BOOK

Second Edition

Also from the publishers of *Rust Book*:

What's it Worth?
All Around the Farm®
Ageless Iron Almanac®
Ag Reference Guide
Shop Reference Guide

Find these and other farm business and
rural lifestyle publications at
successfulfarmingbooks.com.

CONTENT COMPILED BY GREG PETERSON

Successful Farming® Magazine
Ageless Iron Almanac®
Publisher: **Scott Mortimer**
Editor In Chief: **Loren Kruse**
Executive Vice President: **Doug Olson**
Vice President/Group Publisher: **Tom Davis**

Rust Book Second Edition
Editor: **Dave Mowitz**
Art Director: **Kathy Grove**
Copy and Content Manager: **Paula Barbour**
Copy/Production Coordinator: **Pam Garbett**
New Product Manager: **Diana Willits**

Pete's point of view

When I started this business of compiling auction sale price data on used farm equipment 19 years ago, I was one year out of college and just married. The perfect time to take a chance – when you're young and broke!

All I had to go on was faith. Faith in what my father, a third-generation implement dealer, told me. "I think this is really good information that can help folks." OK, Dad. If you say so, I'll take a chance on this gig.

I'm glad I listened.

What a fun ride it's been.

Back in those early days, I always took heart when I'd go to auctions, look around the crowd, and see folks jotting down sale prices on pieces of scratch paper. Hmm, maybe this is a good idea. Now what really makes me smile is when I go to auctions, look around, and see folks armed with our *Rust Book*, leafing through looking for prices.

I had that happen on a recent antique tractor collector auction I attended in central Minnesota. I was at the sale with camera guy in tow, to shoot footage for Season 2 of our *Successful Farming Machinery Show* on RFD-TV. Next thing I hear is …

"Machinery Pete's camera guy bought it!"

Yessiree, he sure did, a Farmall B tractor for $2,250. Lorn, my camera guy, was drawn to the tractor immediately when we arrived at the auction. "I grew up driving a tractor just like that," was his comment. So he shot video with the camera in one hand and used his free hand to secure the winning bid.

Emotion is a powerful thing when it comes to old iron.

It is with great joy that we bring you *Rust Book*. It's filled with more than 9,600 of the latest auction sale prices compiled on all types of farm and construction equipment more than 30 years old.

My hope is that you'll find this book both enjoyable on a personal level (as you browse through the sale prices), and also as a trusted and valuable resource in pursuing your antique iron passion.

See you at the sales.

Greg "Machinery Pete" Peterson
Contributing Editor
Successful Farming *Magazine*

No guesses, just the facts

Not until Greg Peterson showed an interest in antique farm machinery have collectors had anything approximating a *Blue Book* guide for old iron. Sure, other price guides exist. They are speculative, based on an auction bid or two once heard by a guesstimating "expert." These are estimates at best and likely poor ones at that.

Machinery Pete, on the other hand, reports on real sales. Every price listed in these pages was actually paid by someone. Pete's data is also detailed.

This culmination of facts is just the ticket for anyone trying to put a price on old iron. Armed with this book, you can finally put a realistic limit on your final bid or a reasonable value to your collection.

Dave Mowitz
Machinery Director
Successful Farming *Magazine*

Contents

Contents

Abbreviations

air = air conditioning

auto = automatic transmission

aux. = auxiliary

bkt. = bucket

bu. = bushel

CAH = cab, air, heat

CH = chopper

cyl. = engine cylinders

GP = general purpose

GVW = gross vehicle weight

hi/lo = high/low speed transmission

hd = heavy-duty

hp. = horsepower

hyd. = hydraulics

K = 1,000

LP = liquefied petroleum

MFWD = mechanical front-wheel drive

NF = narrow front

OD = overdrive

OH = overhaul

power adj. = power adjustable rear wheels

ps = powershift (transmission)

PS = power steering

pt = pull type

PTO = power take-off

QR = Quad Range

RN = narrow row

ROPS = roll-over protective structure

RW = wide row

sep = separator

SN = serial number

SP = self-propelled

spd. = speed

ss = stainless steel

std. = standard

TA = torque amplifier

tach = tachometer

trans. = transmission

WF = wide front

1 hyd. = one hydraulic outlet

2 hyd. = two hydraulic outlets

1 pt. = one-point hitch

2 pt. = two-point hitch

3 pt. = three-point hitch

2WD = two-wheel drive

4WD = four-wheel drive

Area

The abbreviation under this column represents the region of a state in which the machinery sold. Examples:

NEIA = northeast Iowa

SCCA = south-central California

NWMB = northwest Manitoba (Canada)

Cond. = Condition

The following indicates the overall condition of the machinery as it sold at auction.

P = poor

F = fair

G = good

E = excellent

Allis-Chalmers

The mid-1950s, the glory days of American agriculture and the tractor. Allis Chalmers made its WD-45 tractor during this period, from 1953 to 1957. In 2007 I saw 20 WD-45s sell at auction, ranging from a high sale price of $4,500 for an original 1956 model in good shape sold November 24, 2007, in west-central Illinois, to a low sale price of only $125.

Even with that one clunker for $125, the average auction price of WD-45s was up 4.7% in 2007 to $1,551. Back in 1996, AC WD-45s sold at auction for an average of $1,463.

Allis-Chalmers

Model	Year	Hours	Cond.	Price	Date	Area	Comments
60			F	$500	1/06	WCCA	Canopy, 1 hyd.
15-25	1925		G	$38,000	11/07	WCIL	15-25, running, old restoration/repaint, belt pulley, engine has vacuum line to oil pressure springs that drops oil pressure at idle and increases oil pressure under load, 1,705 units were built, very few around
170			F	$3,000	5/06	SEWI	Gas
170			G	$4,500	12/06	ECIL	Gas, WF, 3 pt., weights, hyd. loader with bale forks
170			G	$6,700	10/06	NCND	Gas, high/low trans., 2 hyd., 3 pt., 540 PTO, low profile, umbrella, radio, good rubber with Allied 600 hyd. loader, 6' bkt.
170		1,760	G	$8,000	3/06	ECIL	Great Bend 440 loader, 6' quick-attach bkt., gas, 3 pt., quick hitch, 16.9-28 tires
170	1969		G	$4,700	12/06	SEIA	3 pt., 2 hyd.
170		9,725	F	$2,400	2/05	WCOK	3 pt., PTO, 1 hyd., diesel, tires poor
170			G	$4,100	9/05	SCON	Canadian sale, 2WD, diesel, 3 pt.
170			G	$4,750	3/05	ECMN	Gas
175	1970		G	$5,100	9/07	WCMI	Diesel, AC 500 hyd. loader, 3 pt., PTO, 2 hyd.
180	1969		G	$5,500	1/08	WCIL	Diesel, open station
180	1967	4,533	G	$4,800	3/07	ECIA	Diesel, no cab, like-new 18.4-28 tires, rear fenders, WF, 3 pt.
180			P	$3,100	11/06	NEIA	Rough, gas, no cab
180			F	$3,700	3/06	NWIL	Diesel
180			F	$4,000	6/06	NCPA	Gas
180	1968	4,280	F	$4,250	10/06	ECSK	Canadian sale, 2WD gas tractor with 64 PTO hp., standard trans., 4,280 hours showing, SN 6007-1 and Great Bend 900 Hi-Master loader and grapple fork
185	1975		G	$8,200	2/08	NCOH	Diesel
185	1971		G	$3,800	12/06	SEIA	Crop Hustler, 3 pt., 2 hyd., front weights
185	1974	2,955	P	$2,800	9/05	NECO	Ezee-On loader, cab, diesel
185	1974		G	$5,400	2/05	SCMN	WF, 3 pt., PTO
185	1974		G	$6,100	8/05	WCMN	Diesel, 540/1,000 PTO, good 18.4-30 tires
190			P	$3,900	6/06	NCPA	Rough, tires poor, jumped out of fourth gear
190	1965		G	$5,200	3/06	NCCO	2 years on OH
190	1970		G	$4,250	11/05	NCIN	Series III
190XT			G	$2,300	1/08	SESD	Diesel
190XT			F	$2,500	1/08	NEMO	Diesel
190XT			F	$2,500	11/07	SEND	Series III, cab
190XT		4,338	F	$4,400	2/07	ECNE	Dual loader, cab, 2 hyd., needed clutch, jumped out of gear
190XT		5,102	F	$5,400	11/07	ECKS	Sold with Farmhand XL 740 loader, 90 hp.
190XT	1965	4,850	F	$2,250	1/07	WCIL	Open station, snap coupler, diesel
190XT			P	$1,500	12/06	NEMO	Disassembled
190XT			G	$6,500	1/06	ECIL	Series III, Westendorf TA-26 loader and 6' bkt., cab, 2 hyd., 18.4-34
190XT	1966	3,271	G	$4,250	12/06	NEMO	Like-new tires, 540 PTO, console control
190XT	1971	4,480	F	$5,300	4/06	SCMI	500 hours on OH, cab, 3 pt., 2 hyd.
190XT			F	$3,100	12/05	WCMN	Series III, with Koyker loader, 3 pt., PTO, 2 hyd.
190XT		5,902	G	$3,300	8/05	NCIA	Diesel, cab, 3 pt., 2 hyd., 18.4-34 duals
190XT		3,092	F	$4,200	1/05	SWIL	
190XT		7,370	G	$5,000	1/05	ECNE	2 hyd.
190XT	1970		G	$3,200	11/05	NCIN	Series III, 2 hyd., 18.4-34
20-35			G	$5,000	6/07	SEID	Full restoration just finished, July 2006 painted, decals

Allis-Chalmers

Model	Year	Hours	Cond.	Price	Date	Area	Comments
200	1975	5,600	F	$5,600	3/08	SWIL	CAH, 3 pt., Allied 595 loader, divorce auction: owner bought this tractor back (some bidders aware of situation; didn't bid)
200			F	$2,600	9/06	NWIL	Diesel, cab, WF, new tires
200			G	$3,500	9/06	NEIN	Diesel, WF, weights, OH
200			G	$4,450	8/06	NCOH	WF, 2 hyd., 3 pt., canopy, front weights, 500 hours on major OH, duals
200			G	$4,950	12/06	WCIL	Diesel
200	1965	4,450	F	$4,000	11/06	WCIL	Diesel, Year-A-Round cab, front and rear weights, 2 hyd., 18.4-34 tires
200	1974	2,585	G	$3,200	1/06	SCMI	WF, diesel, PTO, 3 pt., 6 front weights
200	1974	5,344	G	$5,400	8/06	NCIA	Cab, air, 2 hyd., front rock box
200	1975	6,507	G	$5,200	8/06	WCMN	Diesel, 3 pt., PTO, 2 hyd., 18.4-38
200		7,991	F	$3,600	3/05	SCMI	Canopy, 6 front weights, diesel
200		2,900	G	$5,900	2/05	NCIN	
200	1973		G	$5,800	2/05	NCIN	Diesel, CAH, like-new engine
200	1974		G	$6,200	4/05	SWMN	WF, diesel, Year-A-Round cab, 3 pt., 2 hyd., 16.9-38 tires
210	1972	3,453	G	$6,100	8/07	ECIL	Cab, 3 pt., 2 hyd., weights
210			G	$4,000	12/06	ECIL	
210			G	$6,800	9/06	NEIN	3 pt., PTO, original
220			F	$5,200	11/07	SESD	Loose, may run
220	1970	4,000	G	$3,550	6/06	ECSD	Cab, rock shaft
220			G	$2,500	4/05	NCOH	Cab, 3 pt., 2 hyd., 703 hours on complete OH, 38" axle duals
440	1976	3,291	F	$5,400	3/07	NWIL	4WD, cab, 555 Cummins turbo, 13 speed, 3 pt., 28L-26 tires
5020	1978		G	$2,800	3/07	SEWY	WF, 3 pt. snowblower, 6' belly mower, diesel
5040	1976	3,000	G	$2,500	2/08	WCMN	Open station, 8 speed, 3 pt., PTO, 13.6-28 tires
5040	1978		G	$2,700	9/06	NECO	New tach – 4 hours, WF, 3 pt., 2 hyd., no cab, diesel
5050	1977	2,650	G	$4,500	9/06	NEIN	Utility, diesel
7000			F	$2,000	3/08	NEAR	OROPS, canopy, 18.4-38 duals, 3 pt., PTO, 2 hyd.
7000			F	$3,500	1/08	NEMO	
7000		10,080	G	$4,250	2/08	SEMN	
7000			F	$2,200	12/06	SEMN	
7000			F	$4,450	8/06	NWIL	Diesel
7000			F	$3,900	3/05	NWIL	Diesel, duals
7000			G	$4,400	6/05	SWMN	Diesel, cab, WF, 3 pt., 2 hyd.
7000	1978	8,068	G	$4,800	12/05	NCIL	18.4-38 duals
7030			G	$3,700	4/07	ECSK	Canadian sale
7030			G	$3,500	9/06	NEIN	Diesel, tach shows 1,134 hours, CAH, WF, clean
7030			F	$3,800	3/05	NWIL	Diesel
7040			F	$5,500	8/07	WCSK	Canadian sale, 2WD tractor, 136 PTO hp., PS, cab, air, 18.4×38 duals, dual PTO, new 4-rib front tires
7040	1976		F	$3,300	3/07	WCNE	10-speed manual trans., less than 100 hours on complete OH, 18.4R-38 rear tires with duals, 14L-16 front tires, 2 hyd., 540/1,000 PTO, 3 pt., weak hyd., SN 70405396
7040	1974		G	$3,400	12/06	NWIL	Diesel, cab, new 18.4-38 rear tires, 11:00-16 fronts
7040	1976	8,826	F	$3,900	2/06	ECIL	Weights, 3 pt., 2 hyd., 540/1,000 PTO, duals

Allis-Chalmers

Model	Year	Hours	Cond.	Price	Date	Area	Comments
7040	1976		G	$4,000	11/06	WCIL	Cab, PS, front/rear weights, 540/1,000 PTO, 2 hyd., 18.4-38 tires, diesel
7040			G	$4,500	3/05	NWMN	CAH, 3 pt., hyd., PTO, band duals
7040		4,296	G	$5,500	10/05	ECIL	Cab, air, 2 hyd., 18.4-38 tires, weights, original tractor
7040			G	$5,600	3/05	WCIL	
7040	1977	7,000	F	$4,950	3/05	NCIA	AC, cab, air, PS, 3 hyd., front weights, 1,500 hours on OH, 10-bolt duals
7050			G	$12,000	3/06	NEMO	2WD, cab, Westendorf TA-46 loader
7050	1974	2,025	P	$2,500	4/05	NWKS	Standard trans., 3 pt., PTO, clamp-on duals, broken head
7060	1976		G	$4,200	4/08	ECND	2WD, CAH, gear, 3 hyd., 3 pt., PTO, 20.8-38 duals, hours unknown
7060	1976	7,600	G	$4,700	4/08	ECND	2WD, CAH, PS, 3 hyd., 3 pt., PTO, 18.4-38 singles, shows 7,600 hours
7060	1975		F	$4,400	1/07	WCIL	Cab, Power Director trans., dual PTO, 2 hyd.
7060	1975		F	$4,500	12/07	WCMN	3 pt., front weights, 18.4-38 tires, no toplink, $3,882 in work orders on trans., zero hours since work completed
7060	1975	5,698	G	$8,200	2/06	NWKS	Less than 1,000 hours on OH
7060	1976		G	$5,200	3/06	NENE	2 hyd., 20.8-38
7060	1976	4,430	G	$5,400	12/06	NWIL	Cab, new 20.8R-38 tires, diesel
7060	1978	3,000	G	$6,300	8/06	NCOH	CAH, PS, 3 hyd., duals
7060	1975	5,000	G	$6,200	3/05	NCIA	Cab, air, 2 hyd., 2 PTO, front weights, power director trans., 1,000 hours on OH, duals
B			F	$700	4/08	SCMN	
B			F	$700	5/07	ECND	1-row cultivator
B			G	$850	4/05	ECND	Mid-mount 5' mower
B			G	$1,600	11/07	SESD	Belly sickle mower
B			G	$1,600	6/07	NCIN	2-row mounted cultivator, unrestored, runs
B			G	$2,200	6/07	SCSK	Canadian sale, belt pulley, 21 PTO hp., smaller tires on rear 9.5-24
B			G	$2,300	6/07	SCSK	Canadian sale, belt pulley, 21 PTO hp.
B	1938		G	$900	7/07	WCMN	Hand crank start, Woods mower
B	1939		G	$1,750	9/07	WCIA	Repainted
B	1948		F	$900	6/07	NCIN	Old restoration, runs
B			F	$700	11/06	SCND	WF, PTO, completely restored
B			G	$700	10/06	SEWI	Belt pully, loose, hyd. lift
B			G	$850	4/06	WCTX	
B			G	$900	1/06	NEMO	
B			F	$1,900	3/06	NWOH	Gas, Woods L59 belly mower, WF
B			F	$550	9/05	NEIN	
B			F	$860	4/05	SEND	6' finish belly mower, WF
B			G	$900	9/05	NENE	WF, steel wheels
B			G	$1,075	3/05	SCIL	Electric start
B			G	$1,400	11/05	WCIL	WF, hand start, OH
B			E	$1,525	7/05	ECMN	Buyers premium 8% so final total was $1,647, new paint, new tires
B			G	$2,200	6/05	ECIL	12-volt conversion, new paint
C			G	$1,900	5/08	ECMN	NF, PTO
C	1941		G	$1,700	4/08	NENE	WF, 13.6-26 rear tires, PTO, fenders, Artsway 6' belly mower
C	1947		G	$1,650	5/08	ECMN	Restored, NF, PTO, newer 12.4-24 tires
C	1948		G	$2,000	5/08	ECNE	NF, repainted
C			F	$700	8/07	NWIL	Gas, 48" mower

Allis-Chalmers

Model	Year	Hours	Cond.	Price	Date	Area	Comments
C			G	$750	1/07	WCIL	NF, gas, 72" belly mower
C			F	$775	5/07	ECND	Mower
C			F	$850	8/07	SESD	Running
C			G	$1,025	8/07	SESD	Belly mower
C			G	$1,050	11/07	ECSD	NF, PTO
C			G	$1,200	11/07	WCIL	Running, old restoration/repaint, fenders, PTO, lights, hyd. remote, belt pulley
C	1941		G	$2,000	7/07	ECIL	NF, fenders, new front tires, belt pulley, PTO, has hand brakes, nice restoration
C			G	$1,050	4/06	NWMN	NF, SunMaster 6' belly mower
C			G	$1,050	11/07	ECSD	Belly mower
C			G	$800	2/08	WCIL	NF, fenders, PTO, belt pully
C			F	$1,100	4/05	SCMI	NF, 5' belly sickle mower
C			G	$1,900	11/05	WCIL	WF, converted to 12 volt
C			G	$1,900	11/05	WCIL	WF, converted to 12 volt, restored
C	1944		G	$1,125	8/05	NEIN	NF, sickle mower
CA	1956		G	$4,250	3/08	WCMN	Fully restored
CA	1951		F	$500	12/07	ECNE	NF, 11-24 new rear tires, belt pulley, PTO, oil leak in rear end
CA	1951		G	$3,000	9/07	NWIL	Gas
CA	1953		F	$600	6/07	NCIN	Unrestored, original, runs
CA			G	$1,000	11/07	ECIL	60" Woods belly mower
CA			G	$1,050	8/06	NWOH	NF, gas, loader, restored
CA			G	$1,900	8/07	NWIL	Diesel, WF, belly mower
CA			F	$2,100	6/06	SWMN	Woods belly mower
CA			G	$1,500	11/05	WCIL	NF
CA			G	$1,900	7/05	SESD	WF, planter, cultivator, plow, disk, runs
CA			G	$2,800	11/05	WCIA	WF, good
D10			G	$2,800	11/05	WCIL	Old restoration, repaint, 3 pt.
D10			G	$4,200	11/05	WCIL	WF, unrestored, front rock shaft
D12			G	$3,700	1/08	NCOH	WF, snap coupler, traction booster, PTO
D12			G	$5,100	11/07	NCIL	Series III, original, very good tin, good tires, hyd. leak, not used in couple years, sold by local school district
D12			G	$4,000	12/06	WCIL	WF, restored
D14			F	$1,500	4/08	SCMN	
D14			G	$2,600	1/08	NCOH	NF, snap coupler
D14			G	$3,300	4/08	NEIA	Belly mower, nice
D14	1959		G	$3,300	4/08	NENE	WF, 13.6-28 rear tires, PS, PTO, fenders
D14			F	$2,100	11/07	WCIL	Loader
D14	1957		G	$1,500	2/07	NCNE	WF, PTO, 2 pt.
D14			F	$3,050	8/06	NEKS	Running
D14	1958		G	$2,550	12/06	NWIL	Gas, NF, snap coupler, good paint, PS, like-new 14.9-26 tires
D14			G	$1,700	10/05	WCIL	WF
D14			G	$2,500	11/05	SEND	
D15	1959			$0	4/08	NENE	
D15			F	$3,600	3/07	SWOH	Series II, loader, gas
D15			G	$4,000	9/06	NEIN	Series II, 3 pt.
D15	1964		G	$1,450	12/06	WCIL	Gas, WF, PS, snap coupler
D15			G	$2,900	1/05	SWIL	Row crop, 2 pt.
D15			G	$3,000	11/05	WCIL	WF, gas, old restoration, 3 pt.
D15			G	$3,200	3/05	WCIL	Gas, WF, PS
D15			G	$5,600	9/05	NEIN	LP Series II

Allis-Chalmers

Model	Year	Hours	Cond.	Price	Date	Area	Comments
D17			F	$2,500	5/08	ECNE	Diesel, WF, factory Allis cab
D17			G	$2,700	3/08	SCIL	Series III, WF, PS, weights, snap coupler with 3 pt. adapter
D17			G	$3,000	2/07	NEIN	Gas, WF, 3 pt., 1 hyd.
D17			G	$3,500	2/07	NCNE	WF, 2 hyd.
D17			G	$4,100	3/07	WCIA	WF, no cab, looked rough
D17			G	$4,250	3/07	ECIL	Diesel, 3 pt., PTO, 1 hyd., 16.9-28 tires, restored
D17	1958	5,541	F	$2,750	12/07	WCNE	Gas, WF, 3 pt., dual 325 loader and grapple
D17	1962	3,416	G	$6,000	3/07	ECIL	Series IV, 3 pt., PTO, 1 hyd., factory WF, only 3,416 one-owner hours, rare series
D17	1965		G	$5,500	11/07	WCIL	Series IV, running, restored, PTO, hyd. remote, deluxe seat, 3 pt.
D17	1966		G	$2,250	1/07	WCIL	
D17				$0	6/06	ECSD	Series IV
D17			P	$200	12/06	SEIA	For parts
D17			G	$1,050	11/07	ECSD	Parts tractor
D17			F	$1,450	3/06	NWOH	Gas, 3 pt.
D17			G	$1,850	8/06	NCIA	WF, snap coupler, 3-bottom plow, cultivators and subsoiler
D17			G	$1,950	8/06	NCIA	Cab, WF, snap coupler
D17			G	$2,500	3/06	NENE	WF, gas, 14.9-28 rear tires, Farmhand loader
D17		6,405	F	$2,500	8/06	NWOH	WF, New Idea 504 loader
D17			G	$2,800	6/06	NCPA	NF, gas, not painted
D17			G	$2,900	6/06	ECMN	Gas
D17			G	$3,700	2/08	SEMN	
D17	1958		F	$1,700	10/06	SWNE	WF, 16.9-28 rear tires, factory propane, fenders, snap coupler hitch
D17	1958		G	$4,000	3/06	ECNE	Series III, 3 pt., gas, WF
D17	1959		G	$3,000	8/06	WCMN	One owner, gas, dual 75 loader, PTO, 2 pt.
D17	1962		G	$2,900	11/06	NEKS	WF, syncro, 540 PTO
D17	1965	5,049	F	$6,600	4/06	SCMI	AC 517 loader with 5' bkt., 3 pt., 1 hyd., WF, 5 hours on OH
D17	1966	5,587	G	$4,300	12/06	SEIA	Series IV, snap coupler, 3 pt., WF
D17	1967	6,338	G	$5,500	12/06	SEIA	Series IV, factory 3 pt., 2 hyd.
D17			G	$1,050	12/05	WCOH	Duals
D17			F	$1,575	9/05	SEIA	
D17			P	$1,625	4/05	NCOH	Duals, cab, clutch stuck
D17			P	$1,625	4/05	NCOH	
D17			G	$1,950	12/05	SEND	Farmhand loader
D17			F	$2,000	3/05	NWIL	Loader, WF, gas
D17			G	$2,200	3/05	ECMN	Gas
D17			G	$2,300	11/05	WCIL	WF, gas, fenders
D17			G	$2,500	11/05	WCIL	WF, gas, PS, front weights
D17		4,800	F	$2,600	4/05	SCMI	NF, AC loader, LP
D17			F	$2,600	4/05	NCOH	Gas, WF, PS, wheel weights, 1 hyd.
D17			G	$4,200	12/05	SEMN	WF
D17		5,179	G	$5,000	11/05	ECIL	Series IV, 3 pt., WF, gas
D17	1958		G	$1,300	1/05	NENE	Gas, WF, rock shaft, 540/1,000 PTO, 14.9-28 rear tires
D17	1958		G	$3,000	2/05	NCIN	Diesel, 3 pt., WF
D17	1962		G	$3,175	7/05	SESD	Series 4, WF, snap hitch
D17	1964	4,400	G	$4,900	6/05	ECIL	WF, power adj.
D17	1967		G	$5,700	11/05	SWMN	Series IV, gas, WF, cab, 3 pt., 2 hyd.

Allis-Chalmers

Model	Year	Hours	Cond.	Price	Date	Area	Comments
D19			G	$2,900	1/08	NCOH	WF, gas, runs
D19			G	$4,200	11/07	SESD	LP, WF
D19			F	$2,200	8/07	NWIL	LP, WF
D19	1962	3,203	F	$1,700	1/06	NECO	WF, 3 pt., gas
D19	1963		F	$3,750	4/06	SCMI	Gas, WF, 2 hyd.
D19			F	$1,700	4/05	NCOH	WF, aux. fuel tank, snap coupler, 1 hyd., duals, weak fuel pump
D19			E	$2,300	4/05	NCOK	LP, snap-on duals, runs great, fresh OH on motor
D19			G	$3,000	1/05	SWIL	3 pt.
D21			G	$10,100	1/08	NCOH	Series II, turbo, diesel, runs good but cylinder O-ring is cracked and leaks
D21	1968		E	$12,100	5/08	ECMN	Restored, Series II, turbo charged, newer 18.4-38 tires
D21	1968		G	$10,500	11/07	WCIL	Running, unrestored/original, rear wheel weights, fenders, lights, PS
D21			G	$9,000	9/06	NEIN	OH, original
D21	1968	3,600	G	$19,000	6/06	NCPA	Grill painted white, front lights not original, steering wheel rough, repainted, new front tires, first bid at $7K, 50 hours on complete OH, nice paint job
G	1948		E	$3,000	5/08	ECMN	Restored, newer tires
G	1949		G	$3,700	5/08	ECNE	New rubber, repainted
G	1955		E	$4,200	5/08	ECMN	Restored, Woods belly mower, newer tires
G			G	$2,500	6/07	SCIL	All original, easy restoration, creeper low gear, good sheet metal, runs good
G			G	$3,600	9/07	ECIN	
G			E	$5,400	11/07	WCIL	Good paint, new rubber, great shape!
G	1948		F	$1,900	6/07	NCIN	Original, missing hood, Goodyear tires, runs
G	1948		G	$2,400	6/07	NWIL	Gas, new tires
G	1948		G	$2,800	6/07	SEID	Runs good
G	1948		G	$2,900	6/07	NCIN	Original, new rear tires, runs
G	1948		G	$2,900	6/07	NWIL	Gas, new tires
G	1948		G	$3,000	11/07	SWIN	Restored, new front and rear tires
G	1948		G	$3,150	6/07	NWIL	Gas, new tires, weights
G	1949		G	$2,800	6/07	NCIN	Front weight, old repaint, runs
G	1949		E	$3,000	9/07	WCIA	Redone, like new
G	1949		G	$3,300	6/07	NCIN	Unrestored, original, mounted cultivator, old repaint, new rear tires, runs
G	1949		G	$3,300	11/07	WCIL	Belly mower
G	1953		G	$3,000	11/07	WCIL	Running, old restoration/repaint, fenders, hyd., good brakes, tires and sheet metal, runs good, factory hyd.
G			F	$1,500	3/06	SEPA	Cultivators, fair rubber
G			G	$1,250	9/05	NWOR	Row crop, gas
G			G	$3,250	11/05	WCIL	Diesel, restored, hand start
G	1948		G	$1,300	6/05	ECIL	Electric lift blade, new tires
N/A	1959		G	$2,550	9/06	NECO	AC loader, WF, no 3 pt., gas
L			G	$7,500	8/06	SCIL	Crawler, 6-cyl. gas
RC			F	$1,100	11/05	SWMN	For restoration
U			G	$3,000	6/05	ECWI	
WC			P	$300	5/08	NEMO	For parts, late model
WC			P	$350	5/08	NEMO	For parts, late model
WC			F	$400	11/05	NCOH	Styled, gas, NF, snap coupler

Allis-Chalmers

Model	Year	Hours	Cond.	Price	Date	Area	Comments
WC			G	$1,500	1/08	NCOH	Unstyled, gas, hand brakes and crank, original gas starting tank
WC			G	$425	8/07	SWMN	Gas, NF
WC			F	$700	1/07	NCIL	
WC			G	$950	2/07	NEIN	Gas, NF
WC			G	$1,000	6/07	NCIN	Restored, runs
WC	1935		P	$550	8/07	ECIA	Row crop flat-top gas tractor, needs work
WC	1936		G	$3,000	11/07	WCIL	Running, restored, PTO, belt pulley
WC	1936		G	$3,500	11/07	WCIL	Running, restored, fenders, PTO, belt pulley, new magneto, original fenders
WC			F	$625	9/06	NCIA	Late 1930s or 1940s
WC			G	$700	4/06	WCTX	Full round spoke wheels
WC			G	$950	10/06	SCMN	Reverse loader, 1936 model
WC			G	$1,200	10/06	SEWI	Older restoration
WC			E	$1,925	3/06	NEKS	WF, new rubber, good paint
WC	1935		F	$500	10/06	SWNE	Unstyled, rear spoke wheels, mounted loader
WC	1936		F	$800	8/06	NCIA	Skeleton wheels, NF, partially restored
WC	1937		G	$850	10/06	WCMN	Factory WF, new manifold, new water pump, new tires, new muffler, new paint
WC			F	$525	9/05	NENE	Drawbar, 13-24 tires
WC			F	$600	10/05	NEPA	
WC			G	$750	1/07	WCIL	
WC			G	$1,150	6/05	WCMN	NF, PTO
WC			E	$1,400	1/05	SWIL	Completely rebuilt, new tires
WC	1941		G	$850	12/05	NWIL	NF, good rubber
WD			F	$625	1/08	NCOH	Gas, NF, no serial number
WD			F	$700	2/08	NEMO	NF
WD			F	$800	1/08	SESD	Dual loader
WD			G	$1,250	1/08	SESD	Repainted, WF
WD			G	$1,300	1/08	NCOH	Gas, factory PS, NF, snap coupler, runs, gas
WD			F	$1,300	2/08	WCMN	Reversed, loader
WD			G	$1,500	1/08	NCOH	NF, gas
WD			G	$2,500	6/07	SCIL	WF, PS, snap coupler, good rubber, attached AC loader with bkt. and forks
WD			G	$2,550	3/08	SCIL	Row crop, snap coupler
WD			P	$200	11/07	WCIL	
WD			P	$275	3/07	ECIL	NF
WD			P	$550	8/07	WCIL	Not running
WD			F	$650	11/07	SESD	
WD			F	$775	5/07	WCWI	Loader
WD			F	$800	12/07	ECIN	Mower
WD			G	$1,000	6/07	NCIN	Restored, runs
WD			F	$1,100	11/07	SESD	Dual 250 loader
WD			G	$1,200	1/07	NCIL	
WD			G	$1,300	7/07	WCMN	John Deere 6' deck, turf tires, 12-volt charging system, PS
WD			G	$1,400	11/05	WCIL	Gas, NF, new rear tires
WD			G	$1,600	2/06	WCIL	NF, AC side-mounted 7' sickle mower
WD			G	$1,650	6/07	NCIN	WF, gas, PS, wheels, runs
WD	1950		G	$1,300	9/07	NWIL	Gas, mounted round bale elevator
WD	1951		F	$500	8/07	SWIL	NF, lights, belt pulley
WD	1951		P	$900	8/07	ECIA	Gas, row crop, seized engine
WD	1951		G	$1,500	8/07	WCIL	Gas, restored

Allis-Chalmers

Model	Year	Hours	Cond.	Price	Date	Area	Comments
WD	1953		G	$1,100	9/07	NWIL	Gas, 3 pt., NF, drawbar
WD			P	$250	9/06	NEIN	Original
WD			G	$500	6/06	ECSD	Not running
WD			F	$775	5/06	SWWI	
WD			F	$850	3/06	NENE	WF, gas
WD			G	$1,200	2/06	ECMN	WF, good tin
WD			F	$1,600	3/06	SEPA	2-way 2-bottom mounted plow, NF, good paint/metal/rubber
WD	1951		P	$500	10/06	SWNE	NF, 12-28 rear tires, American 20 loader, engine stuck
WD			P	$200	11/05	ECNE	Loader
WD			F	$450	10/05	NEPA	
WD			F	$475	3/05	NWIL	Gas, NF
WD			G	$700	11/05	WCIL	NF, fenders
WD			G	$850	9/05	WCMN	NF, good paint and rubber
WD			G	$900	11/05	WCIL	
WD			F	$900	2/05	NWWI	3 pt.
WD			F	$925	9/05	SEIA	Blade
WD			G	$1,075	7/05	SESD	
WD			G	$1,250	12/05	ECIN	Rotary mower, gas
WD			G	$1,550	11/05	WCMN	WF
WD			G	$1,600	4/05	NCOH	Styled, loader and manure fork
WD			G	$2,100	11/05	SEND	WF, 3 pt., loader
WD			G	$2,100	6/05	WCMN	NF, PTO
WD	1947		G	$1,350	4/05	NCOK	Loader, bolt-on duals, PS, rebuilt WD-45 motor, WF
WD	1949		G	$625	4/05	NCOK	Gas, ran rough, WF
WD	1949		G	$650	11/05	WCIL	NF, PTO, 3 pt.
WD	1949		G	$1,400	4/05	NEIN	Gas, NF
WD	1949		G	$1,650	8/05	NEIN	Repainted, 3-bottom plow
WD	1950		G	$1,000	6/05	ECIL	6' side-mount twin belt sickle mower
WD	1951		F	$675	12/05	ECIL	
WD	1951		G	$1,050	3/05	NCIA	Fenders, 12 volt
WD	1954		G	$1,550	10/05	SWSD	2 pt., gas, good rear rubber, runs good
WD45			F	$700	2/08	SCMI	NF, 3 pt., gas
WD45			F	$1,050	1/08	ECMI	Gas
WD45			G	$1,075	1/08	SWKY	
WD45			F	$1,500	2/08	NEMO	WF, snap coupler, engine stuck
WD45			G	$1,700	1/08	NEMO	
WD45			G	$2,000	3/07	ECND	WF, 2 pt., PTO, 12-volt connection
WD45			P	$125	3/07	SCNE	For parts
WD45			F	$825	8/07	SESD	WF, gas
WD45			P	$1,000	6/07	SCMN	WF, 12-volt system, loader, not running, needs engine repair, 14.9-28 tires
WD45			F	$1,275	3/07	NWIL	AC 170 loader
WD45			F	$1,400	3/07	SECO	WF, sold, Farmhand loader
WD45			G	$1,425	8/07	SESD	WF, gas
WD45			G	$1,500	6/07	SWMN	
WD45			G	$1,550	8/07	WCIL	Gas, WF, good tires
WD45			G	$1,900	7/05	SESD	3 pt., WF, new tires
WD45			G	$2,150	1/07	NCIL	WF
WD45			G	$2,450	6/07	SCSK	Canadian sale, gas tractor with 3 pt., wheel weights, belt pulley, PTO, 13.6×28 rear tires

Allis-Chalmers

Model	Year	Hours	Cond.	Price	Date	Area	Comments
WD45	1950		G	$900	9/07	NWIL	Gas, no 3 pt.
WD45	1950		G	$1,500	9/07	NWOH	NF, snap coupler hitch
WD45	1953		F	$1,100	11/07	SEND	WF
WD45	1953		F	$2,000	2/07	SEMI	Gas, WF
WD45	1954		G	$1,575	8/07	ECIA	Factory WF, like-new 14.9-28 tires
WD45	1956		P	$450	12/07	WCNE	Gas, WF, PTO, locked
WD45	1956		G	$1,300	9/07	NWIL	NF, factory PS, quick-attach snap coupler, good paint and decals, WF
WD45	1956		G	$2,100	9/07	SCMO	WF, snap coupler hitch, 1 hyd., wheel weights, 14.9×28 new tires, loader
WD45	1956		G	$4,500	11/07	WCIL	Diesel, running, unrestored/original, fenders, PTO, lights, 1 hyd., PS, nice original tractor
WD45			F	$525	7/06	WCOH	
WD45			G	$650	12/06	ECMN	Loader, NF, snow chains
WD45			G	$700	10/06	SEWI	
WD45			F	$700	8/07	NWIL	
WD45			F	$750	3/06	NENE	42" rear tires, Towline tractor, new battery
WD45			F	$950	4/06	SWIN	Needs minor repair
WD45			F	$1,000	4/08	NENE	13.6-38 rear tires
WD45			F	$1,300	3/06	NWIL	Gas, WF
WD45			G	$1,350	1/06	SCMI	NF gas
WD45			F	$1,450	8/06	NWIL	
WD45			F	$1,900	4/06	SCMI	Gas, NF
WD45			G	$1,950	3/06	NEKS	Mounted cultivator
WD45			G	$2,000	4/06	SCMN	New paint
WD45			F	$2,100	8/06	NEKS	Not running, had been a dual fuel
WD45			G	$2,200	3/05	ECMN	Gas, WF
WD45			G	$2,500	9/06	SEIA	WF, 3 pt., 5.5' hyd. AC loader
Wd45			G	$2,500	9/06	NCIA	Factory WF, PS, snap coupler, sold with arch frame loader, snow and manure bkts.
WD45	1952		G	$2,150	12/06	NWIL	Gas, WF, good paint, 14.9-28 tires, crankshaft and pistons
WD45			F	$750	8/05	SCMI	NF, rear blade
WD45			G	$750	10/05	WCMT	
WD45			G	$1,000	11/07	WCIL	
WD45			G	$1,000	11/07	WCIL	NF
WD45			F	$1,350	8/05	NWIL	Gas
WD45			F	$1,600	8/05	WCMN	Loader, WF
WD45			G	$1,600	11/05	SWMN	Loader
WD45			F	$1,800	4/05	NCOH	Gas, WF, PS, trip loader, 1 hyd.
WD45			G	$1,900	7/05	WCMN	WF, 1 hyd., 12 volt
WD45			G	$2,200	6/05	SEND	WF, gas, PTO, wheel weights, PS
WD45			G	$3,100	11/05	WCIL	WF, gas, snap coupler
WD45			E	$4,000	11/05	WCIA	New rubber, WF
WD45			G	$4,100	2/05	SCMN	PS
WD45			G	$4,700	10/05	SWSD	WF, PS, loader and attachments
WD45	1954		G	$700	6/05	ECIL	NF, power adj.
WD45	1955		G	$950	11/05	ECNE	
WD45	1957		G	$1,800	6/05	ECIL	WF, 3 pt., rear weights, 12-volt conversion
WD45	1957		G	$2,550	8/05	NEIN	WF, loader
WF	1948		G	$4,500	11/07	WCIL	Running, restored, fenders, PTO, lights, belt pulley, very nice complete restore with parts list of work done

Case

You know an antique tractor is rare when you start getting up around $20,000 at auction. But $400,000? Wow. $400K is definitely at the top end of sale prices I've run across in my 19 years of tracking sales. The 1913 Case 18-30 that sold for $400,000 September 2007 in northeast Illinois is one of only five left known in existence. Case made 493 Model 18-30s from 1912 to 1916.

Case

Model	Year	Hours	Cond.	Price	Date	Area	Comments
1030			F	$3,100	1/08	NEMO	Workmaster 800 loader
1030			G	$5,000	5/08	ECMN	Restored, Comfort King, diesel, WF
1030	1967		P	$800	3/08	SWNE	For parts, WF, Saddle King fenders
1030	1969		F	$800	3/08	ECNE	Comfort King, 3,230 hours on tach
1030			F	$3,000	7/07	NCND	1,000 PTO, 2 hyd., Allied 580 loader, 18.4-34 tires (70%)
1030		2,765	G	$3,500	12/07	WCMN	3 pt., PTO, WF, new Goodyear 18.4-34 tires, Schwartz loader
1030	1967	5,155	G	$3,500	10/07	NECO	WF, 3 pt., front weights, diesel
1030	1969		F	$2,600	6/07	NCIN	3 pt., 2 hyd., WF, snap-on duals
1030			G	$2,300	9/06	SCMN	WF, 3 pt.
1030	1967		G	$5,000	3/06	SWOH	WF, diesel, Comfort King
1030			F	$2,400	6/05	SWMN	Diesel, WF, new PTO, 3 pt., new 18.4-34 tires
1030			F	$2,650	6/05	SEND	Comfort King cab, 2 hyd., 3 pt., 540 PTO, 18.4-34 singles
1070			G	$8,250	4/08	NEIA	Recent engine/trans. work, cab
1070	1971		G	$2,600	1/08	ECMI	2 hyd., PTO, 18.4-38 tires
1070		8,339	G	$3,700	2/07	NEIN	3 pt., 2 hyd., recent OH
1070			G	$6,250	1/07	WCIL	Yellow, no cab
1070	1971	7,344	G	$8,200	11/07	SWNE	2000 model Great Bend 660 loader, 7' bkt. and grapple, diesel
1070			G	$4,100	1/06	SESD	PS, CAH, yellow cab, diesel
1070			G	$4,100	1/06	ECIL	White, air, 18.4-38, axle duals, 2 hyd.
1070			F	$5,000	3/06	NCMN	
1070		7,490	G	$12,500	4/06	SWSD	Agri King, sold with Farmhand F235 loader with 3-tine grapple and hay basket, 3 pt., 2 hyd., cab, air, radio, 18.4-38, good tires all around, brakes need work
1070	1974	32,200	F	$3,100	11/06	ECIL	
1070	1974	4,735	P	$4,000	2/06	NCCO	PTO clutch out, 2 hyd., 18.4-38
1070	1974	7,951	G	$6,000	2/06	NECO	Diesel, clamp duals, 3 pt., cab
1070	1974	7,951	G	$7,500	2/06	NECO	Diesel, front weights, 3 pt., cab
1070	1974	8,090	F	$7,500	1/06	NCCO	Axle duals, 2 hyd., 540/1,000 PTO, one owner
1070	1977		G	$4,300	6/06	ECSD	White cab
1070	1977	5,454	G	$8,750	11/06	NCKS	Cab, PS, 3 pt., 2 hyd., PTO
1070			F	$4,000	11/05	SCMI	Cab, 6 front weights
1070		4,454	G	$4,300	11/05	NWIL	Diesel, cab, front weights, 3 pt., 18.4-34
1070			F	$4,500	4/05	SCSD	Black Knight
1070			G	$5,100	1/05	NENE	Diesel, 3 pt., WF, 18.4-38
1070			E	$5,600	11/05	SEMI	No cab
1070		4,663	G	$6,200	8/05	SWOH	CAH
1070	1974	4,200	G	$2,600	8/05	NEIN	CAH, duals
1070	1974		F	$3,600	12/05	SEND	12-speed ps, 2 hyd., 3 pt., PTO, 18.4-38 singles
1070	1975	7,500	G	$6,500	4/05	NCCO	One owner, 18.4-38, duals, 2 hyd., 540/1,000 PTO
1090			G	$4,000	8/05	SWIL	
1175		3,100	G	$2,500	2/08	WCMN	2WD, turbo diesel, 2 hyd., 3 pt., PTO, 20.8-38 tires
1175		4,230	G	$5,500	3/07	ECIL	Agri King, CAH, 3 pt., 540 PTO, 2 hyd., 18.4-38 tires and duals, 4,230 hours
1175			G	$9,750	8/07	SWOK	Dual loader

Case

Model	Year	Hours	Cond.	Price	Date	Area	Comments
1175		5,132	G	$12,500	9/07	NCND	2WD, hyd. Artsway loader with grapple, standard trans., Jobber 3 pt., brand-new 18.4-38 tires, 540 PTO, 2 hyd.
1175		4,500	G	$6,300	3/06	NEMO	
1175			G	$3,700	3/05	NWIL	Weights, diesel
1175	1973	6,610	G	$4,000	3/05	SEWY	Diesel, cab, 3 pt.
1175	1976	3,946	G	$9,100	11/05	NCOH	No weights or duals
1210			G	$6,500	12/06	SCMT	
1270			F	$3,200	3/08	ECND	
1270			G	$5,000	7/07	WCSK	Canadian sale
1270	1972	5,376	F	$3,000	6/07	SCND	2WD, CAH, PS, 12 speed, 2 hyd., 3 pt., 540/1,000 PTO, rebuilt heads 150 hours ago
1270	1975	8,556	G	$4,000	4/07	ECSK	Canadian sale, 2WD, 145 PTO hp., PS, 18.4-38 duals, 2 hyd., 8,556 hours showing ($4,600 in recent reconditioning), 135 hp; hyd. 12 gpm @ 2,400 psi; oil pressure 55 psi
1270			G	$4,250	12/06	SEMN	
1270		3,795	G	$6,400	10/06	NCND	Factory 3 pt., PS
1270	1972	4,853	F	$3,200	6/06	NECO	Cab, diesel, slips out of fourth
1270	1977	6,368	G	$4,750	11/06	NEKS	PS, duals, 2 hyd., 1,000 PTO, 3 pt., 10 front weights
1270			F	$1,350	12/05	SCND	Agri King
1270		9,048	G	$4,200	4/05	SEND	451 turbo engine, recent clutch work, 3 pt., 2 hyd., PS, 1,000 PTO
1270			G	$5,500	12/05	SCND	PS, 3 pt., PTO, 2 hyd.
1370			G	$5,750	2/08	WCMN	CAH, 1,000 hours on OH, 2 hyd., 3 pt., PTO, 18.4-38 rear tires, 11×16 front tires, rock box
1370			G	$7,500	4/08	NEIA	Cab, duals
1370			F	$3,500	3/07	NEND	Diesel, PS, CAH, 3 pt., 18.4-38 tires, duals
1370		4,465	G	$5,500	3/07	ECIL	CAH, 2 hyd., 1,000 PTO, like-new 20.8-38 tires, duals
1370			G	$6,200	2/07	WCIL	
1370	1973		P	$2,800	3/07	ECND	2WD, partial PS, cab, 2 hyd., 3 pt., 1,000 PTO
1370	1973	7,314	G	$4,000	12/07	SEWY	Cab, 3 pt., diesel
1370	1973	7,841	G	$4,250	12/07	SCMT	Cab, PS, PTO, one owner, 20.8-38 duals
1370	1975		G	$6,900	4/07	ECSK	Canadian sale, 2WD tractor with 145 PTO hp., 20.8×38 factory duals, 4,162 hours showing
1370	1975	3,409	G	$7,000	12/07	SCMT	3 pt., 1,000 PTO, PS, cab, 20.8-38 duals
1370	1976	3,866	G	$7,500	3/07	ECIL	CAH, PS, 20.8-38 tires and duals, 3 pt., 1,000 PTO, only 3,866 hours, has been in the family since purchased new
1370	1978	4,664	F	$4,000	3/07	ECND	3 pt., PTO, long axle
1370			F	$2,500	8/06	NWIL	WF, gas
1370			F	$2,800	12/06	SEND	
1370			F	$3,100	3/06	NWIL	Duals, weights
1370			F	$4,600	3/06	NCMN	
1370		3,032	G	$5,000	3/06	NWPA	Cab, 2 hyd., 1,000 PTO
1370			G	$7,750	12/06	SCMT	Leon loader
1370	1972	3,675	F	$3,900	2/06	NECO	Duals, 3 pt., cab
1370	1973	8,043	F	$3,000	3/06	NWKS	Agri King, 4 range, PS, 18.4-38, duals, 2 hyd., 3 pt., recent engine work, new turbo, rear wheel weights
1370	1974	3,756	F	$3,900	6/06	WCMN	9-bolt pressed steel duals, PS, 2 hyd., 1,000 PTO, 3 pt., quick coupler, rock box, CAH

Case

Model	Year	Hours	Cond.	Price	Date	Area	Comments
1370	1974	5,600	G	$7,500	2/06	NECO	Diesel, 3 pt., cab
1370	1975		F	$3,600	9/06	NWIL	Cab, WF, Leon 707 loader, diesel
1370	1978	3,850	G	$5,500	11/06	ECND	2 hyd., 3 pt., PTO
1370	1978	4,000	G	$6,500	12/06	NENE	New motor at 3,800 hours, 16 front end weights, 2 hyd., 18.4R-38, 9-bolt duals, quick hitch
1370			P	$1,300	4/05	SCSD	Needs trans. work
1370			F	$3,500	8/05	SEMN	
1370		5,750	F	$4,700	6/05	SEND	2 hyd., 3 pt., 1,000 PTO, 18.4-38 hub duals
1370			G	$5,100	9/05	NWIA	
1370	1972		G	$3,250	2/05	NEIN	Diesel, CAH, patch on trans., no leaks
1370	1972	6,100	F	$3,850	4/05	ECND	PS, 3 pt., PTO, 20.8-38
1370	1974	3,809	G	$5,500	2/05	WCIL	Factory cab, PS, 1,000 PTO, 3 pt., 2 hyd.
1370	1975	5,228	G	$2,250	8/05	NEIN	Cab, new hyd. gasket
1370	1975		F	$2,800	6/05	NESD	2WD, CAH, 12-speed ps, 2 hyd., 1,000 PTO, 20.8-38 singles
1370	1976	4,626	G	$4,000	1/05	NCIA	Cab, 2 hyd., quick coupler, clamp-on duals, front weights
1370	1977	3,397	G	$4,600	9/05	NCIA	Air, front rock box, 2 hyd., quick coupler, duals
1470			G	$2,200	12/06	ECIN	Cab, 4WD, diesel
1470			F	$1,800	3/05	NWMN	4WD
1470			F	$5,100	2/05	SCMN	4WD, 900 hours on OH, 3 pt., duals, no PTO
1570			G	$6,750	1/08	SESD	CAH, 3 pt.
1570		6,400	F	$3,900	2/07	NEIN	Cab, 3 pt., 2 hyd., 8 front weights, snap-on duals
1570	1978	3,620	G	$6,250	10/07	NCMI	Diesel, cab, air, heat, 20.8-38 axle duals, 1,000 PTO, 3 pt., 2 hyd.
1570			F	$4,700	3/06	NCMN	
1570		5,800	F	$6,750	6/06	NWMN	3 pt., hyd., PTO, Leon 808 loader
1570	1977	6,000	G	$9,600	4/06	SWSD	Agri King, diesel, cab, air, radio, 3 pt., 2 hyd., 1,000 rpm, 20.8-38, good rear tires and duals
1570		3,901	F	$2,300	12/05	WCOH	Agri King, faulty trans., 20.8-38, duals
1570			F	$3,600	3/05	WCIL	Diesel, duals
18-30	1913		E		9/07	NEIL	The first Case tractor, 493 made from 1912-1916, only 5 left in existence, 28,500 lbs.
2090	1974	7,229	G	$6,500	2/06	SEIA	Fully-equipped cab, 2 hyd., axle-mount duals
22-40	1920		G	$57,500	9/07	NEIL	Crossmotor tractor, parade ready
2290	1978	4,752	F	$6,000	9/06	NWIL	Cab, WF, weights, diesel
2390	1978	6,727		$0	6/06	ECSD	CAH
2390	1978	4,575	F	$6,700	2/06	WCIL	Fully-equipped cab, 3 hyd.
2470		9,000	F	$2,250	3/08	WCMN	4WD, Agri King, CAH, PS, 2 hyd., 18.4-34 duals, 50% rubber, no 3 pt. or PTO
2470			F	$3,800	4/08	SCMN	4WD, PTO, 3 pt.
2470			G	$5,600	3/08	WCMN	4WD, 10' dozer blade, 2 door, 4 hyd., 3 pt., PTO, 18.4-34 tires
2470			P	$800	8/07	SESD	Salvage
2470			P	$1,900	9/07	NWIL	Diesel, 3 pt., PTO, no duals
2470			F	$3,000	3/07	WCKS	4WD, 3 pt.
2470		4,705	G	$5,300	3/07	ECIL	4WD, 3 pt., PTO, 2 hyd., 23.1-30 tires
2470	1973	4,251	G	$8,250	8/07	ECIA	4WD, one owner, new engine, good 34" duals front and rear
2470	1977	6,150	F	$2,000	11/07	SEND	4WD, 18.4-34 duals

Case

Model	Year	Hours	Cond.	Price	Date	Area	Comments
2470	1978	6,936	G	$6,000	12/07	SEWY	Diesel, 3 pt., PTO, 4 hyd.
2470		5,000	F	$4,400	3/06	NWIL	3 pt., PTO
2470	1973	9,839	G	$3,200	4/06	NWMN	Traction King, CAH, 12-speed ps, 2 hyd., 18.4-34 duals
2470	1976	4,335	G	$4,100	9/06	NECO	Leon 10' dozer, cab, no 3 pt., 2 hyd., diesel
2470			F	$2,100	2/05	SCCA	4WD, cab, 3 pt., 2 remotes, 23.1-30 tires
2470	1973	2,586	G	$8,000	3/05	SWNE	4WD, cab, 3 pt., 4 hyd., quick hitch
2470	1976	6,303	G	$4,300	9/05	NECO	Cab, 3 pt., 4 hyd., diesel
2670			F	$3,500	2/07	ECMI	4WD, 18.4-34 duals, 4 hyd., quick hitch, 5,254 metered hours, hyd. lift blade
2670			G	$4,400	3/07	ECIL	4WD, 3 pt., 2 hyd., no PTO, 23.1-30 tires, duals
2670	1975	3,657	F	$3,500	7/07	NWIL	4WD, diesel, 3 pt., weights, cab, fast hitch, duals
2670			F	$3,200	8/06	NCIA	4WD, cab, air, PTO, 3 pt., 4 hyd., duals
2670			G	$7,000	10/06	WCWI	225 hp., 4WD, duals all around, 3 pt. hitch,1,000-rpm PTO
2670	1976	7,640	G	$8,000	3/06	NECO	4WD, PS, 3 pt., PTO, 4 hyd., 18.4-34, duals, 1,280 hours on engine, trans. OH
2670	1977	4,443	F	$2,500	9/06	NWIL	4WD, 3 pt., OH trans., diesel
2670	1978		F	$6,000	3/06	NWKS	4WD, 1,167 hours on OH, 4 speed, PS, 20.8-38, duals, 3 pt., PTO
2670			F	$3,100	7/05	SESD	4WD, 3 pt., PTO, duals
2670			F	$4,500	2/05	NWWI	4WD, turbo, duals
2670	1976	8,188	F	$3,000	6/05	NWMN	4WD, CAH, PS, 4 hyd., duals
2670	1976	3,340	G	$5,300	12/05	ECIL	4WD, duals, 3 pt., 4 hyd., no PTO
2870	1977	4,130	G	$3,100	3/07	WCNE	4WD, 3 pt., 4 hyd.
2870			G	$4,250	8/06	SCMI	4WD, cab, 3 hyd., 3 pt., bad tranny, tach shows 2,242 hours
2870	1977	4,200	G	$7,000	3/06	SWOH	4WD, PS, no PTO
300	1956		G	$1,900	8/07	WCIL	Gas
300			G	$1,800	11/06	NEND	
300	1957		G	$3,700	6/05	SWMN	Gas, Eagle hitch, restored
400	1958		G	$6,000	5/08	SEMN	WF, Case-O-Matic drive, fenders, low hours on rebuilt engine
400			G	$325	7/07	NCND	Collector tractor, old, straight
400			G	$1,600	4/07	NWMN	Row crop, PTO, American hyd. loader, 18.4-38 rear tires, diesel
400			F	$3,050	3/07	WCKS	
400	1955		G	$2,350	8/07	WCIL	Diesel
400	1958		G	$600	12/07	SEWY	Gas, 3 pt., double front
400			G	$1,000	12/06	SCMT	Super diesel
430			G	$3,550	8/06	NWIL	Gas, loader, 3 pt., PTO
430			G	$4,500	3/06	NWIL	Diesel, WF
430			F	$950	11/05	ECNE	Loader, 14.9-24 tires
500			F	$700	1/07	SWKS	WF, propane, 18.4-34 rear tires, 2 hyd.
500			G	$6,750	11/07	WCIL	Diesel, clean Western original
500	1954		G	$775	9/06	NEIN	Diesel, works good, little antifreeze in oil
500	1959		F	$900	8/06	ECNE	Diesel, front end removed and replaced with Caterpillar maintainer front end, blade and wheel tip, 18.4-30 rear tires, rear wheel weights
500			P	$900	4/05	SCSD	Standard, for parts
500B	1959		F	$1,600	6/07	NCIN	Gas, Eagle hitch, rear weights

Case

Model	Year	Hours	Cond.	Price	Date	Area	Comments
530			F	$2,500	3/08	NCCA	Loader tractor
530			G	$15,000	4/06	WCIA	Military tug
530	1963	2,964	F	$4,000	9/06	NWIL	Hyd. loader and backhoe
540	1963	5,073	G	$1,250	8/05	NEIN	WF, 3 pt., PTO
570			G	$14,000	3/07	SCTX	Loader landscape tractor 4×4, powered by Case 4-390 diesel engine, 70 hp., equipped with 4F/4R shuttle trans., ROPS, loader with general purpose bkt., box, 3 pt., PTO
600	1957		F	$1,200	1/07	SWKS	LP engine, propane, 18.4-34 tires
600			F	$1,800	11/06	ECIL	LP
600	1959	5,000	G	$4,300	3/06	SWOH	WF, gas, Case-O-Matic drive, live PTO
600			P	$1,500	4/05	SCSD	Diesel
600			G	$2,100	9/05	NEIN	Standard, original
630	1962		G	$1,500	2/07	SEMI	Gas, NF 3 pt., 1 hyd., Case-O-Matic
630	1967		G	$1,250	3/07	SEWY	WF, 3 pt.
630			P	$1,900	4/05	SCSD	Eagle hitch, stuck engine, for parts
700	1958	5,032	F	$900	2/06	NECO	3 pt., WF, Johnson Workhorse loader
700	1959	5,032	F	$1,300	6/06	NECO	WF, 3 pt., gas
700			P	$950	4/05	SCSD	For parts
700			G	$3,200	3/05	WCIL	Restored, new motor, new paint, new steering
700			G	$4,800	9/05	NEIN	Standard, original, high crop
700	1959		G	$1,550	8/05	ECIA	LP conversion
730			P	$2,700	3/07	ECIL	3 pt., 540 PTO, 2 hyd., sells with older John Deere loader, entire outfit is very rough
730	1963		G	$5,250	1/07	SEKS	Comfort King, diesel, WF, Case 70 loader with 6' bkt. and bale fork, 2 hyd., 3 pt., 7.50-15 front tires, 15.5-38 rear tires
730	1966		G	$15,000	11/07	WCIL	Running, restored, LP orchard, new front tires, points, cowling, electrical, gauges, rebuilt head, rings, bearings, epoxy primer, paint and clear coated
730			F	$1,175	3/06	SEIA	Gas, hyd. loader
730			G	$2,900	6/06	ECMN	Diesel
730	1960		G	$1,550	2/06	WCMN	2WD
770	1970		G	$4,500	2/08	WCMI	
800			F	$800	8/05	NWIL	Gas, WF
800			F	$1,450	1/06	SESD	Dual loader, diesel
800			P	$825	4/05	SCSD	LP, Eagle hitch, for parts
830		5,843	G	$4,600	3/08	SWOH	Comfort King, 18.4-34 tires
830	1968		F	$3,900	2/08	WCOK	Case-O-Matic, 2 hyd., PTO, rebuilt engine with new crank, alternator and hyd. pump, excellent rubber, Kent 1600 front end loader
830	1962	1,320	F	$1,200	8/07	ECIL	Case-O-Matic tractor, 3 pt., hyd., fenders
830	1964		F	$2,000	4/07	NCSK	Canadian sale, 2WD tractor with 64 PTO hp., diesel, dual-range trans., 540 PTO, 18.4-30 3-rib front tires and Agritec front-end loader with bkt., not running, needs starter, tune-up, 1 back tire
830	1960		G	$5,300	4/06	WCSD	Row crop, PTO, 3 pt., diesel, 63 hp., clean, straight
830			F	$500	9/05	NEIN	LP, Case-O-Matic, Comfort King standard
830			P	$600	4/05	SCSD	Engine stuck, for parts
830			P	$1,000	4/05	SCSD	Standard, diesel, for parts
830			F	$1,550	2/05	NCIN	Gas

Case

Model	Year	Hours	Cond.	Price	Date	Area	Comments
830			F	$1,925	2/05	SCMN	Gas, Comfort King, WF, 3 pt., orange post, antifreeze leak by radiator
830			E	$3,200	11/05	SEMI	Gas, WF, no cab
830			G	$4,200	5/05	SCMN	WF, 3 pt.
830	1965		F	$2,200	8/05	ECIA	Comfort King, factory LP
870			G	$4,000	1/08	NWOH	3 pt., 2 hyd., snap duals, 4 front weights
870			P	$2,950	8/05	NWIL	Loader, 7' bkt., no cab, diesel
870		7,000	G	$4,100	10/05	NCKS	Cab, gas, 3 pt., 2 hyd., dual PTO
900	1958		F	$2,000	5/06	WCKS	Propane
900			P	$800	4/05	SCSD	For parts
900			G	$1,000	11/07	WCIL	Diesel, electric, Wheatland
900			F	$1,650	9/05	NEIN	Standard, original, 900B
900	1958		F	$800	7/05	SESD	Wheatland
900B	1960		G	$1,500	5/08	ECNE	
930			P	$400	1/07	SWKS	WF, propane, cab, 18.4-34 rear tires, 7.50-18 front tires, 540 PTO
930			P	$850	3/07	WCNE	Wheatland, no cab, diesel, WF
930			F	$1,600	4/07	ECND	Diesel, Comfort King, 3 pt., PTO
930			P	$1,800	11/07	NENE	Rough
930			F	$2,000	7/07	SEND	Cab, diesel, 3 pt., singles
930			F	$2,400	3/07	WCIL	Diesel, standard, WF
930	1960		F	$800	1/07	SWKS	Comfort King tractor, WF, propane, cab, 18.4-34 rear tires, 2 hyd., 540 PTO
930	1969		G	$2,000	4/07	NWMN	Diesel, standard, PS, PTO, hyd., 18.4-34 rear tires
930			G	$1,350	9/06	NWTN	
930			F	$1,600	9/06	NEND	Comfort King, PTO, hyd., diesel
930			G	$1,600	12/06	SCMT	
930			G	$2,600	6/06	WCMN	Comfort King, Draft-O-Matic trans., WF, 2 hyd., 3 pt., 540 PTO, Koyker K5 loader
930			G	$2,750	8/07	NWIL	LP, cab, quick hitch
930			F	$3,300	3/06	NWIL	LP gas, WF, cab, Comfort King
930		3,712	G	$4,650	7/06	ECIL	Comfort King, gas, WF
930	1963		F	$1,550	9/06	NWIL	Western, WF, fenders, hyd., diesel
930	1964	5,200	G	$2,100	9/06	NECO	Wheatland, cab, diesel, tach shows 2,197 but 5,200 actual hours
930	1966	5,000	G	$5,300	3/06	SWOH	WF, diesel, Comfort King, Draft-O-Matic drive
930	1967		F	$2,600	9/06	NWIL	Comfort King, open station, WF, diesel
930			P	$800	4/05	SCSD	Blade, for parts
930			P	$900	4/05	SCSD	LP, 3 pt., no PTO, rough
930			G	$1,000	7/05	ECND	PTO, LP, 2 hyd.
930		5,114	G	$2,600	4/05	ECND	Comfort King, 2 hyd.
930	1966		G	$1,500	6/05	ECIL	8 speed, 3 pt.
930	1966		F	$2,150	8/05	ECIA	Factory LP, Comfort King
970		5,600	F	$4,200	3/08	WCMN	2WD, cab, heat, 2 hyd., 3 pt., 540/1,000 PTO, wheel weights, 18.4-38 band duals
970	1972	4,682	G	$5,900	1/08	WCIL	Agri King, diesel, 2 hyd., blower, 18.4-34 tires, weights, original tractor
970	1974		P	$5,000	4/08	SWOH	Rough, diesel, high hours
970		4,913	G	$4,400	12/07	WCNY	Cab, duals
970		8,700	G	$4,750	2/07	NEIA	WF, no cab
970	1969	4,900	F	$3,200	8/07	NWIL	Diesel, cab, good 18.4-34 tires, WF

Case

Model	Year	Hours	Cond.	Price	Date	Area	Comments
970	1973	7,400	F	$4,000	11/07	ECSD	Agri King, cab, heat, PTO, 3 pt., PS, 2 hyd.
970	1977		G	$6,250	1/07	SEKS	WF, 2 hyd., 3 pt., 540 PTO, 6 front weights, 18.4-34 tires
970			F	$1,150	9/06	NEIN	Cab, no brakes, tach shows 3,495 hours
970		6,000	E	$7,100	11/06	NEKS	New tires, good paint, diesel, white hood, AC cab, 12-speed ps, 18.4-34 tires
970	1977	4,617	G	$7,000	9/06	NCIA	Cab, air, PS, front rock box, quick coupler
970	1977	5,069	F	$8,000	2/06	WCNE	Diesel, 3 pt., Great Bend 900 loader and grapple, cab
970			P	$1,300	4/05	SCSD	For parts
970			P	$1,700	8/05	NWIL	Diesel
970			G	$4,100	9/05	SCON	Diesel
970			G	$5,000	6/05	NESD	2WD, cab, 12-speed ps, PTO, 2 hyd., no 3 pt.
970	1973	5,000	G	$4,000	8/05	ECIA	Diesel
970	1975	4,447	G	$6,000	2/05	NWOH	18.4-34
C	1930		G	$2,500	5/08	SEMN	Rear steel wheels
C	1938		G	$3,100	1/08	SESD	Repainted
C	1937		F	$600	8/07	SWIL	Flat-top fenders, flat spoke wheels
C	1937		G	$1,900	8/07	SWIL	Top fenders, flat spoke front and rear steel wheels
C			F	$1,300	10/06	SEWI	Full steel
C			P	$950	4/05	SCSD	For parts
C			G	$1,150	9/05	NENE	Drawbar, 13-24 tires
CC			F	$900	8/07	SESD	
CC			F	$2,500	9/07	WCIL	On rubber, original engine free, third owner, water pump needs repacked
CC	1937		F	$500	8/07	NWOH	Cut-offs
CC	1938		P	$225	10/06	SWNE	Engine stuck
D			G	$4,000	9/07	ECIN	Repainted
D	1949		G	$3,100	11/07	WCIL	Running, restored, fenders, PTO, lights, belt pulley, perfect restoration, new automotive type paint, starts and runs perfect with no smoke, new battery
D	1954		F	$300	6/07	SEID	WF, runs
D	1948		G	$3,100	3/06	SWOH	One owner, WF, gas
D	1951		G	$1,400	4/06	WCIA	Original
D			G	$650	9/05	SCON	Canadian sale
D			P	$750	4/05	SCSD	For parts
DC			F	$800	5/08	WCIA	WF, 12 volt, clamshell fenders, PTO, 1×wheel weights
DC			F	$400	8/07	NWOH	No tag, fenders, lights
DC			G	$1,600	11/07	SESD	
DC			G	$1,750	9/07	ECIN	
DC	1950		P	$300	8/07	SWIL	Not running, rough, NF
DC			G	$400	10/06	SEWI	Old repaint
DC			G	$400	4/06	WCTX	
DC			F	$400	7/06	NWMN	Good rubber
DC			G	$1,300	8/05	NEIN	WF, original
DC			P	$300	4/05	SCSD	Eagle hitch, for parts
DC			P	$400	4/05	SCSD	Eagle hitch, for parts
DC			G	$575	6/05	SEND	540 PTO, new rear tires, horn loader
DC			P	$1,150	4/05	SCSD	Standard, for parts

Case

Model	Year	Hours	Cond.	Price	Date	Area	Comments
DC	1948		G	$1,600	11/05	WCIL	Electric start
DC	1951		G	$1,000	11/05	ECNE	PTO, 16.6-38, DC-4
DC	1952		G	$3,250	6/05	ECIL	Rear rims widened 3"
DC4			F	$1,000	6/07	NCIN	WF, clean sheet metal, engine free, belt pulley, live hyd.
DC4	1950		G	$1,650	8/07	WCIL	Gas
L			F	$1,200	6/07	SEID	No tag, running, on full steel
L	1940		G	$1,200	9/07	WCIA	On rubber
L			G	$300	10/06	SEWI	
L			G	$650	4/06	WCTX	
L			G	$950	8/05	WCIA	
L			G	$1,350	10/06	SEWI	Factory rubber, loose
L			G	$1,600	10/06	SEWI	
LA			G	$2,100	11/05	SEND	WF
LA	1948		G	$3,750	5/08	WCIA	No PTO, no hyd., have OH notes from shop, bored and stroked to 577.6 cubic inches, new sleeves, compression calculated at 8.5 to 1, new valves, connecting rods, new radiator and core installed, new clutch plate
LA			P	$300	6/07	SEID	On rubber, runs, decent tires, new grill
LA			F	$800	1/08	SESD	LP gas
LA			G	$3,400	3/06	ECIN	
LA			G	$500	4/06	WCTX	
LA			G	$1,750	10/06	SEWI	Belt pully, loose
LA			P	$275	4/05	SCSD	For parts
LA			P	$925	4/05	SCSD	For parts
LA	1949		G	$2,000	12/05	WCIL	PS, hyd.
RC	1938		G	$2,300	5/08	SEMN	PTO, mechanical lift, cut-downs on rear, rubber on front, starter, no generator, second owner
RC	1938		G	$1,200	6/07	NCIN	Old repaint
RC	1938		G	$3,600	11/07	WCIL	Running, restored, fenders, belt pulley, hidden fourth gear
RC	1938		G	$1,800	9/05	ECMN	Gas
RC	1939		G	$1,050	9/05	NENE	PTO, drawbar, 10-38 tires
S			G	$1,350	6/07	SEID	Runs, WF, nice, barn fresh
S			G	$2,250	9/07	ECIN	
S	1950		G	$1,750	9/07	WCIL	Older restoration, runs
SC	1944		G	$1,650	5/08	SEMN	Fenders, PTO, new tires front and rear
SC	1950		P	$200	5/08	WCIA	Not running, cultivator attached
SC			F	$700	6/07	SEID	On skeleton steel, front rubber
SC			F	$800	6/07	NCIN	Single front, fenders, unrestored, original
SC			G	$1,700	9/07	ECIN	
SC	1944		F	$500	6/07	NCIN	NF, dual fuel tank, fenders, lights, electric start, unrestored
SC	1950		F	$400	8/07	SWIL	NF
SC	1950		F	$500	11/07	WCIL	
SC	1951		F	$750	6/07	NCIN	Unrestored, original
SC	1954		G	$1,100	11/07	WCIL	Running, unrestored/original, rear wheel weights, PTO, belt pulley
SC			F	$150	6/06	ECSD	Eagle hitch, NF running
SC			G	$350	4/06	WCTX	Eagle hitch, no center link
SC			G	$700	10/06	SEWI	

Case

Model	Year	Hours	Cond.	Price	Date	Area	Comments
SC			P	$425	4/05	SCSD	For parts
SC			P	$550	4/05	SCSD	For parts
SC			G	$1,100	4/05	NENE	540 PTO, Duncan loader
SC	1941		G	$2,700	6/05	ECIL	New paint
SC	1943		G	$1,000	4/05	NEIN	Gas, NF
SC	1948		G	$2,100	12/05	WCIL	
SC	1950		P	$90	11/05	ECNE	PTO, 3 pt.
SO	1942		G	$1,900	6/07	NCIN	Orchard, 1 rear weight, correct air cleaner
SO	1949		G	$1,100	11/07	WCIL	No orchard tin
VA			G	$2,000	9/07	ECIN	
VA			P	$550	7/06	ECND	WF, not running
VA			G	$700	9/05	SCON	Canadian sale
VA			F	$750	3/06	NENE	WF, PTO, drawbar, 12.4-38 tires
VAC	1944		G	$1,350	5/08	ECNE	NF, Woods L59 belly mower
VAC	1948		F	$500	5/08	WCIA	6 volt, good grill and sheet metal
VAC			G	$800	7/07	SEND	NF, 5' belly mower, 12 volt
VAC			G	$1,000	8/07	SESD	3 pt.
VAC			G	$1,250	12/05	ECIN	
VAC			G	$2,250	3/07	SCMI	Gas, NF, 3 pt., restored
VAC	1944		F	$1,150	9/07	WCIA	NF, mower deck, Woods 6' mower
VAC	1945		G	$1,700	9/07	WCIA	WF, sickle mower, repainted, fenders
VAC	1946		F	$3,030	8/07	ECIL	NF, 1 new rear and 1 new front tire
VAC	1947		F	$700	6/07	SEID	Restored, runs, NF
VAC	1948		F	$1,000	6/07	NCIN	Old restoration, rear weights
VAC	1950		F	$850	6/07	NCIN	NF, 3 pt., rear wheel weights, good rear tires, unrestored
VAC	1954		G	$1,700	11/07	WCIL	Running, old restoration/repaint, belt pulley, repainted to two-tone, sat in shed for three years without starting, started 11/13/07 by putting gas in tank and good battery
VAC			F	$600	4/07	NWMN	
VAC			G	$1,500	4/06	WCIA	
VAC			G	$1,500	4/06	WCIA	Air-cooled engine
VAC	1939		G	$1,400	6/06	WCMN	NF, all original, running order
VAC			G	$1,100	1/08	WCIL	NF, new gauges
VAC	1949		F	$700	4/05	SCNE	Runs, 11-28 tires
VAC	1952		G	$1,500	8/05	SCMI	WF
VAH	1948		G	$3,000	5/08	SEMN	High crop, WF, straight drawbar, PTO, low hours on rebuilt engine and clutch
VAH	1949		G	$2,400	11/07	WCIL	Running, old restoration/repaint, fenders, PTO, 3 pt., metal good and complete
VAI			F	$1,300	1/05	ECNE	Woods 59" mower
VAI	1945		G	$5,750	5/08	SEMN	Rear tractor, used during wartime, complete including mechanical lift, cut-downs on rear, rubber on front, starter, no generator
VAO			E	$10,500	11/07	WCIL	Orchard, running, restored, beautiful little tractor with full orchard dress and immaculate restoration
VC			P	$450	4/05	SCSD	For parts

Cockshutt

On May 10, 2008, I attended a very nice antique tractor collector auction in east-central Minnesota featuring 84 tractors. All makes and models. This collector wasn't tied to one color. One tractor on this sale was a 1956 Cockshutt 30 that had been restored and had newer tires. It sold for $2,600.

Standing with me in the crowd that day was a guy from Ireland. He'd flown across the pond to bid on antique tractors on this sale. The extended weakness of our U.S. dollar has opened the door to the antique tractor market over here for European collectors. I look for this trend of more foreign buyers in the antique tractor market to continue to grow.

Cockshutt

Model	Year	Hours	Cond.	Price	Date	Area	Comments
1850			F	$3,000	4/07	SESK	Canadian sale, 2WD tractor with PTO, 2 hyd., 18.4-34 rubber, SN 156040427, MF loader with bkt. and bale fork (original owner)
20			G	$3,300	11/05	WCIL	WF, unrestored, rear wheel weights, fenders, front wheel weights, hyd., 3 pt.
20	1952		F	$3,000	11/05	WCIL	WF, fenders
30	1956		G	$2,600	5/08	ECMN	Restored, newer tires
30			G	$2,400	4/07	ECSK	Canadian sale, 2WD, PTO, hyd., 11-38 rear tires, shedded
30	1947		G	$3,700	7/07	ECIL	NF, new front and rear tires, belt pulley, fenders, deluxe seat, restored
30	1949		G	$3,200	7/07	ECIL	NF, fenders, lights, restored
30			G	$2,200	6/07	SCSK	Canadian sale, antique tractor with 4-cyl. gas, standard trans., 3-rib front, 12.4-38 tires, belt pulley, live PTO
30			G	$2,200	10/06	WCSK	Canadian sale, 4-cyl. gas, standard trans., 3-rib front, 12.4-38 tires, belt pulley, live PTO
30	1947		G	$1,400	10/06	WCMN	NF, new paint
30			G	$1,500	10/05	ECMN	Gas, saw rig
30			G	$2,400	9/05	NENE	Factory WF, live pump, PTO, drawbar, 13.6-38 tires
35	1956		G	$6,500	5/08	WCIA	Blackhawk, 35-1844, restored 1995, great grill, very good sheet metal, live hyd., belt pulley, 2× rear wheel weights, new fronts, original rears
35	1957		G	$1,900	11/07	WCIL	35 Deluxe, running, unrestored/original, fenders, PTO, hyd. remote, deluxe seat, good running, straight sheet metal, big fenders
550			G	$3,500	6/07	NCIN	NF, fenders, old repaint, gas, runs
560			F	$2,400	8/07	SESD	Standard, loose, not running
560	1960		G	$3,000	11/07	WCIL	Running, unrestored/original, fenders, PTO, hyd. remote, PS, off-the-farm tractor
70	1940		G	$8,000	9/07	NEIL	70 Standard, SN 307931, equipped with rare rear wheel extensions

Ferguson

TO-35s were made by Ferguson from 1955 to 1960. One of the nicest TO-35s I've ever seen sold on farm auction in east-central Nebraska in May 2008 for $6,000. It was a 1955 model in excellent condition. Back on September 7, 1996, I saw a 1955 TO-35 with a fresh overhaul in good condition sell for $5,750 on an auction in southeast South Dakota. So it looks like values on TO-35s have held pretty steady over the last decade plus.

Ferguson

Model	Year	Hours	Cond.	Price	Date	Area	Comments
30			G	$2,400	8/07	SWIL	Power-adjusted rear wheels, loader
30	1954		F	$3,700	12/06	WCMI	Gas, Shawnee loader
30			G	$1,700	10/05	WCIL	WF, gas, PTO, lights
N/A			G	$1,500	2/08	NWSC	Gas, turf tires
N/A			G	$1,300	2/07	SCWA	
N/A			G	$2,250	10/07	WCOH	Like-new rubber
N/A			G	$2,400	11/07	WCIL	Restored
TE-20			G	$2,400	11/07	WCIL	Running, restored, fenders, PTO, 3 pt., fresh paint and new rubber all around, runs good, could use a ring and valve job
TE-20			G	$2,350	2/05	NEIN	Loader, gas
TO 20			G	$1,350	8/07	SWOK	
TO 20			F	$1,700	8/06	NWIL	Gas, loader, new tires
TO 20			G	$1,500	11/05	WCIL	
TO 20			F	$1,600	3/05	WCIL	Gas, cab, front blade
TO 20			E	$3,000	4/05	NCCO	A1 paint, new tires, 3 pt., PTO, new rims, strong engine, great paint job, a cutie
TO 30	1954		G	$1,600	2/08	WCOK	
TO 30			G	$1,000	11/07	WCIL	Restored
TO 30			G	$1,400	6/07	NCIN	Old repaint, runs
TO 30			G	$2,050	3/07	SEMI	Gas, 3 pt., good 12.4-28 tires, fresh paint
TO 30	1952		F	$1,400	9/07	NECO	Gas, loader, 3 pt.
TO 30	1953		F	$700	12/07	SEWY	Gas, shop-made pipe frame loader
TO 30			G	$1,000	8/06	SETN	
TO 30			G	$1,425	6/05	NCIA	3 pt., like Ford 8N
TO 30	1953		G	$1,300	8/05	NEIN	
TO 35	1955		E	$6,000	5/08	ECNE	Deluxe, 3 pt.

Ford and Fordson

Ford 8N and 9N models have not enjoyed the same value appreciation over the last 10 to 12 years that I've seen with many of the Farmall and Deere antique tractors. Not sure why. Just the way it is, I guess. Here's the proof. Back in 1997, Ford 9Ns sold for an average of $1,431 at auction. Ten years later in 2007, 9Ns sold for an average of $1,224.

Much the same with 8Ns. In 1997, 8Ns sold for an average of $1,955 at auction vs. 2007, which saw 8Ns selling at an average clip of $1,520 at auction. Perhaps the 8N-9N will make a comeback with collectors in the next 10 years.

Ford

Model	Year	Hours	Cond.	Price	Date	Area	Comments
2000			G	$4,500	3/07	NCOH	Gas, 8×2 trans., 3 pt., new rubber, sharp
2000	1963		G	$5,000	9/07	NECO	Diesel, loader, 3 pt., 6' blade, WF
2000	1964		G	$1,900	1/07	SECO	Enclosed cab, sweeper attachment, King Cutter 3 pt.
2000	1964	2,600	G	$6,750	11/07	WCIL	Offset, running, unrestored/original, fenders, PTO, lights, original paint, gas, runs good
2000	1970		G	$6,000	11/07	WCIL	Running, restored, fenders, PTO, lights, 3 pt.
2000			G	$2,100	9/06	NWTN	
2000			G	$2,150	5/06	ECMN	Gas
2000			G	$3,200	9/06	NWTN	
2000			F	$1,600	2/05	NWKS	3 pt., PTO, gas
2000			F	$3,000	4/05	SWPA	
2000			G	$3,500	9/05	SCON	Canadian sale, gas, ROPS
2000			G	$3,750	6/05	ECMN	Flail mower
2000			G	$4,600	9/05	NEIN	Loader
2000	1963	960	G	$2,600	1/05	SWOH	
2000	1969		G	$2,650	1/05	SWOH	Fresh OH
2000	1974	3,899	G	$3,200	1/05	SWOH	
2600	1977	2,163	G	$3,500	8/07	NWIA	Utility, fenders, 13.6-28 rubber
2N			G	$1,000	7/05	ECND	Gas, 3 pt., PTO
2N			G	$1,900	7/07	ECIL	Utility, PTO, 3 pt., older restoration
2N			G	$2,100	8/07	SWOK	
2N	1947		G	$1,925	3/07	NEIN	Gas, 3 pt., restored, 400 hours on OH
2N			G	$1,900	7/05	WCMN	New tires/paint
2N	1943		G	$1,500	5/06	SCIL	Rubber and steel, hyd. lift
2N	1945		G	$1,650	10/06	SCSD	Gas, repainted, 3 pt., unusual front wheel weights
2N			F	$850	10/05	NEPA	
2N			G	$1,300	6/05	ECWI	
2N			G	$1,600	9/05	SCON	Canadian sale
2N	1943		G	$700	8/05	NCIL	100 hours on OH
2N	1944		G	$1,000	8/05	NEIN	12 volt
2N	1946		G	$3,000	9/05	ECMN	Gas
3000			G	$2,250	2/08	NETX	
3000	1966		G	$4,500	3/08	SWKY	Gas, PS, spinouts
3000			G	$1,600	2/07	SCWA	
3000			G	$2,200	8/07	SESD	Gas
3000			F	$2,250	12/07	WCNY	Gas, fenders
3000		1,869	G	$3,700	1/07	NWNC	
3000			G	$3,700	1/07	NENC	
3000			G	$2,400	9/06	NWTN	
3000			G	$3,500	9/06	NEIN	Loader
3000			G	$5,000	12/06	NWWI	Loader
3000			G	$6,000	3/06	NWIL	Utility with 100 hours on OH, gas, Ford loader
3000	1969		G	$4,700	3/06	SWIN	Gas, 8 speed, power-spread wheels, good tires, weights
3000	1972		G	$3,100	7/06	SEND	2WD, diesel, Select-O-Speed trans.
3000			G	$1,300	12/05	SWMS	Diesel, 3 pt.
3000			G	$1,700	1/05	NWGA	3 pt., PTO
3000			G	$2,500	12/05	ECIN	Gas, 3 pt.
3000			G	$2,700	9/05	NCTN	Front-mounted broom
3000			G	$3,000	10/05	WCTN	

Ford

Model	Year	Hours	Cond.	Price	Date	Area	Comments
3000			F	$3,950	3/05	SWIN	PS, gas
3000			G	$4,000	10/05	WCTN	
3000		2,900	G	$4,150	1/05	SWOH	
3000		1,845	G	$5,300	1/05	SWOH	
3000	1971	3,000	G	$4,400	11/05	WCIL	40 hp., 8-speed trans., live PTO
3400			G	$3,400	4/08	SCMI	Industrial tractor, 730 loader, 3 pt., 2 hyd., gas
3400			G	$3,500	9/07	SENE	Utility, standard shift, 2,000 hours on tach
3400			G	$4,700	8/05	SCMI	Recent OH rebuilt hyd., bkt., 3 pt.
3600	1978		G	$4,600	4/08	ECMN	Utility with canopy and mounted Model A3A sod harvester, diesel, 19.5L-24 tires
4000		1,807	G	$2,600	4/07	ECMA	Tractor loader, gas, direct drive
4000			G	$5,250	1/07	WCIL	Utility, WF with Dunham Lehr loader
4000	1971	1,162	G	$4,900	8/07	SWOH	
4000			F	$1,750	8/06	NWOH	Gas, 3 pt., PTO, NF
4000			G	$2,500	6/06	NEIN	
4000			G	$2,650	3/06	NWIL	Diesel
4000			G	$3,400	8/06	NWOH	Gas, 3 pt., PTO, WF
4000			G	$3,500	9/06	NWTN	Tricycle front end
4000			G	$3,700	10/06	WCMN	3 pt., fenders, gas, Select-O-Speed
4000			G	$4,100	8/06	WCIL	
4000		2,695	G	$5,750	3/06	NCWI	Freeman 3000 fully hyd. loader
4000			G	$8,750	2/06	SWIN	Diesel, hi/lo trans., weights, PS, new paint
4000	1964		G	$4,300	9/06	NCIA	Select-O-Speed, 3 pt., completely restored, 13.6-28 tires, spin-out rims
4000	1968	3,864	G	$4,500	1/06	WCNE	WF, new paint
4000	1974	3,261	G	$4,800	4/06	WCMT	Diesel, Select-O-Speed trans., 3 pt., PTO, 1 hyd., Farmhand F11 loader with 5' bkt.
4000			G	$800	12/05	SWMS	Diesel
4000			P	$800	8/07	SESD	NF, Select-O-Speed, gas
4000			G	$1,500	10/05	WCTN	Select-O-Speed
4000			F	$1,900	11/05	NWIL	Diesel, Select-O-Speed, loader
4000			F	$2,000	1/05	NWOH	Utility with Superior loader, 3 pt., Select-O-Speed
4000			G	$2,500	10/05	WCMT	
4000		5,880	P	$3,200	8/05	NCOH	Loader, diesel
4000			G	$5,200	3/05	ECMN	Gas
4000			G	$5,500	12/05	SWMS	Diesel, 3 pt.
4000			G	$6,100	5/05	SWIN	Gas, 4 speed, PS, Farmhand 22 front-end loader
4000			G	$6,400	8/05	NCIA	Utility, Select-O-Matic, sold with Ford 727 loader, WF, 3 pt., new tires/rims, fenders
4000			G	$6,900	7/05	SEND	Woods dual loader, PS, 3 pt., PTO, hyd., 6.5' bkt.
4000	1969		G	$7,700	7/05	ECMN	8% buyers premium, final total $8,308, Ford 727 loader, tach showed 3,400 hours but not working, PS, 540 PTO, 3 hyd., 8/2 trans., fenders, sheet metal straight, very little rust, tires all need to be replaced
4000	1975		G	$2,450	2/05	NEIN	Gas, WF, Select-O-Speed
4100			G	$7,000	9/06	NWTN	
4400	1972		G	$4,100	9/07	NECO	Tach shows 2,154 hours, gas, cab, 3 pt., PTO, Leland 7' dozer
4500			G	$6,600	9/05	SCON	Canadian sale, loader, backhoe

Ford

Model	Year	Hours	Cond.	Price	Date	Area	Comments
5000		3,152	G	$4,500	2/08	WCMN	2WD, Select-O-Speed trans, 1 hyd., 3 pt., PTO, diesel, 15.5-38 tires
5000		4,500	G	$6,200	2/08	SEMO	Diesel, clean
5000		2,798	G	$7,000	3/08	NCIL	WF, 3 pt., quick hitch, Ford 772 hyd. loader, gas, 16.9-30 rear tires, one owner
5000	1972	5,900	G	$5,100	3/08	SWKY	Diesel, 1 hyd., Dunham loader, cast iron spinouts
5000			G	$3,400	5/07	WCSD	Front diesel, cab, 3 pt., new batteries
5000			G	$3,500	11/07	ECSD	Open station, 1 hyd., 15.5-38 singles (50%)
5000			G	$5,000	12/07	SCMN	Gas
5000	1970		G	$5,250	4/07	NEMI	Diesel, duals, hyd. outlets, 66 hp., one owner
5000			G	$3,950	8/06	SETN	Front loader
5000		1,680	G	$5,000	3/06	NCWI	New tires, 2 hyd.
5000			G	$5,500	9/06	NWTN	Front loader
5000		2,800	G	$7,250	4/06	SWIN	Diesel, good tires
5000	1974	1,972	G	$6,400	2/06	NECO	Diesel, cab, Westendorf WL21 loader, 3 pt.
5000	1975	5,377	G	$7,000	4/06	SEMI	WF, 3 pt., 1 hyd., diesel, dual power, 16.9-38 tires
5000			G	$3,150	8/05	NEIN	Diesel
5000		3,186	G	$4,150	11/05	SCMI	Ford 727 loader, diesel
5000			F	$4,650	10/05	NEPA	
5000			G	$5,000	4/05	NCPA	WF, 1 hyd.
5000			G	$5,000	12/07	SCMN	Roll bar, 3 pt., 540 PTO, 2 hyd., gas, 16.9-34
5000		3,394	G	$6,100	4/05	ECND	Diesel, cab, 3 pt., PTO, 2 hyd., 15.5-38, clean
5000			F	$6,750	5/05	SWIN	Diesel, 6 speed, PS
5000	1962		G	$3,750	2/05	NEIN	Gas, NF
5000	1972		F	$3,400	6/05	NESD	2WD, cab, 1 hyd., 3 pt., 18.4-30 rear tires
5600	1978	721	G	$8,200	8/07	SWOH	Allied 590 loader, 3 hyd.
600			G	$4,000	4/06	WCSD	New rear tires, new battery
600			F	$1,400	2/05	WCIN	Industrial, 3 pt., PTO, hyd. front loader, no hood
600			G	$1,500	12/05	ECIN	Gas, 3 pt.
600			G	$4,300	11/05	WCIL	WF, gas, 134 cubic inch
6000			P	$900	4/08	NCOH	For parts, gas, WF
6000		3,451	G	$4,600	3/08	ECWI	New paint, restored
6000	1967	4,786	F	$2,500	9/07	NECO	Dual loader, gas, WF, Select-O-Speed
601	1959		G	$3,200	11/07	WCIL	Running, restored, fenders, PTO, lights, 3 pt.
601			G	$1,000	8/06	SETN	
601	1961		G	$1,500	12/06	SCMT	4 speed
601			G	$2,000	11/05	WCIL	One-arm loader
601			G	$4,400	9/05	NEIN	New Kelly loader
640			F	$1,200	1/08	NEMO	
640			F	$2,000	1/05	NWOH	
640			G	$2,200	11/07	WCIL	Running, unrestored/original, fenders, PTO, lights, 3 pt.
641			F	$2,500	9/07	WCIA	Original, factory LP, repainted
641	1959	1,030	G	$2,200	2/06	NECO	New tires, 3 pt., gas
641	1958		G	$2,300	8/05	NEIN	Gas
661			G	$2,850	1/05	SWOH	Live power
7000	1973		G	$6,000	4/07	NEMI	Diesel, load monitor, duals, dual range, 83 hp., one owner
7600	1974	4,300	G	$11,000	3/06	SWIN	Diesel, dual power, load monitor, roll bar canopy, 2-speed PTO, weights, radio, good tires

Ford

Model	Year	Hours	Cond.	Price	Date	Area	Comments
7600	1976		G	$8,000	4/05	NCCO	Cab, recent OH, 15.5-38, duals, 2 hyd., 540 PTO, new engine
800	1955		G	$2,900	3/08	NEMO	WF, 3 pt.
800			G	$1,900	2/07	NEIN	Tach shows 285 hours, gas, 3 pt., PTO
800			F	$2,200	6/07	SEID	Backhoe and loader, runs goods, everything works
800			G	$2,900	3/07	SEMI	Gas tractor, 3 pt., PTO, 13. 6-28 tires
800			G	$2,700	3/06	SWOH	WF, gas, utility tractor, PTO, duals, Ford trip bkt., hyd. front loader
800	1953		G	$3,700	3/06	ECMN	Gas
800	1956		G	$2,250	6/06	NECO	3 pt. blade, gas
800	1957		F	$850	7/06	ECND	3 pt., 2 hyd., PTO
800	1957		G	$4,600	9/06	NECO	6' loader, 3 pt. blade, gas
800	1958		G	$2,750	3/06	NWIL	Gas
800			F	$2,200	5/05	SEPA	Gas, WF, rubber 50%
800			G	$2,900	12/05	SWMS	Gas, 3 pt.
800			G	$3,800	8/05	WCIL	Loader
8000			G	$2,600	1/06	NEIN	Diesel, 3208 Cat engine, 5 speed, steel spreader box
8000		9,269	G	$3,750	11/07	ECSD	Dual power, open station, 2 hyd., 3 pt., PTO
8000			G	$4,750	5/07	WCWI	WF
8000	1971	4,701	F	$6,400	3/06	NWIL	Diesel, WF, dual power, rear end OH in 2000
8000			G	$3,400	4/05	SEMN	Diesel, 3 pt., 23.1-34 tires
8000			G	$6,000	2/05	NCIN	Diesel, WF, OH
801			G	$3,300	11/07	SESD	Select-O-Speed, new motor
801	1960		G	$2,350	5/06	SEWY	3 pt., WF
801			G	$3,000	8/05	ECMI	5 speed, live PTO, good rubber, remote hyd., diesel
801			G	$5,500	12/05	NECO	Select-O-Speed, Ford step side loader, 3 pt., PTO, new rubber
841	1959		E	$4,650	12/06	SWMI	Power Master, WF, gas, tractor tie weights, front bumper grill, completely restored in showroom condition
860			G	$2,200	1/07	NWIA	One owner
860	1955		G	$3,950	9/07	NCIA	PS, 5 speed, live PTO, 12-volt alternator, 3 pt., repainted
8600	1976		F	$4,100	3/07	WCIL	No cab, weights, diesel
8600			P	$2,600	4/05	NCOH	Did run, diesel, cab, 38" duals, weights, full power
861			G	$2,300	3/07	WCMI	Gas, 3 pt., 13.6-28 tires
861			F	$2,300	11/07	SESD	Loader, 3 pt.
861	1960		G	$6,750	10/06	SCSD	WF, 5 speed, gas, PTO, new rubber, 3 pt., loader, complete valve job, repainted
861			G	$2,200	2/05	NEIN	Gas, WF
861			G	$2,900	2/05	SCMN	5-speed trans.
8700	1978	5,797	G	$4,300	12/07	SEWY	3 pt., WF, cab, diesel
881	1959		E	$5,600	12/06	SWMI	Select-O-Speed, PS, WF, gas, PTO, all new rubber, completely restored – nicer today than from the factory, everything new, calendar showpiece tractor
8N			G	$1,100	6/07	ECMN	3 pt., PTO
8N			G	$1,800	2/08	NWSC	
8N			P	$600	6/07	NCIN	
8N			F	$650	11/07	SESD	
8N			F	$750	12/07	ECNE	

RUST BOOK

Ford

Model	Year	Hours	Cond.	Price	Date	Area	Comments
8N			F	$900	8/07	SESD	Loader
8N			F	$900	11/07	ECKS	11.2-28 rear tires, 540 PTO, 3 pt.
8N			F	$950	6/07	SWNE	11.2-28 rear tires, side distributor, rpm tach, 3 pt., PTO
8N			F	$950	12/07	WCNY	Fenders
8N			F	$975	8/07	SESD	Side distributor
8N			F	$1,000	12/07	WCNY	No front tire, fenders, front blade
8N			G	$1,100	11/07	WCKS	3 pt. posthole digger, PTO
8N			G	$1,150	12/07	SEIA	
8N			F	$1,200	6/07	NCIN	
8N			F	$1,300	6/07	NCIN	Grill, hood, radiator damage, runs
8N			G	$1,300	8/07	SWOK	
8N			F	$1,400	6/07	NCIN	Good tires, Sherman rear, unrestored, bumper
8N			F	$1,500	11/07	WCIL	
8N			G	$1,500	8/07	SWIL	Front bumper, older repaint, 3 pt.
8N			G	$1,600	8/07	SWOK	
8N			G	$1,600	9/07	SCMO	
8N			G	$1,700	6/07	NCIN	WF, gas, unrestored
8N			G	$1,750	4/07	NWNY	Powered by gas engine, equipped with PTO, 3-point hitch, Bush Hog mower.
8N			G	$1,850	3/07	ECNE	11.2-28 rear tires, 4:00-19 front tires, 3 pt.
8N			G	$1,900	5/07	ECND	PTO, 3 pt., Ford loader, hyd. bucket, tire chains
8N			G	$2,400	4/07	ECSK	Canadian sale, good tires
8N			G	$2,700	4/07	NWMN	3 pt., PTO, side distributor
8N			G	$3,100	7/07	SCMN	All hyd. loader, good rubber, front-mounted hyd. pump
8N			G	$3,100	2/07	ECNE	Straight tin, nice tractor
8N	1947		F	$1,100	6/07	NCIN	Side distributor, unrestored
8N	1947		G	$1,600	6/07	NCIN	Unrestored, runs
8N	1947		G	$1,650	5/07	SEWY	Gas, 3 pt., belt pulley
8N	1948		E	$2,600	4/07	NECO	Gas, 12 volt
8N	1949		F	$1,300	6/07	NCIN	Old repaint
8N	1951		G	$1,100	12/07	SEWY	Gas, 3 pt.
8N	1952		F	$2,500	10/07	NWWI	New rubber and paint but does not run
8N			P	$800	4/06	ECMN	Gas
8N			F	$850	1/06	SEPA	Fair rubber
8N			F	$900	6/07	NCIN	Loader
8N			G	$1,000	6/06	ECSD	Ford loader, good runner
8N			G	$1,150	3/06	NEOH	Loader, blade
8N			G	$1,150	9/06	NEIN	Gas, 3 pt., PTO
8N			F	$1,250	4/06	SCKS	
8N			G	$1,400	4/06	WCSD	WF, 3 pt., buzz saw, belt driven
8N			G	$1,400	4/06	NCOK	Ran great, good tires, PTO, 4 speed
8N			G	$1,500	3/06	NWKS	OD, 11.2-28 tires, 3 pt.
8N			G	$1,600	3/06	SCIN	Loader, Bush Hog mower
8N			F	$1,600	8/06	NWIL	Loader, OD
8N			G	$1,600	8/06	NWIL	Gas, Woods 5' mower
8N			G	$1,600	10/06	SESK	Canadian sale, 3 pt., one new tire
8N			G	$1,800	8/06	SWIA	3 pt.
8N			G	$1,800	8/06	ECIL	Hyd. loader
8N			G	$1,850	5/06	ECMN	Gas
8N			G	$1,850	3/06	SCIN	Woods 7' sickle mower
8N			G	$1,900	8/07	NWIL	Gas

Ford

Model	Year	Hours	Cond.	Price	Date	Area	Comments
8N			G	$2,100	10/06	WCFL	Gas
8N			G	$2,500	4/06	WCTX	6 cyl.
8N			G	$2,800	8/06	NCIA	New tires all way round, OD, buggy top, 12 volt
8N			G	$2,800	2/06	SWNE	4 speed, side-mount distributor, 3 pt., 540 PTO, 11.2-28 tires
8N			G	$3,100	9/06	NEIN	6-cyl. funk conversions
8N	1947		F	$1,400	3/06	NESD	3 pt., OD trans.
8N	1947		G	$1,400	9/06	NECO	3 pt., gas
8N	1947		G	$2,000	2/06	NCKS	3 pt., PTO
8N	1948		F	$1,500	7/06	ECIA	Runs OK, new tires
8N	1948		G	$1,750	10/06	NECO	Gas, converted to 12 volt
8N	1948		G	$3,000	10/06	SCSD	3 pt., gas, WF, 4 speed, lights, restored
8N	1948		G	$3,650	6/06	NECO	Davis loader, 3 pt. blade, gas
8N	1948		G	$5,000	8/06	WCIL	Restored
8N	1950		G	$1,600	9/06	NECO	Bush Hog 6' 3 pt. mower, gas
8N	1950		G	$3,200	3/06	ECMN	Gas
8N	1950		G	$4,000	10/06	SCSD	3 pt., gas, side distributor, 4 speed, repainted
8N	1951		G	$2,250	9/06	SEIA	
8N	1951		G	$4,200	10/06	SCSD	3 pt., lights, side distributor, tach, Dearborn pipe loader, restored, gas
8N			P	$335	10/05	WCIL	Parts tractor, bad
8N			P	$700	8/05	WCIL	Missing parts, bad
8N			P	$700	8/05	WCIL	Gas
8N			F	$700	11/05	WCKS	3 pt.
8N			G	$1,000	1/05	NWGA	3 pt., PTO
8N			G	$1,000	4/06	WCMT	
8N			F	$1,100	2/05	SCCA	3 pt., PTO, 11.2-24 tires
8N			F	$1,100	1/05	SWOH	
8N			F	$1,200	1/05	SWOH	
8N			G	$1,250	1/05	SWOH	
8N			G	$1,350	1/05	SWOH	
8N			G	$1,400	9/05	SCON	Canadian sale, gas, loader, 3 pt.
8N			G	$1,400	12/05	ECWA	4 cyl., flatbed, 4 speed, PTO, 3 pt.
8N			G	$1,550	1/05	SWOH	
8N			F	$1,550	9/05	NEIN	Gas, 12 volt, Sherman trans.
8N			F	$1,600	3/05	WCIL	Gas, new rear tires
8N			G	$1,800	4/05	NCIA	Ford side-mount sickle mower
8N			G	$1,800	9/06	NEIN	
8N			G	$1,900	11/05	WCIL	New tires, gas
8N			E	$2,000	4/05	NCOK	New paint, 12-volt system, runs excellent
8N			G	$2,000	4/05	ECMN	
8N			G	$2,050	1/05	SWOH	
8N			G	$2,500	2/05	NWKS	3 pt.
8N			G	$2,900	6/06	ECMN	Hi/lo trans.
8N			G	$3,000	12/05	NECO	3 pt., PTO, clean
8N			G	$3,000	12/05	NECO	3 pt., PTO, clean
8N			G	$3,150	4/05	NWIL	Loader
8N	1948		G	$2,275	6/05	NEND	4 speed, 3 pt., PTO
8N	1950		G	$1,100	8/05	NEIN	Gas, repainted
8N	1951		G	$950	11/05	ECNE	
8N	1951		G	$2,500	6/05	NEND	4 speed, 3 pt., PTO

Ford

Model	Year	Hours	Cond.	Price	Date	Area	Comments
8N	1952		G	$2,100	11/05	WCIL	Side distributor running, 3 pt., front bumper
8N	1959		G	$2,600	2/05	NECO	3 pt., PTO, 1 hyd., rubber 40%
900			G	$1,700	8/05	ECMN	NF
901			F	$1,900	3/08	NEAR	Diesel, 3 pt., PTO, WF, 13.6/12-28 tires
901	1957		G	$2,400	4/08	NENE	Power Master, WF, 13.6-28 rear tires, 3 pt., fenders
901	1961		F	$950	4/08	NENE	Select-O-Speed, diesel, WF, 13.6-28 rear tires, 3 pt.
901			E	$5,400	10/06	SCSD	Select-O-Speed, NF, 3 pt., PS, live PTO, live power, complete restoration, mint, gas
901	1959		F	$1,600	5/06	SEWY	WF, 3 pt.
901			F	$1,150	7/05	SESD	NF, Select-O-Speed
951	1961		F	$900	12/07	SEWY	Diesel, 3 pt., double front
960	1960		G	$2,700	11/07	WCIL	Running, old restoration/repaint, fenders, PTO, lights, deluxe seat, PS, OH in 2000 by previous owner and repainted at that time
960	1957		G	$4,500	10/06	SCSD	NF, 5 speed, gas, wheel weights, lights, 3 pt., blue color, mint
9600				$0	7/07	WCMN	Factory cab, 2 hyd., 3 pt., PTO, air
9600	1976	3,432	G	$8,000	3/07	WCMI	Cab, diesel, 10 front weights, 2 hyd., PTO, 18.4-38 axle duals, one owner
9600	1976		G	$9,000	4/07	NEMI	Cab, duals, 135 hp., diesel
9600			F	$3,900	8/06	NWOH	Cab, 2 hyd., 3 pt., duals
9600	1974	9,100	G	$7,000	12/06	WCMN	2WD, cab, band duals, recent injector pump and injector work, 2 hyd., 3 pt., PTO
9600			P	$2,700	8/05	NEIA	Rough
9600		5,061	F	$5,000	11/05	SCMI	OH in 1994, cab, 3 pt., 2 hyd., 10 front weights, snap-on duals
9600		7,313	G	$5,500	2/05	WCOH	ROPS, 540/1,000 PTO, 2 hyd., 3 pt., duals
9600	1974		G	$4,400	2/05	NEIN	CAH, diesel
961	1958		G	$6,000	10/06	SCSD	PS, NF, 5 speed, 3 pt., new rubber, lights, diesel
9700	1977	8,823	G	$11,750	3/08	WCMN	2 hyd., cab, air, front weights, 3 pt., 540/1,000 PTO, cast duals, excellent rubber
9700	1978	7,978	G	$10,000	2/08	SCMN	Major OH at 7,200 hours, factory cab, air, 2 hyd., axle duals, rock box
9700	1975		G	$5,500	7/06	ECND	3 pt., PTO, duals
9700	1977		G	$11,200	9/06	NWOH	Dual power, cab, air, front weights, duals, 125 hours on complete OH
971	1960		G	$3,500	3/07	NCNE	Select-O-Speed, 3 pt., dual loader with bkt., runs good
9N			F	$800	2/08	SEAL	
9N			G	$2,200	3/05	ECMN	3 pt.
9N			F	$750	12/07	ECNE	
9N			F	$800	11/07	WCIL	Running, unrestored/original, fenders, PTO, lights, deluxe seat, 3 pt., overrunning PTO clutch, battery cover, PTO cover, side toolbox, rebuilt carburetor, new wiring
9N			F	$1,100	6/07	NCIN	One new tire
9N			G	$1,100	4/07	ECSK	Canadian sale, 2WD
9N			F	$1,200	1/07	NWOH	Original rear tires, used as orchard sprayer, gas
9N			F	$1,200	11/07	WCIL	
9N			G	$1,200	12/07	SCMT	3 pt., PTO, near-new tires
9N			F	$1,250	11/07	NENE	
9N			G	$1,300	2/07	SCWA	

Ford

Model	Year	Hours	Cond.	Price	Date	Area	Comments
9N			F	$1,400	6/07	NCIN	Unrestored
9N			F	$1,400	11/07	WCIL	Bush Hog mower
9N			G	$1,500	12/05	WCMN	Over/under trans., Dearborn loader
9N			G	$1,600	6/07	NCIN	Complete, original, runs
9N			G	$1,700	6/07	NCIN	Original, OD, front hitch
9N	1940		F	$850	12/07	ECIL	Front hitch
9N	1940		E	$1,800	3/07	NWIL	Gas, new paint/tires
9N	1943		F	$650	12/07	SEWY	Gas, needs work
9N			G	$550	12/06	SCMT	Loader
9N			G	$800	12/06	SCMT	
9N			F	$1,025	10/06	SCSD	Gas, original, step up, 3 pt., as is
9N			F	$1,385	4/06	SCMN	12-volt electric system
9N			G	$1,500	10/06	WCSK	Canadian sale, 3 pt.
9N			F	$1,800	3/06	SEPA	Sounder loader, good rubber
9N			G	$1,800	3/06	WCKS	Scoop, 3-pt. blade, one-bottom plow
9N			G	$2,100	10/06	NCKS	3 pt., PTO, hi/lo
9N			G	$2,100	1/06	SESD	3 pt.
9N	1939		G	$1,450	9/06	ECMI	60" belly mower
9N	1940		G	$1,700	9/06	NCIA	Good tires/rims, runs good, Dearborn 19-8A loader
9N	1940		G	$1,800	10/06	WCMN	New tires, repainted
9N	1943		G	$1,800	6/06	NECO	3 pt., gas
9N			F	$800	9/06	NEIN	Gas
9N			F	$900	3/05	NWIL	Gas
9N			G	$1,125	1/05	SWOH	
9N			G	$1,150	1/05	SWOH	
9N			G	$1,350	4/05	NWIL	3 pt., PTO
9N			G	$1,400	12/05	NENE	
9N			G	$1,450	1/05	SWOH	
9N			E	$1,600	4/05	NCOK	New paint, 12-volt system, runs great, 5' mower
9N			F	$1,600	8/05	ECTN	Made in the 1940s, new paint
9N			G	$1,850	7/05	SEND	3 pt., hyd., PTO
9N			G	$2,000	6/05	NWMN	2-speed OD, 3 pt., PTO
9N	1940		F	$950	6/05	NEND	3-speed, 3 pt., PTO
9N	1940		G	$1,600	6/05	NEND	3 speed, 3 pt., PTO
9N	1940		G	$2,500	6/05	NEND	3 speed, 3 pt., PTO, turf tires
9N	1942		G	$1,400	9/05	NECO	3 pt., gas
9N	1943		F	$1,300	9/05	WCNE	Gas, 3 pt., runs good, tin work bent badly
9N	1943		G	$1,700	3/05	NECO	3 pt., PTO, rebuilt, new tires
Jubilee				$0	1/08	NWOH	
Jubilee			G	$2,000	1/08	NWOH	Utility
Jubilee	1954		G	$4,000	5/08	ECMN	Golden Jubliee, 3 pt., unrestored
Jubilee	1954		G	$5,250	5/08	ECMN	Golden Jubilee, OD, 3 pt., restored, new tires
Jubilee			G	$3,400	7/07	SESK	Canadian sale, antique tractor, 3 pt., PTO
Jubilee	1953		G	$2,500	8/07	NWIA	Runs good

Ford

Model	Year	Hours	Cond.	Price	Date	Area	Comments
Jubilee	1953		G	$5,400	11/07	SWIN	Golden Jubilee, restored, live power, new rears
Jubilee			G	$2,100	7/06	WCCA	3 pt., PTO
Jubilee	1953		G	$5,000	4/06	ECND	NAA, extensive reconditioning
Jubilee	1954		G	$1,900	10/06	NECO	Gas, 3 pt., 6' blade
Jubilee	1953		G	$1,100	1/05	NWGA	3 pt., PTO
Jubilee			F	$1,950	1/05	SWOH	
Jubilee			G	$2,300	6/05	ECMN	
Jubilee			G	$2,800	1/05	SWOH	
Jubilee			G	$2,900	1/05	NEMO	Utility, Ford loader
Jubilee			G	$3,000	8/05	ECMI	Rebuilt engine, new gauges and wiring, new paint, good rubber
Jubilee	1953		F	$1,700	9/05	WCNE	Gas, 3 pt.
Jubilee	1953		G	$2,600	9/05	WCNE	Gas, 3 pt. blade, new tires
Jubilee	1953		G	$3,400	6/05	ECIL	New tires, 3 pt., 1 hyd., restored
Major	1958		G	$2,600	10/07	NECO	Diesel, 16" backhoe, 8' dozer
Major	1960		G	$3,100	5/07	SEWY	Great Bend 440 loader and 6' bkt., diesel
N/A	1972		G	$3,100	2/07	ECNC	Ford diesel, 8 speed, PTO, enclosed cab
NAA	1953		G	$3,200	8/07	WCIL	New tires, Ford 5' 3 pt. grooming mower
NAA	1954		G	$1,550	10/07	NECO	Gas, WF, 3 pt.
NAA			G	$1,500	9/06	NWTN	Jubilee
NAA	1954		G	$5,000	10/06	SCSD	Lights, gas, 3 pt., complete motor OH, restored, repainted
NAA			G	$2,400	2/05	NCIN	
NAA	1953		G	$3,600	12/05	NECO	Golden Jubilee, 3 pt., PTO, nice
TW 10	1978	6,792	G	$10,000	3/06	SEMN	2WD, one owner, cab, air, 18.4-38 tires, duals, 3 pt., 2 hyd.

Fordson

Model	Year	Hours	Cond.	Price	Date	Area	Comments
F			G	$2,100		WCIL	Running, old restoration/repaint, early 1920s
Major			G	$1,450		WCMN	Diesel, Ford dual hyd. loader
Major			G	$3,800		WCNE	Diesel, 3 pt., 14-30 rear tires, 7.50-16 front tires, PTO
N/A			F	$1,200		NWOH	Full steel, no fenders, old repaint
N/A			G	$1,800		NCIN	Full steel, good fender, restored, runs
N/A	1922		F	$1,200		WCIA	
N/A	1925		G	$1,600		WCIA	WF, hard tires, gray with red wheels
N/A	1937		G	$6,250		WCIA	Single front wheel, blue, very rare
N/A	1937		F	$6,500		WCIL	Row crop, running, old restoration/repaint, PTO, belt pulley, engine block was repaired, rear rims need work
N/A			F	$700		NWIL	Hybrid, gas
N/A			F	$1,600		NWOH	On steel, gas, WF, did not run
N/A			G	$3,750		WCTX	Rare, NF, older restoration
N/A	1919		F	$2,750		NWIL	On steel, gas
N/A	1921		F	$1,200		NWIL	On steel, gas
N/A	1924		F	$1,900		NWIL	On rubber, gas
N/A			F	$700		ECWI	

IHC/Farmall and McCormick

Old and really old, reliable red tractors continue to draw buyers' interest. First the relatively old, a 1978 IHC 1486 with 4,919 hours sold on a farm auction in north-central Iowa November 30, 2007, for $17,000. That's the highest auction price I've seen on a 1486 in six years and the second highest price since 1998. Note the timing of that sale. Commodity prices began their surge higher in late November 2007 pushing used tractor values up right along with them.

Now for the really old. Farmall/IHC made the venerable M tractor from 1939 to 1952. Back in 1997, M's were selling for an average of $1,175 at auction. By 2007, M's were up to an average auction price of $1,450, an increase in value of 23%. In 2008? M's were up even further, to an average sale price of $1,885.

IHC/Farmall

Model	Year	Hours	Cond.	Price	Date	Area	Comments
10-20			F	$850	8/07	NWOH	Old restoration, rear steel, cutoff fronts, no SN
10-20	1917		E	$50,000	9/07	NEIL	Mogul 10-20, engine rebuilt, new sheet metal, original plow guide
10-20			F	$250	10/06	SEWI	Full steel, missing radiator
10-20			F	$375	10/06	SEWI	Full steel, missing radiator
10-20			G	$400	10/06	SEWI	Cutoffs, rough fenders
10-20			F	$400	1/08	ECNE	On steel
10-20			G	$600	10/06	SEWI	Full steel, fairly complete
100			G	$2,200	6/06	SEID	Cab, 3 pt., dual PTO
100				$2,200	6/06	SEID	Cab, 3 pt., dual PTO
100			G	$2,750	6/06	NEIN	Farmall, repainted
100			F	$2,900	3/06	SEPA	Cultivators, WF, good rubber
100			G	$6,500	2/06	WCME	Hydro, cab, 10 rear weights
100	1974	5,117	F	$6,000	2/06	WCIL	Hydro, factory cab, 2 hyd., weak hydro
100	1974	9,571	G	$7,250	1/06	WCNE	Diesel, cab, dozer
100			F	$900	8/06	WCMN	Farmall, original
100			G	$4,200	6/05	WCMN	Restored, fast hitch, WF, fenders
1026	1970		G	$8,700	3/07	NEOH	Open station, hydro
1066		6,800	G	$6,250	3/08	WCMN	Diesel, 2 hyd., 3 pt., 540/1,000 PTO, 10.00-16 fronts, 18.4-38 axle duals
1066		5,900	G	$8,400	3/08	ECSD	Turbo, 2 hyd., 4-speed hi/lo, 540/1,000 PTO, IHC CAH, axle duals, no 3 pt.
1066			G	$8,800	2/08	SWWI	No cab, flat-top fenders, OH 3 years ago, very good original paint
1066	1973	7,831	G	$8,000	2/08	NEMO	18.4-38 tires, 2 hyd., 540/1,000 PTO
1066	1974		G	$8,200	3/08	SCNE	9-bolt duals and hubs, 540/1,000 PTO, 3 pt.
1066	1974	6,500	G	$11,000	1/08	NWIL	Year-A-Round cab, new clutch, axle hub duals
1066	1975	6,700	F	$7,800	1/08	NWOH	ROPS, OH at 5,800 hours, hub duals
1066		7,375	G	$6,100	3/07	ECIL	Turbo, 1300 Series cab, 540/1,000 PTO, 3 pt., 2 hyd., 18.4-38 tires, duals
1066			G	$6,200	6/07	ECND	Red, CAH, 2 hyd., 3 pt., 14.9-38 hub duals, 80% rubber
1066			G	$6,500	7/07	WCMN	2WD, 2 hyd., rebuilt engine, factory cab, rock box, axle duals
1066		6,771	G	$6,750	7/07	SESK	Canadian sale, 2WD tractor, 116 PTO hp., 2 hyd., dual PTO, 4-rib front, 18.4-38 rear tires
1066			G	$7,600	12/07	SWWI	No cab
1066		1,500	G	$7,700	10/07	WCWI	Dual PTO, 2 hyd., OH, diesel
1066			G	$8,250	7/07	WCMN	Pulling tractor, pulls 12,500, 15,000, 18,500, farm stock, very competitive, new hd TA, new hd clutch, 9-bolt hubs
1066			F	$8,700	8/07	NWIA	Diesel, WF, cab, 18.4-38 near new rubber, 3 pt., 2 hyd., 540/1,000-rpm PTO, recently painted, engine OH
1066			G	$8,800	3/07	NEWI	Diesel
1066		6,442	G	$9,100	6/07	SWMN	Black stripe, cab, 3 pt., 2 hyd., 18.4-38 hub duals, one owner
1066		2,600	E	$16,000	2/07	SEIL	Fender tractor, 2 hyd., weights, quick hitch, 18.4-34 tires, extra sharp
1066	1971	8,677	G	$5,300	2/07	NEND	Farmall cab, 2 hyd., 3 pt., 540/1,000 PTO, twin aux., fuel tanks, front suitcase weights, 18.4-38 hub duals, 11L-15 fronts
1066	1971	9,200	F	$6,500	2/07	NWIL	Cab
1066	1971	5,118	G	$9,600	7/07	NWIL	Diesel, cab, axle duals, dual PTO, 3 pt., new tires

IHC/Farmall

Model	Year	Hours	Cond.	Price	Date	Area	Comments
1066	1972	8,920	F	$3,000	12/07	WCNE	Diesel, WF, no cab, turbo, runs
1066	1972	7,447	F	$4,600	3/07	SEWY	Diesel, WF, 3 pt., cab
1066	1972	6,700	G	$9,500	8/07	NWIL	Dual PTO, 2 hyd., 3 pt., OH
1066	1974		G	$6,000	7/07	NCIA	Red, cab, 3 pt., 2 hyd.
1066	1974		G	$8,000	11/07	WCIL	Turbo, diesel, 18.4-38 rear tires and duals, 10.00-16 front tires, 540 and 1,000 PTO, 2 hyd., tach showing 2,132 hours – not correct, sells with front weights, SN 45944
1066	1974		G	$8,700	12/07	ECIL	18.4-38 tires, turbo, 10 front weights, dual PTO, 2 hyd.
1066	1974	4,400	G	$10,000	12/07	WCIL	200 hours on OH, open station, clamp-on duals, diesel, no cab
1066	1974	1,870	G	$10,000	8/07	SCMN	One owner, Year-A-Round cab, 20.8-38 hub duals, rock box
1066	1975	4,780	P	$1,900	12/07	WCNE	Cab, diesel, turbo, 3 pt., engine locked
1066	1975	16,603	G	$6,500	12/07	ECNE	Hydro, 16.9-38 rear tires, 10:00-16 front tires
1066	1975	4,748	G	$7,750	8/07	WCMN	Second owner, Year-A-Round cab installed in 1991, 2 hyd., 3 pt., PTO, K&M rock box, 18.4-38 tires with 9-bolt duals, 11:00-16 tri-rib front tires, completely restored, always shedded
1066	1976		G	$9,000	3/07	NWIA	No cab
1066	1976	4,226	F	$9,000	2/07	NWIL	Diesel, black stripe, cab, like-new rubber
1066			F	$2,800	12/06	SEND	Factory cab, 3 pt., PTO, 20.8-38 singles, 15% rubber
1066			F	$2,900	12/06	SCMT	
1066			F	$4,400	3/06	NWIL	Diesel
1066			G	$4,400	7/06	NWMN	Turbo, cab, hub duals, 3 pt., 2 hyd., PTO
1066			G	$4,500	3/05	NWMN	Turbo, cab, 3 pt., 2 hyd., PTO
1066			F	$4,600	3/06	NCMN	
1066		8,000	F	$5,500	2/06	NENE	Cab
1066		4,000	G	$5,500	8/06	NCOH	Open station, WF, 540/1,000 PTO, duals
1066			F	$5,900	3/06	NWIL	Diesel, TA out
1066			G	$6,000	3/06	ECIN	Diesel, PTO, 3 pt.
1066		4,252	F	$6,300	8/06	NWOH	1 hyd., 3 pt., duals
1066		3,790	G	$6,500	8/06	WCIA	
1066			G	$7,100	3/06	NCMN	
1066		5,981	F	$7,600	2/06	NENE	2 hyd., no cab
1066		4,260	G	$8,200	8/06	WCIA	Westendorf WL-42 loader
1066		5,111	G	$8,250	3/06	NWPA	Cab, 2 hyd., dual PTO, rebuilt motor
1066			F	$8,500	4/06	NENE	Turbo, cab, 3 pt., 2 hyd., axle- mount duals, front weights
1066		3,500	G	$9,100	1/06	NEIN	Black stripe, front fenders, nice
1066	1969		G	$5,000	11/06	NWKS	3 pt., 2 hyd.
1066	1971		F	$5,900	8/06	NCKS	Wheatland
1066	1972	5,774	G	$6,500	7/06	NCIA	Open station, 2 hyd., 3 pt., 1,000 PTO, less than 1,000 hours on OH and injection pump, 18.4-38 tires
1066	1973	8,964	F	$3,100	9/06	NECO	Cab, 400 hours on OH, diesel
1066	1973	5,668	G	$7,750	2/06	SWOH	2 hyd., 18.4-38, duals, 12 front weights, 2 rear weights
1066	1974	5,890	F	$3,600	5/06	SEWY	Year-A-Round cab, WF, 3 pt.
1066	1974		F	$6,000	1/06	NWKY	No cab, turbo, average
1066	1974	6,981	F	$7,500	2/06	SCNY	2WD, open, 2 hyd., no weights
1066	1975	33,230		$0	1/06	WCIL	CAH, 2 hyd., weights

IHC/Farmall

Model	Year	Hours	Cond.	Price	Date	Area	Comments
1066	1975		F	$5,250	3/06	NCCO	
1066	1975	4,819	G	$5,400	9/06	NEIN	Diesel, CAH, duals
1066	1975	5,500	G	$8,000	2/06	NCCO	Axle duals
1066	1975	3,230	E	$14,250	1/06	WCIL	CAH, 2 hyd., weights
1066	1976	9,662	G	$5,200	6/06	NECO	Cab, diesel, front weights, duals
1066	1976	6,203	G	$6,750	2/06	NECO	Axle duals, diesel, 3 pt., cab
1066	1977	2,736	G	$9,300	2/06	SEIA	Cab, 2 hyd., duals
1066		6,935	G	$4,300	1/05	ECNE	
1066			F	$4,900	4/05	SCSD	
1066		10,035	G	$5,600	2/05	NEIN	Diesel, cab, no doors
1066			G	$6,700	3/05	NENE	
1066		9,045	G	$6,800	2/05	NWKS	1466 engine, 18.4-38, clamp-on duals, 2 hyd., dual PTO, cab, IHC 2350 loader, 6' bkt. with grapple
1066			G	$6,900	7/05	SESD	3 pt., diesel
1066			F	$7,300	4/05	SEPA	Cab, WF, good TA, rubber 40%
1066			G	$7,400	6/05	SWMN	2,484 hours on tach, diesel, cozy cab, WF, 2 hyd., 3 pt.
1066		5,693	G	$7,400	4/05	NCOH	Quick hitch, 2 hyd., 2 PTOs, full weights, axle duals
1066		4,400	F	$8,000	3/05	SCIL	Cab, air, slick tires
1066			G	$8,500	1/05	ECNE	Low hours, nice
1066			G	$8,500	8/05	SEMN	Duals
1066			G	$8,500	9/06	NEIN	Original
1066			G	$8,750	3/05	NCMT	Farmhand 235 loader, 4,300 hours since OH, duals, 3 pt., PTO
1066			G	$9,100	1/05	NENE	Diesel, Koyker K5 quick-tach loader, 3 pt., WF, 3 pt., 18.4-38
1066	1972		G	$5,750	1/05	NEIA	Cab, diesel, 38" tires, 9-bolt axle duals, rock box
1066	1972		G	$8,100	8/05	ECNE	2,000 hours on OH, 3 pt., 2 hyd., 540/1,000 PTO, ROPS canopy, fenders, 18.4-34, rear weights
1066	1973	7,250	G	$8,000	1/05	ECIA	Diesel, WF, 3 pt., 2 hyd., good 18.4-38 tires
1066	1973	3,400	G	$8,500	1/05	ECIA	Diesel, WF, 3 pt., 2 hyd., 3,400 hours on engine, new paint
1066	1974		G	$7,200	3/05	ECNE	1,117 hours on OH, 20.8-38, duals, 3 pt., 2 hyd., 540/1,000 PTO
1066	1974		G	$10,780	3/05	SWIN	One owner, tach shows 4,061 hours but not working, approximately 8,000-9,000 total hours, Hiniker cab, air, turbo, 8/4 hi/lo, 540/1,000 PTO, 2 hyd., wheel weights, 18.4-38, hub duals, quick hitch
1066	1975	6,250	E	$8,800	1/05	SENE	Diesel, weights, duals
1066	1975	3,934	G	$9,300	2/05	WCIL	Diesel, cab, weights, extra fuel tank, fast hitch, 18.4-38 tires
1066	1975	1,988	E	$9,500	4/05	NEIN	Diesel, black stripe, no cab
1066	1976	5,990	G	$6,750	6/05	NEND	2 hyd., 3 pt., 540/1,000 PTO, 18.4-38 duals, 80% rubber
1066	1976	5,900	G	$8,000	2/05	NCIN	Black stripe
1086	1977	9,400	G	$7,250	3/08	ECND	CAH, 3 hyd., 3 pt., 540/1,000 PTO, 18.4-38 duals
1086	1977	7,352	G	$8,500	3/08	WCMN	2 hyd., 3 pt., dual PTO, long axles, K&M step kit, 18.4-38 tires
1086	1977	5,461	E	$14,400	1/08	ECIL	One owner, cab, new 20.8-38 tires, front/rear weights

IHC/Farmall

Model	Year	Hours	Cond.	Price	Date	Area	Comments
1086	1976	9,250	F	$5,500	3/07	WCNE	Cab, 3 pt., 3 hyd., diesel, duals, 540/1,000 PTO
1086	1976	4,729	G	$8,600	12/07	SEWY	Cab, Farmhand 358 loader with grapple, diesel
1086	1976		F	$8,750	11/07	SWNE	Cab, dual 3,100 loader
1086	1977	8,315	G	$9,100	7/07	NWIL	New tires, axle duals, 2 hyd., 3 pt., dual PTO, diesel
1086	1978		G	$6,600	12/07	WCNE	Tach says 3,831 hours, diesel, cab, WF, duals, 3 pt.
1086	1978		G	$7,500	4/07	ECSK	Canadian sale, 2WD tractor with 131 PTO hp., standard trans., dual PTO, 2 hyd., 4-rib front tires, 20.8-38 duals
1086	1978	6,438	G	$8,000	10/07	NCMI	Diesel, CAH, 3 hyd., 3 pt., dual PTO, 18.4-38 tires with new direct axle duals
1086	1978		F	$8,400	12/07	SEMI	Cab Model 4CH7, SN 32619, new torque/engine clutch in 1998
1086	1978	7,800	F	$8,750	3/07	NEOH	Cab
1086	1978	6,500	G	$11,000	3/07	NEOH	Cab
1086	1978	6,000	G	$14,000	7/07	SESK	Canadian sale, 2WD tractor, 131 PTO hp., standard trans., 18.4-38 duals, 4-rib front, 6,000 hours showing, SN U032625 sells complete with Degelman 10' front-mount dozer blade with hyd. lift
1086	1976			$5,000	10/06	SCSK	Canadian sale, IHC 2350 loader, triple hyd., bale spear
1086	1976	5,095	G	$7,100	3/06	NWOH	Cab, 3 pt., 2 hyd., dual PTOs, 8 front weights, snap-on duals
1086	1977	4,273	G	$11,700	1/06	NWIA	IHC cab, air ride seat, WF, 3 pt., 2 hyd., dual PTO
1086	1977	3,270	E	$15,300	3/06	NCIL	CAH, 2 hyd., dual PTO, 18.4-38 axle-mount duals, one owner
1086	1978	7,092	F	$9,900	1/06	NWIL	18.4R-38 hub duals
1086	1978	4,665	G	$14,000	12/06	WCIL	Fully equipped cab, 540/1,000 PTO, 2 hyd.
1086	1978	4,825	G	$14,100	12/06	WCIL	Fully equipped cab, 540/1,000 PTO, 2 hyd.
1086	1976		F	$5,000	3/05	WCIL	Diesel
1086	1976	4,881	G	$9,500	2/05	NCIL	New TA, 3 pt., 2 hyd., 540/1,000 PTO
1086	1976	4,527	G	$15,500	2/05	ECNE	Dual PTO, 6 front weights, near-new Firestone 23 18.4-38R tires, ISO 3 hyd., heavy drawbar, AM/FM, air, chrome pipe
1086	1977	3,749	F	$4,100	9/05	NECO	Cab, 3 pt., duals
1086	1977	4,667	G	$10,750	3/05	NEKS	Weights, cab, air, good 18.4-38 tires, 2 hyd., air ride seat
1086	1978	3,270	G	$9,750	4/05	SEMN	Diesel, 3 pt., 2 hyd., digital dash, air, 18.4-38 tires
1086	1978	5,990	G	$10,100	3/05	SEND	3 pt., 3 hyd., long axles, new clutch, TA and air at 5,820 hours
1086	1978	5,875	G	$10,750	11/05	NCIN	2 hyd., axle duals, new TA in 2003
1086	1978	6,960	G	$10,900	12/05	NCOH	Overhauled at 5,000 hours, CAH
1206	1966		G	$6,000	12/07	WCMN	Wheatland, 2 hyd., no TA, 1,000 PTO, 18.4-34 tires
1206	1967	7,044	G	$6,000	8/07	WCMN	Factory cab, turbo diesel, 2 hyd., 3 pt., dual PTO, 18.4-38 tires, no TA
1206			G	$6,500	8/06	SEMN	
1206			G	$10,000	3/06	NWIL	Diesel, new rear tires, no cab
1206	1966		G	$8,000	8/06	WCIL	
1206	1967	12,103	G	$7,400	4/06	NWMN	CAH, 2 hyd., 3 pt., 540/1,000 PTO, front suitcase weights, 18.4-38 duals
1206			P	$555	3/05	WCIL	Diesel

IHC/Farmall

Model	Year	Hours	Cond.	Price	Date	Area	Comments
1206			F	$3,000	2/05	WCIN	Mechanic's special, 18.4-34 tires, bad TA, clutch, 3rd gear, noise in rear end, does run
1206			G	$4,900	7/05	ECND	2 hyd., 3 pt., 540/1,000 PTO, fenders, diesel
1206			G	$5,600	2/05	NWKS	3 pt., dual PTO, rear weights, 18.4-34, newer 10:00-16 4-rib front tires, 2 hyd.
1206		8,716	G	$8,100	4/05	NWIL	3 pt., PTO, 18.4-38, duals, TA is out
1206	1966	8,242	G	$2,200	6/05	NWMN	Wheatland, diesel, cab, PTO, 2 hyd.
1206	1966		G	$8,800	8/05	SEIA	Diesel, 18.4-38, 56 shifter
1206	1967	4,018	G	$9,300	1/05	NWIL	WF, 2 hyd., 3 pt., 12 front weights, 18.4-38, clamp duals
1256			F	$5,600	1/08	SESD	Trans. problems, 1968 model
1256	1967	9,279	F	$4,800	5/07	SEWY	WF 3 pt., no cab, diesel
1256	1967		G	$4,500	2/06	WCKS	Cab, 3 pt., PTO
1256			F	$5,000	2/05	NWWI	Turbo, little on the rough side
1256			G	$10,000	2/05	SCMN	Low hours
1256	1971		G	$7,750	1/05	NWOH	WF, 3 pt., reconditioned several years ago, very bright
1256	1971	5,300	E	$13,500	1/05	NEMO	
140			G	$2,400	3/08	SWKY	Cultivators
140	1963		G	$6,500	5/08	SEMN	Farmall, high crop, fenders, PTO, low hours on rebuilt engine and clutch
140			G	$2,900	3/07	SEPA	
140	1971		G	$2,000	11/07	WCIL	Running, restored, fenders, PTO, lights, hyd. remote, deluxe seat, new paint, battery box, seats and lights
140			G	$1,500	3/06	SWKY	Cultivators
140			E	$0	12/05	SWWI	No sale at $2,100, new paint
140			G	$1,800	9/05	SCON	Canadian sale, gas
140			E	$2,300	3/05	WCMN	Belly-mount 6' sickle mower
140			G	$3,000	2/07	NEIN	Low hours, 1pt.
140	1959		G	$4,000	6/05	NEND	WF, PS, 4 speed, 2 pt., fast hitch, PTO
1456			G	$8,750	2/08	ECIA	1970 or 1971 model, diesel, no cab, actual hours unknown, WF
1456	1971		E	$12,500	5/08	ECMN	Restored, open station, diesel, 3 pt., 2 hyd., new 18.4-38 tires
1456			F	$4,250	1/07	NWIA	18.4-38 tires
1456		4,139	G	$5,800	3/07	ECIL	Diesel, no cab, good rubber, torque out
1456			F	$3,600	11/06	ECND	Row crop, cab, dual PTO, no TA, 18.4-38 hub duals
1456			G	$8,500	12/06	SCMT	Dual loader
1456	1971		G	$5,900	9/06	SEIA	3 pt., cab, duals
1456	1972		G	$5,300	7/06	WCMN	2 hyd., 3 pt., cab, 540/1,000 PTO, new injection pump, new Goodyear rubber
1456	1971	6,470	P	$1,450	9/05	NECO	3 pt., excellent tires on rear, cab, front weights, for parts
1466			F	$4,000	5/08	ECMN	Open station, square fenders, 3 pt.
1466	1973	4,500	G	$4,750	3/08	ECND	Red cab, CAH, 2 hyd., 3 pt., 540/1,000 PTO, aux. fuel tank, antifreeze in oil
1466			F	$5,000	2/07	NEIN	Tach shows 3,815 hours, cab, 3 pt., 2 hyd., 11 front weights, axle duals
1466		4,186	G	$5,000	6/07	ECND	CAH, 2 hyd., 3 pt., 540/1,000 PTO, 20.8R-38 duals, 75% rubber, showing 4,186 hours
1466		8,042	G	$7,200	3/07	NWIA	Diesel, cab, WF, 3 pt., duals, front/rear weights

IHC/Farmall

Model	Year	Hours	Cond.	Price	Date	Area	Comments
1466		4,592	G	$8,700	3/07	NCNE	Diesel engine (recent OH – less than 1,000 hours), 4-speed high-low trans., TA had recent OH, WF, CAH, 3 hyd.
1466			G	$9,600	10/07	WCWI	Cab, 2 hyd., PTO, diesel
1466	1973		F	$3,900	4/07	SESK	Canadian sale, 2WD, diesel
1466	1974		G	$8,500	7/07	WCMN	No cab, 3 pt., dual PTO, long axles, duals, 9-bolt hubs, 500 hours on new 467 cu. in. engine
1466	1976	3,000	E	$10,000	12/07	SCIA	Cab, very clean, not a scratch, actual hours
1466			G	$2,200	6/06	SEID	Runs but has oil in radiator
1466			F	$2,200	6/07	SEID	Runs but has oil in radiator
1466			F	$4,700	6/06	NCIA	Cab, front tank and weights, 18.4-38 axle duals, 2 PTO, 3 pt., 2 hyd., TA out
1466			G	$6,000	8/06	SEPA	ROPS, turbo, WF, average rubber, weights
1466			P	$6,200	3/06	NEOH	Rough, needed clutch
1466		3,145	G	$7,000	5/06	SEND	CAH, 3 pt., 20.8-38 hub duals, dual PTO
1466	1971		F	$3,200	12/06	SCMT	
1466	1971	9,172	G	$5,250	4/06	NWMN	Cab, air, 2 hyd., 3 pt., 540/1,000 PTO, aux. fuel tank, 20.8-38
1466	1972		G	$10,000	1/06	WCIL	Loader and rear forklift, black stripe, 3 pt., 2 hyd.
1466	1973	7,944	G	$6,400	3/06	ECNE	Diesel, 2 hyd., 3 pt., 2 PTOs
1466	1973		G	$7,500	2/06	WCKS	3 pt., PTO, IHC hyd. loader
1466	1974		F	$4,000	12/06	SCMT	
1466	1975	3,373	G	$10,500	8/06	ECNE	Cab, air, 2 hyd., weights, 20.8-38 duals
1466	1976		G	$7,000	2/06	NCCO	2,000 hours on rebuilt engine, 1,000 hours on new TA, 18.4-38 tires
1466			P	$2,400	4/05	SCSD	Rough
1466		6,173	G	$8,200	1/05	WCOH	2 hyd., 1,000/540 PTO, 3 pt., front weights
1466			F	$8,900	9/05	NENE	High hours, torque out
1466			G	$10,700	3/05	ECMN	Cab, diesel
1466	1971		G	$3,900	6/05	NEND	Turbo, 2 hyd., 1,000 PTO, 18.4-38 duals
1466	1974	4,831	G	$9,000	1/05	ECIA	Diesel, cab, WF, 3 pt., same as new 38" radials, 700 hours on new clutch and TA
1466	1974	5,087	G	$9,300	1/05	WCIL	Diesel, open station, fenders, dual PTO, 2 hyd., 3 pt., 10 front weights, 800 hours on new clutch and TA
1468	1972	6,700	G	$11,500	1/07	WCIL	Diesel
1468	1972	4,666	G	$14,000	9/07	NEIA	Cab, WF, 3 pt., 2 hyd., dual PTO, hub duals, front weights, V-8
1468	1973	5,486	P	$1,500	12/07	WCNE	Cab, 3 pt., engine locked, diesel
1468	1974	5,903	G	$13,500	8/07	WCMN	V-8, 2 hyd., 3 pt., 540/1,000 PTO, new rear 18.4-38 tires, wide axle, restored
1468	1972	6,874	G	$15,250	4/05	SEMN	Diesel, 3 pt., 2 hyd., V-8 diesel, 20.8 radials
1486	1976	7,807	F	$6,200	3/08	SCNE	3 hyd., 540/1,000 PTO, 3 pt., OH at 5,000 hours
1486	1977	4,547	G	$7,500	3/08	ECNE	Diesel, 4,547 hours on tach, 18.4R-38 rear tires, 11:00-16 front tires, 3 hyd., 540/1,000 PTO, air
1486	1978	4,679	F	$8,400	1/08	ECMI	Diesel, needs TA work, 2 hyd., good rubber
1486	1978	3,130	G	$9,250	3/08	ECNE	Diesel, 3,130 hours on tach, 20.8R-38 rear tires, 11:00-16 front tires, 3 hyd., 540/1,000 PTO, 10 front weights, air, AM/FM
1486	1977	3,364	G	$11,500	3/07	ECIL	Diesel, CAH
1486	1977		G	$12,750	3/07	NEOH	No TA, new paint, 540/1,000 PTO

IHC/Farmall

Model	Year	Hours	Cond.	Price	Date	Area	Comments
1486	1978	4,095	G	$13,400	7/07	NWIL	CAH, new 11:00-16 4-rib front tires, diesel, 18.4-38R clamp-on duals, dual PTO, 3 pt., 2 hyd.
1486	1978	4,919	G	$17,000	11/07	NCIA	Cab, air, 2 PTO, 2 hyd., front weights, air ride seat, big steps, intercooler, OH, 18.4-38 duals
1486	1976	4,108	G	$11,600	2/06	WCKS	Cab
1486	1978	4,788	G	$14,250	2/06	NWIL	Fully equipped cab, dual PTO
1486	1977		F	$6,750	6/05	NESD	2WD, CAH, 3 hyd., 3 pt., 540/1,000 PTO
1486	1978		F	$6,500	12/05	WCMN	
1486	1978	9,188	G	$8,100	7/05	NWMN	CAH, 3 hyd., 3 pt., 540/1,000 PTO, 20.8-38 hub duals
1486	1978	8,146	G	$10,000	1/05	ECIA	Black stripe, diesel, cab, WF, 3 pt., new 30.5-32 rear, new 14L-16 front rubber
1566			F	$4,900	4/08	SCMN	Factory cab
1566		4,500	F	$4,900	1/08	WCIL	Red cab, diesel
1566			F	$4,600	3/06	NCMN	
1566	1973		F	$5,500	3/06	NEMI	Cab, duals, low hours on OH
1566	1975	3,695	G	$5,700	2/06	WCIL	CAH, duals
1566	1975		G	$6,250	3/06	ECNE	10-bolt duals, 3 hyd., PTO, less than 50 hours on OH
1566	1975	5,409	G	$6,600	3/06	WCMI	Cab, TA, 2 hyd., dual PTO, axle 20.8-38 duals, diesel
1566	1975		G	$9,500	1/06	WCNE	WF, 3 pt., 2 hyd., 1,000 PTO, duals, 1,000 hours on OH, diesel
1566	1976	6,390	G	$5,200	3/06	WCMI	20.8-38 axle duals, dual PTO, 2 hyd., 3 pt., one owner, diesel
1566	1976		G	$5,400	2/06	WCIL	Black stripe, CAH, duals, new TA and clutch
1566	1975	7,500	G	$6,400	7/05	ECND	CAH, 3 pt., 540/1,000 PTO, duals, front fuel tank
1566	1975	7,800	G	$7,500	4/05	ECND	Factory cab, 3 pt., PTO, 2 hyd., 18.4-38R, duals, good rubber, clean
1566	1976		G	$6,200	3/05	NWMN	Factory cab, 3 pt., PTO, 2 hyd., black stripe
1586	1976	7,305	G	$6,500	3/07	ECND	CAH, 3 hyd., 3 pt., PTO, front weights, 20.8-38 duals
1586	1978	6,085	G	$7,500	4/07	NEND	CAH, 3 pt., PTO, 3 hyd., front weights, 18.4-42 duals
1586	1976	7,552	F	$7,300	3/06	NEKS	2 hyd., 18.4-38, 14L-16.1 front tires
1586	1977		F	$4,000	8/06	NCNE	Duals, $1,500 spent on AC
1586	1977	4,242	G	$8,000	3/06	NWOH	Cab, 3 pt., 6 front weights, 2 hyd., snap-on duals
1586	1976		F	$5,000	12/05	WCMN	
1586	1977		F	$4,100	12/05	SEND	CAH, 3 hyd., 1,000 PTO, no 3 pt., 20.8-38 press steel hub duals (50%), 300 hours on new clutch
1586	1978	2,479	G	$10,600	1/05	WCOH	Duals, 3 hyd., front weights
185	1977		G	$3,100	3/06	SEND	Cadet Lo-Boy, 5' mower
200			F	$800	1/06	NCCO	Cultivator
200			F	$825	1/06	SEPA	NF, fair rubber
200	1955		F	$1,600	2/06	NECO	2 pt.
200			G	$3,100	2/05	ECMI	Farmall, fast hitch, gas
200			G	$6,500	7/07	WCMN	All restored, fast hitch, WF, fenders
230			G	$2,800	8/07	WCMN	WF, fast hitch, unrestored
230			G	$3,250	12/07	SCMN	NF, 2 pt.
230			G	$2,900	3/06	ECND	Reversed, NF, gas, Woods RM90 finishing mower

IHC/Farmall

Model	Year	Hours	Cond.	Price	Date	Area	Comments
230	1957		G	$3,700	10/06	SCSD	NF, gas, fenders, lights, PTO, 11.2-36 tires, restored
230			G	$2,600	7/05	SESD	2 pt.
230			G	$3,400	9/05	WCMN	Fast hitch, WF
230			G	$3,700	11/05	WCIA	NF, Continental mower, good
230	1957		G	$4,000	6/05	NEND	NF, 4 speed, 2 pt., PTO, belt pulley, hyd. front weights, band duals
240			G	$3,550	8/07	SCMN	
240			G	$4,000	8/07	SESD	Utility, Wagner loader
240			G	$6,000	9/07	WCIA	Utility
240	1959		G	$5,000	11/07	WCIL	Running, restored, fenders, PTO, lights, hyd. remote, deluxe seat
240			G	$1,500	11/06	SETN	Gas, 2 pt.
240			G	$1,750	12/06	SCMT	
240			G	$2,500	10/06	WCWI	Industrial tractor with 6' side-mount flail mower
240	1959	2,485	F	$2,300	1/06	NECO	120 Wagner loader, backhoe
240		2,214	E	$5,350	2/05	NEIA	Rear blade, fast hitch
240	1959		G	$4,600	11/05	NWIL	Gas, loader, no PS, good tires
2424	1966		F	$2,800	2/07	NEIL	Loader
2444	1973		G	$6,200	12/07	ECIL	Industrial tires, IHC 2050 loader, 60" bkt., 84" bkt., pallet forks, 540 PTO
2500	1976	3,500	P	$3,000	6/05	WCMN	Utility, loader and grapple fork, one owner, 3 pt., PTO, 2 hyd., hi/low 4 speed, shuttle trans., rough
2544	1972	3,000	E	$4,600	8/05	NEIN	Hydro, restored, loader
2656	1970	5,000	F	$6,300	1/06	NWIL	Industrial tractor, hydrostat, industrial loader
284	1978	1,362	G	$5,000	9/06	NEND	Utility, 3 pt., PTO, 72" belly mount mower
300			G	$2,300	3/08	ECND	Utility, WF, live PTO, fast hitch, hyd. loader
300			G	$2,500	1/08	SWKY	
300			F	$2,600	3/08	NWKS	Utility
300			G	$3,500	3/08	WCMN	WF, gas, fast hitch, live hyd., loader
300	1955		G	$2,500	5/08	SEMN	Farmall, gas, fenders
300	1955		G	$4,900	4/08	SCMI	NF, fast hitch, restored
300			F	$800	6/07	SWNE	Gas, WF, 12.4-38 rear tires, PTO, fenders, 3 pt., front-mount hyd. post pounder
300			F	$975	9/07	SCMI	NF, gas, 3 pt., 2 hyd.
300			G	$1,400	8/07	NWIL	
300			G	$2,250	2/07	ECNE	NF, live power
300			G	$2,250	8/07	WCMN	Factory WF, propane
300			G	$2,600	5/07	WCWI	Utility, industrial loader, 3 pt. blade
300			F	$3,000	6/07	SCIA	Loader, blade, WF, fast hitch, used as snow tractor
300	1955		F	$1,500	11/07	WCIL	Running, unrestored/original, rear wheel weights, PTO, hyd. remote
300	1955		G	$2,250	11/07	WCIL	Running, unrestored/original, PTO, hyd. remote, PS, runs good, good sheet metal
300	1956		G	$2,250	3/07	WCIL	Gas, WF, good paint/tires
300	1956		G	$2,250	3/07	WCIL	Gas, WF, good paint/tires
300	1956		G	$2,400	12/07	SEMI	
300			G	$1,650	8/06	NWIL	TA out, new paint, NF, gas
300			F	$1,700	3/06	NWPA	
300			G	$2,150	1/06	SESD	NF, 3 pt.
300			G	$2,500	2/06	SWPA	New rubber

IHC/Farmall

Model	Year	Hours	Cond.	Price	Date	Area	Comments
300			F	$2,500	6/06	NWMN	Utility, PS, live PTO, 3 pt., hyd., near-new tires, loader
300			G	$2,600	9/06	ECMI	WF, complete, restored
300		3,678	G	$3,000	5/06	NENE	Utility, gas, 2 pt., 13.6-28 tires
300			G	$3,400	8/06	WCIA	WF, fast hitch, Westendorf WL30 loader
300			G	$3,700	3/06	WCMI	Loader/backhoe, gas
300			G	$4,075	1/06	SCMN	Vaughn hyd. loader, 48" and 80" bkt., gas, fast hitch, one owner
300			G	$4,500	1/06	SCMN	Utility, gas, fast hitch, one owner
300			G	$5,100	8/06	SCIL	2 pt.
300	1955		G	$1,500	10/06	SCSD	NF, good TA, PTO, fenders, gas
300	1955		G	$2,300	1/06	NECO	3 pt.
300	1956		G	$1,500	10/06	NECO	Gas, 2 pt., 2 hyd., double front
300	1957		F	$1,800	7/06	ECND	Utility, live PTO, hyd.
300			F	$950	8/05	ECMI	Utility tractor, new rubber and paint
300			F	$1,150	2/05	NCIN	
300			F	$1,300	2/05	WCIN	Gas, NF, 6 volt, belt pulley
300			F	$1,700	3/05	SCMI	Gas, NF
300			F	$1,800	9/05	SEIA	Farmall
300			G	$1,900	2/07	NEIN	WF, TA
300		3,687	G	$2,000	12/05	ECMI	Gas, NF, fast hitch
300			G	$2,600	3/05	NENE	
300			G	$2,700	6/05	ECMN	Farmall, fast hitch, PTO
300			G	$2,800	11/05	WCIL	Industrial, gas, utility tractor, loader
300			G	$2,900	6/05	WCMN	Restored, fast hitch, WF, fenders
300			G	$3,400	9/05	WCMN	Farmall, WF, original
300	1955	8,804	F	$950	2/05	WCNE	Gas, 2 pt., live PTO
300	1955	2,769	G	$2,400	3/05	SEWY	Gas, 3 pt., double front
300	1959		G	$3,500	7/05	NCMN	4 cyl., gas, 3 pt., PTO, PS, loader, blade and sickle mower
300	1959		G	$3,500	7/05	NCMN	4 cyl., gas, 3 pt., PTO, PS, loader
330			F	$1,500	1/07	SWKS	Utility tractor, 3 pt., PTO
340			F	$2,500	8/07	SESD	Single front wheel
340	1959		G	$2,300	11/07	WCIL	Unrestored/original, fenders, PTO, lights, hyd. remote, 3 pt., PS, TA
340	1959		G	$3,200	11/07	SESD	NF, repainted
340			G	$2,700	4/06	NWSD	Gas utility tractor, 3 pt., chains
340	1960	1,300	G	$2,500	3/06	NCCO	Actual hours, PTO, remote, 12.4-36 tires, quick hitch
340	1960		G	$2,600	10/06	SCSD	WF, IPTO, TA, gas
340	1960		G	$2,700	5/06	SEWY	Gas, WF, 3 pt.
350			G	$2,650	4/08	SCMN	WF, 2 pt., gas, PS
350	1958		G	$9,250	5/08	SEMN	Farmall, gas, PS, fenders, rear wheel weights, rear spin out, 2 pt., new tire rear and front, new clutch, steering wheel with center cap, second owner
350			G	$3,200	3/07	NCIA	Farmall with PS, good tin
350			G	$3,950	6/07	ECIL	Fast hitch, gas, 12-38 tires
350	1957		G	$1,900	9/07	NECO	Gas, double front, 2 pt., needs TA
350	1957		F	$2,000	9/07	NWIL	Gas, loader
350	1957		G	$2,300	2/07	SEMI	Gas, NF, fast hitch
350	1957		G	$2,500	3/07	SEWY	WF, 3 pt., gas

IHC/Farmall

Model	Year	Hours	Cond.	Price	Date	Area	Comments
350	1957		G	$4,900	8/07	SCMN	Diesel, straight tin, IHC, WF, 2 pt., PS, one owner
350	1958		G	$2,400	11/07	WCIL	Running, unrestored/original, rear wheel weights, PTO, lights, hyd. remote, PS
350			G	$1,900	7/05	WCMN	WF, 1 hyd., fast hitch
350			G	$2,100	3/06	NWIL	Gas, loader, new tires
350			F	$2,900	8/06	SCMI	Gas, WF, fast hitch, loader
350			G	$3,000	4/06	SEND	Recent OH, gas
350			F	$1,400	9/05	NWOR	Diesel
350			G	$1,800	11/05	WCIL	NF, PTO, PS, fast hitch
350		4,000	G	$1,900	11/05	SCNE	Farmall, TA, 2 pt.
350			G	$6,500	3/05	ECMN	Farmall, gas, tricycle, new paint, rebuilt
350	1957	1,672	G	$2,100	2/05	WCIA	Gas, utility, total restoration, paint, rubber
354	1972	2,622	F	$750	10/06	NECO	3 pt., WF, gas
3588	1978	5,800	G	$4,700	1/08	NEMO	466 cu. in. engine
3588	1978		F	$6,000	1/08	NWOH	2+2, rear duals
3588	1978		F	$5,500	7/06	SEND	2+2, 4 hyd., 3 pt., PTO, 16.9-38 hub duals, unknown hours
3588	1978	4,675	G	$7,900	2/05	NEIN	2+2, diesel
400			F	$1,400	1/08	SESD	2 pt., motor OH, Dual 250 loader
400			F	$2,550	1/08	SESD	Running, diesel
400	1956		G	$5,500	5/08	SEMN	Restored, diesel, WF, rear wheel weights, good fender with extender
400			G	$1,275	2/07	ECIL	Row crop, sold with 2R mounted Massey Harris corn picker
400			G	$1,900	11/07	SEND	Diesel, WF, PS, Farmhand F11 and grapple, PTO
400			G	$2,000	11/07	SESD	Gas
400			G	$2,100	4/07	SCOK	LP gas, WF, rubber 30%, 52 hp.
400			F	$2,150	11/07	SESD	Koyker loader
400			G	$2,500	11/05	WCIL	Gas, NF, good 14.9-38 rear tires, SN 7539
400			G	$2,800	8/07	SESD	NF, gas, 1953 model
400			G	$2,850	8/07	SESD	2 pt., very straight, diesel
400			G	$3,100	3/07	NWIA	Ride and drive, loader
400	1954	6,083	F	$2,200	5/07	SEWY	Diesel, WF, dual loader
400	1955		F	$900	8/07	NWIL	Gas, NF, 3 pt., purchased new without torque
400	1955		F	$1,900	5/07	SEWY	Gas, double front, Farmhand F11 loader and 8' bkt.
400	1955		G	$2,000	12/07	ECIL	Gas, 1 hyd., 12 volt
400	1956		F	$1,900	11/07	WCIL	WF
400	1956		G	$2,300	8/07	SWOH	WF, TA, PS, fast hitch
400	1956		G	$2,500	7/07	ECIL	NF, TA, rear wheel weights, PTO, 1 hyd., PS
400			F	$1,000	8/06	SEMN	
400			F	$1,200	8/06	NECO	
400			F	$1,450	8/06	SEMN	Loader
400			G	$1,550	3/06	WCMI	Farmall, WF, 1 hyd., gas
400			G	$1,600	8/06	NWIL	Gas, new rear tires
400			G	$1,750	6/06	ECSD	WF, dual 300 loader
400			G	$2,200	8/06	NECO	Loader
400			G	$2,200	8/06	SCIL	
400			G	$2,400	11/06	ECIL	NF, gas
400			G	$2,950	8/06	NEIA	NF, PS
400	1955		G	$2,100	9/06	NCIA	Farmall, WF, 12 volt, good metal, clamshell fenders, 13.9-38 tires

IHC/Farmall

Model	Year	Hours	Cond.	Price	Date	Area	Comments
400	1955		F	$3,800	2/06	NWIA	NF
400	1956		G	$1,800	9/06	ECMI	Diesel, good TA
400	1956		G	$2,050	12/06	WCIL	NF, PS, new paint, gas
400				$0	2/05	SCMN	NF, fast hitch, TA, Schwartz loader
400			G	$1,800	11/05	WCIL	NF, gas, TA, rebuilt radiators, oil pump and front end
400			G	$2,150	1/06	SESD	WF, gas
400			G	$3,300	12/05	NCCO	1955 model, Farmall, Dual 325 loader, WF, 2 pt.
400			F	$3,500	3/05	NWMN	WF, PS, Cornhusker 3 pt., IHC 2000 hyd. loader
400			G	$4,000	6/05	WCMN	Restored, fast hitch, WF, fenders
400			G	$4,250	6/05	ECMN	OH, TA rebuilt, front weight set, new paint
400			G	$4,500	12/06	ECMN	Farmall, gas, row crop, WF, new tires, fast hitch, rebuilt
400	1955		F	$2,350	3/05	ECND	WF, TA, single hyd., Cornhusker, 3 pt., PTO
400	1955		G	$3,400	4/05	NCIA	3 pt., WF, Stanhoist loader, hyd. bkt.
400	1956		G	$1,600	9/05	NCIL	LP, gas
400	1956		G	$1,600	9/05	NCIL	LP, gas
404			P	$950	1/06	SESD	Utility, salvage
404		3,145	F	$3,000	3/06	NWPA	Gas, high lift, not running
404	1962		G	$2,800	6/06	NECO	Double front, 2 pt., gas, tach shows 900 hours
404		2,127	G	$3,200	11/05	SCMI	IHC 2000 loader
404			G	$3,350	3/05	NECO	WF, 3 pt., PTO, hyd., 6R front-mounted cultivator
4100	1966	4,787	G	$6,500	8/05	NCOH	4WD, 6 cyl., turbo, diesel, CAH, bareback, 23.1-26 rubber
414			F	$1,600	12/07	WCNY	Fenders, 36 hp., front blade
4166	1977		F	$3,800	1/08	ECMI	4WD
4166			F	$3,350	12/07	ECIN	4WD, 12' snow plow
4166		4,905	G	$7,000	1/07	WCIL	Diesel, 4 speed, AM/FM, PTO, 3 pt.
4166	1973		P	$1,600	4/07	ECND	4WD, diesel, CAH, bad clutch
4166		4,513	G	$3,900	7/06	NWOH	4WD, tires down, cab, 3 pt., major OH at 3,494 hours
4166		2,339	P	$1,600	4/05	SCSD	4WD, for parts
4166			F	$6,100	12/05	SEMN	4WD, blade
424			G	$3,000	5/07	WCWI	Utility, 3 pt.
424			G	$2,200	9/06	NWTN	
424	1966	2,668	G	$4,700	6/06	SCMI	Loader, gas, WF, 1 hyd., 3 pt., PTO
424		2,617	G	$2,700	1/05	SWOH	
424	1966		G	$2,500	8/05	NEIN	Gas, WF, loader
424	1967	3,020	G	$4,750	4/05	NCKS	Gas, 38 hp., PTO, 3 pt., Schwartz 300 loader
4366			F	$3,100	11/07	SEND	4WD, 2 hyd., 18.4-36 duals, no PTO, 3 pt., new engine and turbo at 7,958 hours
4366			G	$5,500	1/05	ECIL	4WD
4366		6,069	G	$7,250	7/05	ECND	4WD, 4 hyd., duals, new batteries, alternator, clutch assembly, center pins and bushings, steering ends, seat and floor mat
4366	1974	7,900	G	$1,400	9/05	NECO	Cab, duals, turbo, diesel
4386	1976		G	$14,250	1/07	ECIL	First one made! SN 501, one owner, 1,500 hours on new engine, 4WD, 3 pt., 3 hyd., 20.8-38 duals
4386	1977	7,425	G	$5,100	2/07	NCNE	4WD, 1,700 hours on new remanufactured IHC 466 engine, cab, air, 18.4-34, duals
4386	1978	4,502	F	$10,500	4/07	SESK	Canadian sale, 4WD, 175 drawbar hp., 18.4-38 duals, 3 hyd., 4,502 hours showing on rebuilt engine, one owner

IHC/Farmall

Model	Year	Hours	Cond.	Price	Date	Area	Comments
4386	1977	6,358	G	$9,000	12/06	WCIL	4WD, 3 pt., 2 hyd.
4386	1977	4,638	G	$4,900	12/05	WCMN	4WD, one owner, 3 hyd., 18.4-34 duals
444		3,202	G	$1,850	2/07	NEIN	Gas, WF
444			F	$1,825	1/05	SWOH	
444	1969		G	$4,500	3/05	NESD	2 hyd., gas, PTO, 3 pt.
450	1957		G	$10,000	5/08	SEMN	Farmall, gas, PS, fenders, 2 pt.
450	1958		E	$8,200	5/08	SEMN	Diesel, WF, 3 pt., fenders, rear wheel weights, pulley, new tires, front and rear
450			G	$1,800	4/07	SCOK	LP gas, NF, rubber 75%, 48 hp.
450			G	$1,800	5/07	WCSD	Gas, loader
450			G	$2,000	11/07	SESD	NF
450			G	$2,100	4/07	SCOK	LP gas, WF, rubber 95%, 48 hp.
450			G	$2,100	4/07	SCOK	LP gas, NF, rubber 40%, 48 hp.
450			F	$2,250	8/07	NWIA	NF, gas, fast hitch, 13.6-38 good rubber, repainted
450			G	$2,700	8/07	SESD	Dual loader
450			G	$3,200	3/05	WCIL	NF, gas, new paint, needs tires
450			G	$5,100	1/05	NENE	Loader
450	1957		G	$1,700	12/07	SEND	5 speed, Syncro
450	1958		E	$9,200	7/07	SWOH	Restored, gas, row crop, looks like new, quick hitch
450		3,478	F	$1,950	3/06	NWOH	Gas, NF, 1 hyd., fast hitch
450		3,807	F	$2,000	3/06	NWIL	LP gas, NF, 12 volt
450			G	$2,000	11/05	WCIL	Farmall, NF, gas
450			G	$2,100	9/06	ECMI	Diesel, WF, all original
450			G	$2,100	6/05	WCMN	Hyd. loader, independent valve
450			G	$2,300	12/06	NWIL	TA, live PTO
450			G	$3,500	8/06	SCIL	LP gas, new paint
450			E	$3,900	8/06	ECNE	Dual loader, straight tin, no welds, gas, 13.4-38 rubber, rear wheel weights, tractor has 3-valve lever so you can run loader, 3 pt.
450			G	$4,250	12/06	SEMN	NF, fast hitch, new tires
450	1957		G	$5,000	2/06	WCIL	NF
450	1958		G	$2,400	11/06	SENE	NF, gas, 15.5-38, 2 pt., cement, rear wheel weights
450			P	$900	7/05	NCMN	Diesel, not running
450			F	$975	7/05	SEIA	Gas, straight bar, hyd., loader sold separate for $10
450			G	$2,600	7/05	SESD	NF, good TA, 3 pt.
450			G	$3,300	12/05	NCCO	Farmhand F11 loader
450			G	$7,000	9/05	ECMN	Farmall, tricycle, fast hitch, new paint, restored, wheel weights
454			G	$4,500	1/08	NEMO	
4568	1976	2,912	F	$5,500	3/07	NWIL	4WD, cab, 4 hyd., bareback, 20.8-38 duals
4586	1977		G	$9,900	12/07	SEMI	4WD, cab
460			F	$1,300	1/08	NEMO	
460			G	$1,700	4/08	SCND	WF, PTO, gas
460			E	$6,000	5/08	SEMN	Restored
460	1961		G	$3,600	5/08	SEMN	Gas, 2 pt., PS, fenders, rear split wheel weights, new clutch and TA, steering wheel with IHC cap, new engine, totally restored
460			F	$1,050	11/07	SESD	NF, running
460			G	$2,200	1/07	NWIA	
460			G	$3,000	8/07	SESD	WF, 2 pt., PTO, repainted

IHC/Farmall

Model	Year	Hours	Cond.	Price	Date	Area	Comments
460			G	$3,200	11/07	SESD	NF, diesel
460			G	$3,200	2/07	SEIL	Gas, row crop, fenders, fast hitch
460			G	$4,400	6/07	ECMN	Gas, WF, 300 hours on complete rebuild, flat-top fenders, fast hitch, new paint
460	1960		F	$1,100	8/07	NWIL	NF, gas, fast hitch, end loader, hyd. bkt.
460	1963		F	$600	3/07	SWNE	NF, diesel, 16.9-34 rear tires, 3 pt. adapter to quick hitch
460	1963		G	$1,400	8/07	SESD	New rubber, motor OH, restored, gas, sharp
460			G	$1,300	9/06	NWTN	Diesel
460			F	$2,750	10/06	SEMN	Fast hitch, WF
460			G	$2,850	7/06	SWWI	NF, fast hitch
460			G	$3,600	3/06	NENE	WF, gas, 15.5-38, original TA in good condition
460	1959		F	$3,000	6/06	NECO	WF, 2 pt., Farmhand F11 loader, gas
460	1960		F	$1,300	2/06	NECO	
460	1961	6,429	F	$2,500	5/06	SEWY	Utility, gas, WF, 2 pt.
460			P	$1,550	12/05	NWIL	Gas, WF, 2 pt.
460			G	$2,050	8/05	ECNE	LP, NF, 2 pt., 1 hyd., front port, 15.5-38, cast weights
460			F	$2,500	12/05	SEMN	Loader
460			G	$2,800	6/05	SWMN	Utility tractor, gas, 3 pt.
460			G	$3,250	4/05	WCWI	Farmall, IHC 2000 hyd. loader, 6' bkt.
460	1959		F	$1,600	11/05	ECNE	2 pt., PTO, utility, 14.9-28 tires
460	1960		G	$1,800	2/05	NEIN	Gas, NF, TA works
464			G	$3,000	12/07	SCNE	Utility tractor, Auburn chain trencher & front angle dozer
464			G	$1,900	8/06	SETN	Gas
464			F	$4,000	4/06	NWKS	IHC 1850 loader, 1,124 hours on tach, not actual hours
464			G	$5,900	9/05	ECMN	Utility
464		2,388	E	$6,200	12/05	ECIL	Loader, gas, WF, PS, 3 pt., new 13.6-28 tires
504	1966		G	$10,500	5/08	SEMN	High-clearance high-crop, 4-cyl. diesel motor, WF, 3 pt., PS, flat-top fenders, new front and rear tires
504	1962		G	$4,200	9/07	NECO	Gas, Farmhand loader, WF, tach shows 3,931 hours
504	1966		G	$14,000	11/07	WCIL	High-clearance utility diesel, running, restored, fenders, PTO, lights, hyd. remote, deluxe seat, TA, one of only 19 diesels built; between 1962 and 1967 there were 53 total built (19 diesels and 34 gas according to IHC archives)
504			G	$900	9/06	NWTN	
504			F	$1,925	3/06	SWOH	Utility tractor, IHC 2000 hyd. front loader
504			F	$2,800	9/06	NCMI	Torque out, diesel, 3 pt., PS, Schwartz WF
504			G	$8,100	11/06	SCIL	Zero hours on OH, clutch, paint, tires and PS, loader
504	1963		G	$3,000	12/06	NWMO	15.5-38 rear tires, 5 speed hi/lo, gas, PS, new seat, new muffler, tach shows 2,391 hours
504			G	$2,050	8/05	ECNE	
504			G	$6,000	8/05	ECIL	Utility, IHC 1709 loader, 5' bkt., new 13.6-28 tires
504	1962		G	$2,200	11/05	NCIN	WF
504	1962		G	$3,900	3/05	ECMI	TA, 1 hyd., 540 PTO
504	1964		G	$1,600	8/05	NEIN	WF
544			G	$3,500	11/07	ECSD	Hydro, 2 hyd., PTO, John Deere 52 loader
544	1968	4,495	G	$5,900	2/07	NCCO	

IHC/Farmall

Model	Year	Hours	Cond.	Price	Date	Area	Comments
544			G	$2,300	5/06	SWCA	PTO, 3 pt., 1 hyd.
544			G	$4,000	2/06	SWPA	Gas, 4 cyl.
560			G	$1,800	2/08	SEMN	WF, diesel
560			G	$3,000	1/08	NEMO	Gas
560	1959		E	$4,750	5/08	ECMN	Restored, row crop, gas, WF, fenders, newer 15.5-38 tires
560		4,208	F	$1,050	2/07	NEIN	Gas, WF, 1 hyd., steel wheels
560			F	$1,350	8/07	SESD	
560			G	$1,450	5/07	WCSD	Diesel, loader
560			F	$1,525	8/07	SESD	Gas, NF
560		4,741	G	$2,000	3/07	NWMO	NF, 1 hyd., 16.9-34 rubber (50%), needs torque, good paint
560			G	$2,300	3/07	NEMI	
560			F	$2,300	10/07	NEKS	Gas, NF, 2 pt., B800 loader
560			F	$2,400	8/07	SESD	Gas, dual 250 loader
560			F	$2,500	1/07	NWOH	Diesel, 2 hyd., fast hitch, snap-on duals
560			F	$2,700	2/07	WCKY	Diesel
560			F	$2,750	11/07	SWNE	Diesel, WF, cab, 2 pt.
560			G	$3,750	11/07	SESD	Gas
560			G	$4,100	1/06	SESD	Gas, Miller loader
560			G	$4,600	2/07	WCIL	Diesel, WF, fenders, new paint, runs good
560	1958		F	$1,400	9/07	NWIL	Gas, NF, fast hitch
560	1959		G	$3,200	8/07	SCMN	Diesel, WF, 2 pt., good rubber
560	1959		G	$4,000	12/07	ECIL	Gas, fast hitch, 1 hyd.
560	1959		G	$4,600	3/07	ECNE	Gas, WF, 15.5-38 rear tires, 6:00-16 front tires, 2 pt.
560	1959		G	$4,700	7/07	SEMN	Fast hitch, NF
560	1960	7,922	G	$4,350	2/07	ECNE	WF, diesel, 540 PTO, TA, 2 pt., fast hitch
560	1962		G	$1,700	8/07	WCMN	WF, 2 pt., 1 hyd., good clutch and TA, diesel
560	1962		G	$3,450	2/07	ECIL	Gas, WF, 3 pt., new 15.5-38 tires
560	1963		G	$3,500	9/07	WCIL	WF, diesel, fast hitch, good TA, sharp, runs, new brakes
560			F	$1,400	3/06	NWIL	NF, gas
560			F	$1,400	3/06	SEIA	Gas
560			G	$1,400	8/07	NWIL	NF
560			F	$1,500	3/06	SEPA	NF, poor rubber
560		5,463	F	$1,700	3/06	NWIL	LP gas, NF
560			G	$1,700	12/06	SCMT	
560		3,245	F	$1,800	3/06	NWOH	Fast hitch, WF, 1 hyd., gas
560			F	$1,850	12/06	SEND	PTO, 2 hyd.
560			F	$1,900	3/06	NWOH	Gas, NF, 1 hyd., fast hitch, snap-on duals, tach says 3,606 hours
560			G	$1,900	12/06	SCMT	
560			G	$2,000	5/07	WCSD	Farmall, no 3 pt., runs OK
560			G	$2,100	7/06	WCCA	2 pt., PTO, 1 hyd.
560		4,411	G	$2,100	6/06	NWIL	WF, cozy cab, new rear tires
560			G	$2,100	6/06	NWMN	Diesel
560			G	$2,100	6/06	NWMN	Gas
560			G	$2,100	8/05	SEMN	
560			F	$2,200	2/06	ECNE	
560			F	$3,300	3/06	NWIL	LP gas, tach says 1,726 hours, WF, Koyker K5 loader
560			G	$3,300	7/06	SCIA	NF, standard drawbar

IHC/Farmall

Model	Year	Hours	Cond.	Price	Date	Area	Comments
560			G	$3,500	12/06	WCIA	Gas, NF, new paint, torque, rear tires
560		8,553	F	$4,250	8/06	NWOH	NF, 1 hyd., diesel, fast hitch, snap-on duals
560	1958		G	$2,900	12/06	NCIA	Gas, fast hitch
560	1958	3,253	F	$3,350	3/06	NWOH	Fast hitch, 1 hyd., WF, gas
560	1959		F	$2,800	7/06	NWIL	Diesel, NF, no hitch
560	1959		G	$3,000	8/06	NCIA	Diesel, WF, fast hitch, OH, new rear tires, 3 pt. adapter
560	1959		F	$3,300	7/06	NWIL	Diesel, NF, quick hitch
560	1963	3,151	G	$2,900	3/06	ECNE	Gas, NF, 15.5-38, 1 set rear weights, 1 hyd., 2 pt.
560			F	$800	7/06	SEND	Wheatland, diesel
560			F	$1,065	9/05	SEIA	Loader
560			P	$1,425	4/05	NCOH	Salvage, diesel
560			F	$1,550	9/05	SCMN	
560			G	$1,750	1/05	ECNE	WF
560		4,300	G	$2,050	1/05	NENE	Propane
560			G	$2,075	2/05	NEIN	Diesel, turbo, WF, TA out
560			G	$2,100	6/05	SWMN	Diesel
560			G	$3,300	4/05	WCWI	Farmall, turbo
560			G	$4,100	2/05	SCMN	
560	1958		F	$2,500	9/05	ECIA	Row crop, gas, high hours
560	1959		G	$2,950	8/05	NCIA	NF, standard drawbar, diesel
560	1959		G	$3,500	6/05	NEND	WF, TA, PTO, 1 hyd., IHC 2000 hyd. loader
560	1960	4,567	F	$1,700	2/05	NWKS	Diesel, 2 pt., 15.5-38 tires, 1 hyd.
560	1961		G	$3,200	1/05	NEIA	Diesel, WF, fast hitch, fenders
560	1962	2,689	G	$3,300	8/05	NEIN	Diesel, NF
560	1962		G	$3,800	8/05	NCIA	NF, fast hitch, repainted, diesel
560	1962		G	$3,950	8/05	ECMI	Farmall, New Idea loader, extra manure bkt.
574		4,350	F	$4,250	3/08	ECWI	Diesel, utility, 16.9-28 tires, 540/1,000 rpm
574		4,595	F	$6,300	4/08	WCWI	Utility, IHC 2250 hyd. loader, 7' material bkt., gas
574		2,892	G	$6,750	3/08	WCMN	Utility, gas, IHC 2050 loader, 3 pt., 540 PTO, 16.9-28 tires
574		2,973	G	$6,200	3/07	ECND	Utility, gas, hyd., 3 pt., PTO, IHC 2250 hyd. loader, 16.9-30 tires, nice
574			F	$6,800	7/07	SWOH	Gas, loader, livestock tractor, well used
574	1975	4,749	G	$9,000	7/07	SEMN	Utility, 2 pt., IHC fully hyd. loader
574			G	$2,600	8/06	WCNC	
574			F	$4,800	3/06	NWOH	Gas, 3 pt., IHC 2050 loader, 1 hyd.
574	1977	2,654	G	$5,000	1/06	NWIL	Utility
574		2,328	F	$3,250	10/05	NEPA	Bkt., good rubber
574	1976	2,581	G	$7,700	3/05	NWMN	Utility, ROPS, canopy, grill guard, 3 pt., PTO, 2 hyd., IHC 2250 loader
600			G	$3,400	7/07	NCND	Diesel, 540 PTO, straight tin, factory fenders
600			F	$750	9/05	NEIN	Standard, original
650			G	$11,000	9/07	WCIA	Diesel, restored, new rubber all around, new gauges, sharp
650			G	$3,000	4/06	NWMN	Diesel, rollover, standard, PTO, 2 hyd., 18.4-34 tires
650	1957	5,438	G	$700	9/06	NEIN	On steel, gas
650			F	$600	9/05	NEIN	LP, standard
650			F	$850	9/05	NEIN	Standard, original
650			F	$1,475	8/05	SEMN	

IHC/Farmall

Model	Year	Hours	Cond.	Price	Date	Area	Comments
650			G	$3,500	11/05	WCIL	Repaint, PTO, lights
656			P	$2,000	4/08	SWWI	Rough
656			F	$2,300	1/08	NEMO	Dual loader
656			G	$4,000	3/08	NWIL	NF
656			G	$4,000	3/05	ECMN	Restored, gas, gear drive, WF, 3 pt., 16.9-34 tires
656	1968		G	$5,400	5/08	ECMN	Restored, hydro, gas, WF, 3 pt., 1 hyd., 16.9-38 tires
656	1970	2,631	G	$6,100	1/08	SENE	Diesel, WF, 2 pt., 3 pt. adapter
656	1972		G	$13,000	5/08	SEMN	High-clearance high crop, 3 pt., flat-top, fenders, new tires, front and rear, WF, totally restored
656			F	$1,750	12/07	WCMN	Hydro, 3 pt., PTO, needs work
656			P	$2,200	2/07	ECNE	NF, gas, 1 hyd., 3 pt., 15.5-38 tires
656			F	$2,600	8/07	SESD	Hydro
656			F	$3,900	8/07	SESD	Gas, WF
656			G	$4,100	7/07	WCSK	Canadian sale, Leon loader
656		6,500	G	$4,100	7/07	WCMN	Gas, 3 pt., 540 PTO, tube grill, long axle
656			G	$4,500	3/07	ECND	Diesel, 2 hyd., 3 pt., PTO, fenders
656			G	$6,750	3/07	ECMN	Loader
656			F	$7,500	1/07	NEIA	Rough, gear shift, gas, Freeman loader, welded
656	1965	4,903	G	$3,900	3/07	SCMI	Diesel, 2 hyd., 3 pt., WF, open station
656	1965	9,442	F	$4,000	5/07	SEWY	Diesel, WF, 3 pt.
656	1966		G	$3,000	9/07	NEIA	WF, standard drawbar, 2 hyd., flat-top fenders, gas
656	1966	7,495	G	$6,000	7/07	NWIL	One owner, NF, diesel
656	1968		F	$3,200	8/07	NWIL	Gas, WF, cab, 3 pt., weights
656			F	$1,100	9/06	NEIN	Hydro, original
656		4,523	F	$4,000	8/06	SCMI	Gas, WF, 1 hyd., fast hitch, snap-on duals
656			G	$4,000	8/07	SESD	WF, 2 pt., diesel
656			G	$4,100	9/06	NEIN	Hydro, original, diesel
656			G	$4,800	1/06	WCIL	Gas, WF, IHC 2000 loader
656			F	$6,000	1/06	NEIA	Gas
656			G	$6,000	2/06	SWPA	WF
656	1966		F	$2,400	5/06	SEWY	Farmhand F11 loader, bale spear, WF, gas
656	1967		P	$1,950	8/06	NWIL	Gas, open station, WF, fender, 3 pt., unknown hours, doesn't run
656	1967	9,477	G	$3,500	11/06	NEKS	WF, diesel, 1 hyd., 540 PTO, 3 pt.
656	1967		G	$3,900	8/06	ECNE	Diesel, WF, 15.5-38, 2 pt., new clutch, pressure plate, TA and throw bearings
656	1969	8,474	F	$2,650	2/06	NECO	No cab, WF, 3 pt., Farmhand F11 loader
656	1969		G	$5,700	2/06	NWIA	Unknown hours, diesel, hydro, NF, 2 pt., 2 hyd.
656	1971	4,358	G	$7,800	3/06	NCIL	NF, hydro, gas, 15.5-38 tires
656	1973	6,631	G	$5,100	1/06	WCNE	WF, 3 pt., 2 hyd., gas
656			G	$0	4/05	NCOH	No sale at $3,750
656		5,424	F	$1,950	9/05	SEIA	
656			F	$2,250	6/05	NESD	Open station, gas, 3 pt., PTO
656			F	$2,950	12/05	SEMN	Diesel
656		10,004	F	$3,000	11/05	NWIL	Gas, NF, 3 pt.
656			G	$3,000	3/08	SCIL	NF, 3 pt., gas
656			G	$3,200	7/05	ECND	WF, 3 pt., 540 PTO
656		5,700	F	$3,600	4/05	SEND	1,000 hours on OH, gas
656		8,854	F	$3,675	9/05	NEIA	Gas, hydro, NF

IHC/Farmall

Model	Year	Hours	Cond.	Price	Date	Area	Comments
656			G	$3,750	12/06	WCIL	Gas, new paint, rings and bearings
656			G	$4,100	3/05	NENE	WF, 3 pt.
656	1965		G	$5,800	2/05	NEIN	Loader, WF, diesel
656	1966	6,476	G	$2,500	8/05	NEIN	NF
656	1967	5,600	G	$3,200	12/05	SEND	Dual loader, clutch replaced in 2004, head worked in 2003
656	1968	6,161	F	$2,000	9/05	NECO	Duals, turbo
656	1968		F	$2,600	6/05	NESD	Gas, 2WD, 540 PTO, 3 pt.
656	1970		G	$6,900	3/05	ECMI	TA, new tires, 2 hyd., 540 PTO, 3 pt.
656	1972		G	$6,000	1/05	WCIL	
660	1960		G	$3,800	5/08	ECMN	Restored, diesel, standard, 18.4-34 tires, no SN
660	1963		G	$4,000	11/07	WCIL	Running, unrestored/original, rear wheel weights, fenders, lights, PS
660			G	$2,900	10/06	SCSK	Canadian sale, 2WD, diesel, 80 PTO hp., 2,090 hours showing, SN 5129, very nice condition
660		7,414	G	$3,800	10/06	SCSK	Canadian sale, 2WD, diesel, 18.4-34 rubber, 80 PTO hp., 7,414 hours showing, Robin front-end loader, second owner
660			G	$3,800	4/06	NWSD	Wheatland, diesel, no 3 pt.
660	1959		F	$2,000	6/05	NEND	Diesel, hyd., PTO, PS, 16.9-34 tires, one owner
660	1963	3,958	G	$2,800	2/05	NEIN	Diesel, WF
664	1972	5,100	G	$7,300	11/05	WCIA	Westendorf loader, diesel, good rubber
666			F	$3,700	8/07	SESD	Cab, 3 pt.
666			G	$5,400	3/07	NEWI	Diesel
666		4,937	G	$5,500	11/07	WCOH	Gas, NF
666	1974		G	$4,900	12/07	ECNE	
666			F	$2,500	6/06	NWSD	Farmhand F25 loader
666			F	$4,800	1/06	SEPA	Hydro, WF, average rubber
666		7,378	G	$5,150	3/06	NEMI	TA, 2 hyd., 800 hours on OH, 540 PTO, less 3 pt. arms
666			F	$5,400	3/06	ECKS	Great Bend 800 loader
666		5,037	G	$6,400	3/06	NWPA	Gas, 1 hyd.
666		6,790	G	$7,000	3/06	ECNE	Gas, WF, 15.5-38, 2 hyd., 540 PTO, 3 pt., rear wheel weights
666	1974	9,644	G	$5,500	7/06	WCMN	WF, gas, 3 pt., loader, 15.5-38 rubber (70%)
666	1976		E	$6,600	4/06	NWOH	Gas, open station, very nice
666			F	$3,200	4/05	SCSD	Gas
666			F	$4,100	12/05	SEMN	
666			G	$5,300	11/05	SCIL	Canopy, 2 pt., TA
666			G	$5,700	12/05	SCMN	Gas
666		7,762	G	$7,000	3/05	NEOH	Completely restored but never an OH, looks like new, diesel, WF, 2 hyd., 3 pt., TA good, one owner
674			F	$3,000	2/08	NWSC	Diesel, 62 hp., 2WD
674			G	$8,300	1/08	NEMO	IHC loader, diesel
674	1975		G	$2,900	12/07	SEWY	WF, 3 pt., gas
674	1976		E	$9,000	3/07	ECIL	Diesel, utility tractor, 3 pt., 540 PTO, 1 hyd., 16.9-30 tires, total hours unknown although 2,000 hours on OH, IHC 2250 loader
674			G	$2,000	7/06	WCCA	3 pt., PTO, 1 hyd.
674			F	$5,000	2/05	NWWI	Early 1970s model, WF, gas, 3 pt., fenders, no-name all-hyd. loader and snow bkt.
674		2,158	G	$5,200	8/06	NCOH	Dunham-Lehr loader, 158 hours on OH
674		3,070	G	$5,900	3/06	NCOH	Woods loader

IHC/Farmall

Model	Year	Hours	Cond.	Price	Date	Area	Comments
674			G	$7,000	11/06	SECA	Front weights, canopy, 3 pt., PTO, rear weights, 1 hyd.
674			G	$4,600	9/05	SCON	Canadian sale, 2WD, 3 pt., hyd.
674		4,834	G	$5,100	3/05	NEOH	Gas, 2 hyd., good paint, 3 pt., PS, 18.4-30 rear tires, one owner
686	1977	2,576	G	$7,250	2/05	NWOH	Gas, 16.9-38 tires
706			F	$2,400	1/08	NEMO	Diesel
706			G	$3,600	4/08	NWMN	Gas, WF, 2 hyd., 3 pt., PTO, Farmhand loader, PTO pump
706	1964		F	$2,500	2/08	NEMO	Diesel, 16.9-34 tires, 2 pt., 540/1,000 PTO
706	1964		G	$4,600	2/08	SCMI	Gas, NF, 1 hyd., 3 pt.
706	1967		E	$5,300	5/08	ECMN	Restored, German diesel, WF, 3 pt., newer 18.4-38 tires
706	1967		G	$5,600	1/08	SENE	German diesel engine, 2 pt., WF
706	1968	6,100	G	$5,000	1/08	ECIA	Diesel, NF, new clutch at 5,600 hours
706		9,149	F	$1,100	3/07	NWIA	3 pt., PS, 2 hyd.
706			G	$1,750	4/07	ECSK	Canadian sale
706			F	$2,000	5/07	WCWI	NF
706			G	$2,100	3/06	NWIL	Gas, new paint, fresh OH, WF
706			F	$2,350	1/07	NWIN	3 pt., NF
706			F	$2,700	7/07	NCIA	Gas, TA out, WF, 3 pt., 2 PTO, 2 hyd.
706			F	$2,700	11/07	WCIL	Running, unrestored/original, rear wheel weights, PTO, factory cab, lights, TA, one owner
706			G	$2,800	8/07	SESD	Diesel, row crop
706			G	$3,000	3/07	NEWI	Diesel
706			G	$3,000	2/07	SEIL	Gas, row crop, fenders, fast hitch, 16.9-34 tires
706			G	$3,100	11/07	SESD	Wheatland, diesel
706			F	$3,100	12/07	ECIN	Gas, WF, 2 pt.
706			P	$3,300	8/07	SESD	German diesel, cab, 3 pt., WF
706			G	$3,350	6/07	ECIL	Fast hitch, gas, 16.9-34 tires
706			P	$3,600	8/07	SESD	As is, 3 pt., WF, diesel, 1965 model
706			G	$3,700	9/07	NCND	2WD, factory 3 pt., diesel, very good 15.5-38 tires, 1 hyd., 540/1,000 PTO
706			G	$4,100	3/07	NWIA	
706			G	$4,600	3/07	NCOH	Diesel, NF, fast hitch
706		7,500	G	$4,800	11/07	SESD	Gas, WF, rear fenders, 3 hours on major OH
706			F	$4,900	2/07	WCKY	Gas, AC loader
706			G	$5,250	1/07	WCIL	Gas, OH, WF
706			P	$7,400	1/07	NEIA	Rough shape
706	1964		G	$2,500	9/07	NEIA	WF, 3 pt., 2 hyd., cab, gas
706	1964		G	$4,500	1/07	WCIL	WF, diesel
706	1965		G	$3,500	12/07	WCIL	Gas, NF, 3 pt.
706			F	$900	6/06	NWSD	Gas
706			F	$2,300	3/06	NWOH	Gas, WF, 1 hyd., fast hitch, dual PTO, tach says 2,071 hours
706			G	$2,400	6/06	ECSD	2 pt., Farmhand F11 loader
706			G	$2,400	7/06	NWMN	Cab, 3 pt., hyd., PTO
706			F	$2,500	3/06	NCOH	WF, broken tach
706			G	$3,700	5/06	ECMN	Gas
706			G	$4,000	3/06	SWMN	WF, fast hitch, 1 hyd., Paulson hyd. loader
706			G	$4,100	2/05	SCMN	
706		3,961	G	$5,000	6/06	NWIL	WF, 3 pt., new tires, cab, fenders, gas
706			G	$6,000	3/07	WCIL	WF, gas

60 RUST BOOK

IHC/Farmall

Model	Year	Hours	Cond.	Price	Date	Area	Comments
706	1963		G	$3,000	3/06	WCMI	Diesel, TA, fast hitch, 1 hyd., dual PTO
706	1964	5,400	F	$1,850	2/06	SCMI	Gas, NF, 3 pt., new 18.4-34 tires, one owner
706	1964		G	$3,100	3/06	ECNE	Diesel, WF, 2 pt., fast hitch
706	1964		F	$3,400	4/06	SEMI	Gas, NF, 3 pt., TA, 18.4-34, tach shows 2,600 hours
706	1964	4,697	G	$4,500	11/06	SENE	NF, diesel, 18.4-34, 1 hyd., 3 pt., 540/1,000 PTO, fenders, front end weight bracket
706	1965	7,671	G	$2,600	9/06	NEIN	Gas, WF, no 3 pt.
706	1965		G	$3,000	7/06	WCMN	WF, gas, 3 pt., 1 hyd., OH 800 hours ago
706	1965	9,761	G	$3,100	8/06	ECNE	Gas, NF, 15.5-38 rear tires, 2 hyd., 3 pt., 540/100 PTO, fenders, rear wheel weights, one owner
706	1965	5,527	G	$5,500	1/06	WCNE	WF, dual 340 loader
706	1965		G	$7,500	12/06	SCMT	Great Bend loader
706	1966	5,572	F	$3,400	2/06	NWIA	NF, 2 pt., 2 hyd., 5,572 hours
706	1967	4,870	E	$11,750	3/06	ECMI	Cab, heat, TA, 310 engine, 1 hyd., dual PTO, 3 pt., diesel, one owner, 15.5-38 tires
706			F	$1,800	7/05	ECND	No 3 pt., 2 hyd., PTO, gas
706			F	$2,200	8/07	NWIL	NF
706			F	$2,200	2/05	NWKS	Diesel, 18.4-38, engine knocks, Farmhand F11 loader, dual PTO, 2 hyd.
706			F	$2,600	12/05	SEMN	
706			G	$2,600	9/06	ECMI	Gas, NF, tach shows 1,619 hours
706		8,360	F	$2,800	8/05	WCMN	German diesel, 3 pt., PTO, band duals, engine oil leak
706		6,341	G	$2,800	9/05	NCIA	Cab, gas, WF, fast hitch
706			G	$3,500	8/05	SEMN	
706			G	$3,550	7/05	SESD	German diesel, WF
706			G	$3,800	3/05	NENE	
706			G	$4,000	12/06	WCIL	Diesel, restored, Western fenders
706		4,682	F	$4,500	1/05	WCOH	Farmall, WF, 3 pt., diesel, older model
706		10,000	G	$4,500	8/05	WCMN	German diesel, Farmall, band duals
706			G	$5,500	12/05	SEND	Cab, heat, filtered air inlet, 3 pt., quick hitch, low hours, new tires, alternator, battery, paint, recent transmission and differential lock
706			G	$5,900	9/05	ECMN	Gas, loader
706		3,524	G	$6,600	11/05	SCNE	Farmall, WF, 2 pt.
706	1965	8,647	G	$1,750	4/05	NEIN	NF, gas
706	1967		G	$3,200	4/05	SENE	Diesel, WF, new tires
756	1969	5,300	G	$7,250	2/08	ECIL	Diesel, 3 pt., 2 hyd., one owner
756			G	$3,900	6/07	ECIL	Gas, WF, 3 pt., 16.9-34 tires
756			G	$4,000	10/07	NEKS	WF, diesel, fair rubber, 2 pt., 2 hyd.
756			G	$5,000	2/07	SEIL	Diesel, fender tractor, WF, 3 pt., 16.9-34 tires
756		6,918	G	$4,700	2/06	ECIL	Fenders, 2 hyd., front & rear weights, quick hitch, good TA, OH
756			F	$5,400	3/06	NCMN	
756	1967	8,566	G	$5,200	1/06	ECNE	WF, 2 pt., 2 hyd., 15.5-38 tires
756			F	$2,600	4/05	SEPA	Gas, WF, 30% rubber
756			F	$2,800	9/05	SEIA	Gas
756			E	$7,300	3/05	WCWI	WF, 3 pt., 2 hyd.
766			G	$4,250	3/08	ECNE	6-cyl. diesel engine, WF, cab, 3 pt., 2 hyd., TA, 18.4-34 tires with 9-bolt hub duals
766	1971		G	$6,400	1/08	SESD	Cab, 3 pt., Farmhand 228 loader
766		5,300	G	$7,600	8/07	SESD	3 pt., original, diesel

IHC/Farmall

Model	Year	Hours	Cond.	Price	Date	Area	Comments
766		6,074	G	$6,100	11/06	WCWI	WF, 3 pt.
766			G	$6,250	3/06	NCMN	Fresh rebuilt engine
766			F	$7,600	1/06	SEPA	WF, good rubber
766			F	$8,100	3/06	SEPA	WF, cab, turbo, average rubber
766	1975	3,833	G	$4,500	1/06	NWIA	WF, 3 pt., 2 hyd., PTO, cab
766		500	F	$1,350	2/05	SCMN	New engine, 18.4-34 duals, 3 pt., 540/1,000 PTO, 2 hyd.
766			F	$4,050	5/05	SEPA	Diesel, WF, no TA, 40% rubber
766			E	$5,200	3/05	NECO	Diesel, 3 pt., PTO, hyd.
766		5,600	F	$9,000	4/05	SEPA	WF, black stripe, 60% rubber
806			F	$4,500	1/08	NEMO	Diesel, new motor and rubber
806		7,151	G	$4,800	3/08	NWMN	IHC cab, 2 hyd., 3 pt., PTO, 23.1-34 singles
806			G	$5,200	3/08	NWIL	Diesel, NF, 1,500 hours on major OH, 3 pt., 2 hyd., heat houser for 806
806			G	$6,500	7/07	WCMN	Factory WF, 4 hyd., 3 pt., PTO, flat-top fenders, rock box, Firestone 18.4-38 tires
806			E	$9,300	4/08	NEIL	Diesel, WF, 2 hyd., dual PTO
806	1964	9,527	G	$4,300	2/08	ECIL	Diesel, ROPS, 2 pt., 2 hyd., dual PTO, 18.4-34 tires
806	1966		G	$5,500	1/08	WCIL	Diesel, quick hitch, WF, good TA, 540/1,000 rpm, 18.4-34 tires
806	1967		E	$9,250	5/08	ECMN	Restored, diesel, 3 pt., no cab, 2 hyd., newer 18.4-38 tires
806	1967	5,232	E	$10,200	1/08	ECIL	Diesel, WF, fender tractor, 3 pt., weights, 18.4-34 tires, nice original tractor
806			F	$2,700	2/07	ECMI	Gas, cab, 3 pt., 2 hyd., dual PTO, 16.9-38 tires
806			G	$3,650	6/07	ECIL	WF, gas, new 18.4-34 tires, 2 pt.
806			F	$3,700	8/07	SESD	WF, 2 pt., diesel
806			G	$3,750	12/07	WCMN	Boom tractor, K&M rear weight box
806			G	$3,900	8/07	SESD	WF, 3 pt., cab, 1964 model
806		6,000	G	$5,200	8/07	WCIL	Diesel, cab, good tires
806			G	$5,500	9/07	WCIA	Factory LP, WF
806			G	$6,500	4/07	SCOK	3 pt., WF, rubber 80%, 94 hp.
806			G	$6,500	7/07	NCIA	Diesel, WF, 3 pt., 2 PTO, 2 hyd.
806			G	$6,600	1/07	NEIA	Diesel, WF, 3 pt., 2 hyd., IHC 2000 hyd. loader, bkt. and bale spear, OH 4 years ago
806			E	$7,300	8/07	NWIA	Major OH front to back, repainted
806	1964		G	$3,100	8/07	SESD	
806	1964	6,700	F	$4,100	9/07	NWIL	WF, fenders, diesel
806	1965		G	$3,500	12/07	WCMN	Wheatland, dual PTO, good torque, 2 hyd., Comfort King cab, Firestone 18.4-38 tires, 85% rubber
806	1966	5,232	G	$3,900	8/07	ECIL	WF, 3 pt.
806	1966		G	$6,400	9/07	NEIA	WF, 3 pt., 2 hyd., flat-top fenders, diesel
806	1967		G	$3,700	8/07	WCMN	Diesel, WF, 3 pt.
806	1967		G	$6,100	2/07	ECIL	Fender tractor, 2 hyd., 18.4-38 tires
806	1967		G	$8,000	7/07	ECIL	Diesel, WF, added turbo, rear wheel weights, lights, PTO, 1 hyd., 3 pt., nice restoration
806		5,523	G	$2,000	9/06	ECMN	Cab, WF, diesel, 3 pt., dual PTO
806		4,983	F	$2,800	4/06	SCMI	Gas, WF, 3 pt., 1 hyd., axle duals
806			G	$3,000	12/06	ECIN	Gas, PTO
806			F	$3,100	4/06	NEIA	Diesel
806			F	$3,250	12/06	NWWI	Duals
806			F	$3,600	5/06	SEND	3 pt., dual PTO, diesel

IHC/Farmall

Model	Year	Hours	Cond.	Price	Date	Area	Comments
806			G	$4,500	10/05	SWWI	WF, 3 pt., 540/1,000 rpm PTO
806			G	$4,600	3/06	NWIL	Cab, diesel
806			F	$4,900	3/06	NWOH	Tach says 2,160 hours, 3 pt., 2 hyd., dual PTO, axle duals
806			P	$5,300	7/06	SCIA	Rough, diesel
806			G	$5,300	3/06	ECIN	Gas, PTO
806			G	$6,500	9/06	NEIN	MFWD
806		6,000	G	$6,700	11/06	SCIL	Diesel, turbo, duals
806	1963	9,697	F	$2,200	11/06	NCKS	Open station, 2 hyd., dual PTO
806	1964		F	$2,500	11/06	SENE	WF, LP gas, 645 hours on OH, 18.4-34, 2 hyd., 2 pt., quick hitch, fenders, 2 sets of rear wheel weights, dual 3000 front-mount loader, 8' bkt., manure teeth, needs work
806	1964		F	$3,700	1/06	NWIL	Diesel, WF, 2 pt., 18.4-34
806	1964		G	$4,600	9/06	NCIA	WF, 3 pt., diesel
806	1965		G	$3,900	3/06	WCMI	Diesel, TA, 2 hyd., dual PTO, good 18.4-34 tires
806	1967	2,320	G	$3,300	8/06	ECMI	Freeman hyd. loader, material bkt., gas, second owner, WF, PTO, 3 pt., 1 hyd., newer 15.5-38 tires, 13.6-38 axle duals, last year made
806	1967	6,788	F	$3,500	9/06	NECO	Diesel, WF, no 3 pt., cab, dual 340 loader
806	1967	9,039	F	$3,700	11/06	NWMN	Diesel, open station, 2 hyd., 3 pt., 540/1,000 PTO, 18.4-38 hub duals
806	1967		G	$4,000	12/06	SCMT	
806	1967	8,200	F	$4,200	7/06	NWIL	Diesel, cab
806	1967	9,611	G	$4,850	1/06	NEMO	
806	1967	6,300	G	$5,600	7/06	NWIL	Diesel, NF, 3 pt.
806			P	$0	2/05	WCOK	No sale at $1,400, wanted $3,500, LP, 3 pt., 1 hyd., tires poor, PTO, no third member
806			P	$750	2/05	WCIL	For salvage, diesel
806			F	$1,700	2/05	NCOK	Wheatland
806			F	$2,400	7/05	ECND	Row crop, diesel, no 3 pt., deluxe fenders, 2 hyd.
806			F	$2,700	7/05	SESD	WF, cab, 2 pt., good TA
806		9,143	F	$2,900	11/05	SCMI	Gas, WF
806			G	$3,500	8/05	SEMN	Diesel, 3 pt., 18.4-38
806			F	$3,900	6/05	WCMN	Cab, diesel, WF, 2 hyd., 2 PTO
806			G	$4,250	2/05	WCIN	2 pt., 18.4-34 tires, duals, 2 hyd.
806			P	$4,400	9/05	NEIA	Rough
806			G	$5,600	4/05	NWIL	WF, Year-A-Round cab, 2 hyd., good TA, diesel
806			G	$5,700	7/05	ECND	Year-A-Round cab, diesel, 3 pt., PTO
806			G	$8,000	12/05	SEND	
806	1963	5,300	F	$3,300	9/05	WCNE	Diesel, 3 pt., cab
806	1966	6,914	P	$2,100	9/05	NECO	Diesel, cab, cracked block, 3 pt.
806	1967	6,815	F	$4,700	1/05	NWIL	WF, dual PTO, 2 hyd., 3 pt., 18.4-34, torque out
8-16	1921		G	$11,000	9/06	ECMI	Steel wheels, restored, runs good, chain-driven wheels
826			F	$4,300	12/07	WCMN	Hydro
826			G	$5,500	12/07	WCMN	2 hyd., 3 pt., WF, diesel, 540/1,000 PTO, 18.4-34 tires, 9-bolt hub duals, rock box
826		5,930	G	$5,600	3/07	NWIA	German diesel, cab, 3 pt., 2 hyd.
826			G	$7,000	8/07	SESD	Hydro, 3 pt., diesel
826			G	$4,000	8/06	SEMN	
826			G	$4,700	9/06	NEIN	Original
826			G	$5,400	8/06	NEIA	Cab, gear drive, WF, diesel

IHC/Farmall

Model	Year	Hours	Cond.	Price	Date	Area	Comments
826			G	$10,000	9/06	NEIN	Gold demonstrator, restored
826	1970	4,050	G	$7,000	12/06	NWOH	Open station, 3 pt., like-new rear tires, duals, weak torque
826	1971	4,505	G	$6,800	1/06	WCIL	
826			G	$5,750	2/05	NEIA	
826		4,771	G	$7,000	4/05	NCOH	WF, 3 pt.
826			G	$12,500	12/05	SWWI	$7K recently spent on engine
826	1970	5,495	F	$5,400	2/05	NWKS	Diesel, 15.5-38, 2 pt., 1 hyd.
826	1970		G	$6,250	9/05	ECIA	German diesel, WF, high hours, cab
856			G	$5,900	4/08	NEIA	No cab
856			G	$4,000	7/07	SEND	Row crop, 2 hyd., 3 pt., PTO, TA, 14.9-38 singles, PTO doesn't work
856			F	$4,200	11/07	SESD	WF, 2 pt., cab
856			G	$5,200	3/07	NWIA	Diesel, WF, 3 pt., dual PTO
856			G	$5,850	1/07	ECIL	Fender tractor, 18.4-34, 2 hyd., PTO
856			G	$6,200	12/07	WCNY	Cab, single tires
856			G	$7,800	10/07	WCWI	WF, 3 pt., 1,000 hours on OH, diesel
856	1967	6,000	G	$4,700	12/07	ECNE	Diesel, 3 pt., 540/1,000 PTO, rebuilt injection pump, 18.4-38 rear tires, 2 hyd.
856	1968	7,033	G	$6,900	9/07	NEIA	Cab, WF, 3 pt., 2 hyd., hub duals, TA out
856	1969		G	$5,000	8/07	SESD	WF, 3 pt.
856	1969	7,413	G	$5,900	1/07	WCIL	Custom, diesel, fenders
856	1969		G	$7,300	11/07	WCIL	Diesel, 18.4R-38 rear tires, 10.00-16 front tires, 540/1,000 PTO, 2 hyd., older cab, no value, 10 front weights, SN 20049
856			G	$4,500	2/06	WCME	Gould & Smith cultivator
856			F	$5,300	2/06	ECNE	2 hyd., 3 pt., 7,800 hours
856		5,500	G	$6,800	11/06	SWIA	Diesel, Year-A-Round cab, 16.9-38 rubber, new injection pump
856			E	$7,700	11/06	NEKS	Diesel, cab, TA, fast hitch, 540/1,000 PTO, near-new 18.4-38 tires
856	1968	1,008	G	$6,100	2/06	NECO	Diesel, cab, 3 pt., WF
856	1968		G	$6,200	8/06	ECNE	Turbo diesel engine, WF, 18.4-38, 11L-15 front tires, 540/1,000 PTO, 2 pt., fenders
856	1970	5,088	G	$8,500	3/06	ECMN	Newer Allied 795 loader, diesel, CAH, +10% buyers premium
856	1971		G	$6,500	2/06	WCMN	2WD, Hiniker cab, tilt steering wheel, Goodyear 18.4-30, 9-bolt hubs, no duals, sharp
856	1971	7,242	G	$6,600	3/06	NCCO	Clutch and TA recently rebuilt 5 years ago, cab, 2 hyd., PTO, quick hitch, 15.5-38, clamp-on duals
856	1971	8,187	G	$8,300	4/06	SEMI	3 pt., 2 hyd., TA, diesel, 18.4-38
856	1973	8,614	G	$7,100	2/06	SEIA	WF, flat-top fenders, 800 hours on OH
856			P	$2,500	12/05	SEMN	Rough
856			F	$5,750	8/05	WCMN	Diesel, 540/1,000 PTO, duals, 2 hyd., 3 pt.
856	1968	5,855	G	$7,300	3/05	SCMI	Diesel, 12 front weights, 150-gal. saddle tanks, duals
856	1968	6,220	E	$11,000	1/05	NEMO	
856	1969	3,504	F	$2,300	9/05	NECO	Cab, 3 pt.
856	1969		F	$5,900	2/05	WCIL	Diesel, cab
856	1969	4,859	G	$6,600	6/05	NCIA	Factory WF, fast hitch, Year-A-Round cab, diesel, 18.4-38 tires
856	1969	4,019	G	$6,700	8/05	SWOH	Diesel
856	1969	5,222	G	$7,900	6/05	SEND	Cozy cab, 3 pt., 540/1,000 PTO, 18.4-38 hub duals, diesel

IHC/Farmall

Model	Year	Hours	Cond.	Price	Date	Area	Comments
856	1970		G	$5,100	12/05	WCMN	
886	1976	9,147	G	$12,000	3/07	ECIA	500 hours on OH, black stripe, diesel, factory cab, 18.4-34 tires, dual PTO, 3 pt.
886	1976	4,131	G	$8,900	8/05	ECIL	One owner, cab, air, 2 hyd., 18.4-34 tires
886	1978	7,806	F	$4,625	3/05	NEKS	ROPS, 2 hyd., good rubber
966			F	$6,000	1/08	NEMO	
966	1974	5,235	G	$5,200	3/08	SCNE	3 pt., diesel, WF, cab, 18.4-38 tires, 2 hyd.
966	1974	4,649	G	$6,050	1/08	ECMI	Cab, duals
966	1974	4,700	G	$9,600	1/08	ECIA	Diesel, OH and clutch at 3,650 hours, WF, cab, Great Bend 900 Hi-Master loader, 18.4-34 tires, duals $200 extra
966			F	$4,800	8/07	SESD	
966		6,170	G	$4,900	3/07	ECIL	IHC cab, 540/1,000 PTO, 2 hyd., 3 pt., 18.4-38 tires, tach shows 6,170 hours, older repaint
966			G	$7,100	11/07	SESD	WF, 3 pt., OH, 2003 Vaughn loader
966	1973		F	$3,500	1/07	SWNE	18.4-38, clamp-on duals, 2 hyd., 540/1,000 PTO, black stripe, cab
966	1973		G	$8,050	2/07	ECIL	ROPS, canopy, 16.9-38, new engine two years ago, quick hitch, attached Leon 808 hyd. loader
966	1973	7,500	G	$8,300	11/07	SESD	One owner, CAH, fast hitch, 18.4-34 tires
966	1974		G	$6,000	12/07	SEWY	Cab, 3 pt., Farmhand 236 loader, grapple, diesel
966		7,319	F	$4,450	9/06	SCMN	Bad torque, cab, diesel, 3 pt., 2 hyd., 540/1,000, bad TA
966			F	$4,750	3/06	SEPA	Good TA, WF, ROPS, good rubber
966			F	$5,400	12/06	NWWI	
966		6,076	G	$6,400	8/06	NWOH	Cab, 1 hyd., 3 pt., 5 front weights
966		6,086	G	$7,600	3/06	NWPA	2 hyd., 2 PTOs
966		4,439	G	$8,000	3/06	NCOH	WF, Year-A-Round cab
966			G	$10,000	8/06	SCMI	MFWD, canopy, 3 pt., 2 hyd., Snapper duals
966			G	$11,000	6/06	NEIN	Black stripe, restored
966		7,800	F	$12,000	1/06	NEIA	
966	1972	6,625	F	$5,100	2/06	WCNE	Diesel, Year-A-Round cab, 3 pt., dual 3000 loader and grapple
966	1972	4,961		$5,100	3/06	NWOH	3 pt., 1 hyd., 4 front weights, dual PTO, snap-on duals
966	1972	1,200	G	$10,200	2/06	NEIA	Cab, new batteries, tires and brakes, bought on August 2005 consignment sale for $6,800
966	1973	5,442	G	$6,200	3/06	WCMI	Diesel, TA, cab, dual PTO, 2 hyd., 3 pt., 16.9-38 axle duals
966	1974	6,500	G	$9,100	3/06	ECNE	3,595 hours on new tach, 18.4-34, 9-bolt duals, 11-15 front tires, 2 hyd., 540/1,000 PTO, 10 front end weights
966	1975	6,835	F	$4,900	7/06	WCMN	2 hyd., cab, rock box, 540/1,000 PTO
966	1976	8,600	F	$5,000	8/06	NWIL	Open station, 3 pt., WF, front/rear weights
966			P	$2,500	6/05	WCMN	Rough, 3 pt., PTO, 2 hyd.
966			P	$3,400	3/05	WCIL	Diesel
966			F	$3,975	9/05	NENE	Year-A-Round cab, 540/1,000 PTO, 2 hyd.
966			F	$4,000	9/05	NENE	
966			F	$5,600	8/05	SEMN	New Idea 319 corn picker
966		3,900	F	$7,150	1/05	WCOH	Duals, 540 PTO, white cab, air
966			G	$8,000	8/05	NWIA	Recent OH, WF
966		4,049	G	$10,000	4/05	WCWI	Front weights
966	1975	3,500	G	$9,500	12/05	SCIA	Cab, one owner
966	1975	1,026	E	$14,500	4/05	SCMN	Hiniker cab, 1 hyd., band duals

IHC/Farmall

Model	Year	Hours	Cond.	Price	Date	Area	Comments
966	1976	7,010	F	$5,000	2/05	WCOK	New paint, 3 pt., PTO, 2 hyd., diesel, cab interior and tires fair
986	1978	5,043	G	$8,900	1/08	ECOK	Cab, air, 3 pt., PTO, 3 hyd., 18.4-38 rubber
986	1978	4,290	G	$9,000	12/07	WCMN	CAH, 2 hyd., PTO
986	1978	7,400	G	$14,000	3/07	WCNE	Diesel, cab, WF, 3 pt., 540/1,000 PTO, 3 hyd., mounted Farmhand 235 loader with bkt. and grapple
986	1973	9,625	F	$5,000	8/06	NWIL	Cab, air, WF, 1,000 and 540 PTO, 2 hyd.
986	1976	5,202	G	$10,300	11/06	SCNE	Diesel, CAH, 3 hyd.
986	1977	8,750	G	$6,300	7/06	SEND	Gear trans., 2 hyd., 3 pt., 540/1,000 PTO, 18.4-38
986	1977	2,327	E	$17,000	3/06	ECMI	2WD, factory CAH, TA, 18.4-38 axle duals, digital tach
986	1977	4,027	G	$11,500	4/05	NEIN	CAH, diesel
A			G	$1,500	6/07	SEID	Farmall, runs
A			G	$1,600	1/07	SCNE	Farmall, WF, new front tires
A			G	$1,600	11/07	SESD	Farmall
A	1939		G	$1,600	7/07	ECIL	Farmall, light utility, new rear tires, fenders, deluxe seat, PTO
A	1941		G	$1,900	11/07	WCIL	Farmall, running, restored, fenders, PTO, lights
A	1941		G	$2,150	8/07	WCMN	Farmall, restored, have all receipts, show-ready, new tires
A	1945		G	$1,800	6/07	NCIN	Farmall, belt pulley, repaint, front weights, runs
A	1945		G	$2,200	8/07	WCMN	Farmall, new paint, tires, fenders, hoses, battery, box, drawbar and hitch
A	1946		P	$175	6/07	SEID	Farmall, for parts
A	1946		G	$3,200	7/07	ECIL	Farmall, light utility, 60" Woods L59 belly mower, PTO, fenders
A	1946		G	$6,500	9/07	NEIL	Farmall Cultivision A, SN 165928
A			F	$1,000	10/06	SEWI	Cultivator lift
A			F	$1,200	3/06	SEPA	Farmall, WF, front blade, good rubber
A			G	$2,200	10/06	SCSD	Farmall, Woods 52" belly mower, gas, turf tires
A			G	$3,000	10/06	WCSK	Canadian sale, McCormick Farmall antique tractor, 4-cyl. gas, 3-rib front, 9×24 rear tires, SN AFAA109294, restored
A			G	$3,000	10/06	SCSK	Canadian sale, McCormick Farmall antique tractor, 4-cyl. gas, 3-rib front, 9×24 rear tires
A			G	$5,500	8/06	SCIL	Cultivator
A	1940		G	$3,150	10/06	SCSD	Farmall, WF, front and rear wheel weights, gas
A	1941		G	$450	1/06	WCNE	WF, PTO
A	1942		G	$875	1/06	WCNE	WF
A	1942		G	$900	1/06	WCNE	WF
A			F	$1,000	12/07	WCNY	Farmall, original paint
A			G	$1,150	9/05	NENE	Farmall Cultivision A tractor, WF, PTO, 9-24 tires
A			G	$1,300	8/05	NEIN	Farmall, gas, WF
A			G	$1,400	10/05	NEPA	Famall
A			F	$2,600	3/05	NESD	Farmall, WF, Woods belly mower
A	1940		G	$800	8/05	NEIN	Gas, WF
A	1940		G	$2,900	6/05	NEND	Farmall, WF, 4 speed, belt pulley, PTO, rear weights, fenders
A	1942		F	$1,400	7/05	NCMN	Farmall, WF
A	1945		G	$1,975	1/05	WCIL	Farmall, Woods 42" belly mower
AV	1944		G	$4,400	5/08	SEMN	High crop, fenders, PTO

IHC/Farmall

Model	Year	Hours	Cond.	Price	Date	Area	Comments
AV	1947		G	$2,800	11/07	SWIN	Farmall, belt pulley, front wheel weights, rear wheel weights
B			G	$2,250	5/08	ECMN	Farmall, tricycle front wheel weights, fenders
B	1941		G	$5,000	5/08	SEMN	WF, high-clearance kit, 1 of 600 built by Bushal Implement, rear wheel guards, fenders
B			F	$900	11/07	WCIL	Farmall, belly mower
B			F	$1,200	6/07	SEID	Farmall, repainted, runs, single front
B			G	$1,600	7/07	ECIL	Farmall, restored, NF, PTO, deluxe seat, fenders, very nice
B	1941		G	$4,600	6/07	SCIL	Farmall, great tires, repainted
B	1942		F	$700	3/07	SEWY	Farmall, gas, single front, front-mount 6' dozer
B	1947		G	$1,600	9/07	WCIA	Farmall, repainted
B			F	$550	4/06	WCTX	
B			F	$600	4/08	ECND	Farmall, NF, 5' mower deck
B			G	$800	3/06	NWOH	Farmall, gas, tricycle front, restored
B			F	$1,200	11/06	WCMN	Farmall, Woods belly mower
B			G	$1,300	7/07	WCMN	Farmall
B			G	$1,300	7/07	WCMN	Farmall, 5' belly mower
B			G	$1,400	4/06	ECND	Cultivision, NF, PTO, Woods 59 belly mower
B			G	$2,000	8/06	SCIL	New tires
B			G	$2,100	6/05	WCMN	Farmall, Cultivision, NF, gas
B			G	$3,200	8/06	SCIL	New paint
B			G	$3,400	8/06	NCIA	Farmall, new tires
B	1944		G	$2,750	10/06	SCSD	Farmall, gas, single front wheel
B	1945		G	$2,400	9/06	NCIA	Farmall, Cultivision
B			P	$450	8/05	ECMN	Farmall, single front wheel, engine stuck
B			G	$1,000	7/05	ECND	Winco 7500W generator
B			G	$1,200	7/05	ECND	Woods 5' belly mower
B			G	$1,450	3/05	ECNE	Farmall, 50 hours on OH
B			G	$1,500	11/05	WCIL	NF, fenders, PTO, lights
C			G	$1,000	3/08	WCMN	Farmall, NF, PTO, 11.2-36 tires
C			G	$1,550	4/08	NEIA	Farmall, belly mower
C			G	$1,900	3/08	SWKY	Farmall, cultivators
C			G	$3,000	2/08	WCMN	Farmall, NF, all new rubber, 11.2R-36 tires, new paint
C	1948		G	$2,100	5/08	WCIA	Farmall, 6 volt, old repaint, Artsway belly mower
C	1953		G	$2,000	5/08	ECMN	Restored, NF, like-new tires
C			F	$600	1/07	NWOH	Farmall, gas, NF
C			G	$1,200	6/07	NCIN	Farmall, NF, oversize tires, belt pulley, IHC draw extension, runs
C			G	$1,300	9/07	NEIA	Farmall
C			G	$1,400	11/07	SEND	Farmall, Woods belly mower
C			G	$1,450	8/07	SESD	Farmall, Woods belly mower
C			G	$1,900	7/05	SESD	Farmall, Artsway belly mower
C	1948		G	$1,350	6/07	NCIN	Farmall, rear weights, front weights, original, fairly straight, runs
C	1949		G	$1,250	6/07	NCIN	Farmall, original, straight
C	1949		G	$3,250	8/07	WCMN	Farmall, restored, have receipts, new tires, show ready
C	1950		G	$1,150	6/07	NCIN	Farmall, NF, fenders, runs
C	1950		F	$1,200	2/07	NWIL	Farmall, Woods 4' belly mower

IHC/Farmall

Model	Year	Hours	Cond.	Price	Date	Area	Comments
C	1950		G	$1,250	3/07	SEWY	Farmall, gas, WF
C			F	$1,100	1/06	SEPA	Farmall, NF, average rubber
C			G	$1,125	8/06	SWWI	Farmall, cultivator, new rubber
C			G	$1,250	3/06	NWOH	Farmall, gas, NF, white, restored
C			G	$1,400	8/07	NWIL	Farmall, gas, belly mower and 5' rotary mower
C			G	$1,900	4/06	WCMT	Farmall, NF, 4-cyl. gas engine, PTO, rear belt drive, paint is poor
C			G	$2,700	12/06	NCIA	Farmall, Woods belly mower
C			G	$2,800	8/06	NCIA	Farmall, Woods 6' belly mower
C			G	$3,300	8/06	SCIL	
C	1948		G	$1,050	7/06	ECND	Farmall, NF, Woods 59 belly mower
C	1948		G	$2,000	10/06	SCSD	NF, lights, gas, wheel weights, motor OH
C	1950		G	$750	1/06	WCNE	Gas, WF, new tires and paint
C	1950		G	$2,000	2/06	WCMN	Farmall, demo
C			F	$1,000	8/06	SEMN	Cultivator
C			G	$1,500	8/05	NCIA	Farmall, belly mower
C			G	$1,600	6/05	ECMN	Tricycle, belt pully
C			G	$1,700	3/05	WCNY	Farmall, repainted but average
C			E	$1,700	3/05	WCWI	Farmall, cultivator
C			G	$1,800	9/06	NEIN	Farmall, Woods belly mower, fenders, new rear rubber
C			G	$2,000	4/05	ECMN	Farmall, belly mower
C			G	$2,700	12/05	SCMN	Farmall
C			G	$2,900	9/05	NENE	Farmall, WF, fenders, PTO, drawbar, 12.4-36 tires, new paint, new tires and rims, completely restored, just OH
C	1948		F	$450	11/05	ECNE	Farmall, PTO, 12.4-36 tires
C	1948		G	$2,075	7/05	NCMN	WF, new tires all around, electric start
Cub			G	$2,300	4/08	WCWI	WF, rear-mount 3' rotary brush mower
Cub	1949		G	$4,000	5/08	SEMN	Farmall, 60" Woods belly mower, hyd. lift, rear turf tires, fenders, new front tires, new PTO shaft and gear, low hours on rebuilt engine
Cub	1959		G	$4,000	2/08	SEAL	Farmall
Cub			G	$1,400	11/05	WCIL	Farmall, Woods 42" mower
Cub			G	$1,900	9/07	ECIN	Farmall
Cub			G	$2,500	3/05	WCMN	Farmall, restored, has all receipts, show-ready
Cub			G	$2,500	11/05	WCIL	Farmall, belly-mount blade, good paint and rubber
Cub	1947		G	$1,700	8/07	SESD	Farmall
Cub	1948		F	$700	11/07	WCIL	Running, unrestored, original, rear wheel weights, fenders, PTO, drawbar, new brakes
Cub	1948		F	$900	8/07	SWIL	Farmall, belly mower
Cub	1948		F	$1,200	5/07	SEWY	4' sickle bar mower, mid-mount, gas
Cub	1948		G	$1,250	6/07	NCIN	Farmall, unrestored, original, belt pulley, runs
Cub	1948		G	$1,700	11/07	WCIL	Farmall, running, restored, rear wheel weights, fenders, PTO, lights, hyd. remote, completely torn down, sandblasted, and reassembled
Cub	1948		G	$2,100	5/08	WCIA	Farmall, sickle mower, original
Cub	1949		G	$1,450	9/07	SCMO	18.3×24 tires
Cub	1955		F	$1,550	8/07	NWIL	Farmall, WF, gas, IHC 4' belly mower
Cub	1957		G	$1,750	8/07	WCMN	9.5-24 rears, new front tires, IHC sickle mower, restored
Cub	1961		G	$3,000	11/07	SWIN	Farmall, new rears, older restoration, belt pulley, underneath exhaust

IHC/Farmall

Model	Year	Hours	Cond.	Price	Date	Area	Comments
Cub	1968		G	$1,600	8/07	WCMN	WF, 9.5-24 rears, rear wheel weights, restored
Cub	1968		G	$3,000	9/07	WCIA	Mower deck, yellow
Cub			P	$100	1/06	WCCA	4-cyl. gas
Cub			F	$500	1/06	WCCA	4-cyl. gas
Cub			F	$750	1/06	SEPA	Farmall, sickle bar mower, average rubber
Cub			F	$1,125	1/06	SEPA	Farmall, cultivators, good paint, average rubber
Cub			G	$1,400	8/06	SETN	Cultivator
Cub			G	$2,250	8/06	SCIL	Lo-Boy, yellow, Woods 60" mower
Cub			G	$3,700	8/06	SCIL	Sickle mower, new paint
Cub			G	$4,000	8/06	SCIL	Cultivator, new paint
Cub			G	$7,250	8/06	SCIL	4WD, compact, yellow
Cub	1949		G	$2,500	10/06	SCSD	Farmall, gas, new rubber, lights, wheel weights
Cub	1949		E	$6,000	10/06	SCSK	Canadian sale, McCormick Farmall Cub antique tractor, 4-cyl. gas, 3-rib front, 9×24 rear tires, underbelly side knife mower, fully restored, painted
Cub	1953		E	$2,150	12/06	WCIL	Farmall, new tires and paint, gas
Cub	1953		G	$3,250	6/06	NWIL	Farmall, one-bottom plow and cultivator
Cub	1974		G	$2,000	2/06	WCIL	60" mower, new paint
Cub			G	$1,100	6/07	ECMN	Needs OH, belly mount grader
Cub			G	$1,250	1/05	SWOH	Farmall, belly mower
Cub			G	$1,300	1/05	SWOH	Farmall
Cub			G	$1,350	9/05	SCON	Canadian sale, Farmall, blade
Cub			G	$1,500	3/05	SCMI	Restored
Cub			G	$1,575	1/05	ECNE	Farmall
Cub			G	$1,600	6/05	ECMN	Farmall, OH
Cub			G	$1,700	10/05	WCIL	WF, restored, Bush Hog mower
Cub			G	$2,000	9/05	SCON	Canadian sale, Farmall, blade/plow
Cub			F	$2,350	12/05	NWIL	Woods 60" deck, rusty
Cub			G	$2,600	6/05	ECMN	Farmall, swept back WF, new paint
Cub	1947		G	$2,600	6/05	NEND	Farmall, WF, 3 pt., PTO
Cub	1947		G	$3,150	9/05	ECMN	Farmall, gas, one-bottom plow and 4' sickle mower
Cub Lo-Boy			F	$1,475	8/07	WCIL	Gas, belly mower
Cub Lo-Boy	1959		G	$2,100	11/07	WCIL	Running, unrestored, original, good running and mowing tractor
Cub Lo-Boy	1969		F	$1,100	6/07	SCOH	Mower deck
Cub Lo-Boy			F	$800	7/06	WCOH	
Cub Lo-Boy			G	$1,600	8/06	SEPA	Belly mower, WF, good rubber, hyd.
Cub Lo-Boy			G	$1,750	11/05	NCCA	Mower
Cub Lo-Boy			G	$1,800	11/05	WCIL	WF, restored, fenders
Cub Lo-Boy	1966		G	$1,400	6/05	ECIL	Auburn trencher and blade
F12	1934		F	$800	4/08	NENE	Farmall, NF, single front wheel
F12	1937		F	$500	2/08	SCMI	Farmall, gas, NF, cracked block
F12	1937		G	$1,000	4/08	NENE	Farmall, NF, single front wheel, hyd. lift, mounted plow

IHC/Farmall

Model	Year	Hours	Cond.	Price	Date	Area	Comments
F12	1933		G	$8,000	11/07	WCIL	Farmall, Fairway 12, running, old restoration, repaint
F12	1934		G	$2,100	7/07	ECIL	Farmall, single wheel front, rear cut-offs, PTO, restored
F12	1936		G	$1,600	6/07	NCIN	Farmall, unrestored, original, cutoffs
F12	1937		F	$900	12/07	WCNE	Farmall, gas, single front, hyd., turns
F12			F	$575	8/06	NCIA	Farmall, skeleton wheels, NF, partially restored
F12			G	$575	10/06	SEWI	Fenders
F12			F	$675	8/06	NCIA	Farmall, power lift, rear skeleton wheels
F12			F	$1,700	8/06	SCIL	2-row cultivator, steel rear wheels and tires in front, engine needs work
F12			G	$1,900	4/06	WCIA	Restored
F12	1934		G	$1,100	9/06	ECMI	Farmall, NF, repainted, runs good
F12	1935		F	$500	8/06	NCIA	Farmall, Hysler trans., tricycle wheels, restored
F12			P	$800	9/05	NENE	Farmall, for parts
F12			F	$875	9/05	SEIA	
F12			G	$1,700	8/05	ECMN	Farmall
F12	1937		G	$700	4/05	NCOH	Farmall on rubber, restored
F14			G	$2,100	5/08	ECMN	Farmall
F14	1939		E	$2,400	2/08	SCMI	Farmall, gas, NF, restored
F14			F	$800	8/07	SWIL	Farmall, flat spoke steel wheels
F14			G	$900	3/06	NWOH	Farmall, gas, NF, restored
F14			G	$2,200	8/06	SCIL	On steel front and rear
F14			G	$2,250	5/08	ECMN	Farmall
F20			G	$1,600	6/05	ECMN	Farmall, restored
F20			G	$2,000	4/05	ECMN	Farmall, restored, round spokes, no SN, newer tires
F20	1939		E	$7,300	2/08	SCMI	Farmall, gas, NF, restored
F20			P	$110	3/07	SWNE	Farmall, for parts
F20			P	$300	9/07	NWIL	Farmall, gas
F20			P	$550	10/07	NEKS	Farmall, rough, motor stuck
F20			F	$600	10/07	NEIA	
F20			F	$600	2/05	WCIL	Farmall, older repaint, runs
F20			F	$1,000	9/07	WCIA	Farmall, new front tires
F20			F	$1,250	9/07	ECIN	Farmall
F20			G	$1,900	2/07	SCKS	Farmall, 12-38 rear tires
F20	1936		F	$325	9/07	NECO	Farmall, gas, double front, wide axle
F20	1936		F	$1,100	8/07	NWOH	Farmall, full steel, rear road bands, old repaint, straight
F20	1936		G	$1,250	11/07	WCIL	Farmall, older restoration
F20	1937		P	$150	8/07	SWIL	Farmall, flat spoke rear steel wheels
F20	1937		F	$1,000	9/07	WCIL	Farmall, factory spokes front and rear Wards rear tires, older restoration, runs
F20	1937		G	$1,400	9/07	WCIA	Farmall, Hysler OD, new tires, cutoffs
F20	1937		G	$1,700	11/07	SESD	Farmall, road grader, older restoration
F20	1937		G	$3,750	9/07	WCIA	Farmall, factory round spokes, restored, cultivator
F20	1938		P	$600	9/07	WCIA	Farmall, no tag, belly pump, fast fourth, engine stuck, new front tires
F20	1938		F	$800	11/07	WCIL	Running, restored, carburetor and mag rebuilt, cleaned and painted
F20			F	$125	4/06	WCTX	Power lift
F20			F	$175	4/06	WCTX	Cut offs, power lift

IHC/Farmall

Model	Year	Hours	Cond.	Price	Date	Area	Comments
F20			P	$180	9/06	NCIA	Farmall, for parts
F20			F	$525	5/06	SEND	Farmall
F20			F	$600	8/06	NCIA	Farmall, belt pulley, hyd. pump
F20			G	$650	10/06	SEWI	
F20			G	$1,050	3/06	NCWI	Farmall, on rubber
F20			G	$1,250	2/06	NCCO	Farmall, restored
F20			G	$2,200	8/06	SCIL	4-cyl. gas
F20			E	$2,400	8/06	SCIL	On rubber, cast rear wheels
F20	1929		G	$975	4/06	NWMN	Farmall, NF, belt pulley PTO
F20	1937		F	$650	2/06	NCIN	Farmall, rear steel flat spokes, front rubber, needs restored, not stuck
F20	1937		F	$2,000	7/06	NWOH	Farmall, original
F20	1939		F	$370	8/06	NCIA	Farmall, partially restored
F20	1939		G	$1,200	8/06	NCIA	Farmall, factory starter, lights, hyd., PTO, older restoration
F20	1939		E	$2,900	9/06	NCIA	Starter, lights, Heisler road gear, 12 volt, 8 speed, parade-ready
F20			P	$25	6/05	WCMN	To be restored
F20			P	$25	6/05	WCMN	To be restored
F20			P	$200	11/05	SWMN	Farmall
F20			P	$225	11/05	SWMN	Farmall
F20			F	$400	3/08	WCMN	Farmall, loader
F20	1939		P	$250	9/05	NCIA	Farmall, motor stuck
F20	1939		F	$700	9/05	NENE	Farmall, factory cast wheels, 11.2-36 tires
F30			G	$1,250	5/08	ECMN	Farmall, restored, round spoke wheels
F30	1936		G	$1,800	5/08	ECMN	Farmall, round spokes, single front wheel
F30			F	$150	10/06	SEWI	Factory round spoke wheels, no radiator
F30			P	$175	11/06	SENE	For parts
F30			F	$175	4/06	WCTX	Front round spokes
F30			P	$225	11/06	SENE	For parts
F30			F	$500	8/06	SCIL	Not been restored
F30			F	$500	4/06	WCTX	Straight
F30			G	$2,500	4/06	WCIA	Narrow tread, new tires
F30			G	$3,500	8/06	SCIL	On steel
F30			G	$1,500	1/05	NENE	Farmall, road gear, 13.6-38 tires
F30	1930		G	$3,250	2/05	NWOH	Farmall, wide tread, older restore
H			F	$500	3/08	NWKS	Farmall, loader
H			P	$700	3/08	WCMN	Farmall, does not run
H			P	$700	3/08	WCMN	Farmall, does not run
H			F	$800	3/08	SEIL	Farmall
H			G	$900	3/08	SWOH	Farmall, NF
H			G	$1,025	3/08	SEIL	Farmall, fenders
H			G	$1,075	7/05	SESD	Farmall, motor OH
H			G	$1,100	3/06	ECND	Farmall, WF, PTO, belly pump, loader
H			G	$1,350	3/08	SEIL	Farmall, fenders
H			G	$1,900	3/08	NCIL	Farmall, NF, live hyd., 13-38 tires, hyd. loader with snow bkt.
H	1939		G	$1,650	3/08	NCIL	Farmall, NF, older repaint, hyd. trip bkt. loader
H	1940		G	$1,800	5/08	ECMN	Farmall, NF, fenders
H	1945		F	$650	5/08	WCIA	Farmall, WF, 6 volt, belt pulley, all original, chains, aftermarket 3 pt. and 6' blade
H	1948		G	$1,400	2/08	SCMN	Farmall, WF
H			P	$100	6/07	WCSD	Farmall, loader, NF

IHC/Farmall

Model	Year	Hours	Cond.	Price	Date	Area	Comments
H			P	$175	3/07	SCNE	Farmall, for parts
H			P	$225	3/07	SCNE	Farmall, for parts
H			P	$400	11/07	SESD	Farmall, NF
H			P	$600	3/07	SEMN	Farmall, needs restoring
H			F	$650	8/07	WCMN	Farmall, NF
H			F	$700	8/07	SESD	Farmall
H			F	$700	3/07	SCNE	Farmall, will use some oil
H			F	$750	5/07	WCWI	Farmall
H			F	$750	3/07	SCNE	Farmall, runs
H			G	$850	4/07	SCKS	Farmall, ran good, PTO
H			G	$850	4/05	ECND	Farmall, NF
H			G	$900	11/07	SEND	Farmall, WF, excellent paint and tires
H			F	$900	8/07	SESD	Farmall, NF
H			F	$900	6/07	NCIN	Farmall, unrestored, original
H			G	$950	1/07	SCNE	Farmall, NF
H			F	$1,000	9/07	WCIA	Farmall, loader, no tag, repainted, 12-volt conversion, aftermarket 3 pt., fenders
H			G	$1,200	8/07	ECIL	Farmall, NF, rear wheel weights, PTO, restored
H			G	$1,300	2/07	NENE	Farmall, repainted, newer rubber
H			F	$1,325	2/07	WCKY	Farmall
H			G	$1,400	11/05	WCIL	Farmall, gas, loader snow blade
H			G	$1,600	8/07	WCMN	Farmall, 11.2-38 rear tires like new, wheel weights, fenders, runs good
H			G	$1,600	8/07	WCMN	Farmall, WF, dual loader, wheel weights, PS
H			G	$1,700	9/07	ECIN	Farmall
H			G	$1,700	10/05	WCIL	Farmall, loader, good paint, average rubber
H			G	$1,900	11/07	SCNE	Farmall, new tires and paint, parade ready
H			G	$2,100	7/07	ECIL	NF, new rear tires, rear fenders, belt pulley, PTO, restored
H			G	$2,300	7/07	ECIL	Farmall, NF, PTO, lights, restored
H			G	$2,975	6/07	ECIL	Farmall, parade restored, new 12.2-38 tires
H	1939		F	$600	11/07	WCIL	Farmall, running, unrestored, original, PTO, belt pulley
H	1939		G	$1,250	8/07	WCIL	Farmall, restored, gas
H	1939		G	$1,400	9/07	NEIA	Farmall
H	1940		G	$800	11/07	ECSD	Farmall, NF, OD trans.
H	1940		G	$1,025	9/07	NEIA	Farmall, OD, loader
H	1940		G	$1,750	9/07	WCIL	Farmall, older repaint, runs, new rubber all around
H	1940		G	$2,600	3/07	NCIA	Farmall, Farmhand loader, 86" snow bkt.
H	1941		P	$200	9/07	WCIL	Farmall, motor apart
H	1941		P	$225	6/07	SEID	Farmall, Farmhand loader, WF, runs, not sure if hyd. works to lift loader
H	1941		P	$350	3/07	NENE	Farmall, for parts
H	1941		G	$2,000	12/07	SEMI	Farmall
H	1942		G	$1,800	12/07	SEMI	Farmall, loader, bkt., 9 speed
H	1943		G	$1,250	8/07	WCIL	Farmall, restored, gas
H	1944		G	$1,550	8/07	NWIL	Farmall, 12 volt
H	1945		G	$1,375	12/07	SEMI	Farmall, new tires and paint in 1996
H	1946		F	$800	11/07	WCIL	Farmall, running, rear wheel weights, PTO, lights, nice old user tractor in farm working condition
H	1946		G	$1,250	12/07	SEMI	Farmall

IHC/Farmall

Model	Year	Hours	Cond.	Price	Date	Area	Comments
H	1947		G	$1,125	1/07	NEIA	Farmall, 12 volt, belt pulley, saw rig
H	1948		F	$700	12/07	ECIL	Farmall, Farmhand loader, wheel weights
H	1948		F	$800	5/07	SEWY	Farmall, WF, gas
H	1949		F	$900	11/07	WCIL	Farmall, running, old restoration/repaint, PTO, lights, hyd. remote, belt pulley
H	1949		F	$1,000	12/07	ECIL	Farmall, one owner
H	1949		G	$1,400	9/07	NWIL	Farmall, NF, 540 PTO, good paint, runs strong
H	1949		G	$1,400	8/07	SESD	Farmall
H	1949		G	$1,400	2/07	WCOH	Farmall, tricycle
H	1949		G	$1,500	12/07	SEMI	Farmall
H	1950		F	$700	11/07	WCIL	Farmall, unrestored/original, rear wheel weights, PTO, belt pulley
H	1951		G	$800	3/07	NWMO	Farmall, runs good, 1 hyd., 12.4-38 rubber (new on left front), 12-volt system, good tin and paint
H	1951		G	$1,200	8/07	SESD	Farmall
H	1951		G	$1,250	6/07	NCIN	Farmall, NF, nice tires, runs
H	1951		G	$1,600	12/07	SEMI	Farmall
H	1951		G	$1,850	12/07	SEMI	Farmall, 9 speed
H	1951		G	$2,200	7/07	NWIL	Farmall, new rear tires, new paint, row crop
H	1952		F	$600	11/07	WCIL	Farmall
H				$0	11/06	NCOH	Farmall, Templeton loader
H			P	$100	9/06	ECMI	Farmall, for parts
H			F	$500	10/06	NCWI	Farmall, gas, WF, uses oil
H			G	$550	11/06	SCCA	Loader, gas, hyd.
H			F	$600	8/06	SEPA	Farmall, average rubber, NF
H			F	$650	12/06	ECIL	Farmall, row crop
H			G	$700	3/06	NWOH	Farmall, gas, NF, restored
H			F	$700	4/06	NENE	Farmall, NF
H			F	$800	6/06	NWMN	Farmall, NF
H			F	$800	8/06	SEPA	Farmall, NF, fair rubber, good tin
H			F	$850	4/06	SWIN	Farmall, runs, restorable
H			F	$875	3/06	NWIL	Farmall, gas, loader, good tires
H			F	$900	3/06	NWOH	Farmall, gas, NF
H			G	$900	3/06	NWOH	Farmall, gas, NF, restored
H			G	$900	10/06	SEWI	1 new tire
H			G	$1,000	8/06	SWWI	Farmall
H			G	$1,000	1/06	NEMO	
H			G	$1,050	8/06	NWOH	Farmall, gas, NF, restored
H			G	$1,050	11/06	NCOH	Farmall, hyd. cylinder kit, belt pulley
H			F	$1,100	4/06	WCTX	
H			G	$1,100	2/06	WCME	
H			G	$1,200	3/06	NEOH	Farmall, fenders
H			G	$1,400	11/05	WCIL	Farmall, good paint
H			G	$1,400	6/06	NWSD	Farmall, runs well, stored inside, excellent tires
H			G	$1,500	2/07	WCOH	Farmall, NF, new rubber, restored in 2000
H			G	$1,750	12/06	NWIL	Farmall, OD trans., new tires
H			G	$1,800	9/06	ECMI	Farmall, NF
H			G	$1,800	11/05	WCIL	Farmall, NF
H			G	$2,100	8/06	SCIL	
H			G	$2,300	8/06	SCIL	
H			G	$2,500	6/07	SCIL	
H	1939		G	$2,500	10/06	SCSD	Farmall, gas, original except for muffler

IHC/Farmall

Model	Year	Hours	Cond.	Price	Date	Area	Comments
H	1940		G	$2,000	9/06	ECMI	WF
H	1942		G	$1,025	8/06	NCIA	Farmall
H	1942		G	$1,200	12/06	WCIL	Farmall, gas, NF, weights
H	1944		F	$1,300	2/06	SCMI	Farmall, gas, NF, good 12.4-38 rubber
H	1945		F	$900	8/06	NCIA	Farmall, new rear tires, OH
H	1945		G	$1,300	9/06	ECMI	Farmall, front end loader
H	1946		F	$750	10/06	NECO	Farmall, Farmhand F10 loader, double front, gas
H	1947		G	$925	10/06	SCSD	Farmall, single front wheel, lights, gas
H	1948		G	$1,100	4/06	ECND	Farmall, WF, PTO, belly pump
H	1948		G	$1,200	11/06	ECNE	Farmall, good old tractor, new rear tires
H	1949			$0	9/06	NCIA	Farmall, 9 speed, 12 volt, PS, clamshell fenders, live hyd.
H	1949		G	$1,400	10/06	SCSD	Farmall, NF, lights, belt pulley, gas, restored, repainted
H	1949		G	$2,300	9/06	NCIA	9-speed, 12 volt, recent restoration
H	1950		G	$2,000	6/06	NECO	Farmall, WF, Workhorse loader, gas
H	1950		G	$2,000	6/06	NECO	Farmall, gas, Workhorse loader, WF
H	1951		F	$1,100	3/06	NWOH	Farmall, gas, NF, IHC front end loader
H	1952		F	$1,100	6/06	NWIL	Farmall, two-way hyd., 2-speed trans.
H	1952		G	$2,000	9/06	ECNE	Farmall, high-speed road gear, 13.6-38 rear tires
H			P	$200	9/05	NENE	Farmall, for parts
H			F	$500	7/05	SESD	Farmall
H			F	$500	9/05	SEIA	Loader
H			F	$540	4/05	SEPA	Farmall, NF, 30% rubber
H			F	$600	8/06	SEPA	Farmall, NF, poor paint, 50% rubber
H			F	$600	3/05	ECCO	Not running
H			F	$650	3/05	ECMI	Farmall
H			F	$650	3/05	SCMI	
H			F	$670	7/05	SESD	Farmall
H			F	$675	4/05	SCMI	Farmall, NF, gas
H			F	$750	7/05	SCMN	Farmall, WF
H			F	$800	7/06	WCOH	Farmall
H			G	$800	1/05	NEMO	Farmall
H			F	$850	10/05	ECMN	Farmall, gas, WF
H			F	$950	1/05	NENE	Farmall, John Deere 7' mower
H			G	$975	1/05	ECNE	Farmall, loader
H			G	$1,000	7/07	SEND	Farmall, factory WF
H			P	$1,025	11/05	NCOH	Farmall
H			G	$1,200	6/05	WCMN	Farmall
H			G	$1,300	11/05	WCIL	
H			G	$1,350	1/05	ECNE	Farmall
H			G	$1,400	11/05	WCIL	NF, restored, belt pully, PTO
H			G	$1,600	7/07	ECIL	Farmall
H			H	$1,700	6/05	ECMN	Tricycle, fenders
H			G	$1,750	4/05	ECMN	Farmall, loader
H			G	$2,500	6/05	ECMN	WF, fenders
H	1940		G	$800	9/05	NCIA	Farmall
H	1940		G	$1,200	4/05	NCIA	Farmall, live pump, fenders
H	1940		G	$1,300	6/05	ECIL	6 volt, rear weights, hyd.
H	1942		G	$2,000	11/05	WCIL	NF
H	1944		E	$1,700	8/05	NCIL	Farmall, restored, new rubber front and rear

IHC/Farmall

Model	Year	Hours	Cond.	Price	Date	Area	Comments
H	1945		F	$1,000	5/05	NCKS	Farmall, PTO, NF, Farmhand F10 loader
H	1945		E	$1,100	2/05	WCNE	Farmall, gas, rear weights
H	1945		G	$1,100	8/05	NCIL	Farmall, live PTO
H	1945		G	$1,150	3/05	ECND	Farmall, Schwartz WF, PTO
H	1946		F	$700	3/05	NECO	Farmall, PTO
H	1947		P	$375	11/05	ECNE	Farmall, new 11.2-38 tires
H	1947		P	$500	9/05	NCIA	Farmall, trans. bad
H	1947		E	$1,050	3/05	WCWI	Farmall
H	1947		G	$2,500	4/05	WCWI	Farmall, one owner, new paint, excellent rubber, duals
H	1948		F	$600	11/05	ECNE	Farmall, new 11.2-38 tires
H	1948		G	$1,075	12/05	ECNE	Farmall, near new 11.2-38 tires
H	1949		P	$550	7/05	NCMN	Not running, missing tin, Farmall
H	1950		G	$1,500	8/05	NEIN	12 volt
H	1951		G	$725	8/05	NEIN	Gas, NF
H	1951		G	$2,400	4/05	SCNE	Farmall, PTO, rear weights, 11-38 tires, runs
H	1952		P	$220	11/05	ECNE	Farmall, for parts, on steel wheels
H	1952		F	$800	2/05	NCKS	Farmall, runs
H	1952		G	$950	1/05	WCIL	Farmall
HV	1948		G	$6,750	5/08	SEMN	Farmall, high crop, wishbone WF, 20" front wheels, original rear nice tires
Hydro 100	1974	6,060	G	$7,600	6/07	SCMI	Canopy, 3 pt., 1 hyd.
Hydro 100	1974	1,200	G	$16,750	3/07	NWIL	Open station, WF, 3 pt., dual PTO, 2 hyd., fenders, 18.4R-38 tires, sharp
Hydro 186	1977	5,486	G	$11,000	3/08	NEMO	Fully equipped cab, 18.4-38 rear tires, 11.00-15 front tires, 2 hyd., 540/1,000 PTO, clean
Hydro 70	1975		G	$10,500	5/08	SEMN	Hi clear, high crop, straight drawbar, WF, PS, flat-top fenders, new front and rear tires, totally restored
M			P	$550	5/08	WCIA	Farmall, 12 volt, all original, lights, hyd.
M			F	$750	3/08	WCMN	Farmall, NF, Paulson trip bkt. loader
M			G	$800	4/08	SCND	Farmall, WF
M			G	$1,400	3/08	NWKS	Farmall, WF, loader, fenders, 1 lift arm
M			G	$1,500	3/08	ECND	Farmall, WF, PTO, belly pump, Koyker hyd. loader
M			G	$1,500	3/06	SWKY	Farmall
M			G	$1,800	3/08	NCIL	Farmall, snap-on duals
M			G	$2,300	4/08	NEIL	Farmall, NF, fenders, live clutch, PTO, restored
M			G	$2,550	1/08	NEIA	Farmall, 12-volt system, ps and fast fourth gear
M	1941		F	$900	3/08	ECND	Farmall, NF, 12-38 tires
M	1941		G	$2,200	3/08	SCKS	Farmall, new starter and generator, runs good
M	1947		G	$2,500	4/08	SCMI	Farmall, gas, NF, restored
M	1950		G	$1,500	4/08	NEIA	Farmall, WF, MM trip-bkt. loader
M	1950		G	$3,700	1/08	ECNE	Farmall, WF, Behlen PS, 540 PTO, Koyker Super K loader
M	1951		G	$1,800	5/08	ECMN	Farmall
M			F	$850	8/07	SESD	Farmall
M			G	$900	9/07	NCND	Farmall, gas, 540 PTO, 12.4-38 tires
M			G	$1,000	5/07	WCWI	Farmall
M			G	$1,050	9/07	NCND	Farmall, WF, gas, 540 PTO, wheel weights
M			G	$1,100	12/07	SCNE	Farmall, Schwartz WF

IHC/Farmall

Model	Year	Hours	Cond.	Price	Date	Area	Comments
M			G	$1,300	3/07	ECIL	Farmall, NF, live hyd., 540 PTO, 1 hyd., older repaint
M			F	$1,400	1/07	NECO	Farmall, gas, 3 pt., double front, 7' rear blade
M			G	$1,450	1/07	NWIN	Farmall, PS, duals
M			F	$1,500	11/07	WCIA	Farmall, NF, no paint, older Paulson loader, ran
M			G	$1,600	1/07	NWOH	Farmall, gas, NF
M			G	$1,675	3/07	NWIA	Farmall, runs good, 9 speed, live hyd. pump, hand clutch
M			G	$1,700	3/07	SEMN	Farmall, WF, runs
M			G	$2,000	3/07	NEOH	Farmall, side-mounted sickle bar mower
M			G	$2,300	1/08	SESD	Farmall, WF, PS, repainted, dual 250 loader
M			G	$2,700	9/07	ECIN	Farmall, repainted
M			G	$2,800	3/07	NCNE	Farmall, WF, Farmhand F11 black stripeloader, 7' bkt. and grapple
M	1939		G	$3,200	7/07	ECIL	Farmall, NF, fenders, lights, belt pulley, PTO, deluxe seat, nice restoration
M	1940		G	$1,300	11/07	WCOH	Farmall
M	1940		G	$1,400	11/07	WCIL	Farmall
M	1941		G	$1,100	8/07	SESD	Farmall
M	1941		G	$2,200	7/07	ECIL	Farmall, NF, lights, PTO, deluxe seat, very nice restoration
M	1942		G	$1,750	11/07	WCIL	Farmall, running, unrestored/original, PTO, lights, good running, straight sheet metal, starting fuel tank, battery box
M	1943		F	$900	8/07	ECIA	Farmall, row crop, gas
M	1945		F	$575	3/07	SCMI	Farmall, NF, gas
M	1946		P	$300	1/07	SEKS	Farmall, NF, PTO, 13.6-38 tires, not running
M	1947		F	$600	9/07	NECO	Farmall, gas, double front, 12 volt
M	1947		G	$800	8/07	WCMN	Farmall, NF, 13.6-38 tires, pulley and bat box included
M	1947		F	$900	12/07	ECNE	Farmall, NF, 12 volt
M	1947		G	$1,000	11/07	WCIL	Farmall, hood in primer
M	1948		F	$1,000	11/07	WCIL	Farmall, older restoration
M	1948		G	$1,400	9/07	WCIA	Farmall, repainted
M	1948		G	$1,500	1/07	SEKS	Farmall, NF, 1 hyd., 13.6-38 tires, 6.50-16 front tires, runs
M	1948		G	$2,400	12/07	SEMI	Farmall
M	1949		F	$900	11/07	SESD	Farmall
M	1950		G	$800	12/07	SEWY	Farmall, gas, WF
M	1950		F	$900	11/07	WCIL	Farmall, running, unrestored/original, rear wheel weights, lights, hyd. remote, belt pulley
M	1950		F	$1,000	7/07	ECIL	Farmall, NF, loader with blade bkt., rear wheel weights, lights, PTO, 1 hyd., PS
M	1950		G	$1,050	11/07	SEND	Farmall, WF, loader
M	1950		G	$1,300	11/07	WCIL	Farmall
M	1950		G	$1,700	11/07	WCIL	Farmall, running, old restoration/repaint, PTO, lights, hyd. remote, belt pulley, very rare dipstick in engine oil pan
M	1950		G	$1,750	12/07	SEMI	Farmall
M	1951		G	$1,650	8/07	WCIL	Farmall, gas, PS, restored
M	1951		G	$1,975	8/07	WCIL	Farmall, gas
M	1951		G	$2,600	5/07	SEWY	Farmall, double front, Farmhand F11 loader and 5' bkt.
M	1952		G	$1,500	8/07	WCMN	Farmall, NF, good rubber, unrestored, second owner

IHC/Farmall

Model	Year	Hours	Cond.	Price	Date	Area	Comments
M	1952		G	$1,675	9/07	NECO	Farmall, gas, Farmhand F11 loader, WF, 12 volt
M	1952		G	$1,700	8/07	WCIL	Farmall, gas, PS, restored, belt pulley
M			P	$500	4/06	NENE	Farmall, NF, not running
M			F	$750	3/06	NEOH	Farmall, loader
M			G	$800	10/06	SEWI	Understrike ignitor
M			G	$900	12/06	ECIN	Gas
M			G	$1,000	7/07	SEND	Farmall, WF, Schwartz front end, PTO pump, hyd., PTO
M			G	$1,100	3/06	NEOH	Farmall, pulley
M			G	$1,200	4/06	WCTX	Older paint
M			G	$1,200	7/05	ECND	Farmall, WF, PTO
M			F	$1,300	8/06	SCIL	
M			G	$1,425	4/06	SCMN	Farmall, Paulson loader
M			G	$1,500	9/06	ECMI	Farmall, WF
M			G	$1,550	1/06	SESD	Farmall, NF, runs
M			G	$1,675	2/06	SEIA	Farmall, WF
M			G	$1,800	3/06	NEOH	Farmall
M			G	$2,000	3/06	NCCO	Farmall, 3 pt.
M			G	$2,500	11/06	SETN	Rebuilt motor
M			G	$2,500	6/07	SCIL	
M			G	$3,750	8/06	SCIL	
M	1940		G	$2,000	3/06	ECNE	Farmall, rear wheel weights, dual 345 loader
M	1941		G	$1,100	7/06	ECND	Farmall, WF
M	1942		F	$800	5/06	SEWY	Farmall, gas, single front, live hyd.
M	1944		E	$1,150	1/06	WCNE	WF, new paint
M	1944		G	$1,300	4/06	NWMN	Farmall, factory WF, 1 hyd., PTO, inside weights
M	1945		F	$650	8/06	ECNE	Farmall, Schwartz WF, 13.6-38, 12 volt
M	1945		F	$1,000	9/06	NWIL	Farmall, gas, 9-speed trans., 2-way hyd.
M	1945		G	$1,000	3/06	ECNE	Towline tractor, single front wheel, 13.6-38, 12 volt
M	1945		G	$1,550	3/06	ECNE	Farmall, WF, Behlen PS, 12 volt
M	1946		F	$700	3/05	NECO	Farmall, needs clutch, 3 pt., gas
M	1946		G	$1,600	3/06	ECMI	Farmall, WF
M	1947		G	$1,550	3/06	ECMI	Farmall, NF
M	1947		G	$2,000	8/06	WCMN	Farmall, new paint, new tires, 12 volt
M	1947		G	$3,000	3/06	NCIL	Farmall, 9 speed, PS, PTO, live hyd., 12-volt system
M	1948		G	$1,000	9/06	NWIL	Farmall, gas, new tires
M	1948		G	$3,050	9/06	NCIA	Farmall, 9 speed, 12 volt, PS, clamshell fenders, live hyd.
M	1949		G	$1,700	3/06	ECNE	Farmall, 13.6-38 rear tires, live hydraulic, belt pulley
M	1950		P	$300	6/06	NECO	Farmall, double front, gas
M	1950		F	$1,000	1/06	WCNE	WF, 12 volt
M	1950		G	$1,000	6/06	NECO	Farmall, single front, gas
M	1950		G	$1,500	8/06	NCIA	Farmall, 12-volt system, belt pulley, NF, restored, 12-38 tires
M	1950		G	$1,800	8/06	ECNE	Farmall, Schwartz WF, PS, 13.6-38, dual 320 loader with 7' bkt. and four-tine grapple
M	1951		F	$900	9/06	NWIL	Farmall, hand clutch, new tires
M	1951		P	$900	9/06	NCIA	Parts tractor, Farmall, 12 volt, 9-speed, PS, not stuck
M	1951		F	$1,000	9/06	NWIL	Farmall, hand clutch, new tires, gas

IHC/Farmall

Model	Year	Hours	Cond.	Price	Date	Area	Comments
M	1951		G	$1,050	10/06	SCMN	Farmall, restored original, looks and runs good
M	1951		G	$1,325	5/06	SEND	Farmall, gas, Schwartz WF, PS, live hyd.
M	1951		G	$1,400	6/06	SCMN	Farmall, NF, fenders
M	1951		F	$1,550	3/06	NCCO	Farmall, loader
M	1952		F	$1,300	2/06	NENE	Farmall, NF
M	1953		G	$0	9/06	NWIL	Farmall, gas
M			P	$200	10/05	NCKS	Parts
M			P	$400	1/05	NENE	Farmall, not running
M			P	$450	4/05	SCSD	Farmall, for parts
M			F	$650	2/05	NWKS	Farmall, WF, 3 pt.
M			G	$650	12/05	SEMN	
M			F	$650	11/05	WCIL	NF, weights, PTO, hyd.
M			F	$900	8/05	SEMN	Mounted picker
M			G	$1,050	7/05	SEIA	Farmall, low hours
M			G	$1,100	7/05	SESD	Farmall, repainted
M			F	$1,100	4/05	SCMI	
M			G	$1,150	9/05	SCON	Canadian sale
M			F	$1,200	11/06	WCMN	Farmall
M			F	$1,250	8/05	NWIL	Farmall, gas
M			G	$1,350	9/05	SCON	Canadian sale
M			G	$1,500	1/05	ECNE	Farmall, new tires and paint
M			G	$1,600	1/05	NENE	Farmall, Farmhand F10 loader
M			G	$3,500	10/06	SCMN	Farmall, loader
M	1941		G	$1,750	3/05	WCIL	Good paint, NF, gas
M	1944		G	$1,425	8/05	ECNE	Farmall, gas, NF, 4" pistons and governor, 13.6-38 tires, wheel weights
M	1944		E	$1,600	9/05	NECO	Gas, WF, new tires
M	1944		E	$2,350	3/05	SCIL	Farmall, PS, live hyd.
M	1946		G	$800	9/05	NECO	Gas
M	1947		G	$825	4/05	SENE	Farmall, gas
M	1947		G	$1,550	4/05	NEIN	Farmall, gas, NF, belt pulley
M	1948		G	$1,800	8/05	NWIA	Farmall
M	1948		G	$2,000	11/05	WCIL	WF, gas, original, carb. OH, belt pully
M	1949		G	$2,500	4/05	NCOH	Farmall, Super M hyd. loader kit and IHC 2000 loader, PS
M	1950		G	$1,500	8/05	NCIL	Farmall, PS, live PTO, hand clutch
M	1950		G	$2,000	3/05	ECMI	Farmall, NF, PTO drawbar
M	1950		G	$2,200	8/05	NCIA	Farmall, 12 volt
M	1951		F	$725	12/05	NWIL	Farmall, gas
M	1951		G	$1,200	1/05	SENE	Farmall, gas, NF, good paint
M	1951		E	$1,300	3/05	WCWI	Farmall, fenders, NF, wheel weights
M	1951		G	$2,375	3/05	NCIA	Farmall, live pump, PS, 12 volt
M	1952		P	$325	11/05	ECNE	Farmall, for parts
M	1952		G	$1,125	2/05	NEIN	Farmall, gas, NF
M	1953		P	$950	7/05	NCMN	Farmall, alternator, electric start
M	1953		G	$3,000	6/05	ECIL	1 hyd., 6 volt, rear weights
MD			G	$1,100	10/06	SEWI	
MD	1954		G	$1,700	6/06	SCMN	Farmall, fenders
MTA			G	$3,200	3/07	ECIL	Farmall, gas, WF, torque out
MV	1950		G	$5,500	5/08	SEMN	Farmall, high crop, single front wheel, fenders, 12 volt, pulley, restored
MV	1948		G	$3,800	10/06	SCSD	Farmall, high crop, new rubber, gas, single front wheels, rear cable winch

IHC/Farmall

Model	Year	Hours	Cond.	Price	Date	Area	Comments
04			G	$7,500	11/07	WCIL	Old restoration, runs, looks good
04			E	$12,500	11/07	WCIL	Orchard, running, restored, excellent profes- sional restoration
04	1941		G	$2,500	6/05	ECIL	Orchard tractor, 12-volt converted, rear weights
Regular			P	$225	3/08	ECNE	Farmall, for parts
Regular			P	$225	3/07	SCNE	Farmall, needs work, duals
Regular	1928		G	$750	4/06	WCIA	Restored, Farmall
Super A			F	$1,400	5/08	ECMN	Farmall, 11.2-24 tires
Super A			G	$2,000	3/08	SWKY	Farmall, cultivators
Super A			G	$2,100	1/07	SWOH	Farmall, cultivators, good rubber
Super A			G	$2,600	11/07	SCNE	Farmall, hyd.
Super A			F	$5,500	9/07	ECMN	Farmall, demonstrator
Super A	1948		G	$2,600	8/07	WCMN	Farmall, industrial, wide tran., new rear 9.5-24 tires, rear wheel weights
Super A	1949		G	$1,800	11/07	WCIL	Farmall, running, unrestored/original, fenders, PTO, lights
Super A	1949		G	$2,200	8/07	WCMN	Farmall, offset WF, restored, rear wheel weights
Super A	1950		G	$2,900	8/07	WCMN	Farmall, demonstrator, new rear 11-2-24 tires, front/rear wheel weights
Super A			G	$2,300	10/06	WCMN	Farmall, new paint, new tires, hyd.
Super A			G	$2,400	7/06	NWMN	Farmall, 1-row cultivator, Woods 59" mower
Super A			G	$2,575	6/06	ECSD	Restored
Super A			G	$3,100	7/07	SCMN	Farmall, new paint
Super A			G	$4,400	8/06	SCIL	
Super A	1948		G	$3,500	10/06	SCSD	Farmall, fenders, front rear wheel weights, motor OH, gas
Super A			G	$2,250	8/05	ECMI	Farmall, restored, starter/generator and carb. rebuilt, new paint, runs good
Super A			G	$2,325	9/05	NCIL	Woods L59 mower
Super A			G	$2,325	9/05	NCIL	Farmall, Woods L59 mower
Super A			F	$2,400	9/05	SEIA	
Super A			G	$2,900	11/05	WCIL	WF, electric valve job, belt pully
Super A			G	$3,000	10/05	SWSD	Farmall, WF, continental rotary belly mower, 5', fairly rare, stored inside
Super A			G	$5,200	11/05	WCIL	WF
Super A	1950		G	$5,400	6/05	WCMN	Farmall, demonstrator, white, restored WF
Super C			F	$1,500	2/08	WCOH	Farmall, weathered paint
Super C			G	$1,500	12/05	WCMN	Farmall, NF
Super C	1951		E	$3,400	1/08	SENE	Farmall, complete OH, 2 pt.
Super C	1953		F	$1,700	3/08	ECNE	Farmall, gas, NF, 11.8-38 rear tires
Super C	1953		G	$3,000	5/08	ECMN	Farmall, restored, WF, wheel weights, rear pulley, newer tires
Super C			F	$1,300	8/07	WCIL	Farmall, Woods 59 belly mower
Super C			G	$1,600	11/07	SESD	Farmall
Super C			G	$2,150	1/07	NWIN	Farmall
Super C			F	$2,600	6/07	SCIA	Farmall, WF, fast hitch, tires good, tin good, paint faded
Super C			F	$2,800	2/07	WCKY	Farmall, cultivators, fast hitch
Super C	1951		F	$950	10/07	NECO	Farmall, WF, 3 pt., gas
Super C	1951		F	$1,400	2/07	NWIL	Farmall
Super C	1951		G	$1,650	9/07	NECO	Farmall, WF, PTO, gas
Super C	1952		F	$800	9/07	NECO	Farmall, gas, single front, PTO, runs
Super C	1952		G	$1,000	9/07	NECO	Farmall, single front, PTO, gas, belt pulley

IHC/Farmall

Model	Year	Hours	Cond.	Price	Date	Area	Comments
Super C	1952		G	$1,050	9/07	NECO	Farmall, single front, PTO, gas, belt pulley
Super C	1952		F	$1,100	6/07	NCIN	Farmall, NF, original, runs
Super C	1952		F	$1,400	6/07	SEID	Farmall
Super C	1952		F	$1,600	9/07	WCIA	Farmall
Super C	1952		G	$2,000	6/07	NCIN	Farmall, NF, wheel weights, belt pulley, runs
Super C	1953		G	$1,775	9/07	NECO	Farmall, WF, PTO, gas, belt pulley, no pulley
Super C	1954		G	$3,100	11/07	WCIL	Farmall, running, unrestored/original, rear wheel weights, fenders, PTO, front wheel weights, lights, 3 pt.
Super C	1954		G	$4,000	11/07	WCIL	Farmall, restored, running
Super C			G	$1,000	9/06	NWTN	
Super C			F	$1,300	12/06	NWWI	Farmall
Super C			G	$1,650	9/06	NCMI	Farmall, gas, NF
Super C			G	$2,200	10/06	WCMN	Farmall, sharp, new paint, WF, 2 pt.
Super C			G	$2,700	3/08	SCIL	
Super C	1952		G	$2,200	2/06	NECO	Farmall, single front, 2 pt., PTO
Super C	1953		G	$1,700	5/06	SEWY	Farmall, WF, 3 pt.
Super C			G	$1,200	11/07	WCIL	NF, gas, weights, fenders, lights, belt pully
Super C			G	$1,850	11/05	NWPA	Farmall, hyd. fast hitch
Super C			G	$2,475	5/05	NEKY	Farmall, cultivator, two-row side dressers, new rubber
Super C			G	$2,500	2/05	NCCO	Farmall
Super C			G	$4,100	9/05	NCND	Farmall, WF, 59" Woods mower, very clean unit
Super C	1953		F	$500	4/05	NCIA	Farmall, NF, fenders, 60" Artsway belly mower
Super C	1953		F	$1,350	9/05	NENE	Farmall, belly-mount mower, PTO, drawbar, 11-36 tires
Super C	1953		G	$4,900	8/05	NCIA	Farmall, factory WF, 12 volt
Super H			G	$1,950	1/08	SWKY	Farmall
Super H			G	$3,400	5/08	ECMN	Farmall, restored, NF, wheel weights, clam fenders
Super H			G	$3,500	4/08	NEIL	Farmall, NF, fenders, rear hyd., PTO
Super H	1953		G	$2,400	4/08	SCMI	Farmall, gas, NF
Super H	1953		G	$3,300	2/08	WCIL	Farmall
Super H			P	$300	3/07	SCNE	Farmall, needs engine
Super H			F	$1,200	11/07	WCIL	Farmall, poor 12.4-38 tires
Super H			F	$1,400	11/07	WCIL	Farmall, decent 12.4-38 rear tires
Super H			F	$1,700	3/07	SCNE	Farmall
Super H			G	$2,750	11/07	ECSD	Farmall, NF, PTO, belt pulley, brand-new rubber all around, straight tin
Super H			F	$3,200	6/07	SCIA	Farmall, dent on hood, paint was brown as can be/faded
Super H	1952		G	$2,500	9/07	WCIA	Farmall, Stanhoist loader
Super H	1953		F	$1,150	8/07	SESD	Farmall
Super H	1953		G	$2,000	12/07	SEMI	Farmall
Super H	1953		G	$2,000	12/07	SEMI	Farmall
Super H	1953		E	$6,100	7/07	NWIL	Farmall, IHC 1701 loader, one owner
Super H	1954		G	$2,800	12/07	SEMI	Farmall
Super H	1954		G	$4,400	11/07	WCIL	Farmall, restored, running
Super H			G	$1,400	11/06	SCNE	Farmall, WF
Super H			F	$1,450	2/06	NENE	Farmall
Super H			G	$2,250	4/06	SCMN	Farmall, fenders
Super H			G	$2,600	8/05	WCIL	Farmall, NF, fenders, restored and one owner
Super H			G	$2,650	3/06	SEMN	Farmall, good tires

IHC/Farmall

Model	Year	Hours	Cond.	Price	Date	Area	Comments
Super H	1952		G	$3,050	10/06	SCSD	Farmall, gas, lights
Super H	1953		G	$1,750	6/06	ECND	Farmall, WF, Schwartz PTO, belt pulley, live hyd.
Super H	1953		G	$1,800	10/06	WCMN	Farmall, fenders, new tires, repainted
Super H	1953		F	$1,850	9/06	ECMI	NF, PTO, all original
Super H	1953		F	$2,200	8/06	ECNE	Farmall, Schwartz WF, PS, 12.4-38 tires, live power & PTO, 2 sets of rear wheel weights
Super H	1954		G	$2,800	9/06	NEIN	Farmall
Super H			G	$800	2/08	WCIL	
Super H			G	$1,650	11/05	SCNE	Live hyd., WF
Super H			G	$1,650	3/05	NWMN	Cornhusker 3 pt., new rear tires
Super H			G	$2,700	6/05	ECMN	Tricycle, new tires, battery under seat, after-market 3 pt. hitch
Super H			G	$3,200	2/05	SCMN	Farmall, Woods mower
Super H			G	$3,300	2/05	WCIA	Farmall, International WF, paint faded, tires 50% and weather checked, tin straight as new
Super H	1953		F	$675	2/05	SCKS	Farmall
Super H	1953		G	$2,500	8/05	NCIA	Farmall, fenders, good tires
Super H	1953		G	$3,400	6/05	WCMN	Farmall, WF, restored, fenders
Super H	1954		G	$2,500	11/05	WCIA	Farmall
Super H	1954		E	$5,250	3/05	NESD	Farmall, NF
Super M	1952		F	$1,700	2/08	SCMI	Farmall, NF
Super M	1952		G	$3,600	1/08	ECNE	Farmall, gas, SN 7055AJ, NF, 12-volt system with electric start, newer 13.6-38 rear tires, (2) sets rear wheel weights, newer front tires, PTO, new paint, amateur restoration, super sharp
Super M	1953		G	$3,600	5/08	ECMN	Farmall, restored, NF, pulley, fenders, newer 13.6-38 tires
Super M	1953		E	$5,000	1/08	ECIL	Farmall, row crop, PS, fenders, hyd., 15.5-38 tires
Super M	1953		E	$14,500	5/08	SEMN	Farmall, custom, 266 IHC V- engine with Holly carb., IHC WF, PS, rear wheel weights, PTO, fenders, Goodyear 16.9-38 radial tires
Super M			F	$900	2/07	NEIN	Farmall, gas, NF, steel wheels
Super M			F	$1,100	12/07	WCMN	Farmall
Super M			G	$1,250	11/07	ECIL	Farmall
Super M			F	$1,500	6/07	SCIL	Farmall, gas, WF, original
Super M			G	$1,850	12/07	SEMI	Farmall
Super M			G	$2,500	11/07	WCKS	Farmall, WF, restorable
Super M	1950		G	$1,250	3/07	SEWY	Farmall, gas, WF, Farmhand F10 loader
Super M	1952		G	$2,700	11/07	WCIL	Farmall, restored
Super M	1952		G	$2,800	11/07	WCIL	Farmall, restored, runs
Super M	1953		F	$1,000	2/07	ECNE	Farmall, NF, 1 hyd., 2-way valve
Super M	1953		F	$1,400	6/07	NCIN	Farmall, unrestored, original, runs

IHC/Farmall

Model	Year	Hours	Cond.	Price	Date	Area	Comments
Super M	1953		G	$1,500	3/07	SCMI	Farmall, NF, gas
Super M	1953		F	$1,700	11/07	WCIL	Farmall, running, old restoration/repaint, rear wheel weights, fenders, PTO, lights, PS
Super M	1953		F	$1,850	8/07	SESD	Farmall
Super M	1953		F	$1,850	9/07	NEIA	Farmall, 12 volt, hyd., OH, WF
Super M	1953		G	$2,900	3/07	ECIL	Farmall, 540 PTO, 1 hyd., like-new 14.9-38 tires, rear weights, nicely restored, SN 50878
Super M	1953		G	$3,750	9/07	WCIL	Farmall, original, belt pulley, hyd., rear weights, clean, runs good
Super M	1953		G	$3,750	9/07	WCIL	Farmall, (Louisville) belt pulley, hyd., solid, all original
Super M			F	$1,100	8/06	SWWI	Farmall, fenders
Super M			G	$1,100	5/06	SWCA	Gas
Super M			F	$1,300	8/06	NEIA	Farmall, loader (rough), PS, WF
Super M			G	$1,350	10/06	WCMN	Farmall, fenders
Super M			G	$1,600	3/06	SEIA	Farmall, PS, 12 volt and live hyd.
Super M			G	$1,750	10/06	SEMN	Farmall
Super M			F	$1,900	11/05	NWIL	Farmall, LP gas, old trip loader
Super M			G	$1,900	11/07	SEND	Farmall, Schwartz WF, PS
Super M			F	$2,050	3/06	SEPA	Farmall, NF, patched block, good rubber
Super M			G	$3,150	3/06	SEMN	Farmall, NF, new tires
Super M	1952		G	$4,900	2/06	NECO	Farmall, Farmhand F11 loader, WF, PS, PTO
Super M	1953		P	$800	6/06	NECO	Farmall, engine frozen, WF, 3 pt., gas
Super M	1953		P	$900	1/06	WCNE	Farmall, gas, WF, not running, for parts
Super M	1953		F	$1,500	3/06	SWOH	Farmall, row crop tractor
Super M	1953		G	$1,650	3/06	ECNE	Farmall, 13.6-38 rear tires
Super M	1953		F	$1,900	8/06	ECNE	Farmall, Schwartz WF, 13.6-38, 5.50-16 front tires, PS, live PTO, shop-built 3 pt.
Super M			F	$1,100	12/05	NWIL	Farmall, gas
Super M			G	$2,000	12/05	WCIN	Farmall, PS, NF, OH, restored
Super M			P	$2,600	10/05	WCMN	Farmall, rough, WF, 3 pt. loader
Super M			G	$3,100	3/05	SCMI	Farmall
Super M			G	$3,500	3/05	ECND	Farmall, gas, WF, PTO, Farmhand F10 loader

IHC/Farmall

Model	Year	Hours	Cond.	Price	Date	Area	Comments
Super M			G	$5,500	3/05	ECND	Farmall, gas, rollover diesel, WF, PTO
Super M	1952		G	$3,200	7/05	NWMN	Farmall, gas, factory WF, 1 hyd., PTO, power steeriing, 2 hyd., bkt. loader
Super M	1952		E	$6,600	9/05	NCIL	Farmall, puller tractor
Super M	1952		E	$6,600	9/05	NCIL	Puller tractor
Super M	1953		F	$2,000	3/05	SCIL	Farmall, fresh dealer OH, PS
Super M	1953		E	$3,900	3/05	WCNY	Farmall, repainted nicely
Super M	1953		G	$4,000	2/05	ECSD	Farmall
Super MD			G	$1,750	6/05	NEND	Farmall, WF, 18.4-38 tires
Super MDTA	1954		G	$9,000	11/07	WCIL	Farmall, restored, running
Super MTA			F	$1,700	1/08	SESD	Farmall, gas, NF
Super MTA			G	$3,100	1/08	SWKY	Farmall
Super MTA			G	$5,000	1/08	SESD	Farmall, gas, NF
Super MTA			G	$5,300	2/08	WCMN	Farmall, gas, WF, good TA, 15.5-38 rear tires, all new rubber, restored, good working order
Super MTA	1954		G	$3,200	3/08	WCMN	Farmall, WF, bad TA
Super MTA	1954		G	$5,250	5/08	ECMN	Farmall, restored
Super MTA	1954		G	$5,600	4/08	SCMI	Farmall, gas, NF, 1 hyd., restored
Super MTA	1954		G	$6,000	1/08	NWOH	Farmall, original, good rubber, good tin, 12 volt
Super MTA	1954		G	$6,500	4/08	SCMI	Farmall, gas, NF, 1 hyd., restored, PS
Super MTA			G	$2,500	8/07	SESD	Farmall
Super MTA			G	$2,600	7/05	SESD	Farmall, gas
Super MTA			F	$3,400	6/07	SCIA	Farmall, TA bad, tires OK, fast hitch, paint bad
Super MTA			G	$5,800	8/07	NEIA	Farmall, original, paint not good
Super MTA	1954		G	$0	2/07	NEIA	No sale at $5,500, Farmall, new rubber
Super MTA	1954		G	$3,100	2/07	SEIA	Farmall, 12-volt system
Super MTA	1954		G	$4,100	12/07	SEMI	Farmall
Super MTA	1954		G	$4,950	11/07	SESD	Farmall, NF, repainted
Super MTA	1954		G	$5,750	11/07	WCIL	Farmall, restored, running
Super MTA	1954		G	$7,300	12/07	SEMI	Farmall

IHC/Farmall

Model	Year	Hours	Cond.	Price	Date	Area	Comments
Super MTA	1954		G	$75,000	9/07	NEIL	Diesel high crop, SN SMDV81642S, 84 of these were ever built
Super MTA			G	$3,100	3/06	NENE	Farmall, WF, gas, 15.5-38
Super MTA			G	$4,450	1/06	ECIL	Farmall, row crop, 13.6-38, nice
Super MTA			G	$4,750	3/06	SEMN	Farmall, good tires
Super MTA			G	$4,900	2/06	NEIA	Farmall
Super MTA	1954		F	$2,000	1/06	NECO	Farmall, Farmhand F11 loader, bent bkt.
Super MTA	1954		G	$3,200	10/06	SCSD	Farmall, gas, fenders, WF, wheel weights, new rubber
Super MTA	1954		G	$4,600	9/06	ECMI	Farmall, complete, good torque
Super MTA			G	$2,000	11/05	WCIL	NF, gas, restored, PTO, lights
Super MTA			G	$2,400	7/05	SESD	Farmall
Super MTA			F	$2,900	9/05	NENE	Farmall, NF, PTO, drawbar, 14.9-38 tires, new clutch and TA, new paint, complete OH
Super MTA			G	$5,100	12/05	WCIN	Farmall, PS, fenders, 3 pt., rebuilt torque, NF, chrome stack, restored, converted to 12 volt
Super MTA	1954		G	$1,300	9/05	NCIL	New Idea horn loader
Super MTA	1954		G	$1,300	9/05	NCIL	Farmall, New Idea horn loader
Super MTA	1954		G	$2,900	9/05	NCIL	Farmall
Super MTA	1954		G	$2,900	9/05	NCIL	
Super MTA	1954		E	$8,900	3/05	WCMN	Farmall, one owner, 12 volt, wide and NF, very nice
Super MV	1953		G	$21,000	9/07	NEIL	Farmall, high crop, 1 of 245 built
Super W6	1954		G	$7,000	11/07	WCIL	Farmall, Super W6TA, standard, running, restored
Super W6	1952		G	$1,525	9/05	NCIL	
Super W6	1952		G	$1,525	9/05	NCIL	
Super W6	1954		G	$2,600	9/05	NCIL	
Super W6	1954		G	$2,600	9/05	NCIL	
W4	1947		G	$1,275	9/07	NWIL	Standard, fenders, belt pulley, standard drawbar, good paint and tires, WF
W4			G	$2,000	10/06	SEWI	
W4	1941		E	$4,150	9/05	NCIL	
W4	1941		E	$4,150	9/05	NCIL	Farmall
W6	1951		F	$1,100	9/07	WCIA	
W6			P	$425	4/06	WCSD	Not running
W6			F	$1,150	3/06	NCCO	LP, dual loader
W6			F	$2,300	10/06	SEWI	New front tires

IHC/Farmall

Model	Year	Hours	Cond.	Price	Date	Area	Comments
W6			G	$1,400	9/05	NCIL	
W6			G	$1,400	9/05	NCIL	
W6	1952		F	$1,600	5/05	NCKS	Engine free but needs tuning, 540 PTO, 1 hyd.
W9			P	$175	1/07	SWKS	WF, propane, 14-34 rear tires
W9			G	$1,050	11/07	SESD	Standard
W9	1947		G	$2,200	9/07	WCIA	
W9	1950		G	$1,600	12/06	WCIL	Gas, WF, good paint
W9	1952		G	$1,200	8/06	NCNE	
W9			G	$1,400	11/05	WCIL	Add-on PS, gas, 4 cyl.
W9	1951		F	$800	9/05	NECO	PTO
WD6			G	$2,550	11/07	SESD	Standard
WD6			F	$900	11/06	ECNE	PTO, 14-30 tires
WD6	1941		G	$2,400	7/05	ECND	2-bottom plow
WD9			F	$1,750	10/06	SEWI	
WK40	1936		G	$4,500	9/07	WCIA	Repainted
WK40			G	$2,600	8/06	ECNE	Factory round spokes, SN WKC3806

McCormick

Model	Year	Hours	Cond.	Price	Date	Area	Comments
10-20			G	$2,400	2/05	NCIN	Repaint, cutoffs, runs
10-20			G	$500	10/06	SEWI	Full steel, side curtains
10-20			G	$550	10/06	SEWI	Cutdowns, no curtains
10-20			G	$600	10/06	SEWI	Full steel, side curtains
10-20			G	$4,000	8/06	SCIL	New gray paint
22-36			G	$2,100	6/07	NCIN	Old repaint, rear round spoke, cut down front, runs
22-36			G	$6,500	8/06	SCIL	New gray paint
W30			F	$650	3/07	WCKS	
W30			F	$475	10/06	SWNE	Complete set of steel wheels, engine stuck
W4			g	$10,000	8/06	SCIL	
W6			G	$1,700	1/07	NWIN	Original
W6			G	$6,750	8/06	SCIL	
W6	1952		G	$1,300	8/05	NEIN	Gas
W9	1948		G	$4,400	5/08	ECMN	Gas
W9			G	$1,900	2/07	SCKS	Ran good, PTO
W9			G	$4,750	9/07	ECIN	Repainted
W9			F	$400	12/06	SCMT	Loader
W9			G	$1,600	4/07	NWMN	PTO, belt pulley
W9			G	$1,700	10/05	WCIL	Gas, electric start, repaint
WK40			G	$4,500	10/06	SEWI	Factory spokes, cast front wheels, loose, good tin

John Deere and Waterloo Boy

The Deere 4020 tractor. Not an antique yet, but I'm already seeing rapidly growing interest from what I'd call collector buyers on 4020s. Deere made the 4020 from 1964 to 1972. For a generation of farm kids, the 4020 was the tractor they grew up using and loving. Now that many of those kids have grandkids of their own, the thought of owning a nice 4020, like the one they remember, is a powerful tug.

Perhaps that explains the 1972 Model 4020 with 2,499 hours I saw sell on an auction in northwest Ohio November 21, 2007, for $31,000! This beautiful 4020 was a diesel, syncro, and was a one-owner tractor. Yes, $31,000 is the highest sale price I've ever seen on a 4020. Can you imagine what a nice original 4020 with less than 1,000 hours would be worth??

John Deere

Model	Year	Hours	Cond.	Price	Date	Area	Comments
1010	1961		E	$9,250	5/08	ECMN	Restored, 3 pt., gas, new 12.4-24 tires, no cab
1010	1965		G	$3,600	3/08	NEMO	Gas, 36 hp., ag tires
1010		3,601	G	$4,450	6/07	WCSD	Gas, 3 pt.
1010	1961		G	$4,900	11/07	SWIN	Restored, new tires, PS, dash shift, PS rears
1010	1963		G	$3,750	11/07	WCIL	Older restoration
1010	1963		G	$4,600	11/07	SWIN	Gas, PS rears, clamshell fenders, restored
1010			F	$1,450	8/06	NCIA	Industrial tractor, loader
1010			P	$2,100	3/06	NCCO	Clutch out
1010			G	$2,300	6/06	ECSD	Side-mount mower
1010	1964	2,203	F	$2,500	12/06	NWMO	
1010			G	$3,700	6/05	ECWI	Remote
1010			G	$3,750	6/05	ECMN	3 pt., PS
1010			G	$5,750	6/05	ECMN	Standard gas, WF, 3 pt., new tires and paint
1010			G	$6,000	6/05	ECMN	Row crop, gas, WF, 3 pt., PTO, lever, new paint, restored
1010			G	$8,800	7/05	SCIA	Standard, restored
1010	1962	764	G	$8,500	2/05	NWIL	WF, 3 pt.
1020	1966	3,687	G	$3,750	4/07	NEMI	WF, 3 pt., hyd. outlets, gas, 12.4-36 tires
1020	1969	8,418	G	$4,300	5/07	SEWY	Gas, WF, 3 pt.
1020	1970	7,642	G	$4,500	9/07	NECO	WF, 3 pt., PTO, gas
1020			G	$4,500	12/06	SCMT	John Deere 47 loader
1020			F	$5,050	3/06	NCCO	Diesel
1020			F	$7,400	3/06	NCCO	Great Bend 332 loader
1020	1967	3,021	G	$5,500	2/06	NCCO	WF, gas, PTO, remote
1020			G	$3,200	1/05	NWGA	3 pt., PTO
1520	1971	2,040	G	$9,000	1/06	SCKS	Diesel, roll guard with canopy, 1 hyd., 14.9-28 tires, front weight bracket, 46 hp., 540 PTO, 3 pt.
1520		6,200	G	$3,750	4/05	WCWI	Utility, gas
1530		3,189	G	$5,000	8/06	SETN	Diesel, sunshade, ROPS
1530			G	$5,550	8/06	NCOH	Utility, 145 loader and quad tach bkt., wheel weights, 700 hours on major OH
1530			G	$8,500	9/06	NEIN	Utility
1530	1974		F	$4,750	10/05	NCOH	1,225 hours on replaced tach, diesel, John Deere front end loader
1530	1975	2,630	G	$10,100	3/05	ECND	8 forward/4 reverse, single hyd., diff. lock, 3 pt., quick hitch, PTO, John Deere 143 loader
1830	1976		G	$8,000	4/07	SEAB	Canadian sale, 2WD, 60 PTO hp., diesel, 3 pt., 2 hyd., 540 PTO, 3-rib front, 16.9-30 rear tires
2010			P	$375	2/08	NCOH	NF, no motor, as is
2010			G	$3,000	1/08	NWOH	Row crop, gas, loader, WF
2010	1962		G	$4,300	5/08	ECMN	Restored, gas, 3 pt., 13.6-38 tires, newer tires
2010		4,780	G	$3,700	12/07	NCOH	Row crop, 1 hyd.
2010			G	$3,750	9/07	WCIA	Factory WF, fenders, 3 pt.
2010			F	$3,750	12/07	ECIN	Gas, loader
2010			G	$4,200	7/07	WCSK	Canadian sale, 4-cyl. gas engine, PTO, 2 hyd.
2010	1962		G	$3,100	10/07	NECO	Diesel, WF, 3 pt., dual loader with 4' dirt bkt.
2010	1962		F	$3,500	11/07	WCIL	Running, old restoration/repaint
2010	1964	4,242	G	$3,950	3/07	NCIA	Gas, open station, NF, 3 pt., 1 hyd., 12.4-36, fenders, bought new November 1965
2010			F	$1,400	3/06	NWIL	Gas
2010			G	$1,400	6/06	NEIN	Midwestern side boom
2010			G	$2,600	12/06	SCMT	Loader
2010			F	$4,350	3/06	NWIL	Gas, 6' bkt.

John Deere

Model	Year	Hours	Cond.	Price	Date	Area	Comments
2010			G	$4,800	3/06	WCSD	Shiny paint, tach doesn't work, new battery, clean, been shedded, no dents, no PS, excellent back tires, gas, 3 pt., 14.9-28
2010			G	$5,200	3/06	ECNE	Gas, QR, 13.9-36 rear, 7.5-15 front, fenders, loader
2010			G	$5,500	10/06	SCSD	WF, diesel, 3 pt., new rubber, motor OH, original
2010	1961	1,565	G	$5,100	2/06	NCCO	WF, gas, one owner
2010	1962		G	$2,850	10/06	SCSD	WF, no 3 pt., gas, repainted
2010	1962		G	$5,600	2/06	NCCO	Low hours, gas, one owner
2010	1963		G	$4,000	7/06	NWOH	Utility, new rear tires
2010	1963	1,250	E	$8,300	7/06	ECIA	Second owner (he bought in 1969), gas, new tires, NF
2010			F	$1,500	9/05	NWOR	Gas
2010			G	$2,500	6/05	ECWI	
2010			F	$3,500	4/05	SCSD	9' sickle bar mower
2010			F	$3,500	10/05	NEPA	Diesel, open cab, good rubber
2010			G	$4,100	6/05	ECWI	Loader
2010			F	$4,700	12/05	NCCO	Loader
2010			G	$5,750	10/05	NWMT	
2010		1,068	G	$10,900	7/05	SCIA	Row crop, NF, fenders, one owner
2010	1961	7,320	P	$1,900	9/05	WCNE	Diesel, 3 pt., smokes
2010	1961		F	$1,950	2/05	NEIN	Gas, tach shows 1,782 hours, WF, 3 pt.
2010	1962		G	$4,200	9/07	NECO	Diesel, new paint, WF
2010	1964	3,110	G	$4,600	8/05	WCMN	Syncro, NF, 3 pt., 1 hyd., second owner
2010	1964		G	$5,700	2/05	NCIN	New paint, good rubber, gas
2020			G	$3,400	9/06	NWTN	
2020			F	$3,900	8/06	SWWI	WF, 3 pt.
2020			F	$5,300	12/06	NWWI	John Deere 146 loader, 13.6-38 tires
2020			G	$6,000	3/06	NCCO	Great Bend 332 loader
2020	1969	4,271	G	$3,150	12/06	WCIL	WF, 3 pt., fenders
2020	1971	8,021	E	$6,000	9/06	NECO	Gas, John Deere 48 loader, WF, 3 pt.
2020		4,661	F	$2,500	8/05	SEIA	Loader, gas, 3 pt., PTO, 2 hyd.
2020			F	$3,200	12/05	NWIL	Gas, WF
2020			F	$4,500	2/05	WCOK	Loader, gas, 3 pt., PTO, 2 hyd., runs fair, tires fair
2020			G	$7,000	10/05	WCMN	Diesel
2020	1966	5,309	G	$5,600	8/05	SWOH	Diesel
2030			F	$3,600	4/07	NWPA	John Deere 145 loader, 600 hours on motor OH, 60 hp.
2030		2,960	G	$8,250	3/07	SEMN	3 pt.
2030	1974		G	$8,400	3/07	SEND	Gas, 2 hyd., 3 pt., PTO, John Deere 145 hyd. loader, 16.9-30 rear tires
2030			G	$3,200	8/06	SETN	Diesel
2030			F	$6,000	3/06	NCCO	John Deere 145 loader
2030	1974		G	$3,700	12/06	WCIL	WF, flat-top fenders, 3 pt., new hyd. pump less than 1,000 hours ago
2030	1974	8,011	G	$4,800	9/06	NEIN	Diesel, WF, 16.9-38 tires
2030			P	$1,500	2/05	SCCA	For parts, WF
2030		3,500	F	$3,400	9/05	SEIA	Brake problem, Koyker loader
2030		4,094	G	$6,900	12/05	NCOH	Utility, diesel, shuttle trans., hi/lo, umbrella, front weights, 1 hyd.
2030			G	$9,300	1/05	NWIL	30 hours on OH, John Deere 145 loader, diesel
2030	1978		F	$4,600	4/05	WCMN	55 hp., 2WD
2120	1975	9,373	G	$7,500	10/06	WCSK	Canadian sale, 2WD, 8 speed, PTO, 3 pt., 7.50-16 front tires, 18.4-30 rear tires, 9,373 hours showing, John Deere 48 loader and bale fork
2130	1976		F	$5,500	4/05	WCMN	55 hp., 2WD

John Deere

Model	Year	Hours	Cond.	Price	Date	Area	Comments
2240	1977		G	$8,250	3/06	NWMN	Open station, ROPS, 8 speed, 2 hyd., 3 pt., PTO, quick hitch, front weights, 14.9-38 singles
2240	1977		G	$7,000	12/05	WCMN	
2340	1976		F	$7,600	8/07	SESD	Farmhand loader
2440	1978		G	$7,300	3/08	NEMO	2WD, John Deere 47 front loader
2510			F	$6,800	1/08	NEMO	Loader, gas
2510			G	$7,400	5/08	ECMN	Restored
2510		5,820	G	$7,500	2/08	ECSD	Gas, syncro, 3 pt., PTO, 1 hyd., 400 hours on OH, new paint, kept inside
2510			G	$8,500	3/08	NCTN	NF, syncro, 1 hyd., front/rear weights, new 15.5-38 tires, low hours
2510		3,100	G	$11,500	1/08	SESD	PS, 3 pt., fenders, John Deere 36A loader
2510	1966		G	$9,000	3/08	WCOH	Restored
2510			G	$5,900	3/07	NCNE	Gas, fenders, 3 pt., hyd., PTO, runs good, may need trans. repair in future
2510			F	$6,100	11/07	SESD	WF, 3 pt., John Deere 48 loader
2510			G	$12,500	8/07	SESD	PS, NF
2510	1966		G	$6,500	11/07	WCIL	Running, unrestored/original, fenders, PTO, lights, hyd. remote, deluxe seat, 3 pt.
2510	1967		G	$3,300	4/07	ECND	Gas, 3 pt., PTO
2510	1966	3,924	F	$4,100	1/06	NECO	Diesel, WF, 3 pt.
2510			F	$5,000	2/05	NCIN	Gas, loader
2510			G	$5,500	2/05	ECNE	NF, 1 hyd., 3 pt., diesel
2510			G	$6,000	12/05	NCOH	Diesel, NF, syncro, row crop, brush guards, wheel weights
2510			G	$7,500	12/05	NCOH	Diesel, NF, row crop, brush guards, wheel weights, PS
2510		1,707	E	$10,000	9/05	WCMN	Diesel, sharp
2510	1966		G	$4,500	6/05	ECIL	Syncro, rock shaft
2510	1966		G	$5,700	7/05	NCIA	Gas, 15.5-38 tires, NF, 2 hyd.
2510	1967		G	$6,300	1/05	NEIA	NF, 3 pt., fenders
2510	1967		G	$7,000	8/05	WCIA	
2510	1968	6,961	G	$5,100	7/05	WCMN	Diesel, cab, WF
2520			G	$6,750	3/08	NCTN	WF, syncro, 1 hyd.
2520	1972		G	$13,750	5/08	ECMN	Gas, WF, 15.5-38 tires, newer tires, 3 pt., fenders, restored
2520			E	$9,500	7/06	NWIL	Running, unrestored/original, PS, 3 pt., good straight tractor
2520			G	$9,500	7/06	WCIL	Running, unrestored, PS, 3 pt.
2520			G	$10,400	8/06	SEPA	Diesel, WF, side console, average rubber
2520	1970		G	$5,500	12/06	NWWI	Gas, WF, console, 2 hyd.
2520	1971	5,211	G	$10,800	3/06	NCCO	One owner, 2 hyd., PTO, 3 pt., one-family owner, no cab, 15.5-38 tires
2520			G	$7,100	7/05	SCMN	Side console, WF, gas
2520			G	$9,500	7/05	SCIA	Syncro, original
2520	1970	4,200	G	$8,900	12/05	SEMN	WF, cab, side console, 3 pt., 2 hyd., rock box
2520	1972	9,516	G	$16,900	4/05	SENY	Side console, PS, 2 hyd., no weights, 15.5-38 tires, straight sheet metal, new engine 250 hours ago
2630	1975	5,309	G	$9,250	2/08	SCMN	Open station, 2 hyd., John Deere 48 loader
2630	1978	2,950	G	$8,000	3/08	WCMN	Diesel, 3 pt., PTO, 15.5-38 tires, hi/lo trans.
2630		12,600	F	$4,500	3/07	WCKS	3 pt., hyd. reverser, John Deere 148 front end loader
2630		258	G	$9,000	8/07	WCIA	WF, diesel engine, Westendorf loader, PTO, 2 hyd., 3 pt.

John Deere

Model	Year	Hours	Cond.	Price	Date	Area	Comments
2630	1975	4,100	G	$8,300	7/07	SEND	MFWD, 2 hyd., 3 pt., John Deere 146 loader, one owner, bale fork and tire chains, 2 sets of weights, 15.5-38 tires on cast iron
2630			G	$4,500	8/06	SETN	ROPS, sunshade
2630	1974		G	$6,200	12/06	ECMO	2WD, 2 hyd.
2630	1974	3,828	G	$7,500	2/06	ECIL	ROPS, 3 pt., 2 hyd., hi/lo trans.
2630	1974		E	$13,500	9/06	NECO	John Deere 148 loader, WF, ROPS, tach shows 434 hours, diesel
2630	1975		F	$7,000	7/06	ECND	3 pt., PTO, John Deere 145 loader
2630	1975	3,650	G	$7,500	12/06	SWIA	Open station, 8 speed, PS, 1 hyd., 3 pt., live 540 PTO, 16.9-28 rears, engine block heater
2630			F	$8,600	11/05	NEIA	John Deere loader, rough
2630			G	$8,900	12/05	NCCO	John Deere loader, rear blade
2640	1978	9,020	G	$7,750	2/08	ECMN	Utility, wheel weights, John Deere 146 loader with 6' bkt., 18.4-30 tires
2640	1976		G	$9,750	6/07	WCMN	2,420 hours showing, WF, 18.4-30 rear tires, PTO, 3 pt., 2 hyd.
2640	1976		G	$9,900	10/07	NEND	2 hyd., 3 pt., PTO, diesel, 16.9-30 tires, weights, 11L-15 front tires
2640	1977		G	$12,250	6/07	NENE	John Deere 146 loader with 5' bkt., hyd. loader controls, WF, collar shift trans., 8 forward gears/4 reverse, 15.5-38 rear tires, 9.5L-15.5L front tires, 2 hyd., 540 PTO, quick hitch
2640	1977		G	$7,750	12/06	SCMT	
2640	1976	3,448	G	$14,300	2/05	WCIL	John Deere 146 loader
2840	1978	10,280		$0	12/07	WCMN	Open station, bar axle, 540/1,000 PTO, 3 pt.
2840	1978		G	$9,500	10/07	SWOH	WF, diesel
2840	1976		G	$13,600	12/05	WCIL	John Deere 146 loader, repainted, good tires, tach reads 850 hours
2840	1978	4,000	G	$7,800	4/05	SCMI	No cab, 2 hyd., diesel, WF, one owner, snap-on duals
300B	1975		F	$2,750	9/07	NECO	Diesel, tach gone, no PTO, 3 pt., loader
3010			F	$3,000	3/08	ECNE	Roll-o-Matic NF, propane, syncro, 15.5-36 rear tires, 6:00-16 front tires, cement rear wheel weights
3010			G	$7,100	3/08	SEMN	Diesel, WF
3010			G	$8,000	3/08	SCMN	Diesel, WF, cab, John Deere 148 loader
3010	1961		G	$7,000	5/08	ECMN	Restored, diesel, NF, 15.5-38 tires, new tires, 3 pt., fenders
3010	1962		G	$4,300	1/08	WCIL	Diesel, 3020 motor, 15.5-38 and 11L-15 tires, syncro, fenders
3010	1962		G	$5,000	3/08	ECND	NF, gas, 1 hyd., 3 pt., tricycle style, 85% rear rubber
3010	1963	4,721	G	$3,700	2/08	WCIL	Gas, NF
3010	1963	7,155	G	$7,800	1/08	SCKS	Propane, 2 hyd., 3 pt., John Deere 148 loader
3010		6,191	F	$3,500	2/07	NEIN	Gas, NF
3010			F	$3,600	8/07	SESD	NF, 3 pt., 1962 model
3010			F	$3,900	12/07	WCNY	Loader
3010		5,800	G	$4,650	2/07	ECIL	Gas, row crop fender tractor, 15.5-38 tires, shedded
3010			G	$4,700	12/07	WCNY	Diesel, fenders
3010			G	$5,200	3/07	SCNE	Diesel
3010			F	$5,300	1/07	NWIN	Utility, John Deere 46 loader
3010			G	$5,700	3/07	SCNE	Farmhand F11 loader
3010			G	$7,000	9/05	ECMN	Loader
3010			G	$8,500	1/07	NCNE	WF, no cab, 1 hyd., extra front fuel tank
3010			G	$8,500	1/07	SCNE	No cab, WF, 1 hyd.

John Deere

Model	Year	Hours	Cond.	Price	Date	Area	Comments
3010	1961	9,000	G	$5,250	7/07	WCMN	Syncro range, 1 hyd., Allied 590 loader, joystick control, 6' bkt., 15.5-30 tires (30%)
3010	1961		G	$5,950	11/07	WCIA	Row crop, gas, 3 pt.
3010	1962		F	$1,500	9/07	NECO	Single front, gas, 3 pt., needs rebuild, tach shows 5,906 hours
3010	1962		F	$2,200	11/07	WCIL	
3010	1962	9,477	G	$3,400	9/07	ECNE	Diesel, Roll-o-Matic NF, syncro range, 16.6-32 rear tires, 6:00-14 front tires, 1 hyd., 540 PTO, 3 pt.
3010	1962		G	$3,700	3/07	NWMN	Gas, open station, 3 pt., PTO, 15.5-38 rear tires, unknown hours
3010	1962		G	$4,750	6/07	NENE	Diesel, syncro, 13.6-38 rear tires, 6:00-16 front tires, 1 hyd., 540/1,000 PTO, 2 sets of rear wheel weights
3010	1962		G	$7,900	7/07	NCSK	Canadian sale, 2WD, 55 PTO hp., syncro, 18.4-30 rubber, approx. 10,000 to 12,000 hours, John Deere 46A loader, 5' bkt.
3010	1963		G	$4,700	2/07	NENE	Diesel, syncro, WF, 3 pt., rear wheel weights, 1 hyd., 13.6-38 rear tires, fenders
3010	1963		G	$8,400	7/07	NCIA	Diesel, factory WF, syncro, 2 hyd., 3 pt., 14.9-38 tires
3010			F	$1,900	3/06	NCUT	Diesel
3010			G	$3,700	6/06	NEIN	LP, WF, original
3010			G	$4,000	3/05	ECMN	Paulson all-hyd. loader, gas, 4 cyl., cab, 3 pt., 15.5-38, +10% buyer's premium
3010			G	$4,400	12/06	ECMO	
3010			G	$5,200	9/06	NCMI	Fresh paint, diesel, turbo
3010			G	$5,500	12/07	WCMN	Diesel, WF, 36A hyd. loader, 3 pt., PTO, 1 hyd., 13.9-36 rear tires (75%)
3010			G	$5,500	11/06	SCMN	
3010		5,250	G	$5,800	6/06	NWIL	WF, 2 hyd., diesel
3010			G	$7,500	4/08	NEIA	
3010	1961	3,050	G	$3,100	11/06	ECIL	NF, no 3 pt., gas
3010	1961		G	$4,500	8/06	NEKS	Diesel, WF, 540 PTO, OH in last five years, 15.5-38, 1 hyd.
3010	1962	4,666	G	$4,500	10/06	NECO	WF, 3 pt., gas
3010	1962	5,068	F	$9,100	2/06	NCCO	John Deere 158 loader, gas, PTO
3010	1963	4,975	G	$4,600	8/06	ECIL	NF, weights, 3 pt.
3010	1963		G	$5,100	6/06	NECO	WF, no cab, Ezee-On loader, diesel, tach shows 853 hours
3010	1963	1,533	G	$5,100	2/06	NECO	Great Bend 850 loader, 3 pt., diesel
3010	1963	5,205	G	$6,900	12/06	WCMI	Diesel, no cab, WF, 1 hyd.
3010			G	$3,100	3/06	NENE	Gas, NF, Roll-o-Matic, 3 pt., drawbar, 13.6-38
3010			G	$3,600	10/05	ECMN	Diesel, John Deere 46 loader
3010			G	$4,100	12/05	WCMN	Gas, WF, 3 pt., PTO, 1 hyd., rock box
3010			G	$4,300	2/05	ECNE	WF, 1 hyd., 3 pt., diesel
3010			F	$4,400	12/05	WCMN	John Deere 48 loader, gas, WF, syncro, 3 pt., PTO, remotes, 18.4-34 new tires
3010			F	$4,650	9/05	WCMN	Gas
3010	1962			$0	9/05	SEMN	Utility, 3 pt., 2 hyd., front weights
3010	1961	5,106	G	$5,600	9/05	WCNE	Diesel, 3 pt., new rear tires
3010	1962	2,979	G	$7,600	2/05	NWIL	Diesel, 3 pt., WF, new rear tires
3010	1962		G	$8,300	9/05	SEMN	Utility, 3 pt., 2 hyd., front weights
3010	1963		F	$4,900	1/05	NENE	Syncro, NF, diesel, one owner, looks rough
3010	1963	1,368	G	$6,100	8/05	SWOH	Diesel, NF
3010	1963		G	$6,200	6/05	ECIL	WF, electric ignition, 46A loader
3020				$0	2/05	SCMN	Side console, low hours, WF

John Deere

Model	Year	Hours	Cond.	Price	Date	Area	Comments
3020		6,407		$0	2/08	NCOH	2 hyd., diff. lock, front and rear weights
3020			F	$3,300	3/08	NWWI	PS, LP, 3 pt., WF, Schwartz loader, diff. lock
3020		5,030	G	$3,500	3/08	WCMN	Gas, NF, open station, 1 hyd., 3 pt., PTO, 15.5-38 tires, 60% rubber
3020			F	$3,500	3/08	NCIL	NF, no 3 pt., gas
3020			G	$3,500	3/08	SEND	Gas, open station, 15.5-38 singles, John Deere 48 loader
3020			G	$4,250	1/08	NWOH	WF, gas, 1 hyd.
3020			F	$5,100	5/08	ECMN	Gas, NF, 3 pt., fenders, 15.5-38 tires, unrestored
3020			G	$5,900	4/08	NEIA	Gas, WF, 2 hyd.
3020		4,800	G	$7,700	4/08	NEIA	Diesel, one owner, NF
3020		9,096	E	$8,000	3/08	WCMN	Gas, 2 hyd., front weights, ROPS, canopy, quick hitch
3020	1964		E	$10,750	5/08	ECMN	Restored, diesel, PS, 15.5-38, newer tires, 3 pt., fenders, WF
3020	1964		E	$27,500	3/08	WCOH	Diesel, PS, restored
3020	1965	10,663	G	$5,000	4/08	SCMI	NF, 1 hyd., 3 pt.
3020	1966		G	$4,100	1/08	NWOH	400 hours on OH
3020	1967	6,407	G	$4,100	2/08	NCOH	2 hyd., front/rear weights
3020	1967		P	$4,600	2/08	NCOH	Rough, 2 hyd.
3020	1969		F	$3,300	2/08	SEAL	
3020	1969		E	$8,500	3/08	NEKS	Farmhand F11 loader and bale fork, diesel, syncro
3020	1970	6,700	G	$10,000	3/08	ECSD	Diesel, syncro, side console, 2 hyd., 3 pt.
3020	1971		E	$14,900	4/08	SCMN	Diesel, WF, low hours, nice
3020			G	$4,100	1/07	SCNE	PS, diesel
3020			G	$5,500	1/07	NWIA	NF, 15.5-38
3020		4,470	G	$5,750	4/07	NWMN	Diesel, PS, PTO, 16.9-34 rear tires
3020			F	$6,600	3/07	WCIL	Diesel, NF, PS
3020			G	$7,750	4/07	NWMN	Diesel, PS, 3 pt., PTO, 2 hyd., fresh OH by John Deere dealer, 16.9-42 rear tires
3020		2,300	G	$7,750	11/07	WCIL	Running, unrestored/original, fenders, PTO, lights, hyd. remote, deluxe seat, PS
3020		5,138	G	$10,000	6/07	WCSD	Diesel, 8 speed, 3 pt., new rear 18.4R-34 tires, 11L-15 front tires
3020			E	$14,500	1/07	NCKY	Rebuilt, new paint, sharp
3020	1964	5,000	G	$6,200	4/07	NCNE	LP, PS, 3 pt., WF, 2 hyd., good tires
3020	1964	10,157	G	$6,900	2/07	NEND	Diesel, syncro, open station, 2 hyd., 3 pt., 540/1,000 PTO, quick hitch, 16.9-34 rear tires
3020	1964		G	$7,000	6/07	SCIL	Diesel, PS, restored, WF
3020	1964		G	$12,000	11/07	SWIN	High crop, diesel, 3 pt., no toplink, new seat, new rear tires, new repo battery boxes
3020	1965		P	$5,000	2/07	ECNE	Rough, diesel, 1 hyd., WF, syncro, 15.5-38 tires
3020	1966		F	$2,700	11/07	SWNE	LP, 3 pt., hyd., PTO
3020	1966		G	$9,500	11/07	SEND	Row crop, diesel, PS, 2 hyd., 3 pt., 540/1,000 PTO, John Deere 46A loader
3020	1966		G	$10,250	11/07	NCIA	Diesel, WF, 3 pt., motor OH and trans., 15.5-38 tires
3020	1966		G	$10,900	9/07	NECO	Tach shows 2,302 hours, cab, 3 pt., Farmhand F258 loader, diesel
3020	1967	6,276	G	$3,700	11/07	SWNE	WF, syncro, gas, 13.9-38 tires, 2 hyd., 3 pt., fenders
3020	1967	6,991	G	$5,800	12/07	ECNE	Gas, PS, WF, 2 hyd., 15.5-38 tires
3020	1968	8,755	G	$5,500	3/07	WCMI	3 pt., 1 hyd., no cab, WF

John Deere

Model	Year	Hours	Cond.	Price	Date	Area	Comments
3020	1968		G	$8,600	11/07	ECNE	5,870 hours on tach, diesel, WF, syncro range, 16.9-34 rear tires, 2 hyd., 540/1,000 PTO, 3 pt., 2 sets of rear wheel weights, fluid in rear tires, new injection pump
3020	1969	10,202	G	$7,800	3/07	WCMI	3 pt., 1 hyd., no cab, WF
3020	1969	7,200	G	$8,250	1/07	NEIA	NF, 3 pt., console, 2 hyd., roll bar
3020	1970	6,700	G	$6,600	2/07	SENE	8-speed syncro, 16.9-34 rear tires, 7.50L-15 front tires, 2 hyd., comfort cab
3020	1970	2,476	G	$11,000	3/07	NWIL	Diesel, open station, WF, 3 pt., 16.9-34 tires
3020			P	$2,400	3/06	NCCO	LP, rear blade
3020			F	$2,550	9/06	NEIA	Gas, OH
3020			G	$4,100	3/06	NCCO	
3020			G	$4,400	12/06	WCMN	Gas, 3 pt.
3020			G	$4,800	12/06	SEMN	9,042 hours, syncro
3020			G	$5,600	4/06	SWPA	Fresh OH, new paint, new tires
3020			G	$6,800	8/06	SEPA	Diesel, PS, WF, average rubber, new paint
3020			G	$7,000	7/06	WCCA	Orchard, syncro trans., 3 pt., PTO, rear weights, 2 hyd.
3020			G	$7,000	12/06	SEND	3 pt., PTO, diesel
3020			G	$8,000	9/06	NEIA	Diesel, John Deere 148 loader
3020		6,100	F	$9,250	3/06	SEPA	WF, average rubber
3020		6,194	G	$13,500	8/06	SCMN	Side console, PS, 3 hyd., 3 pt., new front/rear tires
3020			E	$21,000	7/06	NWIL	High clearance, SN 11T61555, running, restored, PTO, lights, hyd. remote, PS, still working in south Louisiana mid-1990s, John Deere production certificate - shipped 4-10-64 Alexandria, LA
3020			G	$21,000	7/06	WCIL	High clearance, running, restored, PTO, lights
3020			E	$25,000	7/06	NWIL	Lanz standard, restored, fenders, PTO, front wheel weights, lights, PS, 1 of 171, modifications to meet German specs, Bosch lights, kilometers-per hour meter, shipped 3/65 Mannheim, West Germany
3020			G	$25,500	7/06	WCIL	Lanz standard, restored, fenders, PTO, front wheel weights, lights, modifications to meet German specs, Bosch lights, added brake drums for hand brake
3020	1964		G	$6,100	3/06	WCKS	3 pt., PTO, Great Bend 900 Hi-Master loader and bkt., bale fork and grapple fork, gas
3020	1965	6,714	G	$6,350	12/06	WCIL	WF, syncro, 2 hyd.
3020	1965		G	$6,600	11/06	SCSD	PS, 1 hyd., 3 pt., diesel
3020	1965	3,828	G	$9,200	3/06	NWIL	211 hours on OH, NF, PS, diesel, open station
3020	1966		F	$4,900	3/06	NESD	No cab, PS, WF, 1 hyd., Rockshaft, no 3 pt., fenders and dual 320 loader with grapple fork and yellow PTO pump
3020	1966		G	$5,750	12/06	SCMT	Farmhand F11 loader
3020	1966	6,981	G	$7,250	2/06	SCPA	PS, tach not working
3020	1966	4,240	G	$9,400	6/06	NECO	WF, no cab, new tires, diesel
3020	1967		G	$4,300	11/06	WCMN	Schwartz hyd. loader, gas, new engine, rebuilt lift cylinders on loader arms
3020	1967		G	$6,500	10/06	NECO	Koyker loader, WF, 3 pt., gas, tach shows 3,918 hours
3020	1967		G	$10,000	11/06	ECNE	Diesel, 1 hyd., quick hitch, front weights, front fenders, 2 sets rear weights, one owner
3020	1968	4,821	F	$10,500	2/06	NCCO	Diesel, WF, one owner, Year-A-Round cab
3020	1968		G	$11,500	2/06	NECO	3 pt., PTO, 110 hours on OH, diesel, 15.5-38 tires

John Deere

Model	Year	Hours	Cond.	Price	Date	Area	Comments
3020	1969		F	$3,300	3/06	ECNE	Gas, syncro, WF, 1 hyd., 3 pt., 15.5-38, fenders
3020	1969	8,679	G	$8,750	9/06	NECO	Diesel, WF, front weights, excellent cab
3020	1969		G	$9,500	3/06	ECNE	220 hours on new tach and OH, 2 hyd., 3 pt., syncro, console, WF
3020	1970		G	$15,000	8/06	SEPA	WF, fair rubber, PS, diesel
3020	1971	2,105	E	$19,500	1/06	SCKS	Diesel, syncro, 2 hyd., 2,105 actual hours, complete OH at 2,057 hours, WF, full vision cab, R134A air, quick hitch
3020			P	$1,900	9/07	NWIL	WF, diesel
3020		5,800	F	$2,950	8/05	WCIL	Gas, NF, syncro, 1 hyd.
3020			F	$3,050	7/05	SESD	WF, 3 pt., gas
3020			F	$3,300	6/05	ECWI	
3020			F	$3,600	12/05	WCWI	
3020			F	$4,600	12/05	NWIL	John Deere 46A loader, diesel
3020			G	$5,000	12/05	ECIN	Narrow wheel
3020			G	$5,500	12/07	WCMN	Diesel, new tires on rear, quick-tach
3020			G	$5,600	4/06	SWPA	Fresh OH, new paint, new tires
3020			F	$6,250	9/05	SEIA	Diesel
3020			G	$6,450	7/05	SCMN	WF, side console, gas
3020			G	$6,500	9/06	NEIN	Console
3020			G	$7,000	10/05	SWSD	Gas, cab, heater, WF FH 525 loader and grapple, Farmhand F11 loader
3020			G	$9,000	3/05	ECMN	Gas, 148 loader, new paint
3020			G	$9,500	6/05	NWWA	PS, 3 pt., PTO, 2 remotes, 13.6-38 rubber
3020			G	$10,250	3/05	NCAL	
3020	1964		G	$4,900	12/05	NENE	Syncro, 18.4-34 new tires, diesel, NF, 2 hyd., new paint
3020	1964		G	$6,600	2/05	WCIL	Gas, PS
3020	1964	5,393	G	$8,700	7/05	WCMN	John Deere 148 loader, cab, diesel, 2 hyd., QR, average rubber
3020	1965	5,289	G	$5,250	4/05	NCIA	Gas, RC cab, syncro, John Deere WF, 2 hyd., quick coupler
3020	1965		G	$5,900	9/05	NCOH	Diesel, WF, hours unknown, sharp but hood nose dented in
3020	1965	5,509	G	$6,750	8/05	SWOH	Diesel
3020	1965		G	$7,400	1/05	NENE	PS, 2 hyd., WF, 3 pt.
3020	1965		G	$7,900	3/05	SCMI	Diesel, WF, 2,071 hours after OH, 3 front weights
3020	1966	5,780	G	$4,200	11/05	NEIA	Gas, NF, 3 pt., like-new 15.5-38 tires
3020	1967	11,500	F	$2,350	11/05	NEIA	Gas, NF, standard hitch, 15.5-38 tires
3020	1967		F	$4,700	3/05	NESD	Syncro, gas, 16.9-34, 1 hyd., PTO, 3 pt., no cab
3020	1968	6,787	G	$7,900	4/05	SCMI	WF, 1,000 hours on new engine, diesel, square manifold, new rubber, new pressure plate
3020	1968		G	$8,000	7/05	SENE	Diesel, 2 hyd., WF
3020	1969	6,850	G	$7,850	1/05	SCNE	2 hyd., 540/1,000 PTO, gas, repainted four years ago
3020	1969	8,543	P	$9,000	4/05	SENY	Side console, PS, 2 hyd., 15.5-38 tires, poor sheet metal, very dented, no weights
3020	1970	7,811	G	$9,800	3/05	WCNY	Open station, syncro, diesel, 16.9-34, 2 hyd., one owner
3020	1970	3,706	E	$19,000	1/05	NENE	Console, diesel, WF, 15.5-38 rear tires, fenders, front weights, sold by stock auction company
3020	1972		G	$9,400	8/05	WCMN	3,200 hours on OH, diesel, side console, 2 hyd., 3 pt., 540 and 1,000 PTO, new clutch in 2001
320	1956		G	$5,200	9/07	ECMI	WF, 3 pt., 12.4-24 tires

John Deere

Model	Year	Hours	Cond.	Price	Date	Area	Comments
320	1958		G	$10,250	6/07	SCIL	Slant-steer utility, great restoration, very sharp tractor, good tires, rare tractor, two-cylinder production record shows this tractor was shipped in industrial yellow paint, 1 of just 199 built
320			G	$13,000	8/06	WCIL	Running, old restoration, fenders, PTO, lights
320			G	$9,250	11/05	WCIL	Standard, restored, PTO, lights, front weights
320			G	$10,000	1/07	WCIL	Utility, PTO, lights
320			G	$10,100	11/05	WCIL	WF, restored, 3 pt., belt pully, PS
320			G	$10,500	9/07	ECMN	Gas, WF, slant steer, 3 pt., new tires
320			G	$11,000	6/05	WCMN	3 pt., PTO, new tires, fully restored
320	1958		G	$17,500	9/05	SEMN	Standard, 3 pt., fenders, front and rear weights
320S	1957		G	$9,250	11/07	SWIN	320S, restored, nice, fresh OH
320S			G	$13,000	7/06	NWIL	Standard, SN 325410, running, old restoration/repaint, fenders, PTO, lights, 3 pt., nice good running tractor, 1 of 319
320U			G	$7,200	1/08	SESD	WF, PTO, running
330			G	$19,000	7/06	WCIL	Running, old restoration, fenders, PTO, lights
330			G	$25,500	7/06	WCIL	Utility, running, old restoration, painted yellow, no 3 pt.
330			G	$19,750	7/05	SCIA	Standard, good tag, 1 of 844 built, restored
330S			E	$19,000	7/06	NWIL	Standard, SN 330714, running, old restoration/repaint, fenders, PTO, lights, hyd remote, 3 pt., original, nice shape
330U			E	$25,500	7/06	NWIL	Utility, SN 330636, running, old restoration/repaint, no drawbar, originally painted yellow, no 3 pt.
330U			G	$28,500	7/05	NEIA	Restored, 1 of 247 built
40			G	$3,100	11/07	SESD	3 pt., NF
40	1955		G	$4,750	9/07	NEIL	Crawler, SN 68942, older restoration in good mechanical condition
40	1955		E	$28,000	6/07	NWIL	Gas (all fuel), high crop, WF, new paint/tires
40			F	$1,450	1/06	SESD	WF, 3 pt.
40			G	$7,500	7/06	NWIL	SN 60139, running, restored
40	1953		G	$2,350	10/06	SCMN	Tricycle, 3 pt.
40	1954		G	$2,550	11/06	SENE	10-34 rear tires, 3 pt.
40	1954		G	$2,900	2/06	NCCO	3 pt., tri front
40			G	$1,600	2/06	WCIL	3 pt., Model M grill
40			G	$2,650	11/05	SCMI	Gas, WF
40			F	$2,695	2/05	SCMN	WF, 3 pt.
40			g	$4,750	3/05	ECMN	WF, 3 pt., PTO
4000			G	$8,500	3/08	NCTN	WF, syncro, 1 hyd., weights, 18.4-34 tires, diesel, nice original paint
4000			G	$8,500	3/08	NCTN	NF, syncro, 1 hyd., diesel, non-Roll-o-Matic from factory, 15.5-38 tires
4000	1970		G	$9,800	1/07	NWOH	WF, 2 hyd., quick hitch
4000	1970		G	$11,500	7/07	NCIA	Factory WF, syncro, 1 hyd., 18.4-34 tires, great tin
4000	1971	8,600	G	$27,000	3/07	SEPA	1 of 403 made, cab, 2 hyd., dual PTO, differential lock, PS
4000	1972		G	$5,500	11/07	WCIL	Running, unrestored/original, fenders, PTO, hyd. remote, 3 pt.
4000		5,527	E	$10,500	2/06	WCMN	Gas, syncro trans, 3 pt., PTO, 2 hyd., Allied 595 quick-tach loader, 84" bkt.
4000		4,985	G	$12,000	2/06	SCPA	2WD, OH at 4,775 hours, syncro., new front/rear tires, 2 hyd.
4000	1969	8,100	G	$7,000	12/06	SEIA	3 pt., 2 hyd., QR
4000	1969	10,247	G	$11,500	5/06	SEWY	John Deere 148 loader, cab, diesel

John Deere

Model	Year	Hours	Cond.	Price	Date	Area	Comments
4000	1972	8,911	G	$20,500	2/06	SCPA	PS, 1 of about 400 build
4000			F	$6,000	9/05	SESD	Diesel
4000			G	$8,300	3/05	NWIL	WF, console, diesel
4000			G	$11,000	2/05	ECNE	WF, 1 hyd., 3 pt., diesel
4000	1971	6,100	G	$7,250	2/05	ECMI	Side console, diesel, 2 hyd., front weights, one owner
4000	1971		G	$15,250	11/05	SCKY	
4010		3,600	G	$4,000	2/08	SCIL	NF, gas, only 3,600 original hours, original paint, 18.4-38 rears
4010			G	$4,500	2/08	WCMN	Diesel, 2WD, cab, 3 pt., PTO, 16.9-38 tires
4010			G	$5,200	3/08	NEMO	Wheatland, diesel
4010			G	$5,350	2/08	SCMI	Diesel, NF, 3 pt., PTO, syncro, weights, 1 hyd.
4010		5,500	G	$5,700	2/08	SEMO	Diesel, clean
4010			G	$7,000	3/08	ECWI	Diesel, pulling tractor, 3 pt., NF, 16.9-38 tires
4010	1961	4,803	G	$5,800	4/08	SCMI	NF, 1 hyd., 3 pt.
4010	1961		G	$7,500	5/08	ECMN	Diesel, WF, canopy, new style step, 18.4-38 tires, restored
4010	1962	9,302	G	$6,000	2/08	WCOK	Diesel, PTO, 1 hyd., 18.4-34 tires
4010	1962		G	$6,700	1/08	WCKS	3 pt., PTO, syncro, WF
4010			F	$2,250	4/07	ECAB	Canadian sale, 2WD, 81 PTO hp., diesel, PTO, 2 hyd., 18.4-34 rear tires
4010		3,400	F	$3,600	12/07	WCMN	WF, gas
4010			G	$4,000	2/07	SCWA	Syncro, 15.5-38, 3 pt., PTO, 2 hyd.
4010			G	$4,650	8/07	SESD	WF, 3 pt., gas
4010			G	$5,500	8/06	SCIL	Diesel, front weights, original condition, rear weights, 3 pt. with toplink
4010			G	$8,000	11/07	NEND	Open station, dual 3500 loader with 7' bkt., 2 hyd., PTO, 3 pt., 14.9-38 singles
4010			G	$12,500	8/07	NENE	High crop, 8.25-20 front tires, 15.5-38 rear, 3 pt., rock shaft, PTO, diesel, fenders, rollbar, only 170 made!
4010	1961	4,523	F	$2,800	1/07	NECO	Diesel, Wheatland, WF, PTO
4010	1961		F	$4,600	2/07	ECIL	Diesel, 15.5-38, 1 hyd., 540 PTO, restoration in progress
4010	1961		G	$4,600	9/07	ECNE	1,357 hours on tach, diesel, WF, syncro range, 16.9-38 rear tires, 11L-16SL tires, 2 hyd., 540 PTO, 3 pt., cement rear wheel weights
4010	1961		G	$6,400	4/07	NECO	Diesel, 3 pt., WF, front weights, no cab, tach shows 2,031 hours
4010	1962		P	$2,000	4/07	NECO	LP gas, 3 pt., WF, no cab, tach shows 1,369 hours
4010	1962		F	$4,500	12/07	SEWY	Diesel, cab, WF, 3 pt.
4010	1962		G	$5,200	6/07	NENE	Diesel, Roll-o-Matic NF, syncro, 15.5-38 rear tires, 6:00-16 front tires, 2 hyd., 540/1,000 PTO, 3 pt.
4010	1963	6,794	G	$4,000	11/07	SWNE	WF, gas
4010	1963		F	$5,100	4/07	NECO	Diesel, 3 pt., WF, front weights, no cab, tach shows 3,798 hours
4010	1963	8,655	F	$6,200	8/07	SWSK	Canadian sale, 2WD, John Deere 46A loader and 5' bkt., 80 PTO hp., 8F/2R trans., 8,655 hours, 2 hyd., PTO
4010	1963		G	$6,500	1/07	NCIA	Diesel, open station, factory WF, 2 hyd., quick coupler, syncro
4010	1963		E	$7,200	12/07	WCIL	Diesel, 80 hours on OH, 2 hyd., front weights, new paint
4010	1963	7,706	G	$7,800	9/07	NECO	Diesel, no cab, 3 pt., new rear tires, Workmaster loader

John Deere

Model	Year	Hours	Cond.	Price	Date	Area	Comments
4010	1963		F	$8,500	8/07	SWSK	Canadian sale, 2WD, John Deere 46A loader and 5' bkt., 80 PTO hp., 8F/2R trans., 18.4-34 rubber, 2 hyd., PTO
4010	1963		G	$10,200	1/07	SENE	John Deere 158 loader, grapple, new battery and starter
4010			F	$2,650	3/06	NCUT	Diesel
4010			F	$3,900	12/06	NWWI	Diesel, WF, Allied 660 loader, 18.4-34 tires
4010			G	$4,500	4/06	NWSD	Farmhand loader
4010			G	$4,500	2/08	WCMN	1 hyd., PTO, no 3 pt., 60 Series step, 14.9-38 singles
4010			G	$5,500	4/06	WCSD	Farmhand F10 loader with grapple and scoop, rock shaft, no 3 pt.
4010		2,900	G	$6,200	11/06	NCKY	Tricycle front, shaver post driver mounted
4010			G	$6,400	8/06	SEPA	WF, diesel, average rubber
4010			G	$7,200	8/06	NCIA	Diesel, WF, 3 pt., repainted
4010	1961		F	$3,600	3/06	SWNE	2 hyd., diesel, 18.4-34, rear wheel weights
4010	1961		F	$4,000	7/06	ECND	3 pt., PTO
4010	1961	2,476	G	$4,250	1/06	NECO	Diesel, WF, 3 pt., excellent cab
4010	1961		F	$4,800	3/06	SWOH	Diesel, WF, no cab
4010	1961		G	$5,000	2/06	ECKS	
4010	1961		E	$6,750	1/06	WCNE	WF, no cab, rebuilt
4010	1962		G	$4,100	6/06	WCMN	3 hyd., 3 pt., 540/1,000, rock box, cab, 12-volt system, 18.4-34
4010	1962		G	$9,200	8/06	NWIA	Diesel, John Deere WF, no cab, nice
4010	1963	3,365	G	$6,300	1/06	SCMI	WF, cab, 3 pt., 2 hyd., diesel
4010	1963		G	$6,700	4/06	NWOH	Diesel
4010	1963	7,132	G	$6,800	3/06	ECNE	Diesel, syncro, 1 hyd., 3 pt., quick hitch, ROPS canopy
4010	1963	3,643	G	$9,000	3/06	NWKS	Diesel, syncro, 18.4-34, 11:00-16 front tires, 2 hyd., 3 pt., fenders, John Deere roll guard canopy, John Deere 158 loader with grapple fork, one owner
4010			F	$3,000	7/05	SCMN	Loader, diesel
4010			F	$3,400	3/05	SEIA	Industrial, loader
4010			F	$3,750	1/05	NENE	Diesel, WF, 3 pt., weak clutch
4010			F	$3,900	12/06	NWWI	Diesel, early model, 3 pt., 540/1,000 PTO, syncro
4010			G	$4,100	3/05	ECMI	Gas, WF, syncro, 2 hyd., 3 pt., PTO, 15.5-38
4010			G	$4,250	9/05	WCCO	Diesel, PTO, 3 pt.
4010			F	$4,400	12/05	WCWI	LP
4010			F	$4,500	4/05	SEND	500 hours on hyd. pumps, 2 hyd., 3 pt., syncro
4010			F	$5,000	4/05	WCIN	High hours, diesel, NF, good rubber
4010			G	$5,600	4/05	WCWI	Cab, 4020 pistons, clamp-on duals, diesel
4010		3,700	G	$5,900	2/05	WCOK	One owner, bareback, PTO, 1 hyd., tires good, diesel
4010			G	$6,100	10/05	ECMN	Diesel, extra fenders, cab, 3 pt.
4010			G	$6,200	2/05	WCIA	Diesel, WF, roll bar
4010			G	$8,000	12/05	WCIN	WF, syncro, 1,000 hours on OH
4010		4,311	G	$9,200	1/05	SENE	1 hyd., front weights, syncro
4010			G	$20,000	7/05	SCIA	High crop, diesel, good tag, restored
4010	1961		F	$3,600	7/05	ECND	1 hyd., 18.4-34
4010	1961	4,642	F	$4,400	8/05	SEMI	1 hyd., quick hitch
4010	1962		F	$4,600	3/05	NECO	High hours, John Deere 158 loader, grapple
4010	1962		G	$5,200	7/05	NEIA	Diesel, Schwartz, WF, 1 hyd.
4010	1962		F	$5,600	12/05	NWIL	WF, 3 pt., diesel
4010	1962		G	$6,600	3/05	ECND	WF, syncro, 3 pt., PTO
4010	1963		G	$3,500	3/05	ECID	Diesel, syncro, 2 hyd., 540 PTO, 3 pt.

John Deere

Model	Year	Hours	Cond.	Price	Date	Area	Comments
4010	1963	8,118	G	$6,700	12/05	ECMN	Diesel, cab, 2 hyd., 3 pt., quick hitch, band duals, 12-volt system, 14-7-38
4010	1963		G	$9,850	9/05	SEMN	MFWD, hyd. level front, front weights, 3 pt.
4020				$0	3/08	ECMN	Side console
4020			G	$6,000	3/08	SCMN	Year-A-Round cab, gas, clean
4020			G	$8,800	1/08	SESD	WF, 3 pt., PS, Farmhand 235 loader
4020			F	$9,100	4/08	WCWI	Cab, diesel, weights
4020			G	$9,300	3/08	ECSD	8 speed, PS, cab, John Deere 158 loader, 1 hyd., 3 pt.
4020			E	$11,000	1/08	SWKY	
4020	1964	7,119	G	$5,000	2/08	ECMN	Syncro, 2 hyd., 540/1,000 PTO, no 3 pt., new 18.4-34 tires
4020	1964		G	$5,600	2/08	WCOK	LP, PTO, 1 hyd., dozer blade
4020	1965	5,734	F	$3,500	3/08	ECNE	Propane, syncro, 18.4R-34 rear tires, 7.50-18 front tires, 1 hyd., 540/1,000 PTO, John Deere 148 loader with 8' bkt. and 4-tine grapple fork
4020	1965	5,734	F	$4,200	3/08	ECNE	WF, diesel, PS, 5,734 hours on tach, 18.4R-34 rear tires, 11L-15 front tires, 1 hyd., 540/1,000 PTO, Farmhand F25 loader with 4-tine grapple fork
4020	1965		G	$7,000	2/08	WCOK	Cab, diesel, 3 pt., PTO, 2 hyd.
4020	1965	2,759	G	$12,750	3/08	ECWI	WF, 3 pt., 18.4-34 tires
4020	1966		G	$8,500	1/08	NWOH	
4020	1967		F	$6,500	3/08	ECWI	PS, Generation step, M&W turbo, 18.4-38 (70%) on long axles, Pioneer tips, front/rear weights
4020	1967	8,066	F	$9,850	3/08	NWKY	Hub duals
4020	1967	6,500	G	$11,000	2/08	SCKS	Diesel, canopy, 8 speed, 2 hyd., 3 pt., 18.4R-34 tires, 10:00-16 front tires, 75 hours on complete OH
4020	1968	7,005	F	$6,600	3/08	ECNE	WF, diesel, QR, 7,005 hours on tach, 18.4R-34 rear tires, 10:00-16 front tires, 3 hyd., 540/1,000 PTO, 3 pt., missing rear window
4020	1968	8,123	G	$8,500	2/08	SWMN	Syncro, 3 pt., PTO, 2 hyd., rock box
4020	1968		G	$8,700	2/08	SEMN	Diesel, NF
4020	1968		G	$8,800	2/08	SEMN	Diesel, WF
4020	1968	9,854	G	$9,600	2/08	WCOK	Diesel, 3 pt., PTO, 2 hyd., John Deere 148 loader
4020	1969		G	$9,500	3/08	ECMN	Side console
4020	1969		G	$10,900	4/08	ECND	Diesel, 3 pt., PTO, 2 hyd., side console, John Deere 148 loader, 18.4-34 tires
4020	1969	6,877	G	$11,000	1/08	SCNE	Syncro, 18.4R-34 rear tires, 4 front weights
4020	1970		F	$7,500	4/08	WCIL	WF, no cab, weights, diesel, new clutch and trans.
4020	1970	8,836	G	$7,900	2/08	WCOK	3 pt., PTO, console
4020	1970		G	$9,500	3/08	SCNE	70 hours on OH, WF, syncro, 18.4-34 tires, 3 pt., repainted
4020	1970		G	$14,750	2/08	SEIA	WF, side console, 2 hyd., ROPS, 3 front weights and bracket sold separate for $475, quick hitch sold separately for $125
4020	1971		E	$15,750	5/08	ECMN	Restored, diesel, side console, PS, 3 pt., new 18.4-38 tires
4020	1972	5,868	G	$13,000	4/08	SCMI	2 hyd., 3 pt.
4020	1972	5,770	E	$17,000	1/08	WCIL	Diesel, Hiniker cab, WF, syncro, side console, 3 pt., dual PTO, front rail weights, 18.4-38 tires, nice
4020			F	$5,300	8/07	SESD	
4020		9,300	F	$5,600	1/07	WCIL	WF, no cab, syncro, 2 hyd.

John Deere

Model	Year	Hours	Cond.	Price	Date	Area	Comments
4020			G	$6,000	8/06	WCMN	WF, diesel, 1 hyd., 3 pt., PTO, 18.4-34 tires, 85% rubber, hours unknown
4020		1,075	G	$6,500	11/07	ECSD	WF, cab, syncro, 2 hyd., 3 pt., PTO, 18.4-34 singles
4020			G	$6,500	3/07	NEMI	Diesel, PS, 15.5-38 tires, rebuilt motor
4020			G	$7,000	8/07	SESD	LP, WF, 3 pt., cab
4020			F	$7,100	4/07	NCNE	Early model, 1 hyd., poor tires
4020			G	$7,700	3/07	NWIA	
4020			G	$8,100	8/07	SESD	Diesel
4020		3,230	G	$8,100	1/07	ECIL	Diesel, WF, weights, 2 hyd., 850 hours on OH, 16.9-34
4020		5,772	G	$8,400	1/07	SCNE	No cab, 2 hyd.
4020		5,772	G	$8,400	1/07	NCNE	Diesel, no cab
4020			G	$10,000	11/07	NEND	2 hyd., PTO, 3 pt., safety-guard cab, diesel, 18.4-34 tires
4020			G	$10,000	12/07	WCNY	PS, good paint, fenders
4020	1964		G	$6,000	9/07	NECO	Industrial, diesel, WF, no PTO, 1 hyd.
4020	1964		G	$6,100	3/07	WCMN	Open station, turbo diesel, syncro, 2 hyd., 3 pt., PTO, 18.4-34 singles (70%)
4020	1964		G	$7,000	9/07	ECND	PS, open station, PTO pump, Farmhand F11 hyd. loader with bkt. and grapple, no 3 pt., hd standard axle
4020	1964	10,660	G	$7,100	3/07	WCMI	3 pt., 1 hyd., no cab, WF
4020	1964		G	$10,000	4/07	SCOK	John Deere 148 loader, rubber 75%
4020	1965	5,098	G	$6,000	3/07	NEMI	Diesel, PS, 94 hp. PTO, 2 hyd., 16.9-34, hours on tach, not actual
4020	1965		G	$7,500	12/07	SCMN	John Deere 48 loader
4020	1965		F	$7,600	3/07	SWOH	Diesel, WF, no cab, OH 10 years and 1,200 hours ago, hub duals
4020	1965		G	$11,000	2/07	ECNE	Diesel, PS, new tach, 16.9-34 rear tires, 11L-15 front, 1 hyd.
4020	1965	8,547	G	$11,500	9/07	ECNE	PS, turbo, hd WF, 2 hyd., 4,000 hours on OH
4020	1966		F	$3,700	4/07	NECO	Diesel, 3 pt., WF, cab, tach shows 479 hours
4020	1966	6,600	F	$4,600	3/07	ECND	Great Bend 800 loader, grapple fork, gas, PS, 3 pt., PTO, 2 hyd., 14.9-38
4020	1966	7,903	F	$5,050	12/07	WCNE	Diesel, WF, no cab, 3 pt., runs
4020	1966		G	$8,000	12/07	NCIA	Diesel, cab, PS, WF, 1 hyd., rebuilt motor with 479 hours, front tank
4020	1966		G	$8,500	6/07	SCIL	Diesel, PS, showing 1,678 hours on the Killam farm, good tires, no fenders, all original, rear weights
4020	1966		G	$9,400	11/07	SCNE	Diesel, 3 pt., 2-way Fasse hyd. valve, 18.4-34 tires
4020	1966		E	$10,200	2/07	SCKS	3 pt., 2 hyd., slick
4020	1966	7,800	G	$10,750	11/07	ECIL	John Deere 148 loader, diesel, WF, 2 hyd., front/rear weights, syncro, 18.4-34 tires (90%)
4020	1967	7,490	F	$4,600	1/07	SWKS	LP, syncro, WF, 2 hyd., 3 pt., 15.5-38
4020	1967		G	$5,100	10/07	NECO	Wheatland, diesel, Great Bend 900 loader, 8' bkt.
4020	1967		G	$5,300	2/07	ECIL	Diesel, 16.9-34, 1 hyd., extra front fuel tank, front weights, dual PTO
4020	1967		F	$5,650	8/07	SESD	PS, fenders, 3 pt., 600 hours on motor OH
4020	1967		G	$5,700	9/07	ECND	2 hyd., 3 pt., diff. lock, PTO, fenders, diesel, 18.4-34 singles
4020	1967	5,700	G	$6,900	4/07	SEND	Diesel, open station, fenders, 18.4-34 singles, cast wheel weights, one owner
4020	1967	5,341	G	$7,700	8/07	NWIA	Diesel, WF, 3 pt., 18.4-34 rubber
4020	1967		G	$8,100	8/07	SESD	Diesel, PS

John Deere

Model	Year	Hours	Cond.	Price	Date	Area	Comments
4020	1967		G	$8,500	3/07	WCKS	Great Bend 900 loader, bkt., small bale fork and spear
4020	1967		G	$8,750	6/07	ECND	Diesel, row crop, syncro, 3 pt., PTO Farmhand F27 loader with dirt and snow bkts., bale spear and manure fork, 18.4-34 tires
4020	1967	4,306	G	$11,300	8/07	NEIA	Diesel, console, PS, NF, all original
4020	1967	6,400	G	$16,000	8/07	ECIL	WF, 1 hyd., syncro, 18.4-34
4020	1968		G	$6,000	3/07	WCMN	2WD, syncro, 2 hyd., new front tires, Hiniker cab, Farmhand 22 loader
4020	1968		F	$6,000	11/07	NEMO	Diesel, syncro, 2 hyd., 18.4-34 and 10.00-16 tires, front weight pads
4020	1968	6,800	G	$6,400	4/07	SCKS	Diesel, PS, 3 pt., PTO
4020	1968		G	$6,400	4/07	SCKS	PS, diesel, 3 pt., PTO
4020	1968		G	$6,900	3/07	WCKS	Cab, PTO, 3 pt.
4020	1968	8,167	G	$7,050	11/07	NEMO	Diesel, syncro, 2 hyd., 18.4-34 & 10.00-16 tires, factory John Deere ROPS, nice straight tractor
4020	1968	7,704	G	$9,250	6/07	NENE	NF, diesel, syncro, 7,704 hours tach, 18.4R-34 rear tires, 7.5L-15 front tires, 2 hyd., 540/1,000 PTO, rear wheel weights
4020	1968	6,395	G	$9,800	3/07	NCIA	One owner, WF, syncro, quick hitch
4020	1969		G	$7,250	1/07	SWKS	Console, syncro, 2 hyd., rear wheel weights, cab, 18.4-34
4020	1969		G	$7,600	2/07	SWIL	Side console, Year-A-Round cab, fenders sold separately for $850
4020	1969	9,194	G	$7,800	3/07	NEIA	Diesel, side console, no cab, NF, 3 pt., good 18.4-34 rubber, front weights sell separately
4020	1969	11,071	F	$8,000	4/07	NCIA	4,000 hours on OH, side console, WF, syncro, 2 hyd., no cab
4020	1969		G	$8,500	6/07	SCIL	Diesel, fully restored, WF, 3 pt., 1 hyd.
4020	1969	6,865	G	$9,000	3/07	ECIL	Diesel, syncro range, 540/1,000 PTO, 2 hyd., 3 pt., 18.4-34 tires, tach reads 6,865 hours, nicely restored
4020	1969		G	$12,300	3/07	NCIA	Diesel, open station, 2 hyd., syncro, quick coupler, factory WF
4020	1969	6,500	G	$12,900	1/07	NEIA	WF, 3 pt., dual console hyd.
4020	1969		G	$16,250	4/07	SCOK	John Deere 158 loader, rubber 95%
4020	1970		G	$6,100	2/07	ECIL	Diesel, side console, 1 hyd., dual PTO, 16.9-38, extra front fuel tank, front weights
4020	1970	8,750	G	$7,000	12/07	SEWY	Cab, WF, 3 pt., diesel
4020	1970	5,017	G	$10,000	2/07	ECNE	Gas, PS, 18.4R-34 rear tires, 11L-15 front tires, 2 hyd.
4020	1970		G	$10,500	6/07	NENE	WF, syncro, 3,330 hours on tach, 18.4R-34 rear tires, 11L-15 front tires, 2 hyd., 540/1,000 PTO, 3 pt. with third link, 2 sets of rear wheel weights, fenders
4020	1970	4,700	F	$11,100	6/07	SCIA	PS, weak engine, good 34" tires
4020	1970	4,700	G	$14,750	6/07	SCIA	PS, fair 34" tires
4020	1971		G	$14,200	1/07	SCNE	Syncro, console hydraulic
4020	1971	4,800	E	$22,100	3/07	SEIA	Diesel, very nice original tractor
4020	1972	2,499	E	$31,000	11/07	NWOH	Diesel, syncro, 18.4-34 tires, 2 hyd., one owner, nice!
4020		6,354	P	$4,000	11/06	NEIA	Really rough, gas
4020			F	$4,900	8/06	SEPA	Diesel, PS, average rubber, WF
4020			G	$5,200	3/08	NWIL	LP, WF
4020			G	$5,900	8/06	NWOH	2 hyd., 3 pt., turbo, 2 front weights, rear weights

John Deere

Model	Year	Hours	Cond.	Price	Date	Area	Comments
4020			G	$6,000	9/06	SCNE	WF, PS, 2 hyd., 3 pt., 18.4-34 rear, new starter, new injection pump
4020			F	$6,000	12/06	NWIL	WF, cab, diesel
4020			G	$7,500	7/06	NWIL	LP, console, WF, long axle, 6,100 hours
4020		6,500	G	$7,500	8/06	SCMN	
4020			G	$8,200	12/06	NWIL	John Deere 48 loader, quick hitch, diesel
4020			G	$9,000	11/06	NCKY	Diesel
4020			G	$9,100	10/06	SEMN	New rubber
4020			G	$9,450	3/06	SEIA	Koyker loader
4020			G	$9,500	8/06	SCFL	PS
4020			G	$10,500	2/06	WCMN	Diesel, PS, side console, 3 pt., 2 hyd., diff. lock
4020			F	$11,300	2/06	NEIA	High hours, recent OH
4020		5,268	G	$12,500	3/06	SEMN	PS, cab, WF, John Deere 148 loader
4020		7,462	G	$14,000	8/06	SCFL	PS
4020	1964	4,121	F	$3,400	5/06	SEWY	3 pt., gas
4020	1964		G	$5,100	3/06	NCND	Standard, open station, 2 hyd., Faul 3 pt., PTO, OH, 18.4-34 singles
4020	1964		F	$5,250	12/06	SCMT	Farmhand F11 loader
4020	1964		F	$5,750	7/06	ECND	3 pt., ROPS, cab, 2 hyd., syncro
4020	1964		G	$6,100	4/06	ECND	Open station, turbo diesel, syncro, 2 hyd., 3 pt., PTO, 18.4-34 singles (70%)
4020	1964		G	$7,500	3/06	NECO	Diesel, cab, 3 pt., PTO, 2 hyd., 15.5-38, no duals
4020	1964	7,178	G	$7,700	1/06	NECO	WF, 3 pt., Year-A-Round cab, 38" rear tires
4020	1964		G	$8,000	11/06	NWKS	3 pt., 2 hyd.
4020	1965	6,578	P	$3,400	2/06	NCIL	2 hyd., needs clutch
4020	1965		F	$4,000	11/06	ECND	PS, 500 hours on OH, hyd. pump
4020	1965		G	$6,900	10/06	SCMN	PS, WF, cab, diesel, 540/1,000 PTO
4020	1965		G	$10,400	3/06	NCIA	Diesel, syncro, factory WF, 2 hyd., 3 pt., quick coupler, new 18.4-34 tires, waffle weights
4020	1965	7,300	G	$12,000	3/06	NENE	200 hours on OH, 18.4-34 rear tires, 3 pt., fenders, repainted
4020	1966	9,000	P	$3,850	12/06	SCMN	Water in engine oil
4020	1966		F	$4,000	11/06	SCND	Gas, cab, 3 pt., PTO
4020	1966		F	$5,700	12/06	WCMN	New starter, new radiator, good tires and paint
4020	1966	9,225	G	$5,750	9/06	NENE	9,255 hours on new tach, syncro, diesel, positive track wide & NF, 15.5-38 rear tires, 11L-15 front, 2 hyd., 540/1,000 PTO, 2 sets of rear wheel weights, one owner, shedded
4020	1966	7,100	G	$5,900	3/06	ECND	Open station, 1 hyd., 3 pt., PTO, 18.4-34 singles
4020	1966		F	$7,000	3/06	SWOH	No cab, diesel, WF
4020	1966	3,204	G	$7,000	2/06	NECO	Diesel, Farmhand 258 loader, 3 pt., WF, axle duals
4020	1966	7,749	G	$7,000	12/06	NWMO	18.4-34 rear tires (70%), 1 hyd., new battery, no toplink
4020	1966	8,454	G	$7,200	4/06	ECND	Diesel, syncro, open station, 2 hyd., 3 pt., PTO, diff. lock, 12-volt conversion
4020	1966		G	$7,600	12/06	ECIL	PS, weights, quick hitch, 18.4-34
4020	1966		G	$8,000	11/06	WCMN	Diesel, syncro, 2 hyd., 3 pt., PTO
4020	1966	4,500	F	$10,700	2/06	NCKS	John Deere 146 loader with dirt bkt., 3 pt., dual PTO, 1 hyd., syncro, no cab, tractor used oil, tin bent up, side panel missing, several oil leaks, loader welded, bkt. bent
4020	1967		P	$3,600	11/06	SCND	Low hours on OH, 2 hyd., PTO, 23.1-30 tires, John Deere 148 loader and bkt.
4020	1967		F	$5,500	12/06	SCMT	Dual loader
4020	1967	6,400	E	$6,000	2/06	SEMI	Syncro, gas

John Deere

Model	Year	Hours	Cond.	Price	Date	Area	Comments
4020	1967		G	$6,700	3/06	NEND	Diesel, cab, 2 hyd., 3 pt., PTO, diff. lock, 18.4-34 band duals
4020	1967	6,887		$7,000	8/06	SCMI	3 pt., 1 hyd., WF
4020	1967	5,140	G	$7,650	4/06	SWOH	Diesel, WF, Year-A-Round cab
4020	1967	10,600	F	$7,750	8/06	ECNE	Weights, less than 1,000 hours on complete OH, 2 hyd., good 18.4-34 tires, less than 1,000 hours on piston sleeves, 5-gal. oil pan
4020	1967		F	$8,500	3/06	SENE	Loader, diesel
4020	1967	6,406	E	$10,500	3/06	ECND	Open station, 8-speed gear trans, 2 hyd., 3 pt., PTO, 14.9-38 singles, one owner
4020	1968		F	$5,600	5/06	SEWY	Cab, front weights, air
4020	1968		F	$6,900	3/06	NCCO	Canopy
4020	1968		G	$7,000	11/06	ECND	Row crop, 2 hyd., 2 hyd. aux, 3 pt., PTO, diesel, 18.4-38 singles
4020	1968		G	$7,500	12/06	NCNE	Syncro range, diesel, Farmhand F11 loader
4020	1969		G	$8,000	4/06	WCWI	Diesel, console, factory cab, factory WF, syncro
4020	1969	6,890	G	$9,700	9/06	ECNE	WF, 3 pt., 1 hyd., no duals, fair paint
4020	1969		G	$10,800	11/06	SCSD	PS, 2 hyd., 3 pt.
4020	1969		G	$11,250	11/06	WCMN	Diesel, syncro, side console, 2 hyd., 3 pt., PTO, one owner
4020	1969	6,628	G	$11,700	2/06	WCIL	Side console
4020	1969	5,300	G	$14,200	11/06	SCNE	Console hyd., PS, 2 hyd., WF, 18.4-34 tires
4020	1969	5,500	E	$15,400	8/06	ECIL	Quick hitch, 2 hyd., good rubber, diesel, 2 hyd.
4020	1970		G	$5,900	12/06	NCNE	Syncro, gas, Farmhand F11 loader
4020	1970	6,000	G	$7,250	7/06	WCMN	Syncro, 1 hyd., side console, 18.4-34
4020	1970		G	$7,400	9/06	WCWI	10,000 hours, 3 pt., WF, diff. console, 2 hyd., new style John Deere suitcase weights
4020	1970		G	$7,600	9/06	NEIN	PS, factory cab, diesel, original
4020	1970		G	$8,500	9/06	NCIA	Diesel, syncro, duals, 3 pt., 2 hyd., wheel weights, new batteries
4020	1970		G	$8,800	9/06	NEIN	Syncro, 3 hyd., original, diesel
4020	1970		F	$8,800	3/06	SWOH	No cab, diesel, WF, new style
4020	1970	8,321	G	$8,900	12/06	NWIL	One owner, WF, roll bar side console
4020	1970	6,800	G	$9,800	9/06	WCWI	6,800 hours, 3 pt., WF, diff. console, 2 hyd.
4020	1970	7,700	E	$9,800	2/06	WCMN	2WD, side console
4020	1970		G	$10,000	4/06	WCWI	Diesel, console, factory WF, syncro
4020	1970	8,410	P	$12,600	3/06	WCMN	John Deere 158 loader, rough, guard cab, 3 pt., PTO, 2 hyd., diesel, 18.4-34 tires
4020	1970		E	$21,200	11/06	SCSD	MFWD (factory original), PS, 2 hyd., 3 pt.
4020	1971	9,108	G	$10,000	12/06	SEIA	3 pt., 2 hyd., QR, front weights, repainted
4020	1971		G	$10,600	8/06	NEIL	Side console, syncro, like-new 18.4-38 tires, duals
4020	1971		G	$15,500	9/06	NEIN	PS, WF, 2 hyd., original
4020	1972		G	$9,000	4/06	NEND	Turbo diesel, PS, roll guard, CAH, side console, 3 pt., PTO, 2 hyd., 18.4-34 singles
4020	1972	8,344	F	$11,250	3/06	NWOH	3 pt., 2 hyd., weights, axle duals
4020	1972	2,818	E	$21,500	2/06	SWNY	PS, restored, 2 hyd., no rear weights, 4 front weights, 18.4-38 tires
4020		4,151	F	$2,250	9/05	SEIA	Gets water in oil
4020			F	$5,300	7/05	SCMN	PS, diesel
4020			G	$5,750	6/05	ECMN	Row crop, WF, syncro, fenders, 3 pt., PTO, rebuilt
4020			F	$5,750	9/05	SCMN	
4020			F	$5,800	8/05	WCIA	
4020			G	$5,900	3/05	NWIL	Diesel, NF
4020			G	$6,200	6/05	ECIL	3 pt., John Deere WF, syncro, 500 hours since OH, good paint

John Deere

Model	Year	Hours	Cond.	Price	Date	Area	Comments
4020			F	$6,750	2/05	WCWI	2 hyd., Hiniker cab, 3 pt., WF, diesel
4020		6,286	G	$7,100	8/05	SWMN	Diesel, 3 pt., PTO, 2 hyd., diff. lock, syncro, Year-A-Round cab
4020			G	$7,250	12/05	WCMN	Diesel, side console, 2 hyd., 3 pt., PTO, roll guard cab, 18.4-38
4020			G	$7,300	1/05	ECNE	2 hyd., WF
4020			G	$7,700	7/05	ECMN	Diesel
4020			G	$7,800	9/05	NENE	Diesel, WF, 3 pt., drawbar, 18.4-34
4020			G	$7,800	2/05	SWNE	Diesel
4020			F	$8,000	10/05	NEPA	Open cab
4020			G	$9,250	11/05	WCIL	Diesel, console, new paint
4020			G	$9,500	9/05	NEIN	Console
4020			G	$12,900	1/05	SCMN	Console, complete rebuild
4020		5,300	E	$14,400	11/05	SEMI	Duals, weights
4020			E	$28,000	7/05	NWIL	High crop, 1 of 121 made, diesel, restored, new tires, WF, side console ps (possibly the first side console ps off the line), deluxe seat, PTO
4020	1964		P	$3,700	8/05	NWIL	Rough, gas, WF, 2 hyd., good 15.5-38 tires
4020	1964	5,004	F	$4,100	4/05	SCMI	Cab, quick hitch, diesel, WF, good rear rubber, snap-on duals
4020	1964		G	$6,100	12/05	NENE	PS, 18.4-34 near-new rear tires, WF, 2 hyd., fenders
4020	1964		G	$6,200	12/05	WCMN	Diesel, syncro, 2 hyd.
4020	1964	8,517	G	$13,900	12/05	NWIL	Diesel, PS, WF, 2 hyd., John Deere 48 loader
4020	1965		P	$2,400	12/05	NWIL	Parts tractor, diesel, 18.4-34
4020	1965		G	$4,100	4/05	NEIN	2,896 hours on tach, NF, 18.4-34 like-new tires
4020	1965		G	$7,400	7/05	ECND	2 hyd., 3 pt., engine major OH recently
4020	1965		G	$7,750	2/05	NEMI	Diesel, syncro, 2 hyd., 3 pt., PTO, 18.4-34
4020	1966		G	$3,750	6/05	ECIL	WF, 3 pt., 2 hyd., front/rear weights
4020	1966		F	$4,250	2/05	NWIL	NF, 3 pt., 2 hyd., gas, 15.5-38 tires
4020	1966	4,964	F	$4,300	8/05	SEMI	3 pt., 1 hyd.
4020	1966		G	$5,500	1/08	WCIL	Gas, 2 hyd., fenders, WF, 3 pt.
4020	1966	4,909	G	$6,400	8/05	NWIL	Gas, NF, open station
4020	1966	6,988	G	$8,000	4/05	SCMI	WF, diesel, 2 hyd., canopy, new 18.4-34 rubber, new head, radiator, trans., pump and injectors, clean
4020	1966	12,200	E	$8,500	4/05	NCOK	Crown fenders, heavy-duty front axle, John Deere 158 loader, 7' bkt. and round bale spike attachment, PTO, no 3 pt., one-owner family
4020	1966		E	$9,200	8/05	NCIL	1 hyd., restored, syncro, 18.4-34
4020	1966	6,000	E	$10,100	3/05	WCWI	2 hyd., WF, 3 pt., front weights, diff. lock, new paint
4020	1966		G	$11,200	2/05	NCIN	2 hyd., new paint and good rubber, PS, diesel
4020	1967		F	$3,900	12/05	NENE	Syncro, WF, 18.4-34
4020	1967		G	$5,000	2/05	NEIN	Diesel, NF, 500 hours on OH
4020	1967		F	$6,000	9/05	NCOH	Hours unknown, diesel, WF, looked fair
4020	1967		G	$6,700	3/05	ECSD	Diesel, syncro, WF, 18.4-34
4020	1967		G	$8,000	3/05	NCCO	ROPS, canopy, 2 hyd., 3 pt., 540/1,000 PTO, 1,600 hours on OH
4020	1967		G	$8,000	3/05	SENE	High hours, WF, diesel
4020	1967	6,603	G	$9,500	7/05	ECIL	
4020	1967	6,686	G	$10,000	3/05	NCIA	Factory WF, 2 hyd., quick coupler, hyd. pump OH, diesel, 18.4-34
4020	1967	6,826	F	$12,100	4/05	SENY	PS, 2 hyd., 18.4-34, good sheet metal
4020	1968	6,920	F	$4,400	7/05	NEIA	Gas, WF, 3 pt., 1 hyd.
4020	1968		G	$7,300	11/05	NEIL	Diesel

John Deere

Model	Year	Hours	Cond.	Price	Date	Area	Comments
4020	1968		G	$7,500	2/05	NENE	Syncro, 18.4-34 tires, 2 hyd., 3 pt. WF, new paint, bought new
4020	1968	6,600	G	$7,700	4/05	NCOH	Diesel, WF, syncro, 34" tires, 2 hyd., add-on tractor step and handrail
4020	1968	5,910	G	$8,300	8/05	SWOH	Cab, diesel
4020	1968		G	$11,750	2/05	ECIA	3 pt., 2 hyd., new PS and clutch, WF, 200 hours on complete OH, Farmhand F258 high-lift loader with bkt.
4020	1968	15,000	G	$13,500	2/05	ECSD	
4020	1969		G	$0	12/05	NCCO	No sale at $10,900, PS, one owner, Year-A-Round cab, side console, axle duals, OH about 3,000 hours ago
4020	1969		G	$5,700	2/05	NCIN	PS, shift console
4020	1969	7,048	F	$6,700	11/05	SCIA	Hiniker cab, WF, 2 hyd., 18.4-34
4020	1969		G	$8,700	11/05	ECNE	Side console, CAH, 3 hyd.
4020	1969		G	$8,750	8/05	NEIA	Recent OH, average paint, no cab, console shift
4020	1969		G	$9,100	8/05	NEIA	Recent OH, Hiniker cab, console shift, average paint
4020	1970		P	$6,600	8/05	NECO	Bad cab, smokes, console, rough
4020	1970	9,571	G	$6,850	1/05	NEMO	
4020	1970	8,485	G	$7,900	7/05	NEIA	Diesel, side console, syncro range, 3 pt., 1 hyd., NF, newer clutch
4020	1970		G	$9,000	3/05	NWMN	Side console, PS, John Deere roll guard cab, 3 pt., hyd., PTO
4020	1970		G	$9,900	12/05	WCMN	Console, factory cab, one owner, diesel
4020	1970	10,250	F	$10,250	11/05	WCIL	New tires, side console
4020	1970	8,015	G	$11,600	2/05	NWIL	1,350 hours on OH and new clutch, Hiniker cab, front/rear weights, WF
4020	1970	7,808	G	$15,000	3/05	WCNY	Open station, syncro, 18.4-38, diesel
4020	1970	4,096	E	$15,000	3/05	SWNE	3 pt., PTO, 2 hyd.
4020	1971	5,583	F	$7,250	4/05	NEKS	PS, 3 pt.
4020	1971		G	$9,100	7/05	NCIA	Diesel, factory cab, 18.4-34 duals
4020	1971		E	$9,750	3/05	NESD	Syncro range, diesel, 3 hyd., PTO, 3 pt., no cab, 18.4-38
4020	1971	6,220	G	$10,700	1/05	NWIL	Cab, syncro, WF, 2 hyd., new paint, diesel
4020	1971	5,933	F	$10,800	5/05	SEMN	QT cab, Year-A-Round cab, diesel, John Deere 146 loader
4020	1971	6,110	F	$12,500	3/05	WCNJ	PS, WF, Year-A-Round Hiniker cab, front end welded, repainted but poorly
4020	1971		G	$14,250	3/05	SCKS	3 pt., PTO, 2 hyd., new paint
4020	1972		F	$10,400	4/05	WCIN	High hours, poor rubber, console, ROPS with roof
4020	1972	7,123	G	$12,000	12/05	NWIL	One owner, sharp, 3 pt., 2 hyd., M&W turbo, new radiator, 18.4-38
4020	1972	5,430	F	$13,000	3/05	WCNJ	No cab, PS, WF, 2 hyd., poor tires, repainted, hour meter not working for 10+ years
4020	1972	8,042	G	$15,300	2/05	SCKS	PS, 2 hyd., roll guard, may have been one of the last 1972 models made
4030			F	$7,750	1/08	NWOH	ROPS, QR
4030	1976	2,241	E	$19,200	3/08	ECIL	Hiniker cab, 1 hyd., front fenders sold for $250 extra, rear fenders $2,600
4030	1977	1,345	E	$24,500	3/08	ECIL	Open station, 2 hyd.
4030			G	$8,500	12/07	WCMN	2WD, CAH, 3 pt., 3 hyd., 540/1,000 PTO, 18.4-34 tires
4030	1973	5,339	G	$12,000	3/07	ECIA	1,000 hours on OH, diesel, no cab, side console shift, real good 16.9-38 tires, rear fenders, WF, 3 pt.

John Deere

Model	Year	Hours	Cond.	Price	Date	Area	Comments
4030	1975	5,703	G	$7,700	2/07	SEMI	No cab, 3 pt., 2 hyd., syncro range, 1,000 hours on OH
4030		6,100	F	$7,800	5/06	SWWI	WF, like-new tires, clamp-on duals
4030			E	$8,500	3/06	NEKS	Cab
4030	1973		G	$7,200	12/06	ECMO	2WD, 4 post, syncro
4030	1974	6,761	F	$7,300	2/06	SCPA	2WD, QR, 4 post
4030	1974	5,300	G	$11,750	4/06	SWMB	Canadian sale, cab, standard trans., 2 hyd., 540 PTO, 18.4-34
4030	1974	5,300	G	$11,750	4/06	WCMN	Canadian sale, cab, standard trans., 540 speed, diesel
4030	1975	3,065	G	$12,000	12/06	WCIL	Fully equipped cab, QR, 540/1,000 PTO, 2 hyd., front weights
4030	1975	6,960	G	$15,700	2/06	NCCO	15.5-38, axle duals, second owner
4030	1975	2,800	G	$21,000	12/06	NCIA	CAH, QR, 18.4-38
4030	1976	6,253	G	$16,700	2/06	NCCO	One owner, 15.5-38, axle duals
4030			F	$5,000	4/05	WCWI	Front weights, diesel
4030			G	$7,000	3/05	SCIL	Open station, 2 hyd., 1973 model
4030		4,695	E	$18,000	11/05	SWMN	2WD, 540/1,000 PTO, 3 pt., 2 hyd., one owner, 18.4-34
4030		1,966	E	$18,500	1/05	NENE	QR, WF, 3 pt., CAH, shedded, new cab interior, not used in 4 years
4040	1978	6,670	G	$17,000	3/07	SEND	QR, 2 hyd., LSO conversion, 3 pt., 540/1,000 PTO, 18.4-38 on cast
4040	1978	6,615	G	$18,800	8/05	WCMN	Duals, 2 hyd., 3 pt., PTO, QR, CAH, original rear tires
40S	1953		F	$2,500	11/07	SWIN	Original, no draw
40T	1954		F	$2,000	6/07	SCIL	Straight tin, long axle, 3 pt., extra wide rear axles, gauges not John Deere, no tach, no dents in tin, easy restoration
40T	1954		G	$4,300	10/06	SCSD	Single front wheel, 3 pt., lights, new 11.2-34 tires, 12 volt, belly-mounted 9' sickle mower, very nice restoration
40T			F	$1,700	8/05	ECMI	NF, 3 pt., no arms
40T			G	$3,600	6/05	ECWI	New rubber
40T	1953		G	$7,500	9/05	SEMN	WF, front and rear weights, fenders, 3 pt.
40U			G	$4,000	3/08	NWIL	SN 40U-60762, running, old restoration/ repaint, PTO, lights, 3 pt.
420	1958		G	$2,300	2/08	ECMT	Single front, PTO
420			G	$2,750	9/07	ECMI	WF
420		2,717	G	$9,000	2/07	NEIN	MFWD, roll bar, John Deere 420 loader with 5' bkt., 3 pt., PTO
420			G	$1,950	10/06	WCWI	
420			G	$4,400	7/06	WCIL	Running , unrestored, original, fenders, PTO, lights, 3 pt., 4 speed
420	1956		G	$3,600	2/06	NCCO	Gas, 3 pt.
420	1956		G	$8,200	10/06	SCSD	WF, gas, 3 pt., Touch-O-Matic, nice restoration
420			F	$2,800	6/05	ECMN	Gas
420			G	$3,000	11/05	WCIL	WF, running, unrestored
420			G	$4,000	3/05	ECMN	Row crop, WF, 3 pt., PTO, fenders, new tires, power adjustable wheels
420			G	$7,000	9/05	ECMN	Standard gas, WF, new tires, 3 pt., PTO
420			G	$9,750	6/05	WCMN	3 pt., PTO, new tires, fully restored, WF
420	1956		G	$4,400	6/05	SWOH	Gas, WF, utility, no 3 pt.
420	1957		G	$3,000	11/05	WCIL	
420H	1956		G	$14,500	11/07	SWIN	High crop, older restoration, nice, Series 1, all green
420S	1956		G	$5,000	11/07	SWIN	PS rears, Series II, new tires, 3 pt., no toplink, older restoration, drawbar, 4 speed

John Deere

Model	Year	Hours	Cond.	Price	Date	Area	Comments
420T	1957		F	$3,500	9/07	WCIL	5 speed, 3 pt., slant steer, nice original, 28" rear tires, tach, front weight, live PTO, runs
420T	1957		G	$3,900	6/07	SCIL	Single front wheel, all original, 3 pt., lights, good tires
420T	1958		G	$7,000	6/07	SCIL	1 of 234 made, dual fuel, single front wheel, slant steer, valves ground, new radiator core, rebuilt carb., 3 pt., no toplink
420T	1958		E	$17,000	6/07	NWIL	LP, PS, 3 pt., WF, restored
420T	1958		G	$4,700	6/05	ECIL	5 speed, 3 pt., new rears
420U	1956		G	$3,400	11/07	WCIL	Older restoration
420U	1958		G	$12,250	9/05	SEMN	5 speed, reverser, air stack, front and rear weights
4230			P	$3,750	3/08	NEAR	Open station, 3 pt., PTO, 1 hyd., 18.4-38 singles
4230	1973	8,844	G	$8,600	2/08	SCMI	No cab, 2 hyd., quick hitch
4230	1973	7,135	G	$10,100	1/08	WCIL	No cab, 3 pt., 2 hyd., 6 front weights, 18.4-34 rear tires, clamp-on duals
4230	1973	7,100	G	$13,500	3/08	NEMO	18.4-38 tires
4230	1975		G	$12,000	1/08	NWOH	Cab, air, QR, updated 2 hyd., duals
4230	1977	2,000	G	$16,500	3/08	WCMI	2WD, CAH, QR, 2 hyd., stone bkt., 16.9-38 direct-axle duals
4230		7,786	G	$14,000	6/07	SWMN	CAH, QR, 3 pt., 2 hyd., PTO, 14.9-38, rock box, one owner
4230			E	$17,500	3/07	WCSD	Koyker 565 loader and bale spear, cab, 3 pt., good paint, new seat, like new
4230	1972	8,580	G	$8,500	2/07	NCNE	Diesel, cab, QR, 3 pt., 2 PTOs, 2 hyd., 16.9-38 (60%) 116 hp., weights sold separate
4230	1973	9,104	G	$10,200	3/07	NEIA	Diesel, factory cab, 3 pt., 18.4-34
4230	1973	7,395	G	$12,500	1/07	NEIA	Cab, new air, heat, QR, 2 hyd. with power beyond and joystick, 1,900 hours on OH, West-endorf WL42 loader with bkt. and bale spear
4230	1973	6,500	G	$13,500	12/07	ECIL	John Deere 148 loader, QR, John Deere turbo, 18.4-34 duals, 2 hyd., quick hitch, dual PTO, K&M cab step, 6' bkt.
4230	1973	10,576	F	$15,000	7/07	NCSK	Canadian sale, 2WD, 80 PTO hp., QR, dual PTO, 2 hyd., 18.4-34 rubber, 10,576 hours showing, SN 002903R, John Deere 148 loader and 5' bkt.
4230	1973	8,721	G	$15,000	12/07	SCMN	QR, 2 hyd.
4230	1973	8,600	G	$16,000	11/07	ECNE	PS, 18.4R-38 rear tires, 10:00-16 front tires, 2 hyd., 3 pt., 3 sets of rear wheel weights, single large front end weight, Sound Guard cab, R134 air
4230	1973	4,851	G	$18,000	7/07	SCSK	Canadian sale, QR, 2 hyd., 18.4-34 duals
4230	1973	6,112	G	$25,000	4/07	ECSK	Canadian sale, 2WD, 100 PTO hp., QR, 2 hyd., dual PTO, 10.00-16 3-rib front tires, 18.4-34 rear tires, 6,112 hours showing, SN 4230H008286R, John Deere 158 loader
4230	1974		G	$9,600	1/07	NWIA	QR, 16.9-38
4230	1974	8,407	G	$13,750	6/07	SWMN	CAH, PS, PTO, 3 hyd., rock box, one owner, 16.9-38 tires
4230	1974	8,522	G	$14,500	3/07	WCSK	Canadian sale, 100 hp., dual PTO, QR, 2 hyd., 18.4-34 tires, wheel weights, John Deere 148 loader and bkt., shedded
4230	1975	8,618	G	$10,250	6/07	SCMN	CAH, QR, rock box, 18.4R-38 tires
4230	1975		G	$16,500	7/07	SEND	CAH, PS, 2 hyd., 3 pt., 540/1,000 PTO, 18.4-34 rears, 10:00-16 fronts
4230	1976	6,633	G	$12,000	2/07	ECIL	Cab, air, new 18.4-34 tires, 2 hyd., shedded
4230	1976	7,431	G	$13,200	9/07	NECO	Rear weights, diesel

John Deere

Model	Year	Hours	Cond.	Price	Date	Area	Comments
4230	1976		G	$15,200	11/07	NEND	QR, 3 hyd., 3 pt., quick hitch, PTO, Agri-Power, 14.9-38 tires (60%), OH
4230	1977	8,635	G	$12,000	10/07	NEND	QR, 2 hyd., power beyond, 3 pt., PTO, new AC system, 18.4-38 singles (90%), mounts for John Deere 740 loader with joystick
4230	1977	5,376	G	$16,000	9/07	ECNE	Diesel, WF, QR, 18.4R-34 rear tires, 11L-15 3-rib front tires, 540/1,000 PTO, 3 pt., front weight bracket with slab weights, cab, air, AM/FM radio
4230		9,500	G	$8,700	2/06	SWNE	Diesel, WF, QR, 4-post roll guard canopy, 3 pt., 2 hyd., 540/1,000 PTO, 18.4-38
4230			G	$9,500	11/06	SWCA	120 hp., diesel
4230			G	$9,600	3/06	NCUT	PS
4230			E	$10,000	3/06	NEKS	
4230			G	$11,500	12/06	SEMN	300 hours on major OH, PS, 18.4-34
4230		5,590	G	$11,800	5/06	SWWI	Cab, heat, axle-mount duals, PS, turbo
4230		5,150	G	$14,200	6/06	NWIL	5,150 hours
4230		7,000	G	$16,750	2/06	NECO	QR, Koyker 565 loader, 3,660 hours on OH, 1977 model
4230	1973	10,188	F	$8,200	10/06	NECO	WF, cab, 3 pt., diesel
4230	1973		G	$11,400	7/06	SEND	QR, 2 hyd., 3 pt., rear wheel weights, unknown hours.
4230	1974		G	$8,600	2/06	NCCO	Cab, QR, 18.4-38
4230	1974		G	$9,000	12/06	SCMT	
4230	1974	5,338	G	$9,750	6/06	NWKS	2 hyd., PTO, 3 pt., QR, John Deere 148 loader
4230	1975	6,300	G	$11,250	8/06	NWIL	Diesel, cab, weights, duals
4230	1976		F	$6,250	2/06	NWIN	Syncro
4230	1976	8,408	F	$13,500	3/06	NWOH	Cab, 3 pt., 2 hyd., John Deere 148 loader, snap-on duals
4230	1976	7,350	E	$17,750	2/06	WCIL	2 hyd., PS, 197 hours on a complete OH, 18.4-38
4230	1977	7,425	G	$12,600	8/06	NWIA	Good clean tractor
4230	1977	5,296	G	$14,600	8/06	NEIL	One owner, QR, 18.4-34 tires
4230	1977	5,236	G	$17,000	3/06	ECNE	QR, 18.4-34, 11L-15.5 front tires, 2 hyd., front weights, Sound Guard cab
4230	1977	2,714	G	$19,500	11/06	NCIN	CAH, QR, 18.4-34
4230			P	$3,000	2/05	SCCA	Front weights, canopy, syncro, 3 pt., PTO, 1 hyd., rear weights, 14.9R-46 tires
4230		7,020	F	$8,900	4/05	SCMI	CAH, 3 pt., 2 hyd., 18.4-34, snap-on duals
4230			G	$8,900	10/05	SWWI	
4230			G	$10,500	11/06	SWCA	Front weights, canopy, 3 pt., PTO
4230			G	$11,300	2/05	WCWI	2 hyd., CAH, QR
4230		2,600	G	$12,600	4/05	NWIL	PS, 2 hyd.
4230		6,700	G	$12,600	4/05	WCWI	Cab, clamp-on duals
4230			G	$12,800	9/05	SCMN	
4230			G	$13,000	1/05	ECNE	
4230			G	$15,500	4/05	SCSD	QR, John Deere 740 loader
4230	1973		G	$7,000	4/05	SWPA	Cab, weights, 2 hyd.
4230	1973	8,187	G	$12,250	1/05	SCNE	3 hyd., 540/1,000 PTO, 6 front weights, good mechanically, poor paint
4230	1973		G	$15,000	9/05	SWOH	WF, cab, QR, front weights, 1,000 hours on Redpath, OH
4230	1975	7,025	G	$9,900	3/05	ECMI	105 hp., cab, air, 2 hyd., QR
4230	1976	6,997	G	$12,500	7/05	ECIL	
4230	1976	3,846	G	$17,200	8/05	NCIA	Cab, air, QR, 2 hyd., front weights, quick coupler
4230	1977	9,013	G	$12,750	6/05	WCMN	CAH, PS, 2 hyd., 3 pt., PTO, duals, one owner

John Deere

Model	Year	Hours	Cond.	Price	Date	Area	Comments
4240	1978	4,659	G	$22,500	2/08	SENE	Open station, 8 speed ps, 4,659 hours, 18.4R-38 rear tires, 2 hyd., 3 pt.
4240	1978	7,700	G	$19,750	2/07	NCNE	Farmhand 248 loader, 3 pt., 2 hyd., joystick, loader controls, 110 hp.
4240	1978	7,976	G	$26,000	4/07	NEAB	Canadian sale, 2WD, 110 PTO hp., PS, 3 pt., dual PTO, 3 hyd., 11.00-16SL front tires, 20.8-38 Goodyear radial rears
4240	1978		F	$18,000	2/06	NCIL	CAH
4240	1978	5,273	G	$18,500	12/06	SCMN	QR
4240	1978	3,229	G	$26,000	3/06	SWNE	PS, 3 hyd., PTO, John Deere 158 loader with grapple fork, 18.4R-34
4240	1978	8,000	G	$14,000	6/05	NEND	2WD, CAH, QR, 3 pt., PTO, 3 hyd., front weights, duals
4240	1978	7,730	F	$14,700	2/05	WCOK	3 pt., PTO, 2 hyd., PS
4240	1978	7,457	G	$17,000	7/05	WCMN	Diesel, factory cab, QR, 2 hyd., good rubber and duals
4240	1978		G	$18,500	3/05	SEIA	CAH
4240	1978	4,126	G	$19,500	11/05	NEIA	Diesel, fully equipped factory cab, WF, 3 pt., front weight bracket, original 18.4-38 tires
430				$0	3/08	NCTN	WF
430			G	$4,400	8/06	SCIL	WF, loader, hyd. bkt., good tires, been restored
430	1958		G	$4,600	5/08	ECNE	3 pt., older restoration
430	1958		G	$7,700	5/08	ECMN	Restored, gas, 3 pt., PTO, wheel weights, front weights
430			G	$5,000	12/07	SCMN	
430	1959		G	$2,800	11/07	WCIL	Tag says T version, configured as S version
430	1959		G	$3,650	6/07	NWIL	LP, WF, no 3 pt., good tires
430	1959		G	$4,500	9/07	WCIL	WF, 3 pt., no toplink, just repainted, last year received $2,500 OH
430	1960		G	$4,500	9/07	ECMI	NF, PTO, tricycle, 3 pt.
430			G	$7,250	7/06	WCIL	Runnng, restored, fenders, PTO, lights, front weights, 3 pt.
430			G	$4,250	6/05	ECMN	Row crop, gas, WF, fenders, 3 pt., PTO, 1 hyd.
430			G	$13,750	7/05	SCIA	High crop, LP, good tag, restored, 1 of 183 built, PS, live PTO
430T	1959		G	$9,100	9/05	SEMN	Front and rear weights, fenders, 3 pt.
4320			G	$9,500	1/08	SESD	WF, 3 pt.
4320		8,327	G	$16,000	3/08	SCWI	2 hyd., 3 pt., WF, roll guard canopy
4320	1971	6,081	G	$8,250	4/08	SCMI	Cab, 2 hyd., 3 pt.
4320	1972		E	$12,000	1/08	SWKY	Show tractor
4320	1972		E	$13,750	1/08	SWKY	
4320		8,870	F	$5,600	1/07	WCIL	Syncro
4320			G	$8,100	12/07	WCMN	WF, diesel, open station, 1 hyd., 3 pt., PTO, new Firestone 18.4-38 tires, front weights
4320			G	$8,800	1/08	SESD	3 pt., 2 hyd., no cab
4320		3,400	G	$9,000	7/07	WCSD	Turbo, 2 hyd., 3 pt., starts and runs good, 18.4R-38 tires
4320	1971		G	$8,200	1/07	ECIL	Hi-Master loader and pallet forks, 18.4-38 tires
4320	1971		G	$10,000	2/07	NWKS	WF, diesel, syncro, 4,120 hours on OH, 15.5-38 tires with 13.6-38 duals, John Deere 158 loader and 7' bkt., grapple fork, 2 hyd., 540/1,000 PTO, 3 pt., CAH
4320	1971	6,580	G	$11,500	3/07	NWMO	Syncro, 2 hyd., Year-A-Round cab, 20.8-34 rubber (50%), one owner
4320	1972		F	$9,000	2/07	NENE	WF, cab, tin work fair
4320	1972		G	$9,500	7/07	SEND	John Deere roll guard, 3 hyd., 3 pt., PTO, 20.8-34 singles

John Deere

Model	Year	Hours	Cond.	Price	Date	Area	Comments
4320	1972		G	$9,900	10/07	SWOH	WF, diesel
4320			F	$4,400	3/06	NCUT	
4320			G	$9,900	9/06	NEIA	Factory cab, air
4320		80	G	$22,500	3/06	SEPA	MFWD, John Deere 400 loader, ROPS, warranty
4320	1971	8,311	G	$7,250	6/06	NECO	Cab, WF, rear weights, diesel
4320	1971	7,580	F	$7,400	2/06	NCIL	2 hyd.
4320	1971	9,000	G	$7,700	6/06	SWMN	Cab, 2 hyd.
4320	1971	6,080	G	$10,500	2/06	NCIL	
4320	1971		E	$12,750	8/06	ECIA	Actual hours unknown, diesel, syncro, WF, no cab, 3 pt., 2 hyd., new 18.4-38R, sharp
4320	1971	3,487	E	$18,750	2/06	SWNY	Restored, roll bar, 18.4-38 Firestone tires, syncro, 6 front weights, rear weights
4320	1972	6,299	G	$6,000	4/06	NEND	Factory cab, air, side console, 2 hyd., 540/1,000 PTO, no 3 pt., 20.8-34 singles
4320	1972		G	$9,250	2/06	WCMN	2WD, John Deere cab, clean
4320	1972		G	$9,300	2/06	WCMN	2WD, 18.4-38 singles, sharp
4320	1972	6,992	G	$9,500	10/06	NECO	WF, cab, 3 pt., diesel
4320	1972		G	$9,750	9/06	NEIN	Repainted, diesel
4320	1972	8,630	F	$11,750	2/06	SCNY	2WD, 1 hyd., syncro, no weights, 18.4-38
4320	1972	6,989	G	$13,100	9/06	NECO	Farmhand XL1140 loader, diesel, axle duals
4320			F	$5,400	5/05	SEPA	WF, poor paint, 20% rubber
4320			F	$5,900	2/05	SCMN	
4320			G	$7,500	6/05	ECWI	4WD, diesel, cab
4320			G	$7,750	11/05	WCIL	Standard, diesel, unrestored, fenders, PTO
4320			G	$9,100	2/05	WCWI	Weights, 2 hyd., 3 pt., console
4320		8,102	G	$9,200	3/05	ECNE	Syncro, 2 hyd., duals
4320			F	$9,600	7/05	SCMN	
4320			G	$10,000	2/05	WCIN	18.4-34 tires, duals, 3 pt., 2 hyd.
4320			G	$10,750	2/05	NEIA	711 hours on OH, cab
4320			G	$11,500	2/05	NEIA	1,000 hours on OH, fender, quick hitch
4320			G	$12,650	2/05	ECSD	Syncro, 3 hyd., 18.4-38
4320	1971		G	$9,000	3/05	ECNE	2 hyd., 3 pt., 540/1,000 PTO, WF, 18.4-38, clamp-on duals, rear weights
4320	1971	8,942	G	$9,250	2/05	SWIN	6,518 hours on major OH, 3 pt., front/rear weights, open station
4320	1971	6,194	G	$12,100	9/05	WCIL	Diesel, 2 hyd., fenders, syncro range console shift, weights
4320	1972	7,450	F	$8,500	2/05	WCOK	Hiniker cab, Farmhand F236 loader, 3 pt., PTO, 2 hyd., duals (fair), interior poor
4320	1972		F	$9,500	4/05	NWOK	Ansel cab, 2 hyd., 3 pt., PTO, 8 speed, R134A air, loader (nice loader), diesel
4320	1972		G	$10,250	2/05	SCMN	Factory cab
4320	1972	6,000	G	$10,300	2/05	SCMN	Factory cab
4320	1972	7,400	G	$12,500	7/05	WCMN	500 hours on factory rebuilt engine, diesel, cab, QR, 2½ hyd., average rubber, duals
435			G	$6,750	3/08	NCTN	WF, 3 pt., power adjust wheels, no fenders
435	1959		G	$7,500	11/07	WCIL	GM diesel, running, restored, rear wheel weights, PTO, lights, hyd. remote, deluxe seat, 3 pt., new gauges, steering wheel, seat, new paint and decals, tach and hour meter-digital
435	1959		G	$12,500	11/07	WCIL	Restored
435			G	$12,500	9/06	NEIN	Restored
435	1959	3,096	G	$6,250	11/06	NEKS	1 hyd., 540 PTO, 3 pt., fenders, diesel, 13.6-28 rear tires
435	1960		G	$8,900	10/06	WCMN	Diesel, 44 hp., 2-53 Detroit engine, 3 pt., PTO, PS, WF, lights 13.6-28 tires, good rubber, dynoed on 9-9-06, new paint

John Deere

Model	Year	Hours	Cond.	Price	Date	Area	Comments
435			G	$13,600	7/05	NEIA	Restored, 435D
435			G	$15,500	6/05	ECMN	Row crop, diesel, 3 pt., PTO, rebuilt, painted
4400	1975	3,217	F	$875	9/05	SEIA	Diesel
4430		7,557	G	$12,000	3/08	NWWI	QR, Sound Guard cab, air, duals, 3 pt.
4430		6,000	G	$15,250	3/08	SCMN	Diesel, QR
4430		1,830	G	$17,000	3/08	NEMO	CAH, QR, Westendorf loader, near new rubber
4430	1973		F	$10,700	2/08	NCOH	CAH, QR, 2 hyd., one owner
4430	1973		G	$11,000	2/08	SWMN	3 pt., PTO, 3 hyd., quick coupler, rock box, QR
4430	1974	7,603	F	$8,500	3/08	ECND	CAH, 2WD, syncro, 2 hyd., 3 pt., 540/1,000 PTO, 16.9-38 tires
4430	1974	10,738	F	$8,750	2/08	WCMN	QR, 2 hyd., 3 pt., 540/1,000 PTO, 20.8-34 duals (25%), 11.00-16 front tires (65%)
4430	1974	10,738	F	$8,750	2/08	WCMN	2WD, CAH, QR, 2 hyd., 3 pt., 540/1,000 PTO, 20.8-34 duals, 25% rubber
4430	1974		G	$10,750	2/08	WCMN	2WD, CAH, 3 hyd., 3 pt., 540/1,000 PTO, Goodyear 18.4-38 (70% rubber)
4430	1974	7,500	G	$15,500	2/08	WCMN	2WD, CAH, PS, 2 hyd., 3 pt., 540/1,000 PTO, 18.4-38, axle duals, new front tires
4430	1975	8,510	F	$7,250	3/08	NECO	Diesel, front/rear weights
4430	1975	9,229	G	$10,500	4/08	ECND	CAH, 3 pt., PTO, QR, 18.4-38 band duals
4430	1975	10,000	G	$13,500	3/08	ECND	CAH, QR, 3 hyd., 3 pt., PTO, 18.4-38 hub duals (70%)
4430	1975		G	$14,100	1/08	NEMO	Tach is wrong
4430	1976	8,369	G	$10,400	2/08	WCOK	Cab, air, 3 pt., PTO, 2 hyd., QR, diesel
4430	1976		F	$12,750	3/08	SWKY	MFWD, John Deere 158 loader, bkt., hay spear, 3 hyd., 18.4R-38 tires
4430	1976	7,650	G	$16,800	1/08	WCIL	300 hours on OH, QR, 2 hyd., one owner, aux. tank, weights, diesel
4430	1976	7,000	G	$24,000	3/08	SCKS	John Deere 260 loader, 2 hyd., 3 pt., 540/1,000 PTO, 18.4-38, axle duals, full weights
4430	1977	8,115	G	$15,750	2/08	WCMI	MFWD, CAH, QR, 3 hyd., 3 pt., 540 PTO, 18.4-38R duals
4430	1977	7,402	G	$17,000	2/08	NCOH	CAH, QR, weights, adjustable front axle, 3 hyd.
4430	1977	10,317	G	$17,000	3/08	NECO	Diesel, front/rear weights
4430	1977		G	$19,000	3/08	WCMI	2WD, CAH, QR, 3 hyd., quick coupler, hd 4440 kit, 18.4R-38 axle duals, stone box
4430	1977	5,000	G	$22,500	1/08	NEIA	QR, 2 hyd.
4430	1977	2,940	E	$25,500	1/08	WCIL	Fully equipped cab, QR, dual PTOs, 3 pt., 8 front weights, 18.4-38 tires, sharp!
4430		11,000	G	$0	1/07	NWIN	No sale at $12,600
4430			P	$6,400	4/07	NEIA	Not starting, rear end problems, PTO not working
4430			F	$9,000	8/07	SESD	Diesel
4430			G	$10,000	11/07	SEND	CAH, QR, 3 pt., hub duals
4430		7,093	G	$11,000	3/07	ECIL	CAH, 540/1,000 PTO, 2 hyd., 18.4R-38 tires, QR
4430			G	$11,500	7/07	WCMN	QR, 2 hyd., 18.4-38 singles
4430		7,093	G	$12,000	3/07	ECIL	CAH, 540/1,000 PTO, 2 hyd., 18.4R-38 tires, QR
4430			G	$12,000	1/07	WCIL	Open station, loader
4430			G	$12,800	11/07	SESD	CAH, QR, 3 pt.
4430			G	$13,000	3/07	SECO	John Deere 158 loader, recent OH, QR
4430			G	$13,250	8/07	NENE	QR, cab, air, long axle, 18.4-38 tires, dual PTO, 3 pt.
4430			G	$13,500	11/07	NEND	Front weights, QR, 2 hyd., PTO, 3 pt., 18.4-38 band duals
4430			G	$14,000	2/07	SWIL	
4430		7,553	G	$14,500	3/07	SCNE	QR, changed over to ISO

John Deere

Model	Year	Hours	Cond.	Price	Date	Area	Comments
4430			G	$15,500	2/07	SCWA	PS, 16.9-38, 3 pt., PTO, 2 remotes, 8 front weights
4430		6,599	G	$17,600	8/07	NENE	3 pt.
4430			G	$18,500	4/07	NCNE	Recent OH, good tires
4430		3,730	G	$19,000	3/07	NCIA	One owner, 2 hyd., 18.4-38 rear tires, duals, front tank, quick hitch, front weights
4430			G	$20,500	3/07	WCSD	3 pt., good rubber, cab, quick hitch, 500 hours on recent engine and trans. OH, Koyker 565 quick-tach loader with grapple
4430	1973	10,340	G	$8,500	4/07	ECND	Diesel, CAH, 3 pt., PTO, 18.4-38 tires, QR
4430	1973		G	$9,000	4/07	NEND	CAH, 3 pt., PTO, 2 hyd., unknown hours, 16.9-38 tires
4430	1973	7,871	G	$12,000	3/07	NWMO	Diesel, QR, 2 hyd., 3 pt., cab, air, 18.4-38 rubber (50%), wheel weights, one owner
4430	1973	7,150	G	$13,000	2/07	ECIL	Cab, air, 2 hyd., weights, 18.4-34 tires, shedded
4430	1973	4,842	G	$13,500	3/07	WCMI	Cab, QR, 3 pt., 2 hyd.
4430	1973	6,915	G	$14,000	6/07	SCSK	Canadian sale, 2WD, 126 PTO hp., QR, 3 hyd, 18.4-38 factory duals, new front tires, Beline aux. fuel tank, 6,915 hours showing
4430	1973	6,170	G	$15,300	2/07	ECNE	2 hyd., QR, new cab kit, good paint
4430	1973	8,800	E	$18,200	2/07	NENE	Hub duals, front tank, weights
4430	1974		G	$8,750	12/07	SEWY	Diesel, WF, cab, 3 hyd., tach shows 3,010 hours
4430	1974		G	$9,000	3/07	ECND	CAH, 3 pt., PTO, 2 hyd., QR, 18.4-38 duals
4430	1974		G	$10,500	3/07	NEMI	PS, 2 hyd., new 16.9-38 rears
4430	1974	8,171	G	$11,000	3/07	NEMI	QR, 125 hp., PTO, 2 hyd., 16.9R-38
4430	1974	10,230	G	$11,500	3/07	WCMI	No cab, 3 pt., 2 hyd.
4430	1974		G	$14,250	7/07	SEND	HMFWD, CAH, PS, 2 hyd., 3 pt., quick hitch, PTO, rear weights, OH, 20.8-38 singles
4430	1974	7,288	G	$15,800	12/07	SCNE	Syncro, 2 hyd., CAH, power beyond, dual lift-assist cylinders
4430	1974		G	$21,500	4/07	ECSK	Canadian sale, 2WD, 126 PTO hp., QR, 3 pt., dual PTO, 2 hyd., 18.4-38 clamp-on duals, John Deere 158 loader
4430	1975		G	$9,000	4/07	SCOK	Cab, tires 80%
4430	1975		G	$9,780	3/07	WCIL	New paint and tires, cab, tach shows 5,000 hours
4430	1975		G	$12,750	4/07	SCOK	Cab, duals, tires 70%
4430	1975	5,257	G	$13,700	1/07	ECIL	PS, OH at 4,000 hours, duals
4430	1975		G	$14,000	2/07	NENE	High hours, WF, good rubber
4430	1975	8,424	G	$15,750	8/07	SESK	Canadian sale, 2WD, 3 pt., QR., 2 hyd., dual PTO, 20.8-38 rear with snap-on duals
4430	1975	9,182	G	$16,000	12/07	SCNE	PS, 18.4-38 tires, 9-hole duals, 2 hyd., CAH
4430	1975	4,952	G	$16,000	2/07	WCIL	QR
4430	1976	6,005	F	$8,000	1/07	WCIL	Cab, QR, 540/1,000 PTO, 2 hyd.
4430	1976	7,376	G	$9,300	12/07	SEWY	Diesel, WF, cab, 3 hyd.
4430	1976	7,643	G	$9,900	12/07	SCMN	Syncro
4430	1976		G	$10,250	3/07	NCNE	John Deere 260 loader, 3 pt., 3 hyd., older style grapple fork and 7' bkt.
4430	1976	10,725	G	$11,750	3/07	WCMN	CAH, QR, 2 hyd., 3 pt., quick hitch, PTO, front tank and weights, rear weights, 18.4-38 press steel hub duals, 90% insides
4430	1976		G	$12,100	3/07	WCKS	Great Bend front end loader and grapple fork, new rubber, good condition, new rubber
4430	1976	4,832	G	$21,000	4/07	SWIA	3 hyd., PS, 18.4-38, axle duals set for 30" rows, 6 front weights
4430	1977			$0	11/07	SEND	QR, 3 hyd., John Deere 148 loader
4430	1977		G	$10,000	4/07	SCOK	Cab, tires 65%
4430	1977		G	$10,000	4/07	SCOK	Cab, tires 70%

John Deere

Model	Year	Hours	Cond.	Price	Date	Area	Comments
4430	1977		G	$10,900	1/07	NWIA	QR, 18.4-38
4430	1977	8,600	G	$12,500	1/07	SCNE	
4430	1977	10,000	G	$12,800	11/07	SEND	CAH, QR, 3 hyd., 3 pt., PTO, 18.4-38 singles
4430	1977	10,000	G	$13,100	11/07	ECND	CAH, QR, 2 hyd., 3 pt., PTO, front fuel tank, 18.4-38 hub duals
4430	1977	4,900	G	$15,000	3/07	SEPA	PS, no weights, repainted, 18.4-38 duals
4430	1977		G	$15,250	8/07	SESD	CAH, PS, 3 pt.
4430	1977	6,590	G	$15,500	1/07	WCIL	CAH, hub duals, 2 hyd., QR, 3 pt., weights
4430	1977	4,959	G	$16,750	1/07	NWOH	Cab, air, QR, 2 hyd., axle duals, 6 front weights
4430	1977	5,240	G	$17,200	12/07	NWOH	CAH, power quad
4430	1977	5,411	G	$18,000	1/07	SEKS	PS, 18.4-38 axle-mount duals, 3 hyd., inside rear tires new, new 11L-16 front tires
4430	1977	7,200	G	$19,000	12/07	SWIN	One owner, 125 hp., CAH, 16-speed QR, 18.4R-38 hub-mount duals, 2 hyd., wheel weights, 540/1,000 PTO, quick hitch
4430	1977	6,719	G	$20,250	2/07	NCCO	OH at 5,500 hours, axle duals
4430	1977	4,780	G	$24,500	2/07	SCNE	Dual 3100 loader, QR, duals
4430			F	$6,500	8/06	SCNE	Syncro, open station, 3 pt., PTO, 2 hyd., 24.5-32 tires
4430		5,033	G	$8,000	11/06	SWCA	130 hp., diesel, PTO, 3 pt.
4430			G	$8,000	8/06	SETN	
4430			F	$8,900	8/06	SEMN	
4430			G	$10,750	12/06	NWIL	Diesel, cab
4430		9,932	G	$11,000	3/06	SWMN	CAH, QR, 3 hyd., 3 pt., PTO, rock box
4430		14,000	F	$11,500	4/06	SCKS	John Deere 158 loader, 3 pt., PTO
4430			G	$11,750	7/06	WCOH	Diesel, cab, 1,545 hours on OH
4430		4,778	G	$12,000	10/06	NCKS	QR, 3 pt., PTO, 2 hyd., like-new rubber
4430		6,039	G	$12,300	2/06	ECMN	QR, 3 pt., 2 hyd., quick hitch, one owner
4430		5,234	G	$13,500	4/06	NEIA	QR, one owner
4430		9,870	G	$13,750	3/06	WCMN	QR, 2 hyd., rock box
4430		10,000	G	$14,300	11/06	SWIA	10,000+ hours, PS, hyd., 20.8 rubber
4430			G	$15,000	12/06	SWIL	Cab, air, 3 hyd., repainted, good cab interior, hours unknown
4430			F	$15,500	1/06	NEIA	
4430			E	$20,000	3/06	SEND	John Deere loader, joystick control, quick-tach bkt., 4-tine grapple, CAH, QR, 1,000/540 PTO, 2 hyd., 2,700 hours on major OH and complete check over, heavy 9-bolt rims
4430		3,357	G	$23,500	1/06	ECIA	
4430		8,200	G	$25,500	3/06	WCSD	Farmhand F258 loader, 8' bkt., PS, diesel, CAH, radio, 3 pt., 3 hyd., new batteries, nice 20.9-34 tires, shedded
4430	1973		F	$8,100	2/06	NCKS	QR, dual PTO, 3 pt., 2 hyd., uses lots of oil, 18.4-38 (40%)
4430	1973	7,680	F	$8,500	5/06	SEWY	Cab, rear weights, diesel
4430	1973		G	$11,000	8/06	WCMN	QR, CAH, 2 hyd., rock box, 18.4R-38
4430	1973		G	$11,250	8/06	WCMN	No cab, QR, diesel, 2 hyd., 3 pt., 18.4-34, rock box, sharp
4430	1973	9,520	G	$11,250	3/06	ECNE	QR, 18.4-38, 10:00-16 front tires, 2 hyd., rear weights, 6 front weights, Sound Guard cab
4430	1973	8,887	G	$11,750	2/06	NEIN	CAH, 2 hyd., new tires with inner cooler, 3,000 hours on OH
4430	1974		P	$5,900	1/06	NECO	3 pt., PTO, PS
4430	1974		G	$10,600	7/06	WCMN	QR, 2 hyd., 38" rubber, fresh clutch and hyd. pump, 38" rubber
4430	1974	10,263	G	$11,500	8/06	WCMN	QR, 2 hyd., 18.4-38
4430	1974	4,478	G	$14,500	1/06	NEIN	New engine, cab, air, QR

John Deere

Model	Year	Hours	Cond.	Price	Date	Area	Comments
4430	1974	3868	E	$19,000	3/06	NEMO	Sound Guard cab, one owner
4430	1975	1,408	G	$11,250	3/06	ECND	QR, 3 hyd., power beyond, 540/1,000 PTO, front suitcase weights and fuel tank, 18.4R-38 duals
4430	1975	11,800	G	$11,500	4/06	NWMN	CAH, QR, 2 hyd., 3 pt., quick hitch, 540/1,000 PTO, OH at 9,800 hours, 18.4-38 tires
4430	1975	5,125	G	$16,000	2/06	NCIN	18.4R-38 factory duals, QR, quick hitch
4430	1976		F	$8,800	3/06	NWOH	Trans. problem, cab, 2 hyd., 3 pt., 6 front weights, QR, snap-on duals
4430	1976		F	$9,750	2/06	WCMN	2WD, CAH, 1,562 hours on OH, 18.4-38
4430	1976	9,100	G	$10,500	7/06	NEND	QR, 2 hyd., 3 pt., 540/1,000 PTO, front fuel tank, 18.4-38 band duals
4430	1976	10,809	G	$11,000	11/06	ECND	QR, 3 pt., quick hitch
4430	1976	8,800	G	$11,500	7/06	SEND	QR, 2 hyd., 3 pt. quick hitch, 540/1,000 PTO, new hub duals, one owner
4430	1976	10,725	G	$11,750	4/06	ECND	CAH, QR, 2 hyd., 3 pt., quick hitch, PTO, front tank and weights, rear weights, 18.4-38 press steel hub duals, 90% insides
4430	1976	6,459	G	$12,100	3/06	WCIL	QR, 18.4-38, shedded
4430	1976		G	$13,250	1/06	SCMI	1,500 hours on OH, QR, cab, 3 pt., 2 hyd., axle duals
4430	1976	2,100	G	$13,500	2/06	WCIA	Duals
4430	1976	5,342	G	$15,200	8/06	NEIL	QR, one owner, Shoup steps
4430	1976	4,858	G	$16,750	3/06	ECIL	Cab, air, 2 hyd., QR, front/rear weights, quick hitch, new 18.4-38, axle duals
4430	1976	4,140	G	$17,250	2/06	SWOH	2 hyd., 18.4-38, duals, ext. lights, quick hitch, 8 front weights, 4 rear weights
4430	1976		E	$17,500	8/06	ECIA	Diesel, QR, WF, factory cab, 3 pt., 2 hyd., ISO couplers, 18.4-38 rear tires, sharp
4430	1976	6,053	G	$17,750	3/06	NESD	PS, 3 pt., 2 hyd., new 134 air and compressor, hub duals
4430	1976	6,659	G	$19,000	1/06	SCKS	QR, 18.4-38R, 9-bolt duals, 3 pt., 2 hyd., 1,000 lbs. rear weights, R134A
4430	1976	3,330	E	$26,000	4/06	SEMN	3 pt., 2 hyd.
4430	1977	4,300	G	$15,800	8/06	NWIL	Diesel, cab, weights, duals
4430	1977	5,435	G	$15,900	12/06	WCIL	Fully equipped cab, QR, 2 hyd.
4430			G	$1,973	3/05	NCMT	Sound Guard cab, QR, unknown hours, 3 pt., 3 hyd., PTO, 14.9R-46 tires, 125 hp.
4430			F	$6,800	8/05	NECO	
4430			F	$7,000	3/05	SEIA	
4430			F	$8,200	3/05	NWIL	Cab, diesel
4430		9,377	F	$8,700	12/05	WCOH	18.4-34, duals
4430			F	$9,000	4/05	WCMN	Sound Guard cab, 2 hyd., diesel, high hours, jumping out of gear
4430		8,029	F	$9,000	4/05	NCOH	CAH, 2 hyd., 38" duals
4430			F	$9,500	1/05	ECMI	2WD
4430			F	$9,500	1/05	ECMI	2WD
4430			F	$9,600	7/05	SCMN	
4430			G	$10,900	2/05	SCMN	
4430			G	$12,000	3/05	SWCA	Front weights, PS, 3 pt., PTO, 2 hyd.
4430			G	$12,000	11/05	NCCA	Saddle tanks, cab
4430			F	$12,200	4/05	SCSD	QR
4430		7,753	G	$12,400	2/05	ECIL	QR, 2 hyd.
4430		8,700	G	$14,000	11/05	NEIA	OH, 2 hyd., QR, 18.4-38 tires
4430			F	$14,500	12/05	SWWI	QR
4430			G	$14,800	3/05	NECO	Farmhand F258 loader, good tires
4430		6,708	G	$15,200	12/05	WCIN	Loaded, CAH, QR, hub-mount duals, 700 hours on lower end

John Deere

Model	Year	Hours	Cond.	Price	Date	Area	Comments
4430		1,875	G	$19,000	11/05	NWOH	Cab, QR, 3 pt., PTO, 18.4-38 duals
4430	1973		F	$6,700	2/05	NEIN	Diesel, CAH
4430	1973	9,282	E	$7,250	4/05	NCOK	Snap-on duals, PTO, no 3 pt., front weights, rock shaft less arms, wide swing drawbar, 8-speed ps, 2 hyd.
4430	1973		P	$8,000	2/05	NWIL	Year-A-Round cab, 2 hyd., less than 1,500 hours on major OH, 18.4-38
4430	1973		G	$8,500	3/05	ECID	MFWD, QR, 2 hyd., 540/1,000 PTO, 3 pt., 16.9-38
4430	1973		G	$8,500	3/05	ECID	MFWD, QR, 2 hyd., 540/1,000 PTO, 3 pt., 16.9-38
4430	1973	9,100	G	$11,000	3/05	SEWY	Diesel, cab, 3 pt.
4430	1973		G	$12,100	8/05	WCMN	1,956 hours on OH, rock box, wheel weights, 3 pt., 2 hyd., 540/1,000 PTO, 18.4-38, duals
4430	1973	5,000	G	$13,600	3/05	NCCO	3 hyd.
4430	1973	6,745	G	$14,200	1/05	NEIA	Fully equipped cab, deluxe step, 38" near-new tires, good rubber all around
4430	1973	4,304	G	$14,500	8/05	NCIA	Cab, air, QR, front tank and weights, 2 hyd., quick coupler, 18.4-38 duals
4430	1973	7,000	G	$16,000	3/05	ECCO	Koyker 565 loader, 3 pt., PTO, 2 hyd., duals
4430	1973	5,550	E	$17,600	1/05	NENE	One owner, QR, axle duals
4430	1974	7,613	P	$4,500	1/05	SWOH	MFWD, canopy
4430	1974	10,000	F	$9,600	4/05	NWKS	QR, 2 hyd., 3 pt., PTO
4430	1974		F	$10,500	7/05	SCMN	20.8-38R tires, very clean
4430	1974	10,500	G	$12,000	7/05	NCMN	2WD, QR, cab, 3 pt., 4 hyd., quick hitch, band duals
4430	1974		G	$12,000	7/05	NCMN	QR, cab, 3 pt., 4 hyd., quick hitch
4430	1974	4,419	G	$14,000	10/05	NCND	CAH, 20.8-38R tires with 18.4-38 band duals
4430	1974	1,625	E	$19,250	8/05	SEIA	QR, 2 hyd., 6 front weights, 18.4-38 tires
4430	1975		G	$8,000	6/05	ECIL	Rice/cane, quick hitch, syncro, front/rear weights
4430	1975		G	$13,100	7/05	SESD	CAH, PS
4430	1975		G	$13,500	3/05	ECNE	PS, 2,400 hours on OH, 2 hyd., 3 pt., 540/1,000 PTO, 5,000-pound rear weights, R134 air, 18.4-42R rears, 11:00-16 fronts, front fenders
4430	1975	7,583	G	$13,900	11/05	SCIA	QR, CAH, 2 hyd., 18.4-38
4430	1975	5,113	G	$14,400	4/05	NCIA	Cab, air, QR, quick coupler, 2 hyd., front weights, extra steps
4430	1975		F	$14,750	6/05	NEND	QR, 3 hyd., 540/1,000 PTO, no 3 pt., 18.4-38 singles, 90% rubber
4430	1975	3,695	G	$16,600	8/05	NWIL	
4430	1976		P	$4,100	12/05	NWIL	Project tractor, rough, 18.4-34
4430	1976		F	$5,300	2/05	NEIN	ROPS, diesel
4430	1976	9,732	F	$6,100	8/05	SEMI	No cab, 2 hyd., 3 pt.
4430	1976		G	$8,500	3/05	ECID	MFWD, QR, 2 hyd., 540/1,000 PTO, 3 pt., 14.9-38 tires
4430	1976	9,800	F	$12,500	2/05	WCIA	QR, 18.4-38
4430	1976	6,980	E	$13,750	4/05	NCKS	Cab, air, QR, 2 hyd., 540/1,000 PTO, 8 front suitcase weights, extra rear weights, 20.8-38 rears with clamp-on duals, underhaul at 4,000 hours, 40 Series rear end and radiator
4430	1976	8,900	E	$15,000	3/05	SCIL	CAH, QR, 2 hyd., duals, chrome stacks
4430	1976		G	$16,000	3/05	ECND	2WD, CAH, QR, 2 hyd., elec. splitter, 3 pt., QH, 540/1,000 PTO, front fuel tank, 18.4-38 press steel duals (50% rubber)
4430	1976	10,465	G	$16,000	4/05	NENE	OH at 9,200 hours, 2 hyd., lift assist, power beyond, quick hitch, 18.4-38, 10:00-16 front tires

John Deere

Model	Year	Hours	Cond.	Price	Date	Area	Comments
4430	1976		G	$16,000	12/05	NEIA	PS, 2 hyd., air, straddle duals, 3 pt.
4430	1976		G	$16,200	2/05	NWKS	QR, 2 hyd., John Deere 158 loader, 8' bkt., grapple fork, 18.4-38, rear weights, cab
4430	1976	5,081	E	$18,100	12/05	NCOH	Diesel, WF, CAH, 2 hyd., 3 pt., syncro, super clean
4430	1976		G	$18,750	3/05	NECO	1,702 hours on OH, cab, 3 pt., PTO, 2 hyd., 18.4-38, duals
4430	1977		F	$7,900	9/05	ECND	QR, 3 hyd., front fuel tank, 540/1,000 PTO, 14.9-38 tires
4430	1977		G	$15,500	9/05	WCIL	Diesel, cab, 2 hyd., QR, front weights
4430	1977		G	$16,750	3/05	NWMN	2WD, CAH, QR, 3 pt., 540/1,000 PTO, 3 hyd., band duals, push tank and 60 Series steps
4430	1977	6,591		$17,100	1/05	SEIL	Cab, air, 2 hyd., quick hitch, heavy rear end, 18.4-38
4430	1977	5,600	E	$17,500	2/05	WCIL	Diesel, aux. fuel tank, QR, 18.4-38R, cab
4430	1977	4,200	G	$24,300	2/05	SWIN	3,875 hours on major OH, CAH, QR, 3 pt., duals, wheel weights and front weights
4430	1977	2,252	E	$24,500	4/05	SEMN	QR, one owner, always shedded, 20.8-38, duals, full set front weights, new style step, 3 pt. quick hitch, 2 hyd., 540/1,000 PTO
4440	1978		G	$14,000	3/08	ECND	MFWD, 3 pt., PTO, 3 hyd., 14.9-46 tires
4440	1978	6,870	G	$21,325	2/08	WCIL	Cab, QR, duals
4440	1978	7,450	F	$9,000	4/07	NWPA	MFWD, cab, 130 hp., QR, 540/1,000 PTO, 2 hyd.
4440	1978	12,694	F	$13,500	4/07	ECAB	Canadian sale, 2WD, 130 PTO hp., QR, dual PTO, 3 hyd., 20.8-38 rear, 12,694 hours showing, has approx. 2,300 hours on rebuilt engine
4440	1978	10,200	G	$15,500	7/07	SEND	CAH, QR, 3 hyd., 3 pt., 540/1,000 PTO, front weights, 16.9-38 band duals
4440	1978	8,950	G	$18,000	11/07	ECNE	QR, 18.4R-38 rear tires with 10-bolt duals and hubs, 5" spacers, 11:00-16 front tires, 2 hyd., 3 pt., 8 front weights, Sound Guard cab, R134 air, AM/FM/cassette
4440	1978	7,138	G	$18,700	6/07	SCMN	PS, CAH, 3 pt., 2 hyd., rock box, 18.4R-38, 600 hours on OH
4440	1978	3,397	G	$25,000	3/07	NWIA	One owner, 2 hyd., 18.4-38, no duals
4440	1978		F	$10,000	9/06	SEIA	QR, 2 hyd.
4440	1978		F	$15,000	12/06	NWWI	PS, $18K spent on reconditioning in 2003, 20.8-38 tires
4440	1978	9,825	G	$16,250	3/06	ECMI	CAH, 500 hours on John Deere OH, 2 hyd., PS, 18.4-38
4440	1978	8,492	F	$17,000	3/06	NCCO	8-speed ps, 2 hyd., PTO, purchased as a demo when it had 1,200 hours on it, 18.4-38, axle duals
4440	1978	6,357	G	$18,500	2/06	ECIL	QR, 18.4-38 duals
4440	1978	6,727	G	$21,000	4/06	WCMT	CAH, 130-hp. diesel engine, 16-speed QR, 3 pt., dual PTO, 2 hyd., 18.4-38 duals, 12 front weights
4440	1978	5,600	G	$22,500	4/06	SEMI	Cab, PS, 3 pt., 2 hyd., 20.8-38
4440	1978		G	$24,750	2/06	SENE	PS, 3 hyd., 3 pt., 1,157 hours on complete engine and trans. OH, front fenders, 6 front weights, 18.4R-38 (70%), 50 Series front end lights
4440	1978	2,743	G	$25,000	1/06	WCIL	Duals, 8-speed ps, 2 hyd., 3 pt.
4440	1978	6,611	E	$28,500	12/06	WCIA	OH at 6,034 hours, PS, 18.4R-38, axle duals
4440	1978		G	$14,000	3/05	NCMT	Sound Guard cab, QR, rebuilt motor, 3 pt., PTO, 3 hyd., 18.4-38 tires, 130 hp.
4440	1978	8,797	G	$18,750	1/05	ECNE	1,200 hours on OH, PS, 18.4-38, 3 hyd., power beyond, R134A air, AM/FM
4440	1978	6,344	F	$19,250	1/05	NWIL	QR, power beyond

John Deere

Model	Year	Hours	Cond.	Price	Date	Area	Comments
4440	1978	7,144	E	$19,250	12/05	ECIL	
4520			F	$5,800	1/08	NEMO	One owner
4520	1969		G	$8,500	1/08	NWOH	CAH, front weights, duals, 2 hyd., front fenders
4520	1970		G	$6,500	4/08	NEIA	Diesel, side console, John Deere 4630 engine, Sound Guard cab
4520			F	$4,250	11/07	SESD	Koyker loader
4520	1969		G	$4,900	7/07	WCMN	Cab, 18.4-38 duals, 3 pt., quick hitch
4520	1969	5,766	G	$6,500	1/07	ECIL	Fender tractor, 20.8-38, duals
4520	1970	8,426	G	$5,500	7/07	WCMN	Syncro, 3 hyd., cab, heater, no air, 1,000 PTO, 20.8-38 duals
4520	1970	6,620	G	$6,700	1/07	WCIL	John Deere 4620 engine, cab, new engine, 2 hyd., 3 pt., 1,340 hours on new engine
4520			F	$7,500	4/06	SWPA	700 hours on OH, 2 hyd., needed new tires
4520		5,735	G	$7,900	10/06	SEMN	Open station, side console, 3 pt., 2 hyd.
4520	1969	3,589	G	$4,700	11/06	NEKS	WF, syncro
4520	1969	5,807	G	$4,900	7/06	NCIL	Syncro, side console, 2 hyd., 18.4-38 clamp-on duals (duals $175)
4520	1969	5,000	G	$6,900	7/06	SEND	Cab, air, 3 hyd., 3 pt., PTO, aux. fuel tanks, single tires
4520	1970		G	$12,500	6/06	SCMN	Open station, axle duals, 3 pt., 2 hyd., 1,000 PTO
4520	1969		G	$5,250	2/05	NCIN	4,620 updates, 500 hours on engine complete OH, diesel
4520	1970		G	$13,500	2/05	SCKS	3 pt., PS, roll guard, 4620 motor
4620		11,032	G	$6,200	4/08	NCND	8-speed syncro, 2 hyd., on-side console
4620	1971	8,315	G	$6,900	2/08	WCMN	2WD, CAH, syncro, 2 hyd., 3 pt., 540/1,000 PTO, Firestone 20.8-38 duals (40% rubber), 11.00-16 fronts (70%), rock box
4620	1972	5,784	G	$7,200	3/08	ECNE	Cab, syncro, 5,784 hours, 18.4R-38 rear tires, 10:00-16 4-ribbed front tires, 2 hyd., 540/1,000 PTO, quick hitch
4620			F	$3,100	7/05	SESD	
4620			G	$7,100	8/07	NWIA	
4620			G	$7,100	8/07	NWIA	
4620	1971	4,328	G	$12,900	2/07	WCIL	Cab, syncro, 18.4-42 tires
4620	1972		G	$4,300	4/07	NCNE	
4620	1972		G	$11,900	1/07	ECIL	PS, John Deere 148 loader, forks, duals
4620		7,847	G	$7,000	8/06	SCMN	
4620	1970	8,997	G	$6,500	3/06	NEND	Side console, syncro, 3 hyd., added hd 3 pt., quick hitch, 1,000 PTO, diesel, 20.8-38
4620	1972	6,747	G	$8,500	7/06	NCIL	Syncro, $8,536 spent on OH, John Deere CAH, 2 hyd., 20.8-38 axle duals
4620	1972		G	$12,500	7/06	NWIL	Running, restored, PTO, factory cab, lights, deluxe seat, 2 hyd., big and beautiful, ready for parade, show or go back to field, very strong, new hoses and belts, new John Deere batteries, new OEM seat
4620	1972		G	$12,500	7/06	WCIL	Running restored, PTO, factory cab, cab, lights, deluxe seat, 2 hyd., new hoses and belts
4620			F	$4,400	4/05	SCSD	
4620			G	$8,500	12/06	WCWI	
4620	1971	9,079	F	$3,700	3/05	SEWY	Cab, diesel, no 3 pt.
4620	1971		G	$7,200	2/05	NEIN	Diesel, WF, no cab
4620	1971	6,320	G	$9,450	3/05	ECNE	PS, 3 hyd., 3 pt., 1,000 PTO, diesel, PS, cab, console, 18.4-38, 11:00-16 fronts
4630			F	$5,000	3/08	NEAR	Cab, air, weights, 18.4-38 tires, 3 pt., PTO, 2 hyd.
4630			F	$6,400	1/08	SESD	Open station, 3 pt., syncro, duals

John Deere

Model	Year	Hours	Cond.	Price	Date	Area	Comments
4630			G	$11,000	1/08	WCIL	CAH, PS, diesel, 250 hours on OH, weights
4630		9,681	G	$14,000	3/08	ECSD	About 2,000 hours on OH, 3 hyd., 1,000 PTO, 8-speed ps
4630		6,909	G	$14,500	2/08	SCMN	Syncro, 8 speed, CAH, 2 hyd., quick hitch, rock box, 1,000 PTO, hub duals
4630		5,500	G	$21,000	2/08	NCOH	Cab, 3 pt., 1,000 PTO, 3 hyd., duals
4630	1975		F	$8,600	3/08	SCNE	QR, 2 hyd.
4630	1975		G	$12,500	3/08	NEMO	CAH, axle duals, QR, 2 hyd.
4630	1975		G	$20,500	1/08	WCKS	QR, duals, 3,000 hours on motor OH, 3 pt., quick hitch, dual 3000 loader
4630	1976		F	$8,500	2/08	WCOK	Syncro, 2 hyd.
4630	1976	3,505	G	$13,500	3/08	ECND	CAH, 2WD, QR, 3 hyd., 3 pt., 1,000 PTO, 18.4-38 singles
4630	1976		G	$15,000	3/08	NEMO	CAH, clamp-on duals, QR, 2 hyd.
4630	1977	6,399	F	$4,100	1/08	SCGA	Cab (busted window), 2 hyd., 18.4-38 tires (30%), needs cab kit
4630	1977	9,100	G	$10,750	2/08	ECMN	QR, 2 hyd., 3 pt., PTO, 20.8-38 duals (80% rubber)
4630	1977		G	$12,000	3/08	SCKS	4,300 hours since OH, 3 hyd., 18.4-38 duals, 3 pt., 1,000 PTO, radar
4630	1977	6,335	G	$18,250	1/08	NWIL	2 hyd., OH at 3,000 hours, QR, 6 front weights
4630			G	$4,500	4/07	ECSK	Canadian sale
4630			G	$4,500	4/07	ECSK	Canadian sale
4630			G	$8,750	11/07	SEND	CAH, 3 hyd., diff. lock, 1,000 PTO, 20.8-38 singles (70%)
4630			G	$17,250	3/07	WCSD	3 hyd., 3 pt., syncro, duals, front weights, quick hitch, low hours on OH, Koyker 565 quik-tach loader and grapple
4630		6,000	G	$18,750	4/07	ECSK	Canadian sale, 2WD, QR, duals, approx. 6,000 hours showing (2,600 hours on new motor)
4630	1974		F	$5,800	3/07	WCIL	Sound Guard body, front fuel tank, weights, syncro, 18.4-38 duals
4630	1974	5,690	F	$7,500	3/07	SEWY	Diesel, duals, rear weights
4630	1974	5,692	F	$8,100	5/07	SEWY	Duals, diesel
4630	1975	12,000	G	$7,750	3/07	WCKS	3 pt., duals, OH and new clutch at 7,000 hours
4630	1975	5,176	F	$8,500	3/07	SEWY	Weights, diesel
4630	1975		F	$8,500	1/07	NWIA	QR, 18.4-38
4630	1975	5,185	G	$8,900	5/07	SEWY	Quick hitch, diesel
4630	1975		F	$9,250	5/07	SEWY	MFWD, Koyker 600 loader, diesel
4630	1975	5,077	G	$17,200	11/07	NWIL	Axle duals, 3 pt. hitch with quick hitch, 10 front weights
4630	1975	6,375	G	$17,500	6/07	NENE	PS, 18.4R-38 rear tires, 11-16 front tires, 2 hyd., 1,000 PTO, 3 pt., quick hitch, cab, air, 2 front end weights
4630	1976		G	$7,500	10/07	NECO	Diesel, duals, front weights, quick hitch
4630	1976		F	$7,500	3/07	SEND	PS, 3 hyd., quick coupler, PTO, front rock box, 18.4-38 hub duals
4630	1976		G	$8,000	9/07	NECO	Duals, front weights, diesel, tach shows 2,020 hours
4630	1976		F	$9,000	4/07	NECO	WF, duals, front weights, diesel, new tach
4630	1976	7,165	G	$9,500	1/07	WCIL	CAH, 3 hyd., QR, 3 pt., weights
4630	1976	9,220	G	$13,250	2/07	ECNE	PS, 18.4R-38 rear tires with cast duals, 10:00-16 front tires, 2 hyd., quick hitch, 8 front end weights
4630	1977	9,000	G	$9,000	7/07	SEND	QR, 3 hyd., 3 pt., quick hitch, PTO, 20.8-38 duals

John Deere

Model	Year	Hours	Cond.	Price	Date	Area	Comments
4630	1977	10,183	G	$9,750	1/07	SWNE	QR, 18.4R-38, 10-bolt axle-mount duals, 3 hyd., quick hitch, 1 set of inside rear wheel weights, 14 front end weights
4630	1977		G	$10,000	9/07	WCIL	42" rubber, nose fuel tank, runs, used to mow road ditches
4630	1977	7,071	G	$10,200	2/07	ECMI	Cab, 14.9-46 duals, QR, 3 hyd.
4630	1977	8,600	G	$12,100	2/07	SWIL	1,600 hours on OH
4630	1977		G	$12,750	3/07	SEWY	Great Bend loader and grapple, diesel
4630	1977	8,700	G	$15,000	1/07	SCNE	QR, 3 hyd., 20.8×38 tires & duals, 950 hours on OH, 1,200 hours on trans.
4630	1977	7,608	G	$20,000	1/07	NEIA	Cab, heat, QR, 3 hyd., front fuel tank, duals, quick hitch
4630			F	$7,200	8/06	SEMN	Open station
4630			G	$8,000	12/06	ECIN	Diesel, PTO, 3 pt., WF
4630			G	$8,500	12/05	ECIN	Diesel, PTO, 3 pt., WF
4630		6,945	F	$9,400	6/06	NCIL	1976 model, PS, 2 hyd., 18.4-38 axle duals, quick hitch
4630			G	$9,900	3/06	NWIL	Diesel
4630		8,136	G	$10,500	11/06	SCSD	PS, 3 hyd., 3 pt., rock box, 20.8-38 tires
4630		6,225	G	$13,500	11/06	ECIL	QR, 2 hyd., 10-bolt hubs
4630			G	$13,700	12/06	WCMN	7,234 hours, cab, QR, rock box, 480/80R38 duals
4630			G	$18,000	12/06	SCMN	PS, 18.4-42 duals 90%
4630			G	$18,000	12/06	SCMN	
4630	1973	9,535	F	$7,400	7/06	SEND	3 hyd., 3 pt., PTO
4630	1973	5,143	F	$7,600	7/06	SEND	CAH, PS, 3 hyd., 3 pt., 1,000 PTO, duals, front fuel tank
4630	1973	10,320	G	$8,000	11/06	NWKS	
4630	1973		P	$8,300	2/06	WCIL	ROPS, PS, IHC loader, no cab
4630	1973	4,652	G	$16,000	3/06	ECND	QR, 3 hyd., power beyond, 1,000 PTO, front weights, 14.9-46 duals
4630	1974	9,630	F	$6,800	11/06	ECND	CAH, QR, 2 hyd., 3 pt., PTO
4630	1974	9,000	G	$7,700	8/06	WCIL	QR, 2 hyd.
4630	1974	8,211	G	$11,000	3/06	NCND	CAH, 8-speed ps, 3 hyd., no 3 pt., 1,000 PTO, aux. fuel tank, new hyd. pump, 14.9-46 singles, one owner
4630	1974	3,906	G	$13,500	3/06	NWKS	18.4R-38, cast duals, 14L-16.1 front tires, 2 hyd., 3 pt., quick hitch, belly fuel tank, rear wheel weights, 10 front end weights, AM/FM radio, air, one owner
4630	1974	4,207	G	$16,300	8/06	NEIL	QR, Shoup steps, 18.4-38 rubber
4630	1975	8,000	G	$8,500	12/06	WCMN	QR, duals, 2 hyd., PTO, 3 pt., diff. lock, front weights
4630	1975	8,000	G	$8,700	12/06	WCMN	Duals, 3 pt., PTO, 2 hyd., QR, CAH
4630	1975	9,700	G	$8,900	8/06	WCMN	Complete OH, QR, 3 pt., PTO, 3 hyd.
4630	1975		F	$9,700	2/06	NCCO	PS, 5,687 hours on rebuilt engine and trans., 18.4-38, axle duals
4630	1975	10,000	G	$10,000	2/06	SWNE	8-speed ps, 3 pt., quick hitch, 3 hyd., 1,000 PTO, 18.4R-42 duals
4630	1975	5,357	G	$10,250	4/06	NEND	CAH, 8-speed gear, 2 hyd., 1,000 PTO, no 3 pt., 60 Series step, 20.8-38 band duals (90%)
4630	1976	8,700	G	$6,500	7/06	NEND	QR, 2 hyd., no 3 pt., 1,000 PTO, front fuel tank, 20.8-38 cast duals
4630	1976	10,000	F	$7,500	10/06	NCND	CAH, 2 hyd., 1,000 PTO, 20.8-38 band duals
4630	1976		F	$7,750	12/06	NEKS	Diesel, 8-speed syncro, Year-A-Round CAH, 3 pt., 1,000 PTO, 20.8-38, approx 9,700 hours, OH 1,200 hours ago
4630	1976	11,800	G	$8,400	7/06	SEND	2 hyd., 3 pt., quick hitch, duals

John Deere

Model	Year	Hours	Cond.	Price	Date	Area	Comments
4630	1976	8,700	G	$10,000	8/06	WCMN	QR, 20.8-38 duals, 2 hyd., 3 pt., quick coupler, 1,000 PTO, CAH
4630	1976	5,886	F	$10,200	2/06	ECIL	QR, CAH, weights, 2 hyd., 3 pt.
4630	1976	8,636	G	$10,750	6/06	WCMN	20.8-38 10-bolt duals, 1,000 PTO, CAH, QR, 2 hyd., rock box
4630	1976	5,900	G	$17,100	1/06	NEIN	Cab, air, QR
4630	1977		G	$12,000	5/06	SEWY	Great Bend 770 loader and grapple, diesel
4630	1977	7,064	G	$13,500	3/06	NWOH	Cab, 2 hyd., 10 front weights, PS, axle duals
4630	1977	6,200	G	$16,500	6/06	SWMN	Cab, duals, 2 hyd.
4630	1977	3,403	G	$16,800	3/06	ECIL	Cab, air, 2 hyd., QR, front/rear weights, 20.8-38, axle duals
4630	1978		G	$15,700	2/06	SEIN	18.4-34R duals, QR, 500 hours on remanufactured engine
4630			F	$4,600	2/05	SCMN	PS
4630			P	$5,700	12/05	SEMN	Blown engine
4630		10,387	F	$6,000	9/05	NWOR	Diesel, duals
4630			G	$8,100	3/05	NWIL	Cab, air, diesel
4630			G	$9,750	3/05	NWIL	Weights, diesel
4630			G	$9,900	3/06	NWIL	Diesel
4630		7,209	G	$11,300	9/05	NWIL	PS, 2 hyd., 3 pt.
4630			F	$14,500	4/05	SCSD	QR, MFWD
4630	1973	4,800	G	$9,250	2/05	WCIL	CAH
4630	1974		G	$13,100	1/05	ECIL	PS, front/rear weights, quick hitch, front fuel, new duals, OH, 18.4-38, duals
4630	1974		G	$15,250	3/05	SEWY	1,738 hours on newer tach, diesel, cab, 3 pt.
4630	1975	6,708	G	$11,000	8/05	NCIL	QR, 2 hyd., 18.4-38 duals, 500 hours on major OH
4630	1975	12,490	E	$12,500	4/05	NCOK	QR, front weights, 2,500 hours on reconditioned tranny, 2 hyd., quick coupler, PTO, 3 pt., axle duals
4630	1976		G	$9,300	7/05	SCMN	QR, rubber 85%
4630	1976	6,336	G	$10,000	6/05	NWMN	CAH, 8-speed ps, 3 pt., PTO, 3 hyd.
4630	1976	8,316	G	$11,250	9/05	NWIA	2 hyd., 3 pt., diesel
4630	1976	10,000	G	$13,000	7/05	NCMN	QR, CAH, 3 pt., PTO, 4 hyd., quick hitch, band duals, 4,000 hours on new motor OH
4630	1976	3,500	G	$13,000	7/05	NCMN	CAH, 3 pt., PTO, 4 hyd.
4630	1977	5,700	F	$8,000	12/05	SCIA	QR, no forward gears
4630	1977	5,131	G	$11,800	8/05	NWIL	Cab, air, quik hitch, 10 front weights, good 20.8-38R tires, clamp-on duals, 2,000 hours on major OH
4630	1977	4,700	G	$13,300	9/05	WCIL	Diesel, cab, weights, 2 hyd., QR, quick hitch
4630	1977		G	$15,500	3/05	SEWY	977 hours on new tach, diesel, cab, 3 pt.
4640	1978	12,000	F	$16,000	3/08	WCKS	PS, 3 pt., PTO, duals, Ezee-On loader
4640	1978	1,709	G	$16,500	3/08	ECND	2WD, CAH, QR, 3 hyd., 3 pt., 1,000 PTO, 14.9-46 singles
4640	1978	6,263	G	$17,500	12/07	SEIA	Cab, PS, 2 hyd.
4640	1978	5,750	G	$18,000	12/07	SWIN	Second owner at 1,600 hours, 156 hp., CAH, 16-speed QR, 18.4-42 hub-mount duals, 3 hyd., 10 front weights, OH at 4,700 hours, paperwork available
4640	1978	4,516	G	$19,500	8/07	NWIA	Diesel, PS, 14.9R-46 rubber with axle duals, 2 hyd., 3 pt. quick hitch, 1,000 rpm
4640	1978		F	$8,800	7/06	WCMN	MFWD, QR, duals, quick hitch, 1,000 PTO
4640	1978	8,775	G	$10,000	11/06	ECND	MFWD, CAH, QR, 3 hyd., 3 pt., PTO, 20.8-38 band duals
4640	1978		F	$12,700	1/06	SCNE	QR, 3 hyd., 6 front weights, new 18.4-42, duals, accessory powerstrip

John Deere

Model	Year	Hours	Cond.	Price	Date	Area	Comments
4640	1978	11,620	G	$13,500	2/06	NECO	QR, duals, 2,500 hours on OH, 11,620 total hours
4640	1978	6,778	G	$18,000	8/06	NWIA	Cab, 18.4-38 axle-mount duals, OH at 5,000 hours, nice
4640	1978	7,647	G	$21,000	12/05	WCIA	PS, 3 hyd., fuel tank
4640	1978	4,647	G	$22,000	1/05	NWIL	1,968 hours on complete OH, cab, 2 hyd., quick hitch, 12 front weights, 18.4-38 hub duals
4840	1978		G	$10,000	9/07	NECO	Duals, front weights, diesel, tach shows 864 hours
4840	1978		G	$12,000	10/07	NECO	Diesel, front weights, quick hitch, duals
4840	1978	7,900	G	$15,750	3/07	NCNE	18.4R38 (60%) with duals (70%) with cast inserts, set of 20 front weights., good running tractor, 180 hp.
4840	1978	5,454	G	$18,500	3/07	SEND	PS, 3 hyd., PTO, quick hitch, front fuel tank, 18.4-42 cast hub duals, 90% rubber
4840	1978	8,500	G	$19,750	12/07	WCMN	Duals, PS, rock box, 3 pt., PTO, 3 hyd.
4840	1978	6,600	G	$22,000	12/07	ECIL	Engine and rear axle OH in 2003 at 4,420 hours, 3 hyd., 20 front weights, 8-speed ps, quick hitch, 4 inside rear weights, new cab liner and radio
4840	1978		F	$10,000	7/06	SEND	CAH, PS, 3 hyd., 3 pt., 1,000 PTO, wide space duals, aux. lights
4840	1978	9,440	G	$13,200	11/06	NWKS	Recent OH
4840	1978	3,129	F	$13,500	3/06	SWNE	PS, 2WD, 18.4R-38, cast duals, rear wheel weights
4840	1978	4,652	F	$14,000	12/06	ECIL	PS, 18.4R-38, axle duals, front/rear weights, quick hitch
4840	1978	8,717	F	$12,000	4/05	NWOK	PS, duals, 3 pt., no third member, PTO, 2 hyd., good interior, no front weights
4840	1978	9,578	F	$13,000	2/05	NWKS	PS, 18.4-42R, cast duals, 16.5L-16.1 front (4 rib), 3 hyd., 3 pt. with lift assist, quick hitch, 20 rear end weights, large 1,000 PTO
4840	1978	6,800	F	$14,600	8/05	WCMN	PS, 3 hyd., duals, 3 pt., PTO
4840	1978	6,925	F	$17,000	3/05	WCNJ	PS, 3 hyd., 18.4-42, duals, 12 front weights, quick hitch, no park
4840	1978		G	$19,100	1/05	ECNE	PS, 18.4-42R, 10-bolt duals, 3 hyd., power beyond, 18 front weights, rear weights, R134A air, AM/FM, engine and trans., OH
50	1953		G	$5,900	3/08	WCOH	Restored
50	1955		G	$3,500	5/08	ECMN	Restored, row crop, gas, NF, 12.4-38 tires, newer tires
50			G	$2,250	11/07	WCIL	Running, unrestored/original, fenders, PTO, hyd. remote, live power
50			F	$2,350	10/07	NEIA	Loader
50			G	$2,600	5/07	WCWI	Roll-o-Matic, NF
50	1954		P	$900	9/07	WCIL	No live power, like-new fronts, stuck
50	1954		G	$1,500	12/07	SEWY	Gas, 3 pt., single front
50	1955		F	$1,750	8/07	SESD	PS, gas
50	1955		G	$5,000	9/07	SCMO	14" bottom plow combination
50	1955		G	$5,000	6/07	SCIL	
50	1956		F	$3,300	1/07	NCIL	PS, fenders, not running
50	1956		E	$5,800	6/07	NWIL	LP, NF, 3 pt., fenders
50	1956		E	$9,100	6/07	NWIL	Gas (all fuel), single front, PS, 3 pt., 42" tires, long axle
50	1960		G	$4,500	9/07	ECMI	NF, new 15.5-38 tires, PTO, PS
50			F	$1,000	10/06	WCWI	
50			G	$2,050	6/06	ECSD	PS, new rubber, restored
50			F	$2,300	3/06	SWMN	NF, 1 hyd.

John Deere

Model	Year	Hours	Cond.	Price	Date	Area	Comments
50			G	$2,600	1/06	NEIN	Single front, 3 pt., original
50			G	$3,500	7/06	NWIL	SN 5024662, running, unrestored/original, rear wheel weights, the tractor is a straight original John Deere 50, the Barber-Greene 550 loader requires restoration
50	1953		G	$2,250	2/06	NCCO	
50	1953		G	$3,600	3/06	WCSD	Looks nice, WF, 3 pt., runs good, good paint, live hyd., good seat, shedded, no PS
50	1953		G	$4,100	10/06	SCSD	NF, lights, gas
50	1955		G	$1,850	9/06	NEIN	Gas, NF, PS
50	1959		G	$1,650	12/06	ECMO	Tricycle
50			P	$400	6/05	ECWI	For parts
50			F	$1,050	9/05	NENE	PTO, drawbar, 12.4-38 tires, tire chains, John Deere 45 loader with hyd. bkt. and lift cylinder
50			F	$1,100	9/05	NENE	NF, Roll-o-Matic, PTO, drawbar, 12.4-38
50			F	$1,450	1/05	ECNE	$650 extra for 3 pt.
50			F	$1,500	9/05	NENE	NF, Roll-o-Matic, live PTO, drawbar, PS, 12.4-38 tires
50			F	$1,650	7/05	SCIA	
50			F	$2,000	6/05	ECWI	Painted
50			G	$2,100	1/07	SWOH	
50			G	$2,300	9/05	NEIN	All fuel repainted, 3 pt.
50			F	$2,500	9/05	NENE	Factory WF, factory PS, PTO, drawbar, new crank, needs assembly, 13.6-38 tires
50			G	$2,500	11/05	WCIL	NF, lights, PS, manual and parts
50			G	$3,250	6/05	ECMN	Row crop, gas, tricycle, new tires and paint
50			G	$5,500	3/05	ECMN	Row crop, gas, WF, 801 hitch, fenders, new paint, front weights, PS
50			G	$9,000	7/05	NEIA	LP, expo, restored, 1 of 731 built
50	1953		F	$2,050	1/05	NEIA	
50	1953			$4,000	12/05	NECO	Single front, 3 pt., PTO, new rubber, restored
50	1953		G	$4,200	9/05	SEMN	NF, front and rear weights, fenders, 3 pt.
50	1954		G	$2,550	8/05	ECMI	NF, new rubber
50	1954		G	$2,800	7/05	NCIA	PS, John Deere 45 loader with hyd. bkt.
50	1955		G	$2,150	6/05	WCMN	Clean, original
50	1955		F	$2,650	1/05	NEIA	PS
5010			G	$6,700	3/08	NEMO	Wheatland, diesel
5010	1963		F	$4,000	5/08	ECMN	Unrestored, diesel, 24.5-32 tires
5010	1964		G	$6,000	8/07	WCMN	M&W turbo kit, no cab, 2 hyd., 3 pt., PTO, rock box, 18.4-38 band duals, shows 1,172 hours, repainted, new clutch, fresh OH
5010	1963		F	$79,100	12/06	NEOK	SN 1000, first John Deere 5010 made, on eBay, 114 bids, rusty, seller's uncle bought from construction company, used to pull sheep foot roller
5020			F	$3,400	3/08	NEAR	Open station, 3 pt., PTO, 2 hyd., 18.4-38 singles
5020	1966		G	$7,000	3/08	SWNE	Wheatland, syncro, 2 hyd., dual 3100 loader with grapple fork, recent work done
5020	1968		G	$6,900	5/08	ECMN	Restored, diesel, 3 pt., wheel weights, 18.4-38 newer tires
5020			G	$4,250	11/07	SESD	Very original
5020	1966		G	$4,000	3/06	SWNE	Wheatland, diesel, 2 hyd., 1,000 PTO
5020	1967		G	$3,700	12/06	SCMT	
5020	1969		G	$6,000	4/06	NEND	Turbo diesel, row crop, Sound Guard CAH, 3 pt., PTO, 2 hyd., 18.4-42 singles and fronts, press steel hub duals, 60 Series exhaust and step
5020			G	$3,400	12/05	NWIL	Diesel, front blade

John Deere

Model	Year	Hours	Cond.	Price	Date	Area	Comments
5020			G	$3,750	12/05	SWMS	Diesel, canopy, 3 pt.
5020			F	$3,750	9/05	NWOR	Diesel
5020			G	$5,250	9/05	SCON	Canadian sale, diesel
5020			G	$9,250	6/05	ECWI	PTO
5020	1967		G	$4,800	11/05	ECNE	Syncro, 2 hyd., 18.4-38
5020	1967		G	$5,700	6/05	ECIL	Quick tach, 2 hyd., 162 hp., dyno, 24.5-32 rears, 12-volt conversion
520		4,195	G	$5,450	1/08	SESD	Factory WF, 3 pt., 1956 model
520	1957		E	$5,700	5/08	ECMN	Restored, gas, row crop, rock shaft, 13.6-38 tires, newer tires
520	1957		E	$7,000	5/08	ECMN	Restored, gas, NF, 3 pt., clamshell fenders, air stack, 12.4-36 tires, newer tires
520	1956		F	$3,500	6/07	SCIL	1 of 764 made, frame broken (has been repaired), PS, 3 pt., fair sheet metal
520	1957		G	$4,100	1/07	NCIA	PS, repainted, new manifold
520	1957		G	$4,600	9/07	NWIL	LP gas, WF, 3 pt., PS, original condition
520	1958		G	$2,700	9/07	ECMI	NF
520	1958		E	$9,250	11/07	SWIN	Older restoration, PS, 3 pt., clamshell fenders, new rears
520			F	$1,300	3/06	SEPA	NF, average rubber
520			G	$2,600	7/05	SESD	NF
520			G	$4,250	7/06	NWIL	LP, SN 520LP9868, running, restored, PTO, deluxe seat, Oregon tractor
520			G	$4,500	2/08	WCMN	35.5 hp., NF, PS, IPTO, lights, 3 pt., Roll-o-Matic, 12.4-36 tires, good rubber, newer paint, dynoed on 9-9-06
520			G	$6,200	7/06	NWIL	Restored, fenders, PTO, lights, 3 pt., belt pulley, PS
520			G	$6,200	2/07	WCIL	Restored, PTO, lights, 3 pt., belt pully, PS
520	1957		G	$2,850	10/06	SCMN	Single front wheel, good rubber, looks and runs good
520	1957	5,163	G	$4,500	9/06	NECO	Gas, single front, 3 pt.
520			G	$2,100	8/05	SEMN	
520			F	$2,500	8/05	ECMI	NF, Roll-o-Matic, PS, fenders, newer paint
520			F	$2,850	10/05	NEPA	
520			G	$3,300	6/05	ECMN	WF
520			G	$3,450	6/05	ECWI	Black dash, painted
520			G	$3,500	6/05	ECMN	Row crop, gas, tricycle, PTO, 1 hyd.
520			G	$3,900	6/05	ECMN	WF, 3 pt., fenders
520			G	$4,900	3/07	WCIL	New restoration, 77th of 520 produced, deluxe seat, Rollan engine
520			G	$6,000	8/06	WCMN	Gas, PS, NF, original, new rear tires
520	1957		G	$4,500	9/05	NWIA	Gas
520	1957		G	$8,850	9/05	SEMN	NF, 3 pt., front and rear weights, fenders, air stack
520	1958		G	$2,900	4/05	NCCO	PTO, 3 pt., remote, complete OH and restore
530			G	$6,900	3/08	NEMO	Gas
530			G	$7,750	3/08	NCTN	NF, 1 hyd., PS, new 12.4-38 tires, flat-top fenders
530	1958		G	$6,000	5/08	ECNE	NF, 3 pt., original
530	1959		E	$7,000	5/08	ECMN	Restored, gas, NF, 3 pt., air stack, square fenders, 13.6-38 tires, newer tires
530			G	$4,100	3/05	ECMI	NF
530			G	$6,700	8/07	SESD	NF, fenders, PS, restored
530			G	$6,900	6/07	NWIL	Gas, PS, WF, 3 pt., fenders

John Deere

Model	Year	Hours	Cond.	Price	Date	Area	Comments
530			G	$7,250	7/06	WCIL	Running, unrestored/original, fenders, PTO, lights, hyd. remote, deluxe seat, 3 pt., PS, live hydraulic and PTO, covered for 90% of its life
530	1958		E	$13,500	9/07	WCIA	WF, fenders, 3 pt., like new
530	1959		G	$4,750	6/07	SCIL	Original fenders 50%, 3 pt., float ride seat, new battery box and manifold, hood 90%, rebuilt carb.
530	1959		G	$5,100	9/07	NWIL	NF, PS, PowrTrol, gas
530	1959		G	$5,250	6/07	SCIL	NF, PS, flat-top fenders, restored
530	1959		G	$7,000	6/07	SCIL	WF, flat-top fenders, PS, great tires, restored
530	1959		E	$7,750	11/07	WCIL	Running, unrestored/original, lights, 3 pt., really an excellent original Nebraska tractor, all paint is original, only newer parts are one grille screen and seat cushions
530	1959		G	$8,500	6/07	NWIL	LP, single front, 3 pt.
530	1960		G	$7,700	6/07	SCIL	Vegetable, PS, long axles, carburetor rebuilt, float-ride seat, single front wheel (new tire), A4577R rear wheels, restored, 3 pt., sharp
530	1960		G	$8,000	6/07	SCIL	LP, row crop, all original, unusual 3 pt., great original tractor ready for restoration, 1 of 420, 1 hyd., air stack with precleaner, sheet metal 95% complete, Roll-o-Matic, runs good
530			G	$3,900	9/06	NEIN	WF, 3 pt., fenders, 2 hyd.
530			G	$5,800	9/06	ECMI	Single front wheel, tricyle, restored, complete
530			G	$6,700	6/06	SWMN	Restored
530	1956		G	$7,750	9/06	ECMI	WF, PS, restored, complete
530			F	$4,500	4/05	SEPA	Gas, NF, nice tin, 30% rubber
530			G	$5,250	6/05	ECIL	PS, NF
530			G	$6,500	3/05	ECMN	Row crop, WF, PTO, hyd., new paint
530			G	$6,750	3/05	ECNE	
530			G	$7,750	6/05	ECMN	Row crop, gas, WF, fenders, 3 pt., PTO
530			G	$8,000	7/05	SCIA	
530			G	$8,000	11/07	WCIL	NF, gas
530			G	$8,500	6/05	ECMN	Row crop, new tires, 1 hyd., new paint, OH
530			G	$9,100	7/05	NEIA	LP, WF, 3 pt., 1 of 417 built
530	1958		G	$8,600	8/05	WCIA	
530	1959		G	$9,800	9/05	SEMN	Air stack, 3 pt., fenders, front/rear weights
530	1960		G	$6,700	7/05	NCIA	PS, 3 pt.
60			G	$2,100	3/08	NCIA	John Deere 45 loader, hyd. bkt.
60			G	$2,500	6/05	ECWI	NF
60	1953		G	$3,600	4/08	NENE	Standard, restored
60	1953		G	$8,000	3/08	WCOH	Restored
60	1954		G	$2,200	3/08	NECO	Gas, single front, 3 pt., dual 325 loader
60	1954		E	$5,000	5/08	ECMN	Restored, gas, NF, 14.9-38 tires, newer tires
60	1954		E	$6,200	5/08	ECMN	Restored, gas, standard, 18.4-30 tires, newer tires
60			G	$1,500	9/06	ECMI	Tricycle
60			F	$1,600	3/05	WCIL	All original, runs
60			F	$1,600	11/07	SWIN	Orchard, all fuel, incomplete, for parts or restore, turf tires
60			G	$2,200	11/07	WCIL	Running, unrestored/original, fenders, PTO, lights, hyd. remote, live power
60			G	$3,250	2/07	NENE	Roll-o-Matic, NF, runs great
60			G	$3,500	11/07	SESD	NF
60	1952		G	$2,700	9/07	SCMO	13.6-38 rear tires, NF, PTO, 2 hyd., hand clutch
60	1952		G	$4,600	9/07	SCMI	NF, gas, restored, rear weights, 1 hyd.
60	1953		G	$2,200	10/07	SWOH	Row crop, diesel

John Deere

Model	Year	Hours	Cond.	Price	Date	Area	Comments
60	1953		G	$3,900	6/07	SCIL	Restored, all new tires
60	1953		E	$5,900	6/07	NWIL	Restored, WF, fenders, gas
60	1954		F	$1,100	6/07	SCIL	Standard, live PTO, 1 hyd., low seat, ready to restore
60	1954		F	$1,400	11/07	SWIN	Orchard, all fuel, incomplete, for parts or restore, fenders
60	1954		G	$1,500	6/07	NENE	Gas, Roll-o-Matic NF, 13.6-38 rear tires, 6:00-16 front tires, 1 hyd., PS
60	1954		G	$2,000	8/07	SESD	Gas
60	1954		F	$3,750	11/07	SWIN	Orchard, propane, original, rough sheet metal
60	1954		G	$4,900	4/07	NEIA	NF, gas, good tires
60	1955		F	$1,250	11/07	SWIN	Orchard, propane, somewhat complete
60	1955		F	$1,400	11/07	SWIN	Orchard, all fuel, incomplete, for parts or restore, one new rear tire
60	1955		F	$1,500	11/07	WCIL	800 hitch
60	1955		F	$1,700	8/07	SESD	Original
60	1955		F	$2,200	9/07	WCIL	Electric start, oval hole rear cast, fairly straight, runs, did head gasket in 1997
60	1955		G	$2,900	11/07	WCIL	Running, unrestored/original, PTO, lights, deluxe seat, 3 pt., PTO, PS, new seats, just had $1,300 put into repairs (PS seals, starter parts, plugs, battery, voltage regulator)
60	1955		G	$4,800	9/07	NWOH	Power block, NF
60	1956		F	$1,900	11/07	SWIN	Orchard, all fuel, incomplete, for parts or restore, fenders
60			F	$775	5/07	WCWI	
60			F	$850	10/06	WCWI	
60			F	$1,350	10/06	WCWI	
60			G	$2,100	1/06	SESD	Row crop, WF, fenders
60			G	$2,500	11/06	SETN	WF, PS, 3 pt.
60			G	$4,100	3/06	NWOH	Gas, NF, restored
60			G	$4,300	10/06	SENY	Hi seat, restored
60			G	$4,500	10/06	SENY	Lo seat, restored
60			G	$7,500	10/06	SENY	227 mounted corn picker, restored
60			E	$11,500	7/06	NWIL	SN 6050014, running, restored, fenders, PTO, lights, PS, this tractor has been beautifully restored to its original condition
60	1954		G	$1,800	9/06	NENE	Gas, Roll-o-Matic NF, 13.6-38 rear tires, live power, hyd. cylinder, 3 pt., rear wheel weights, one owner, shedded
60	1954		G	$4,500	9/06	ECMI	3 pt., WF, all original, complete
60			P	$800	1/05	ECNE	Stuck, as is, NF
60			F	$1,000	4/08	NENE	Dual loader, gas, PS, PTO, drawbar, 13.6-38 tires
60			F	$1,100	7/05	SCMN	
60			F	$1,225	7/05	SESD	NF
60			F	$1,400	9/05	NEIN	PS
60			G	$1,600	8/06	NWIL	Gas
60			F	$1,600	3/05	WCIL	NF, unrestored, PTO, lights
60			G	$1,650	6/05	WCMN	WF, one owner
60			F	$1,800	6/05	ECWI	Painted
60			G	$1,800	9/05	NENE	NF, good tin
60			F	$2,000	4/05	SEPA	Gas, NF, 30% rubber
60			G	$2,300	11/05	WCIL	NF, PS
60			G	$3,700	10/05	NWMT	
60			G	$3,750	6/05	ECWI	Standard, low seat, new tires
60			G	$3,800	6/05	ECWI	New rubber, restored

John Deere

Model	Year	Hours	Cond.	Price	Date	Area	Comments
60			G	$3,800	8/05	SWMN	PTO, NF, PS
60			G	$3,900	8/05	SWMN	PTO, WF, PS
60			G	$4,000	3/05	ECMN	Row crop, new tires and paint, 3 pt., hyd.
60			G	$4,100	7/05	SCIA	Row crop, restored
60			E	$57,000	7/05	SCIA	LP, standard, restored, 1 of 25 built
60	1952		G	$1,900	2/05	NEIN	Gas, NF
60	1952		F	$3,700	3/05	NECO	Dual loader, WF, 3 pt., PTO, hyd.
60	1953		P	$1,225	1/05	NEIA	Rough
60	1953		G	$3,600	4/05	NEIN	Gas, NF
60	1954		G	$4,850	9/05	SEMN	Low seat standard PTO, PowrTrol, 2 hyd.
60	1955		G	$4,800	9/05	SEMN	Row crop, front and rear weights, 800 Series, 3 pt.
60	1955		G	$4,800	9/05	SEMN	Standard, front and rear weights, 800 Series, 3 pt.
60	1956		F	$2,450	1/05	NEIA	PS
6030			E	$16,000	2/08	SCIL	CAH, original paint, near perfect
6030	1976		E	$17,500	5/08	ECMN	Restored, diesel, 3 pt., canopy, 20.8-38 tires
6030			F	$5,900	3/07	WCIL	Open station, 2 hyd., PTO, 3 pt., 20.8-38 duals
6030		5,500	G	$12,500	1/07	NWIN	Fender tractor
6030	1973	3,118	G	$11,200	1/07	ECIL	8-speed syncro, duals, weights, front tank
6030	1973		G	$16,500	11/07	SESD	3 pt., factory duals, cab, restored
6030	1976	8,000	G	$9,400	11/07	SEND	Cab, air, PTO, 24.5-32 duals, no 3 pt.
6030	1972	9,925	F	$4,800	2/06	NECO	Axle duals, PTO, cab
6030	1972		F	$6,600	3/06	SWOH	Diesel, WF, new style cab, 819-cubic-inch engine, hub duals
6030			G	$8,000	2/05	NCIN	
6030			G	$8,750	9/05	NEIN	Original, factory cab
6030			G	$14,000	3/05	SEIA	
620	1958		G	$8,500	3/08	WCOH	Restored
620			G	$2,900	6/05	WCMN	Gas, Roll-o-Matic, NF, John Deere 45 loader with snow bkt.
620	1956		G	$8,000	11/07	SWIN	Standard, PS, 3 pt., no center link, new rear tires, older restoration
620	1957		F	$2,300	11/07	WCIL	LP, 3 pt.
620	1957		G	$2,800	6/07	SCIL	Nice original tractor, hood 90%, radiator hood 75%, excellent runner, PS
620	1957		G	$4,000	9/07	WCIA	NF, 3 pt., repainted
620	1958		G	$2,750	9/07	ECMI	NF
620	1958		G	$3,500	6/07	SWMN	LP, NF, 13.6-38 tires, 6,671 hours showing, John Deere 45 loader
620	1958		G	$3,500	6/07	SCIL	NF
620	1958		E	$20,000	11/07	SWIN	Orchard, gas, restored, PS, new fronts, complete reproduction sheet metal
620	1958		E	$26,000	11/07	SWIN	Orchard restoration will be complete by sale time, turf tires
620	1958		E	$29,000	11/07	SWIN	Orchard, propane, restored, nice, PS, new front tires, rear turf tires, complete orchard sheet metal reproduction
620			G	$1,950	6/06	ECSD	NF, 3 pt.
620			G	$2,500	8/07	SESD	LP, NF, restored
620			G	$2,500	8/06	NWOH	NF, gas
620			G	$2,900	6/06	ECSD	WF, 3 pt.
620			G	$3,000	7/06	NWIL	LP, SN 6209649, running, old restoration/repaint, PTO, hyd. remote, deluxe seat, 3 pt., belt pulley, PS
620			G	$3,300	6/06	ECSD	NF

John Deere

Model	Year	Hours	Cond.	Price	Date	Area	Comments
620			G	$3,500	3/08	WCMN	Paulson loader, 3 pt., WF, new paint on hood and grille, new radiator
620			G	$16,750	10/06	SENY	620LP, 1 of just 37 built, restored, adjustable front end, PS
620	1957		F	$2,000	9/06	ECNE	NF, new rubber, some rust
620	1957		F	$2,600	9/06	ECMI	NF, has cracked block
620			F	$1,700	3/05	WCMN	NF, original
620			F	$2,000	8/05	SEIA	
620			G	$2,700	6/05	WCMN	PS, NF, new tires, gas
620			G	$2,750	8/05	WCMN	Runs good, good tin
620			G	$3,100	2/05	NEIA	Newly repainted
620		3,854	G	$3,100	6/05	WCMN	WF, 1 hyd., PS, duals, dual 325 hyd. loader
620			G	$3,250	11/05	WCIL	NF, gas, 3 pt., belt pully, PS
620			F	$4,000	8/05	NECO	Farmhand F11 loader, 3 pt.
620			G	$4,000	3/05	ECMN	Row crop, 3 pt., PTO, 1 hyd., fenders, new paint
620			G	$5,000	5/08	ECMN	Row crop, 3 pt., PTO, 1 hyd., fenders, new paint
620			G	$7,250	6/05	ECMN	Standard gas, PTO, 1 hyd., new paint
620	1956		G	$3,900	3/05	WCNJ	Nice tractor
620	1956		G	$9,500	9/05	SEMN	Standard, front and rear weights, 3 pt., air stack
620	1958		G	$2,300	4/05	NENE	NF, 13.6-38 tires
620	1958		G	$9,850	9/05	SEMN	NF, 3 pt., front/rear weights, air stack
630			G	$4,750	3/08	NCTN	NF, 1 hyd., original 3-piece front weights, 3 pt., PS, good tires
630	1957	4,500	G	$7,600	1/08	ECIA	NF, factory 3 pt.
630	1958		G	$8,200	3/08	ECWI	Gas, John Deere WF, flat-top fenders, 3 pt., new 14.9-38 tires, SN 6300900
630	1959		G	$4,500	4/08	SCMI	Gas, WF, 3 pt.
630	1959		E	$6,500	5/08	ECMN	Restored, gas, row crop, WF, air stack, 3 pt., square fenders, 13.6-38 tires, new tires
630	1960		E	$6,700	5/08	ECMN	Restored, gas, NF, 3 pt., square fenders, new 14.9-38 tires
630			G	$0	8/07	NENE	No sale at $5,000
630			G	$3,750	8/06	SCIL	No tag, original, single front fenders
630			G	$4,750	9/07	WCIA	Schwartz WF, fenders, 3 pt.
630			G	$4,900	2/06	NEIA	Restored, new rubber
630			G	$6,100	2/07	NENE	Gas, NF, 3 pt., good tin, one owner
630			G	$10,000	9/07	WCIA	WF, 3 pt., factory LP, repainted
630			E	$13,500	6/07	NWIL	Gas, WF, PS, fenders, new paint and tires
630			G	$14,000	4/07	NEMO	Standard, fresh restoration
630	1958		G	$2,450	9/07	ECMI	NF
630	1958		E	$4,400	7/07	NWOH	SN 6300809, gas, Roll-o-Matic, PS, 3 pt., nearly show-ready
630	1958		G	$6,000	9/07	ECMI	NF
630	1958		G	$6,500	6/07	SCIL	Row crop, original, sheet metal is with the tractor, less than 200 hours on complete OH done in 1987, 3 pt. (no toplink), main hood and radiator in primer, new clutch drive plate, all new clutch disks, new throttle linkage ball-joint
630	1958		G	$9,000	9/07	WCIA	WF, 3 pt., fenders, repainted
630	1959		G	$2,700	6/07	NWIL	LP, single front, no 3 pt., fenders
630	1959		G	$2,750	11/07	SWIN	Propane, single front wheel, precleaner, repainted
630	1959		G	$8,000	6/07	SCIL	WF, PS, flat-top fenders, 3 pt., good tires, restored
630	1959		G	$12,500	11/07	WCIL	LP, standard, clone, 3 pt.

John Deere

Model	Year	Hours	Cond.	Price	Date	Area	Comments
630	1959		E	$14,000	6/07	SCIL	Standard, restored, super sharp, new tires, 1 set rear weights, PTO, air cleaner, 1 hyd.
630	1960		G	$8,000	11/07	WCIL	Running, restored, fenders, PTO, lights, hyd. remote, deluxe seat, PS, unit has been restored to original condition using original John Deere parts, only thing not original is the twin chrome stacks (one exhaust, one intake)
630			G	$1,950	9/06	NEIN	LP, 3 pt., original
630			G	$3,750	3/06	NWIL	SN 6316229, running, restored, fenders, lights, hyd. remote, PS, nice restoration
630			G	$4,000	6/06	ECSD	NF, original
630			G	$4,500	7/06	WCIL	Running, old restoration, PTO, lights, 1 hyd.
630			G	$4,500	3/06	NWIL	LP, SN 6304272, running, old restoration/repaint, fenders, PTO, lights, hyd. remote, deluxe seat, 3 pt., belt pulley, PS
630			G	$4,750	7/06	WCIL	Running, old restoration, hyd. remote, 3 pt., new radiator core, batteries, seat
630			G	$4,750	7/06	NWIL	SN 6309872, running, old restoration/repaint, PTO, lights, 1 hyd., 3 pt., PS, new radiator core, batteries, seat, muffler, clutch, rebuilt carburetor and water pump
630			G	$5,500	9/06	NEIN	WF, 3 pt., fenders, original
630			G	$6,250	1/06	SCKS	13.6-38 rear tires, Roll-o-Matic front end, 3 pt., 1 hyd., 3 sets rear weights, shedded, 50 hours on OH, runs great
630	1958		G	$6,700	9/06	ECMI	NF, PS, complete, rebuild
630	1959		G	$4,650	3/06	NCIA	Gas, NF, PS, fenders, good tin work
630	1959	5,741	G	$6,200	11/06	NEKS	WF, gas, 13.6-38 rear, 1 hyd., 540 PTO, 3 pt., new paint
630	1959		E	$7,500	12/06	WCIL	LP, 3 pt., PS, new paint
630		4,273	G	$3,400	6/05	WCMN	NF, PS, new rubber
630			G	$3,750	6/05	ECMN	Tricycle, float ride seat
630			G	$4,500	12/06	ECMN	Row crop, WF, deluxe fenders, 3 pt., front weight set
630			G	$5,500	3/05	ECMN	Row crop, tricycle, 3 pt., PTO, 1 hyd.
630			G	$5,900	6/05	ECWI	3 pt., WF
630			G	$6,100	7/05	SCIA	Row crop
630			E	$6,300	11/05	NEIA	NF, restored, flat-top fenders
630			G	$10,500	9/07	ECMN	Standard, gas, PTO, 1 hyd
630			G	$14,000	6/05	ECMN	Standard, gas, 3 pt., PTO, new paint, rebuilt
630	1958		G	$3,700	9/05	WCNE	Gas, 3 pt., 2,267 hours on tach
630	1958		G	$6,600	8/05	WCIA	
630	1958		E	$20,000	6/05	NEOK	LP standard, 3 pt., adjustable front, Rice Special restored
630	1959		G	$4,350	12/05	NCCO	WF, 3 pt.
630	1959		G	$9,250	9/05	SEMN	NF, air stack, 3 pt., front/rear weights
630	1959	5,039	G	$9,450	1/05	NENE	NF, PS, live power, factory 3 pt., gas, 13.6-38 tires, rear wheel weights
630	1960		G	$3,200	9/05	NENE	Gas, Roll-o-Matic, 13.6-38 unused tires
630	1960		G	$4,200	6/05	ECIL	New rear tires, rock shaft, PS
70			G	$3,000	1/08	ECOK	NF, diesel, PS, 2 cyl.
70	1954		E	$4,600	5/08	ECMN	Restored, row crop, gas, NF, 13.6-38 tires, newer tires
70	1954		E	$12,250	5/08	ECMN	Restored, standard, gas, wheel weights, 18.4-30 tires, newer tires
70	1955		G	$4,500	4/08	NENE	Standard, diesel, WF, pony start, 3 pt.
70	1956		G	$4,100	2/08	NCOH	Roll-o-Matic, pony start
70			P	$1,500	3/07	NWIL	Diesel, NF

John Deere

Model	Year	Hours	Cond.	Price	Date	Area	Comments
70	1952		G	$2,300	9/07	NEIL	Unrestored
70	1953		G	$2,700	12/07	ECNE	Roll-o-Matic, NF, electric start
70	1953		G	$3,650	9/07	SCMI	NF, gas, restored, 1 hyd.
70	1954		G	$3,750	4/07	NCNE	LP, WF, good tin
70	1954		G	$5,250	6/07	SCIL	Nice original 70, fenders, great runner, used with loader on the owner's farm
70	1955		G	$1,900	12/07	SEWY	Gas, 3 pt., single front
70	1955		G	$2,000	12/07	SEWY	3 pt., WF, diesel
70	1955		G	$2,000	8/07	ECIA	Row crop, NF, LP conversion
70	1955	8,843	G	$3,000	9/07	NECO	WF, diesel, 3 pt., PowrTrol
70	1955		G	$3,400	10/07	NECO	Diesel, pony start, WF, 3 pt.
70	1955		E	$4,400	3/07	NWIL	Pony start, diesel, PS, live PTO, Roll-o-Matic, NF, new tires and paint
70	1955		G	$5,500	9/07	ECMI	WF, diesel, PTO, nice tractor, PS, 14.9-38 tires
70	1955		G	$5,750	11/07	WCIL	Running, restored, rear wheel weights, fenders, PTO, lights, hyd. remote, deluxe seat, 3 pt., belt pulley, PS
70	1956		F	$1,350	6/07	SCIL	Gas, to restore, PS, alternator, tach, noise in engine
70	1956		G	$3,250	11/07	WCIL	Diesel, electric start conversion, square WF
70	1956		G	$4,000	6/07	SCIL	Pony start, 1 hyd., tin 95%, Wheatland, very nice original tractor, live PTO, PS
70			P	$235	9/06	ECMI	For parts
70			G	$1,200	11/06	SCCA	Gas
70			G	$2,000	10/06	WCSK	Canadian sale, running, row crop tractor, gas
70			G	$2,500	7/06	NWIL	SN 7008772, running, old restoration/repaint, PTO, lights, belt pulley, exellent working and running tractor
70			G	$2,800	5/06	SEND	Diesel, pony engine, 3 pt., WF, PS, recent OH
70			G	$3,200	9/06	NEIN	Standard, diesel
70			G	$8,000	10/06	SENY	Restored
70	1953		G	$2,700	9/06	ECMI	NF, PS, complete, clamshell fenders
70	1954		F	$1,500	4/06	NEND	Gas, WF, PTO, 1 hyd., 14.9-38 singles
70	1954		G	$3,500	4/06	ECND	Gas, WF, open station, electric start, 1 hyd., PTO, rock shaft
70			P	$300	2/05	ECNE	Salvage tractor
70			G	$2,000	11/05	WCIL	John Deere adjustable WF, diesel, pony, PTO, lights
70			G	$2,300	6/05	ECWI	PS, painted
70			F	$2,550	9/05	SEIA	
70			G	$2,600	8/05	WCIL	NF, diesel, pony
70			G	$3,000	11/05	WCIL	NF, diesel
70			G	$3,050	6/05	ECWI	PS, painted
70			G	$3,800	6/05	ECMN	Gas, rockshaft
70			G	$4,100	6/05	ECMN	Row crop, tricycle, fenders, PTO, hyd.
70			G	$4,100	8/06	WCIL	Std, restored, PTO, lights, PS, diesel
70			G	$4,500	4/05	WCWI	Standard
70			G	$4,500	12/06	ECMN	Row crop, tricycle, fenders, PTO, hyd.
70			G	$4,500	12/06	ECMN	Row crop, gas, WF, new paint
70			G	$5,600	7/05	SCIA	Gas, standard, restored, 1 of 1,035 built
70			G	$7,700	7/05	NEIA	LP standard, restored
70	1949		G	$5,100	6/05	ECIL	New paint, 45-degree tires, implement-mount brackets
70	1953		G	$4,000	3/05	SEIA	
70	1954		F	$1,400	7/05	NEIA	Gas, 13.6-38 tires, 1 hyd., not running
70	1954		G	$2,100	2/05	NENE	Gas, Behlen PS, 13.6-38, 1 hyd.
70	1954		F	$2,200	7/05	NEIA	1 hyd., gas, 13.6-38 tires

RUST BOOK

John Deere

Model	Year	Hours	Cond.	Price	Date	Area	Comments
70	1954		G	$5,000	7/05	NCIA	PS, 13.6-38 tires
70	1954		G	$5,200	9/05	SEMN	Row crop, front and rear weights, 800 Series, 3 pt.
70	1954		G	$5,250	9/05	SEMN	Standard rear weights, PTO, PowrTrol
70	1955		G	$2,300	4/05	SCMI	NF, gas, PS, fenders, original condition
70	1955		G	$5,600	9/05	SEMN	Pony start, front and rear weights, 3 pt., fenders
70	1955		G	$6,500	9/05	SEMN	Standard, pony start, PTO, PowrTrol
7020			F	$4,000	3/08	WCMN	4WD, 2 hyd., 1,000 PTO, 18.4-34 duals, cab, good runner
7020			G	$7,000	4/06	WCSD	4WD, Degelman dozer, OH approximately 650 hours ago
7020			G	$9,000	4/06	SCMN	8 new 18.4-34 tires, bareback, sharp
7020	1972		F	$5,700	1/06	SCMI	4WD, cab, 3 pt., 3 hyd., PTO, duals, front blade, 18.4-34, diesel, tach shows 1,857 hours
720			E	$5,250	5/08	ECMN	Restored, diesel, pony, clam fenders, air stack, 16.9-30 tires, newer tires
720	1957		E	$5,500	5/08	ECMN	Restored, diesel, standard, pony, air stack, PTO, 18.4-34 tires, newer tires
720			G	$2,900	11/05	WCIL	LP, 3 pt., square WF
720	1956		F	$1,350	9/07	ECMI	NF
720	1956		G	$7,750	11/07	WCIL	Running, unrestored/original, fenders, PTO, lights, hyd. remote, deluxe seat, PS, very clean, original Western tractor, factory 3 pt., 100% original paint and decals
720	1957		G	$3,000	11/05	WCIL	Diesel, pony, 3 pt., sqaure WF
720	1957		G	$4,000	11/07	WCIL	Diesel, running, unrestored/original, PTO, lights, hyd. remote, deluxe seat, 3 pt., belt pulley, PS
720	1957	1,520	G	$4,900	11/07	SEND	Diesel, Roll-o-Matic, pony motor start, factory 3 pt., 15.5-38 rear tires (90%)
720	1957		E	$8,200	11/07	SWIN	Standard, pony start, diesel, 1x wheel weight, 3 pt., no center link, precleaner, PS, older restoration
720	1958		G	$2,900	9/07	ECMI	
720	1958		G	$4,750	8/07	WCMN	Diesel, WF, PS, new tires, new paint, new battery, $1,500 in recent work orders (available), new rubber all around, restored, fuel gauge and tach not working
720	1958		E	$5,100	4/07	WCSK	Canadian sale, gas., restored, good running condition
720			G	$3,600	10/06	NCND	Dual loader, PowrTrol, gas pony engine start
720			G	$4,000	4/06	NWSD	Diesel, 3 pt., comfort cover, WF, weights, Johnson loader, grapple and corn bkt., new tires, hay head
720			G	$10,900	4/06	WCSD	Standard, factory LP, 3 pt. (missing third member), 52 hp.
720	1956		G	$4,400	9/06	ECMI	Pony motor, diesel, pony motor, runs excellent, all original
720	1957		G	$4,800	10/06	SCMN	Electric start, factory WF, sharp original condition
720			F	$2,050	3/06	SEPA	Diesel, tricyle, good tin, poor paint, 50% rubber
720			g	$3,100	11/05	WCIL	Diesel
720			G	$3,800	9/05	NEIN	LP, standard, 3 pt.
720			G	$3,800	6/05	WCMN	Original, NF, PTO, hyd.
720			G	$4,000	6/05	ECWI	LP, 2 hyd., 3 pt., new rubber
720			G	$4,100	7/05	SCIA	LP all-fuel row crop, 42" rubber, single front wheel, 3 pt., 2 hyd., fenders, 5-piece weights, 1 of 413 built

John Deere

Model	Year	Hours	Cond.	Price	Date	Area	Comments
720			G	$4,900	3/07	WCIL	WF, diesel, pony, PTO, lights, PS
720			G	$5,400	8/05	SWMN	Diesel, WF, PS, 3 pt., PTO, restored
720			G	$5,500	7/05	NEIA	Pony, new tires, clamshell fenders, diesel
720			G	$5,600	3/05	WCIL	NF, OH, new front tires
720			G	$6,000	3/07	WCIL	Diesel, pony, PTO, lights
720			G	$6,000	6/05	ECMN	Gas, WF, 3 pt., 1 hyd.
720			G	$6,750	3/07	ECMN	Row crop, LP, WF, 1 hyd., 3 pt., new paint
720			G	$7,250	6/05	ECMN	Diesel, pony start, 3 pt., new paint and tires
720	1956		G	$10,250	9/05	SEMN	Standard, pony start, rock shaft, rear weights, air stack
720	1957		G	$3,600	2/05	NCIN	Diesel, NF, 3 pt., duals
720	1957		G	$7,500	6/05	ECIL	Restored John Deere 45 loader, 3 pt.
720	1958	39,150	G	$2,500	9/05	WCNE	New paint, double front, 3 pt.
720	1958		G	$3,900	2/05	WCIL	Diesel, pony motor, good rubber, 3 pt., PS
720	1958		G	$4,600	2/05	NCIN	Diesel, NF, rebuilt hyd.
720	1958		F	$4,750	9/05	SCMI	Diesel, electric start, NF
720	1958		G	$8,650	9/05	SEMN	3 pt., air stack, fenders, front weights
720	1959		G	$4,950	3/05	NWMN	WF, PS, pony start, diesel, one owner
730			G	$2,700	9/05	NCTN	LP, 3 pt., PS
730			G	$8,000	9/05	NCTN	Diesel, NF, 2 hyd., electric start, diesel, PS, flat-top fenders
730	1958		G	$11,000	3/08	WCOH	Restored, diesel
730	1959		E	$9,250	5/08	ECMN	Restored, row crop, NF, air stack, square fenders
730	1959		E	$9,500	5/08	ECMN	Restored, diesel, air stack, new tires, electric start
730	1960		G	$3,600	3/08	ECWI	John Deere square WF, 3 pt.
730	1960		E	$11,700	5/08	ECMN	Restored, diesel, standard, electric start, air stack, PTO, 18.4-38 new tires
730			G	$2,000	4/07	ECSK	Canadian sale, 2WD, diesel, PS, PTO, hyd.
730			G	$3,400	11/07	WCIL	LP, 3 pt., square WF
730			G	$4,400	7/06	WCIL	Diesel, electric start, square WF
730			G	$4,400	6/07	NWIL	LP, WF, PS, 3 pt.
730			G	$4,600	2/07	WCIL	Diesel, elecric start, square WF
730			G	$4,900	6/07	NWIL	Diesel, NF, 3 pt., pony, fenders, 2 hyd.
730			G	$5,700	7/05	ECND	Gas, row crop, WF, John Deere rock shaft, Jobber 3 pt., live PTO, 15.5-38 tires, showing 2,297 hours
730			G	$6,000	3/07	WCIL	WF, 3 pt., restored
730			E	$22,000	11/07	WCIL	High crop, Argentine tractor, rare foot clutch option, excellent restoration, super nice
730	1958		G	$5,600	6/07	SCIL	Diesel, 24-volt electric start, square WF, fenders, 3 pt. (no toplink), tin at 90%, needs battery box, good runner, used on the Killam farm to mow
730	1958		G	$6,500	7/07	NCIA	Diesel, cracked block found on sale day, no fenders, factory WF, 3 pt., PS, good tin
730	1958		G	$14,000	11/07	WCIL	LP, standard, older restoration, 3 pt.
730	1959		G	$4,250	9/07	WCIL	LP, 3 pt., runs, used to pull wagons
730	1959		G	$5,000	6/07	SCIL	LP, all original gauges, 1 hyd., PTO, 3 pt., runs good
730	1959		G	$5,000	6/07	SCIL	Diesel, Wheatland, pony start, all original, straight, complete engine OH, pony motor has new coils, points and condensor, PTO, 1 hyd.
730	1959		G	$5,250	11/07	WCIL	Electric start, original
730	1959		G	$7,000	8/07	ECIA	Row crop, factory LP, NF
730	1959		G	$10,500	11/07	WCIL	Gas, 3 pt., square WF, weights, fenders

John Deere

Model	Year	Hours	Cond.	Price	Date	Area	Comments
730	1959		G	$11,000	6/07	SCIL	Diesel, WF, flat-top fenders, 3 pt., all decked out with the options, restored
730	1959		G	$14,750	11/07	WCIL	Running, restored, fenders, PTO, lights, hyd. remote, deluxe seat, PS, 1 of 292, shown at expo
730	1960		F	$2,500	11/07	WCIL	LP, 3 pt., square WF
730	1960		G	$5,200	3/07	WCIL	Diesel, PS, pony start
730	1960		E	$6,000	4/07	WCSK	Canadian sale, diesel, restored, good running condition, this is the tenth-last John Deere 730 built
730	1960		G	$6,500	11/07	WCIL	Diesel, standard, pony start, restored
730	1960		G	$8,000	9/07	ECMI	WF, gold tag, 15.5-38 tires
730	1960		G	$9,500	6/07	NWIL	Gas, WF, PS, 3 pt.
730	1960		G	$10,000	4/07	NEMO	Standard, electric start, hyd., lights, steps, tires 60%, PTO, straight drawbar, good restoration
730			F	$2,800	12/05	NWIL	Diesel, SN 7314097, running, old restoration/repaint, fenders, PTO, lights, hyd. remote, 3 pt., belt pulley, PS
730			G	$2,900	11/05	WCIL	Running, unrestored, PTO, deluxe seat, 3 pt., belt pully
730			G	$3,400	11/07	WCIL	Running, old restoration, PTO, deluxe seat, PS
730			G	$3,400	12/05	NWIL	LP, SN 7327059, running, old restoration and repaint, fenders, PTO, lights, deluxe seat, belt pulley, PS, painted
730			G	$3,450	10/06	WCWI	
730			G	$4,300	7/06	NWIL	LP, SN 732716, running, restored, factory cab, lights, PS, direction reverser, belt pulley, TA
730			G	$4,500	7/06	WCIL	Unrestored, fenders, PTO, lights, hyd., PS
730			G	$4,500	3/06	NWIL	SN 7326042, running, unrestored/original, fenders, PTO, lights, hyd., PS, tractor runs good, drives good, and trans. works fine, brakes and lights work, pony needs work, only 2,213 of these tractors produced
730			G	$5,500	7/06	NWIL	SN 7311935, running, restored, fenders, PTO, lights, hyd. remote, deluxe seat, PS, new rings and reground valves, new wiring, new seat, rebuilt carb. and steering wheel, nice restoration
730			G	$6,000	8/06	WCMN	Diesel, WF, new tires, 3 pt., PTO, hyd.
730			G	$6,550	12/06	SWIL	Diesel, row crop, electric start, 3 pt., hyd., repainted
730			G	$6,700	3/06	SEMN	NF, air stack, 3 pt., White 1610 quick-tach loader, original
730			G	$7,400	9/06	NEIN	Pony, WF, 3 pt., fenders, original
730			G	$8,250	7/06	NWIL	Diesel, SN 7314097, running, old restoration/repaint, fenders, PTO, lights, hyd. remote, 3 pt., belt pulley, PS
730			G	$8,750	7/06	NWIL	SN 7315591, running, old restoration/repaint, PTO, lights, 3 pt., complete OH in 1996 and stored ever since, new wiring, harness, lights, and flywheel, hydraulic hoses and fittings
730			G	$9,500	7/06	NWIL	SN 731155, running, restored, rear wheel weights, fenders, PTO, lights, hyd. remote, front weight, deluxe seat, 3 pt., belt pulley, PS
730			G	$9,500	7/06	WCIL	Running, restored, rear wheel weights, fenders, PTO, lights
730			G	$10,200	8/06	NECO	Diesel, pony motor start, factory WF, factory 3 pt., factory fenders, 15.5-38 rubber
730			G	$11,000	7/06	NWIL	SN 311983, running, unrestored/original, this is an Argentina high crop

John Deere

Model	Year	Hours	Cond.	Price	Date	Area	Comments
730			G	$11,000	1/08	WCIL	Running, unrestored, Argentina high crop
730			G	$13,000	8/06	WCIL	Running, unrestored, Argentina high crop
730			G	$13,000	7/06	NWIL	SN 7302686, running, restored, rear wheel weights, fenders, deluxe seat, 3 pt., PS, excellent 730 gas row crop, many options
730			G	$13,000	7/06	NWIL	SN 300565, running, unrestored/original, this is an Argentina high crop
730			G	$15,500	7/06	NWIL	SN 7307144, running, restored, fenders, PTO, lights, hyd. remote, front weight, deluxe seat, 3 pt., belt pulley, PS
730	1959		G	$10,000	7/06	WCIL	Gas, standard, running, old restoration, old dealer repaint
730	1959		G	$10,000	7/06	NWIL	Running, old restoration/repaint, old dealer repaint, good straight tractor
730	1960		F	$5,500	1/06	NCCO	Gas, WF, 3 pt.
730	1960		G	$5,500	9/06	ECMI	Diesel, PS, WF
730	1960		G	$8,500	8/06	WCMN	Diesel, WF, 3 pt., fenders, 15.5-38
730	1960		G	$9,600	8/06	WCMN	Diesel, NF, 3 pt., fenders, 15.5-38
730		3,428	G	$3,700	6/05	WCMN	NF, PS
730			G	$3,750	12/06	WCIL	WF, standard, magneto out of pony motor
730			G	$4,500	8/05	SEMN	Standard, electric start, straight
730			G	$4,700	11/05	WCIL	Old restoration, repaint, PS
730			G	$4,900	3/07	WCIL	NF, diesel, pony, M&W pistons
730			G	$5,500	3/05	ECMN	Row crop, electric start, WF, fenders, 3 pt., 1 hyd., 4 wheel weights
730			G	$6,000	6/05	ECMN	Row crop, gas, tricycle, 3 pt., PTO, 1 hyd., toplink
730			G	$6,000	2/05	WCIN	Diesel, 15.5 tires, NF, belt pulley
730			G	$6,500	3/05	ECMN	Row crop, 3 pt.
730			G	$6,750	3/05	NWIL	LP, NF, PS, new paint
730			G	$7,750	6/05	ECMN	Row crop, WF, diesel, PTO, welded U-tie rods, new tires, fenders
730			G	$7,750	6/05	ECMN	Row crop, electric start, WF, fenders, 3 pt., 1 hyd., 4 wheel weights
730			G	$7,750	6/05	ECWI	3 pt.
730		3,800	F	$8,100	4/05	SEPA	Gas, 3 pt., PS, 90% rubber
730			G	$9,700	7/05	SCIA	LP, high crop, diesel, restored, no tag
730			G	$16,000	7/05	SCIA	Argentine, diesel
730			G	$17,000	7/05	SCIA	Restored, gas
730			G	$46,000	7/05	SCIA	Diesel, high crop, good tag, restored, 1 of 78 built
730	1958		G	$4,400	11/05	WCIA	Diesel, good rear rubber, electric start, front weights sold separately for $800 and $400
730	1958	5,590	G	$4,700	4/05	SCMI	NF, diesel, electric start, clam fenders, 38" rubber, 5,590 hours on tach
730	1958		E	$5,250	7/05	SENE	Gas
730	1959		G	$4,700	2/05	NENE	Diesel, 16.6-38 tires, direct start, factory 3 pt.
730	1959		G	$6,500	2/05	NWOH	Gas, original, WF, 3 pt., fenders
730	1959		G	$7,100	8/05	WCIA	
730	1959		E	$82,500	6/05	NEOK	LP high crop restored, rare, 1 of 28 built, SN 7312419
7520	1973		G	$14,000	1/08	WCKS	4WD, duals, 4,000 hours on OH, PTO, syncro, John Deere 12' dozer blade
7520	1974		G	$9,000	2/08	NCOH	4WD, ROPS, cab, 3 pt., PTO, 3 hyd.
7520		6,975	G	$1,000	6/07	WCMO	4WD, John Deere diesel engine, QR, rear duals, 1,000-rpm shaft, 3 pt., 18.4-34 tires
7520			P	$1,600	9/07	ECND	4WD, 3 hyd., PTO, 18.4-38 duals, trans. work needed

John Deere

Model	Year	Hours	Cond.	Price	Date	Area	Comments
7520	1973		G	$8,000	3/07	WCIL	4WD, cab, 2 hyd., 3 pt., 1,000 PTO, 18.4-34 duals, runs good
7520	1974	10,075	F	$4,500	3/07	ECND	4WD, CAH, PTO, 3 hyd., 5,000 hours on 50 Series engine, 20.8R-34 duals
7520	1975		G	$5,500	12/07	SEND	4WD, 3 hyd., 1,000 PTO, 23.1-26 insides, 18.4-34 duals
7520	1975	4,011	G	$7,700	4/07	NECO	4WD, diesel, 3 pt., PTO, duals, John Deere 12' dozer
7520	1972	6,313	G	$14,750	12/06	ECNE	4WD, syncro, 16 speed, factory rebuilt 691 engine, 18.4R-34, 2 hyd., 1,000 PTO, 3 pt., quick hitch, differential lock rear end, R134A air
7520	1973		F	$7,000	3/06	SWOH	4WD, cab, diesel
7520			F	$3,700	8/07	SESD	4WD
7520			F	$6,900	12/05	WCWI	4WD, bad PTO
7520			G	$8,000	8/05	NWIL	4WD, diesel, duals
7520			F	$8,700	3/05	NWIL	4WD, diesel, 3 pt., PTO, duals
7520	1972		G	$17,800	9/05	SEMN	4WD, singles, cab, air, 3 pt., 4 hyd., quick hitch, PTO
80			G	$6,500	8/07	SWIL	Diesel, PS, covered flywheel
80		811	G	$15,000	10/06	SENY	Diesel, pony start
80			G	$5,900	11/05	WCIL	Standard, diesel, restored, fenders, lights
80			G	$9,000	3/05	ECMN	Standard diesel, pony start, no PTO, 2 hyd.
80			G	$13,750	6/05	WCMN	1956, diesel, new tires, PS, WF
80	1956		G	$8,000	11/05	SCKY	
80	1956		G	$8,900	9/05	SEMN	Standard, pony start, PS, front/rear weights, strap hitch, 2 hyd.
820	1958		F	$3,700	5/08	ECMN	Unrestored, diesel, standard, pony, air stack
820			G	$6,200	8/07	SESD	
820	1957		G	$3,750	6/07	SCIL	Original condition, pony start, both engines run good, all original John Deere gauges, air stack precleaner, 1 hyd., live PTO, PS
820	1957		G	$4,000	11/07	WCIL	Running, unrestored/original, fenders, lights, hyd. remote, PS, runs
820	1958		G	$3,750	11/07	WCIL	2 cyl.
820	1958		G	$5,500	6/07	SCIL	Diesel, black dash, PS, 2 hyd., Wheatland, live PTO, 3 sets rear wheel weights, fenders have some dents, tin 80%, pony start, big engine, complete OH, older repaint
820			G	$5,300	8/06	SETN	#75 front loader
820			G	$6,000	3/07	WCIL	Pony, fenders, PTO, lights, PS
820			G	$6,250	6/05	ECMN	Diesel, new start, pony start, 1 hyd.
820			G	$8,100	8/05	WCIL	2 cyl., needs paint and tires
820	1956		G	$4,900	11/05	WCIL	Pony start, factory step, PTO, hyd., PS
820	1958	4,467	P	$1,350	9/05	NECO	Runs, pony motor, needs work, diesel
820	1958		G	$11,500	9/05	SEMN	Standard, pony start, PTO, 2 hyd.
830			G	$9,000	3/08	NEMO	Pony motor, diesel
830	1959		E	$11,500	5/08	ECMN	Restored, standard, diesel, pony, air stack, 23.1-26 tires, new tires
830	1974		E	$9,700	1/08	ECOK	John Deere 145 loader, 3 pt., PTO, 1 hyd., 14.9-28 rubber, approx. 489 actual hours
830			F	$3,800	6/07	SCIL	Electric start, PTO, 1 hyd. Wheatland remote, 2 sets rear wheel weights, tin 40%, fenders have holes, engine needs major work, tractor sold with replacement crankshaft, piston, and complete gasket set, not running since 1995
830			G	$13,500	6/07	SCIL	Rice tires, restored, good tractor, hyd.
830	1959		E	$10,100	3/07	NWIL	Diesel, pony start, PS, water pump, 540 PTO, 2 hyd., new 18.4-34 tires

John Deere

Model	Year	Hours	Cond.	Price	Date	Area	Comments
830	1959		G	$12,750	9/07	ECMI	Diesel, WF, pony, gold tag, new paint, sharp, parade tractor, PS
830	1960		G	$8,000	11/07	WCIL	Pony start, original
830			G	$5,000	3/05	WCWA	
830			G	$5,700	3/05	NWIL	Gas, NF, new paint
830			G	$8,000	11/07	WCIL	Diesel, pony, PTO, lights, PS
830			G	$8,900	6/05	ECIL	Pony start, new paint, 2 hyd.
830			G	$10,000	6/05	ECMN	Standard, diesel, 2 hyd., new tires, 6 wheel weights
830			G	$12,250	6/05	ECMN	Standard, diesel, 2 hyd., new paint
8430	1976		F	$4,500	3/08	WCMN	4WD, 3 hyd., PTO, 18.4-38 duals, needs trans. work
8430	1976	9,100	G	$7,500	3/08	SCND	4WD, 2 hyd., 20.8-38 cast duals, poor tires
8430	1976		G	$16,100	1/08	WCKS	4WD, air, 3 pt., PTO, 3,500 hours on 50 Series motor, duals, all new rubber
8430	1977	6200,	G	$7,000	3/08	WCMN	4WD, 3 hyd., 3 pt., PTO, 800 hours on remanu-factured engine, 23.1-30 rubber
8430	1975	4,700	G	$8,100	1/07	NECO	4WD, diesel, Degelman dozer, bareback, PTO, duals
8430	1976	10,792	G	$7,400	4/07	ECND	4WD, CAH, 3 hyd., PTO, 20.8-34 duals, good rubber, 150 hours on rebuilt head
8430	1976	5,080	G	$9,000	6/07	SWMN	4WD, CAH, approx. 1,000 hours on OH, QR, 2 hyd., 3 pt., PTO, quick hitch, 20.8-34 hub duals
8430	1976	9,807	G	$9,000	10/07	NECO	4WD, diesel, 3 pt., PTO, duals
8430	1976	8,058	G	$9,500	4/07	NWMN	4WD, CAH, PTO, 3 hyd., 20.8-34 duals, 2,500 hours on 50 Series engine
8430	1977	7,032	F	$8,200	6/07	NCIN	4WD, cab, 3 hyd., quick hitch, QR, duals
8430	1978		G	$9,000	11/07	SEND	4WD, CAH, 16 speed, 3 hyd., 3 pt., 1,000 PTO, 18.4-38 duals
8430	1978	3,335	G	$20,250	3/07	NWIA	4WD, John Deere 50 Series engine, PTO
8430			G	$8,000	9/06	NCIA	4WD, duals, 2,000 hours on OH
8430		7,349	G	$8,800	6/06	NCIA	4WD, 18.4-38 axle duals, 3 hyd., PTO, 3 pt., quick coupler, QR
8430	1976		F	$6,500	12/06	NWWI	4WD, 3 hyd., 12' dozer blade, 18.4-34 duals
8430	1976	3,700	G	$11,000	3/06	SCIN	4WD, 3 pt., quick hitch
8430	1977	11,900	G	$7,000	7/06	NEND	4WD, QR, 3 hyd., rock shaft, 1,000 PTO, 18.4-38 press steel hub duals
8430	1978	5,900	G	$9,250	2/06	WCIA	4WD, QR, 3 pt., PTO, duals
8430	1978	6,477	G	$18,000	2/06	ECIL	4WD, CAH, QR, PTO, 3 hyd., 4 inside new tires, duals
8430		7,710	G	$7,500	12/05	SCND	4WD, 3,700 hours on rebuilt engine, PTO, 20.8-38 duals (90%)
8430	1977		F	$6,250	9/05	ECND	4WD, 2,000 hours on engine and trans. OH, PTO, 3 hyd., 20.8-34 tires
8430	·1978	12,370	G	$6,000	4/05	ECND	QR, PTO, duals
8430	1978	6,595	G	$8,300	6/05	WCMN	4WD, 3 hyd., 16/4 speed, average rubber
8430	1978	9,200	G	$10,000	6/05	NEND	4WD, CAH, QR, 3 pt., PTO, 3 hyd., duals, low hours on engine, very clean
8430	1978	6,498	G	$10,500	6/05	NENE	4WD, QR, cast duals, radar, 3 hyd., 3 pt., quick hitch, PTO
8630			F	$5,500	3/08	SEWA	4WD, 24.5R-32 duals, 3 pt., PTO, 2 hyd., QR
8630	1976	10,800	G	$7,500	3/08	SCND	4WD, 2 hyd., PTO, 20.8-38 cast insides, 20% rubber
8630	1976		G	$17,000	2/08	NCOH	4WD, CAH, decals, 3 hyd., no PTO, QR
8630	1976		G	$20,000	1/08	WCIL	4WD, 3 hyd., QR, 3 pt., quick hitch, 1,000 PTO, new 20.8-34 tires, duals
8630	1978	1,979		$0	2/08	NCOH	4WD, CAH, 3 hyd.., PTO, 3 pt., duals
8630	1978	1,979	G	$20,500	2/08	NCOH	4WD, CAH, 3 hyd., PTO, 3 pt., duals

John Deere

Model	Year	Hours	Cond.	Price	Date	Area	Comments
8630		1,366	G	$7,500	3/07	SEND	4WD, 50 Series engine, Goodyear Dyna Torque 20.8-42 cast hub duals with band triples
8630		7,200	G	$8,700	8/07	SESD	4WD, 3 pt., PTO
8630			G	$9,000	7/07	WCSK	Canadian sale, 4WD, 18.4-38 duals
8630	1975	5,530	G	$10,250	11/07	SEND	4WD, CAH, QR, 3 hyd., 24.5-32 band duals, 50 Series engine, always shedded
8630	1975		G	$16,000	12/07	NWOH	4WD, CAH
8630	1975	10,959	G	$16,000	3/07	WCMI	4WD, 198 hours on John Deere 50 Series engine, cab, 3 pt., 3 hyd., PTO
8630	1975	11,553	G	$19,500	3/07	WCMI	4WD, 1,199 hours on OH, John Deere 40 Series engine, cab, 3 pt., 3 hyd., PTO, duals
8630	1976	7,127	G	$7,700	10/07	NCMI	4WD, diesel, CAH, 3 pt., 1,000 PTO, quick hitch, 20.8-38 tires, duals
8630	1976	6,594	G	$12,600	11/07	ECIL	4WD, one owner, Degelman 12' front dozer blade, 3 years on rebuilt engine, quick hitch, 18.4-38 duals
8630	1977	5,404	G	$6,250	6/07	SWMN	4WD, CAH, QR, 3 hyd., 3 pt., PTO, 30.5L-32 inside, 18.4-38 outside
8630	1978		G	$5,500	7/07	SEND	4WD, 3 hyd., PTO
8630	1978	4,281	G	$12,250	12/07	SCMN	4WD, PTO, duals
8630			P	$200	1/06	WCCA	For parts
8630			F	$6,000	2/08	WCMN	4WD, PTO, 18.4-38
8630		7,080		$8,000	9/06	SEIL	4WD, 3 pt., quick hitch, 18.4-38 axle duals, 3 hyd.
8630	1975	7,756	G	$12,100	8/06	WCMN	4WD, CAH, 16-speed quad, 3 hyd., PTO, 23.1-30 tires, out of frame major, 50 Series engine
8630	1976	5,577	F	$7,000	3/06	NWOH	4WD, 3 pt., 3 hyd., cab, duals
8630	1977	6,100	G	$11,250	11/06	SCSD	4WD, 50 Series engine, new 18.4-38 tires, PTO, 3 pt., 3 hyd.
8630	1978	9,870	P	$8,800	2/06	NCKS	4WD, articulated diesel, QR, 3 hyd., 3 pt. with category 3, recent valve job
8630	1978		G	$12,750	11/06	NWKS	4WD
8630	1978	3,496	E	$20,000	3/06	SEIA	4WD
8630		7,000	G	$0	3/05	NCOH	No sale at $15,750, 50 Series motor, 3 pt., quick hitch, PTO, duals
8630			P	$3,750	9/05	SESD	4WD
8630		9,818	F	$6,100	8/05	WCMN	4WD, PTO, 3 pt. hitch, 3 hyd., 2,000 hours on complete OH
8630		5,200	G	$7,750	8/05	WCIA	4WD, 50 Series engine
8630			G	$11,600	3/05	WCIL	4WD, 1,000 hours on new 50 Series engine
8630	1975	6,492	G	$16,000	1/05	NWIL	4WD, 3 pt., 3 hyd., PTO, 18.4-38R
8630	1976	10,011	G	$6,500	2/05	ECND	4WD, QR, 20.8×34 tires
8630	1976		G	$7,300	4/05	NEKS	4WD, duals
8630	1976	7,480	G	$8,100	9/05	NWIA	4WD, QR, 3 hyd., duals all around, diesel
8630	1977		G	$4,250	7/05	ECND	50 Series engine, PTO
8630	1977		G	$4,250	7/05	ECND	50 Series engine, PTO
8630	1977		G	$18,000	2/05	NENE	4WD, QR, 1,670 hours on 50 Series engine, 18.4-42, 10-bolt duals, 3 pt., quick hitch, 3 hyd., PTO
8640	1978	7,412	F	$8,500	8/05	SEMI	4WD, cab, 3 hyd., 3 pt., duals
950	1978		G	$3,700	12/07	SEWY	WF, 3 pt., diesel
950	1978	2,200	P	$700	8/05	SEIA	Bad engine, weights sold separately for $536
A			P	$650	3/08	SCMN	Stuck engine, PS, flat tires
A			P	$1,000	6/07	SCMN	Stanhoist loader, runs, very poor tires
A			F	$1,000	3/08	NWKS	
A			E	$3,100	5/08	ECMN	Restored, newer tires
A	1934		G	$7,200	4/08	NENE	NF, open fan shaft, mostly original, runs

John Deere

Model	Year	Hours	Cond.	Price	Date	Area	Comments
A	1936		G	$1,600	4/08	NENE	Unstyled, NF, hyd. lift
A	1936		G	$2,500	4/08	NENE	Unstyled, NF
A	1936		E	$3,500	5/08	ECMN	Restored, round spokes, unstyled, shutters, newer tires
A	1937		G	$1,600	4/08	NENE	General-purpose tractor, unstyled, NF, cutoff round spoke rear wheels, round spoke front wheels
A	1937		G	$1,900	4/08	NENE	Unstyled, NF
A	1937		G	$2,100	4/08	NENE	Unstyled, NF, road gear
A	1937		G	$2,650	1/08	SESD	Motor OH
A	1937		G	$2,900	4/08	NENE	Unstyled, NF, flat spoke rear wheels, spoke front wheels
A	1938		G	$2,200	4/08	NENE	Unstyled, NF, factory flat rear wheels, hyd. lift, restored
A	1938		G	$2,600	4/08	NENE	Unstyled, NF, cast iron split rear wheels, road gear
A	1938		G	$5,100	4/08	NENE	Unstyled, NF, spoke rear wheels with spade lugs, spoke front wheels
A	1944		F	$475	4/08	NENE	Styled, NF, electric start
A	1945		P	$550	4/08	NENE	For parts, styled, NF
A	1946		F	$500	4/08	NENE	Styled, NF, electric start
A	1949		E	$2,800	5/08	ECMN	Restored, WF, styled, newer tires
A	1951		F	$1,200	4/08	NENE	Styled, Roll-o-Matic NF, spoke rear wheels with spade lugs, spoke front wheels
A	1951		G	$2,200	4/08	ECMO	Roll-o-Matic
A	1952		G	$2,950	3/08	NCIL	NF, repaint, 12.4-38 tires
A			P	$375	5/07	ECND	For parts
A			F	$620	6/07	WCSD	NF
A			G	$975	8/07	NWIA	New M&W oversize pistons, new lights and gears
A			G	$1,300	11/05	WCIL	
A			G	$1,300	11/05	WCIL	Clamshell fenders
A			E	$1,500	4/07	SCKS	Unstyled, restored, ran great
A			G	$1,550	1/06	SESD	Roll-o-Matic
A			G	$1,550	6/07	NCIN	Late model, styled, pressed frame, square axle
A			G	$2,000	11/05	WCIL	Electric start, older restoration 7 to 8 years ago, long hood, pressed steel front and rears, runs
A			G	$4,500	6/07	SCIL	Restored, all new tires
A	1935		F	$1,000	9/07	WCIL	Brass tag, open fan shaft, cutoffs, no motor
A	1936		P	$350	9/07	WCIL	For parts, cast rear
A	1936		P	$400	9/07	WCIL	For parts, cut off rears, pressed steel fronts, rough
A	1936		G	$2,500	7/07	ECIL	Unstyled, pulling tractor, NF, round spoke front wheels, rear on cut-offs, PTO, older restoration
A	1936		F	$3,200	6/07	SCIL	Unstyled, old repaint, flat spokes, flat-top fenders, double stick trans., dual fuel, rear steps, hand start, hood 95%
A	1936		G	$3,600	12/07	SEMI	
A	1936		G	$4,000	4/07	NEMO	Unstyled, Van Zante silk-screened hood, round spokes front and rear, tires 60%, good repaint, nice straight tractor
A	1937		F	$3,000	1/07	NCIL	Spoke wheels, John Deere 290 corn planter, complete, not running
A	1937		F	$3,250	9/07	WCIL	All steel, rear skelton, no carb., not running, not stuck
A	1937		F	$3,300	9/07	WCIL	All steel front spokes, skeleton rears, flat back, unrestored, no carb., no exhaust or magneto

John Deere

Model	Year	Hours	Cond.	Price	Date	Area	Comments
A	1937		F	$3,600	9/07	WCIL	All steel, skeleton rear, flat back, not stuck or running
A	1938		G	$3,000	9/07	WCIL	Flat back, front F&H cast, factory flat spoke rear, older restoration, runs
A	1939		G	$1,300	6/07	SCIL	Styled, round rear spokes, 12-spline axle, dual fuel, styled, 4 speed, hand start, straight sheet metal
A	1940		P	$500	6/07	SCIL	Styled, for parts, Behlen road gear box, high-speed road gear, hand start, spring seat with ratchet adjustment
A	1940		G	$1,600	11/07	SESD	
A	1941		F	$600	6/07	SCIL	Styled, electric start, restorable
A	1941		F	$900	8/07	WCMN	NF, Roll-o-Matic
A	1941		F	$975	9/07	SCMI	NF, gas
A	1941		G	$1,700	6/07	NCIN	Styled, NF, fenders, good rear tires, electric start, older restoration
A	1941		G	$1,850	2/07	SCKS	NF, 12-38 rear tires, 6:00-16 front tires
A	1942		G	$1,000	8/07	SESD	Starter and lights, restored
A	1942		G	$1,625	1/07	ECNE	Gas, NF, 11-38 tires
A	1943		G	$1,000	8/07	SESD	Restored
A	1944			$0	8/07	SESD	Runs
A	1944		F	$750	8/07	SESD	
A	1945		F	$1,300	3/07	SEWY	Gas, single front
A	1946		G	$1,600	11/07	WCIL	Running, restored, PTO, lights, hyd. remote, belt pulley, new rings, ground values, new voltage regulator and generator, new starter switch
A	1946		G	$3,800	6/07	NCIN	Electric start, restored, runs
A	1947		F	$700	4/07	NCNE	A good restoration project
A	1947		G	$1,375	11/07	SESD	NF, OH
	1947		G	$2,100	2/07	SCKS	Electric start, 12.4-38 rear tires, 6:00-16 front tires, engine free
A	1948		F	$500	2/07	SCKS	Roll-o-Matic, NF, 13.6-38 rear tires
A	1948		F	$900	9/07	WCIL	Late style, round axle, runs
A	1948		F	$1,050	9/07	SCMI	NF, gas, fenders
A	1948		G	$1,200	8/07	WCMN	NF, like-new 13.6-38 tires, unrestored, runs good
A	1948		G	$1,550	5/07	WCWI	
A	1948		G	$1,700	8/07	ECNE	Good rubber and paint, 3 pt. sold seperately for $500
A	1948		G	$1,750	6/07	SCMI	Gas, NF
A	1949		F	$500	8/07	WCMN	NF, 13.6-38 rears, 5-50×16 fronts
A	1949		P	$575	6/07	NCIN	Late style, pressed frame, round axle, good rear tires
A	1949		G	$1,000	12/07	SEWY	PowrTrol, gas
A	1949		F	$1,100	6/07	NCIN	Unrestored original, late style, round axle
A	1949		G	$1,950	9/07	NECO	Gas, single front, 3 pt., PowrTrol
A	1949		F	$3,750	6/07	SCIL	Styled, loader, open flywheel, electric start, original
A	1950		G	$2,100	9/07	WCIL	Late style, square axle, clamshell fenders, fair original, runs, just rebuilt carb.
A	1951		G	$1,300	11/07	WCIL	
A	1951		G	$1,300	12/07	SEWY	Gas, 3 pt., single front
A	1951		G	$1,500	9/07	WCIA	Repainted
A	1951		F	$1,575	1/07	NEOH	Styled, no 3 pt.
A	1951		G	$1,700	11/07	WCIL	
A	1951		G	$1,900	8/07	WCIL	Restored
A	1952		P	$650	9/07	WCIL	Water pump

John Deere

Model	Year	Hours	Cond.	Price	Date	Area	Comments
A	1952		E	$5,000	6/07	NWIL	Gas, WF, live hyd., no 3 pt., new paint and tires
A			F	$275	6/06	ECSD	Slant dash, not running
A			P	$300	3/06	WCMN	1 hyd., PTO
A			F	$525	11/07	NCIA	12 volt
A			F	$700	10/06	WCWI	
A			G	$700	4/06	WCTX	Skeleton rears, power lift
A			P	$725	8/06	NCOH	Head cracked
A			F	$800	3/06	NCWI	Roll-o-Matic NF, fenders
A			G	$1,250	3/06	SEMN	Styled
A			G	$1,275	3/06	SEMN	Styled
A			G	$1,400	10/06	WCMN	Styled, lights, starter
A			F	$1,800	5/06	SEWI	
A			G	$1,900	8/06	SEPA	NF, good rubber and paint
A			G	$2,000	4/06	WCTX	Rear round spokes, front cutoffs, flat back
A			G	$2,200	7/06	NWIL	SN 453553, unrestored/original, PTO, belt pulley
A			G	$2,200	11/07	WCIL	Unrestored, PTO, belt pully
A			G	$2,300	12/06	NWIL	SN 464109, running, old restoration/repaint, starts and runs good, motor and trans. good, tires poor
A			G	$2,500	6/06	NEIN	Open fan, full steel
A			G	$2,600	7/06	NWIL	SN 459212, running, restored, good paint job, new hood, very nice looking tractor
A			G	$2,850	6/06	NWIL	Good rubber, hand start
A			F	$3,100	3/06	SEPA	Unstyled, spoke wheels, fair rubber
A			G	$4,250	8/06	SCIL	WF, rear fenders
A			G	$4,750	8/06	SCIL	
A			G	$5,500	7/06	WCIL	Running, restored, PTO, belt pully
A			G	$5,500	7/06	NWIL	SN 474579, running, restored, PTO, belt pulley
A	1935		G	$3,300	9/06	SCNE	General purpose, open shaft, spoken wheels, restored
A	1936		E	$3,250	10/06	WCMN	Spoke steel wheels all around
A	1937		F	$1,100	11/06	WCMN	Unstyled, on rubber, no running
A	1937		F	$1,600	8/06	NCIA	
A	1937		G	$2,050	12/06	WCIL	Gas, hand start
A	1937		G	$2,300	9/06	NEIN	Gas, unstyled, repainted
A	1937		G	$3,350	1/06	SCKS	14.4-36 tires on round spoke rims, SN 459607
A	1938		F	$850	9/06	ECMI	Unstyled, hand start
A	1939		G	$1,400	6/06	WCMN	On rear double steel
A	1939		G	$1,675	10/06	SCMN	Styled, all fuel, hand start, new rubber
A	1941		G	$1,500	9/06	SCNE	Complete OH
A	1941		G	$2,300	11/06	ECIL	
A	1943		G	$1,250	3/06	SWKS	NF, new paint
A	1943		G	$1,300	11/06	NEKS	Styled, WF, diesel, 12.4-38 rear, 5.50-16 front tires, 3 pt., John Deere 435 loader
A	1946		P	$300	10/06	SCSD	Complete, not running, gas
A	1947		F	$750	3/06	NCNE	11-38 rear tires
A	1947		F	$2,500	1/06	NCCO	WF, OH, only needs paint to finish restoration
A	1949		G	$550	9/06	NEIN	Gas, on steel, NF, styled
A	1949		G	$1,000	9/06	ECMI	Styled
A	1950		F	$825	8/06	NCIA	Styled, PS, restored
A	1950		F	$950	10/06	WCMN	NF, all original, dual fuel, square rear axle, water pump
A	1950		G	$1,400	8/06	NCIA	New water pump
A	1951		G	$1,700	10/06	SCMN	Gas, electric start, all original
A	1951		G	$2,100	9/06	ECNE	NF, repainted

John Deere

Model	Year	Hours	Cond.	Price	Date	Area	Comments
A	1951		G	$2,250	9/06	NENE	Gas, Roll-o-Matic NF, 12.4-38 rear tires, 1 hyd., rear wheel weights, one owner, shedded
A	1952		F	$1,500	2/06	NECO	Gas, 3 pt., double front
A	1952		G	$1,650	9/06	NCMI	Gas, electric start, rear blade
A	1952		G	$2,000	10/06	SCSD	Split pedestal Roll-o-Matic, gas, lots of new parts (radiator, gauges, carb.)
A			F	$700	3/05	SEIA	
A			F	$750	8/05	ECMI	Good sheet metal, needs work, no spark
A			P	$750	6/05	ECWI	For parts
A			F	$900	12/05	WCWI	
A			G	$1,100	1/08	WCIL	NF, hand start crank, restored
			G	$1,100	1/08	WCIL	NF, all fuel, PTO, lights, belt pully, styled sheet metal, hyd. lift
A			G	$1,300	6/05	ECWI	9 bolt
A			G	$1,350	7/05	SCIA	
A			F	$1,500	8/05	ECMN	Older loader
A			G	$1,600	1/05	NENE	PS, NF, clean
A			G	$1,650	6/05	ECWI	
A			G	$1,700	10/05	WCIL	NF, M&W pistons
A			G	$2,600	7/05	SCIA	Unstyled, open fan shaft, restored
A			G	$2,800	11/05	WCIL	NF, PTO, lights
A			G	$3,100	9/06	ECMN	Tricycle, PTO, fenders, new tires, water pump
A			G	$3,250	6/05	ECMN	2R corn picker mounted, 51-A, tricycle, new tires and paint
A			G	$5,000	6/05	ECWI	Painted, rear round spokes
A			G	$7,800	7/05	SCIA	High crop, good tag, restored, 1 of 246 built
A	1936		G	$1,250	1/05	NENE	NF, unstyled, Farmhand F10 loader
A	1936		G	$1,800	4/05	SCNE	Unstyled, 12.4-38 tires on spoke rims, rear wheel weights, spoke front wheels, runs
A	1936		F	$2,500	1/05	NWIL	Unstyled
A	1936		G	$3,500	2/05	NWOH	Unstyled, fenders, shutters
A	1936		G	$7,250	9/05	SEMN	Unstyled factory rounds, rear weights
A	1937		G	$1,750	1/05	NEIA	Unstyled
A	1941		G	$1,300	12/05	NWIL	Gas
A	1943		P	$600	7/05	NEIA	Hand start, gas, not running
A	1943		F	$800	6/05	WCMN	Original
A	1946		F	$1,000	2/05	NWOH	Buzz saw, original
A	1946		F	$1,100	2/05	WCOK	LP, ran, row crop, styled
A	1947		G	$1,300	2/05	NEIN	Gas, NF, good rubber
A	1947		F	$1,300	10/05	WCIL	
A	1947		G	$1,700	2/05	NENE	Slant dash, electric start
A	1948		P	$400	2/05	NWOH	Parts tractor
A	1948		G	$1,000	1/05	SENE	Gas, NF, PS, Roll-o-Matic
A	1949		F	$500	3/05	NCIA	New rear tires
A	1949		G	$1,400	12/05	ECIL	
A	1949		G	$4,500	1/05	NENE	NF, 2R front-mount cultivator, one owner
A	1950		P	$550	7/05	NEIA	1 hyd., gas, not running
A	1950		G	$1,800	7/05	NEIA	1 hyd., 2 gas tanks, 11-38 tires
A	1951		G	$1,250	2/05	NEIN	Gas, repainted
A	1951		G	$2,250	7/05	SCIA	Row crop, restored
A	1951		G	$3,650	1/05	NENE	NF, 12.4-38 tires
A	1951		G	$6,700	11/05	SCKY	
A	1952		G	$1,350	2/05	SCMN	
A	1952		G	$1,550	4/05	NENE	12.4-38 tires, fuel burner
AO			G	$8,250	10/06	SENY	Older restoration
AO			G	$8,000	8/05	NEIN	Unstyled, restored, shutters

John Deere

Model	Year	Hours	Cond.	Price	Date	Area	Comments
AO			G	$10,000	9/06	NEIN	Unstyled, restored, factory front round spokes
AOS			G	$5,500	9/06	NEIN	Started restoring, sheet metal in primer
AOS			G	$7,750	7/05	SCIA	Restored
AR	1953		G	$2,850	4/08	NENE	Styled, WF, electric start
AR	1937		F	$2,100	6/07	SCIL	Unstyled, very straight hood, fenders have a few dents, cut offs, older paint, runs good, 4 speed, off-set radiator cap
AR	1938		F	$1,700	9/07	WCIL	Teardrop drawbar, cutoffs, stuck
AR	1952		G	$1,600	11/07	SWIN	Straight sheet metal
AR	1952		G	$2,300	11/07	SWIN	Fairly straight
AR			F	$1,700	5/06	SEND	WF, PTO, 6 speed, hyd.
AR			G	$2,900	6/06	ECMN	Gas
AR			G	$3,750	9/06	NEIN	Styled
AR			G	$3,900	9/06	ECMI	WF
AR	1936		G	$3,350	10/06	SCMN	Restored, new rubber, very sharp
AR	1937		G	$7,000	10/06	WCMN	
AR	1952		G	$5,300	10/06	SCMN	Restored, original, lights, very sharp
AR			G	$1,700	7/05	SCIA	
AR			G	$1,850	9/05	SCON	Canadian sale
AR			G	$2,250	7/05	SCIA	Restored
AR			G	$4,600	6/05	ECMN	Gas, PTO, 1 hyd., new tires, new paint
AR			G	$5,100	6/05	ECWI	Painted
AR	1937		G	$3,200	9/05	SEMN	Unstyled, flat rear spokes, shutters
AW	1951		F	$1,700	6/07	SCIL	Lights, round tube WF, single stick trans., tin good with muffler cut out, rock shaft, no 3 pt., cast rear wheels, creeper first gear
AW	1946		G	$3,000	10/06	WCMN	WF, eagle beak, original
AW	1948		G	$1,600	10/06	WCMN	WF, eagle beak, original
AWH	1949		G	$4,500	5/08	ECNE	WF, frame off restoration, new rubber
AWN			G	$2,000	5/08	ECNE	Rebuilt motor, original
B			F	$1,225	2/08	NCOH	New rubber and radiator, styled
B			F	$1,250	2/08	WCOH	New rubber
B			G	$1,700	3/08	ECWI	Unstyled, new rear tires, flat spoked front and rear
B			E	$2,400	5/08	ECMN	Restored, styled, newer tires
B			G	$2,600	6/05	ECMN	Restored, newer tires
B			G	$2,600	6/05	ECMN	Restored, newer tires
B	1935		G	$3,250	4/08	NENE	Unstyled, NF, round spoke front & rear wheels
B	1935		G	$5,500	4/08	NENE	Unstyled, NF, Texas sand wheels
B	1936		G	$3,200	4/08	NENE	Unstyled, NF, hand crank start, spoke rear wheels
B	1937		F	$800	4/08	NENE	Unstyled, NF, flat spoke rear wheels, engine stuck
B	1938		F	$1,600	4/08	NENE	Unstyled, NF, engine stuck
B	1938		G	$1,950	3/08	SEMN	
B	1938		G	$2,150	4/08	NENE	Unstyled, NF, flat spoke rear wheels, hand crank start
B	1940		E	$3,700	5/08	ECMN	Restored, styled, slant dash, half round fenders, 12.2-38 tires, newer tires
B	1941		G	$1,500	4/08	NENE	Styled, NF, hand crank start
B	1942		F	$650	4/08	NENE	Restored, styled, NF, spoke front wheels, hand crank start
B	1942		F	$750	4/08	NENE	Styled, NF, hand crank start
B	1945		E	$2,200	1/08	ECIL	6' Woods belly mower
B	1946		F	$800	4/08	NENE	Styled, NF, pressed steel rear wheels
B	1949		G	$2,750	4/08	NENE	Styled, WF, electric start, restored
B	1950		G	$1,100	4/08	NENE	NF, rear tires

John Deere

Model	Year	Hours	Cond.	Price	Date	Area	Comments
B	1951		G	$1,700	3/08	SCKS	Runs good
B			P	$400	10/07	NEIA	For parts
B			P	$500	10/07	NEIA	For parts
B			F	$950	11/07	SESD	
B			G	$1,100	7/07	SEND	NF, reconditioned
B			G	$1,500	9/07	NCND	Gas, NF, good tires, 540 PTO, wire winder
B			G	$1,800	9/05	NENE	Roll-o-Matic, NF, runs great
B			G	$2,100	4/07	NEMI	Single wheel front, front-mounted buzz saw
B			G	$3,600	11/07	WCIL	Styled, hand start
B			G	$4,100	6/07	SCIL	1964 California tractor, 42" special dish wheels, rims and tires, adjustable WF, single-stick trans.
B	1935		G	$3,600	11/07	WCIL	Running, old restoration/repairt, PTO, loop drawbar, runs well
B	1935		G	$8,500	9/07	WCIL	Unstyled, brass tag, full steel, older restoration, rear skeleton steel, flat back, runs
B	1936		G	$1,500	8/07	WCMN	Unstyled
B	1936		G	$3,000	11/07	WCIL	Running, old restoration/repaint
B	1937		G	$2,100	9/07	NWOH	Unstyled, 36" rear tires, front-mount cultivator for John Deere A
B	1937		G	$2,500	11/07	WCIL	
B	1937		G	$2,900	9/07	WCIL	Front and rear cutoffs, older restoration, flat back, runs, rebuilt in 1999
B	1937		G	$5,250	6/07	SCIL	Unstyled, factory rear round spokes, new clutch plates, new paint, many items rebuilt, many new parts
B	1938		G	$2,200	7/07	ECIL	Unstyled, NF, rear round spoke wheels with 85% tread, PTO, nice restoration
B	1938		G	$5,600	9/07	WCIL	Front and rear spokes, flat back, runs
B	1939		G	$1,000	8/07	SWIL	High post, round spoke front wheels, flat spoke rear wheels, open flywheel
B	1939		G	$1,350	11/07	WCIL	Running, unrestored/original, PTO, belt pulley
B	1939		G	$2,800	9/07	SCMO	Restored, real good rubber, 11.2×38 tires, SN 70510
B	1940		P	$150	6/07	SCIL	For parts, pan seat, rear axle 12-spline, 9-hole rim
B	1940		G	$2,200	11/07	WCIL	Hand start
B	1941		F	$1,100	2/07	SCKS	Electric start, 11.2-38 rear tires, 5:00-15 front tires
B	1941		F	$1,200	6/07	NCIN	Electric start, fenders
B	1941		G	$1,400	8/07	SESD	Shedded
B	1943		F	$900	9/07	WCIL	On cutdowns front and back, hand start, flat back, 1 new rear, older repaint, runs, rebuilt carburetor and magneto
B	1944		F	$980	12/07	SCNE	
B	1944		F	$1,300	6/07	NCIN	Electric start, wrong seat, grill damaged
B	1944		G	$1,600	9/07	WCIA	Hand start, older repaint
B	1945		F	$500	6/07	SCIL	Styled, flat-top fenders, good tires, electric start
B	1945		F	$750	5/07	WCWI	
B	1946		P	$400	9/07	WCIL	Long hood, stuck
B	1946		F	$900	10/07	NEIA	Electric start
B	1947		F	$650	6/07	NCIN	NF, unrestored, one rear tire missing, nice tin
B	1947		G	$1,300	9/07	WCIA	Original
B	1947		G	$1,750	8/07	WCIL	Gas, restored
B	1947		G	$3,000	9/07	ECMI	NF
B	1947		G	$4,100	7/07	WCSD	Restored, 12 volt, new tires

John Deere

Model	Year	Hours	Cond.	Price	Date	Area	Comments
B	1948		P	$725	8/07	SESD	Good rubber, motor loose
B	1948		F	$1,600	1/07	NCIL	Not running
B	1949		F	$600	6/07	SCIL	Styled, to restore, decent tires, flat-top fenders, partially disassembled, all parts present
B	1949		F	$1,000	6/07	NCIN	NF, patch hood, tires nice but old
B	1949		G	$1,400	8/07	ECIA	Gas, row crop
B	1949		F	$1,500	9/07	WCIL	Older restoration, late style, fairly straight, uncut hood, runs, did carburetor work two to three years ago
B	1949		G	$2,050	9/07	NWIL	Gas, PowrTrol, cylinder, belt pulley, 2 hyd., NF, Roll-o-Matic
B	1949		G	$3,000	3/07	NWIL	Gas, power lift, good original tires, pressed steel rear wheels, new paint
B	1950		P	$675	8/07	SESD	Runs
B	1950		G	$2,250	4/07	NEMO	Good repaint, good tires, used around the farm, great runner
B	1950		G	$3,500	11/07	WCOH	3 pt., fenders
B	1951		G	$1,800	2/07	NCCO	
B	1951		G	$2,000	6/07	SCIL	Runs good, older restoration, good rubber
B	1951		G	$2,100	6/07	SCIL	Styled, cultivators, lights, good tires, single stick trans., Roll-o-Matic, hood has muffler cut, quick-tach, rolling shields, nice old original
B	1952		G	$2,500	9/07	WCIA	Repainted, fenders
B			P	$200	10/06	WCWI	To restore or for parts
B			P	$200	10/06	WCWI	To restore or for parts
B			F	$375	8/06	SCMN	Electric start, lights, needs work
B			P	$525	10/06	WCWI	To restore or for parts
B			F	$850	5/06	SEND	6 speed, running
B			G	$1,000	3/06	NWOH	Gas, NF, restored
B			P	$1,100	3/06	NWOH	Did not run, gas, NF
B			G	$1,250	10/06	SEWI	Long hood, electric start
B			G	$1,650	3/06	SEMN	Styled
B			G	$1,750	6/06	SWMN	Restored
B			G	$2,100	5/08	ECMN	Gas, NF
B			G	$2,200	3/05	ECMN	Gas, WF
B			G	$2,300	7/06	SCIA	
B			G	$2,500	8/06	SEPA	NF, average rubber, 3 pt.
B			G	$3,000	3/08	SCIL	Unicycle
B			G	$4,200	10/06	SENY	Mounted 6R cultivator, high clearance, single tire front
B			G	$4,250	8/06	SCIL	WF
B			F	$4,750	8/06	SCIL	WF
B			G	$5,000	7/06	NWIL	SN 1252, running, restored, completed and shipped December 1934, originally held by factory
B	1936		G	$1,350	8/06	NEIA	Runs good, straight tin, needs paint
B	1936		G	$1,900	9/06	ECMI	Unstyled, hand start
B	1938		P	$725	5/06	SEND	Unstyled, engine stuck
B	1939		G	$1,000	10/06	SCMN	Hand start, complete, original
B	1941		G	$900	10/06	SCMN	Hand start, complete, original
B	1941		G	$2,550	12/06	NCIA	6 speed
B	1942		F	$900	9/06	NENE	NF, 10-38 rear tires, spoke front rims, hyd. cylinder, one owner, shedded
B	1942		G	$1,750	2/06	NCCO	Tri front, rear blade
B	1944		G	$1,200	10/06	SCMN	Hand start, complete with John Deere #5 mower
B	1948		P	$395	5/06	SEND	Needs magneto

John Deere

Model	Year	Hours	Cond.	Price	Date	Area	Comments
B	1948		F	$800	9/06	SCNE	NF, 11.2-38 tires
B	1948		G	$2,000	5/08	ECNE	Good tires, PowrTrol
B	1949		G	$1,800	10/06	SCMN	Electric start, all original
B	1950		G	$0	4/06	NWSD	No sale at $2,500, NF, new tires, runs good
B	1951		F	$1,600	7/06	ECIA	NF, ran OK, very straight, easy restore project, 2R cultivator
B			P	$150	8/05	SCMI	For parts
B			P	$650	7/05	SCIA	
B			F	$800	9/05	NENE	Later model, one of two on same sale
B			G	$900	11/05	WCIL	Styled, gas
B			F	$985	9/05	NENE	Later model
B			G	$1,125	8/05	NWIL	Gas
B			F	$1,400	3/06	NWIL	Gas, full steel
B			G	$1,400	6/05	ECWI	Square axle, restored
B			G	$1,800	6/05	ECMN	Tricycle, repainted
B			G	$1,900	5/08	ECMN	Tricycle, fenders
B			G	$1,900	7/07	ECIL	New tires all around, factory
B			G	$2,000	11/05	WCIL	Unstyled, NF, all fuel, PTO
B			G	$3,100	6/05	ECIL	Long frame, expo restoration, everything new or repaired, new tires all around
B			G	$3,600	6/05	ECWI	
B			G	$4,700	11/05	SCKY	
B			G	$13,000	11/05	SCKY	
B	1935		G	$1,900	11/05	WCIL	Unstyled, NF, gas
B	1936		F	$1,450	6/05	WCMN	Stuck PTO
B	1936		G	$3,400	11/05	WCIL	Hand start, spokes, belt pully
B	1937		G	$2,100	1/05	NEIA	Unstyled
B	1937		G	$2,150	2/05	NWOH	Restored on hard rubber,
B	1937		G	$3,850	9/05	SEMN	Unstyled, factory round spokes front and back, rear weights
B	1937		G	$9,000	6/05	ECMN	Gas, tricycle, round spokes front and rear, new paint, OH
B	1938		G	$1,400	4/05	SCNE	Unstyled, 10-36 tires, spoke wheels, runs
B	1938		G	$2,000	11/05	WCIL	NF, gas, cutoffs
B	1938		G	$2,100	7/05	NCIA	
B	1939		G	$1,650	1/05	NEIA	
B	1939		G	$2,000	11/05	WCIL	NF, hand start, spokes
B	1940		G	$1,400	11/07	WCIL	Motor OH, good tin, gas
B	1941		G	$1,700	7/05	SESD	
B	1943		G	$2,400	1/05	NENE	NF, John Deere #5, 7' mower, 11-38 tires
B	1946		G	$1,100	9/05	NECO	3 pt., new rear tires, gas
B	1947		G	$5,900	7/05	SCIA	42" rears, single front, restored
B	1948		G	$1,200	11/05	NCIA	
B	1948		G	$1,400	1/05	NEIA	
B	1948		G	$2,600	9/05	SEMN	Styled, rear weights, fenders
B	1948		G	$2,700	8/05	NCIA	Single front wheel, repainted
B	1948		G	$2,900	6/05	ECIL	New paint, hyd., shutters, Roll-o-Matic
B	1948		G	$3,100	8/05	ECMI	Wide axle, 42" wheels, new tires
B	1949		F	$1,000	7/05	NEIA	1 hyd., gas, 12.4-38 tires
B	1949		G	$2,050	7/05	SCIA	Single front
B	1949		G	$2,600	2/05	NCKS	Restored, new rubber, electric start
B	1950		G	$1,525	1/05	NEIA	
B	1950		G	$1,650	8/05	SWOH	
B	1950		G	$1,900	12/05	SEMN	Electric start, restored
B	1951		F	$900	9/05	WCNE	Gas, 3 pt., single front

John Deere

Model	Year	Hours	Cond.	Price	Date	Area	Comments
B	1951		F	$1,200	11/05	NEIA	Gas, NF, 10-38 tires, starts and runs but seldom used
B	1951		G	$1,600	7/05	NEIA	1 hyd., gas
B	1951		G	$3,300	2/05	NWOH	Fenders, good rubber
B	1952		G	$1,100	1/05	NEIA	
B	1952		G	$1,350	2/05	SCMN	Restored
B	1952		G	$2,150	10/05	WCIL	New paint, needs trans. work
BI			G	$27,000	7/05	NEIA	Round spoke fronts, 1 of 183 built
BO	1937		E	$5,200	5/08	ECMN	Restored, orchard, newer tires
BO			G	$9,250	8/05	NEIN	Restored, round spokes front and rear
BO			G	$9,900	8/05	NEIN	Restored, factory 7" lights
BO			G	$9,900	9/06	NEIA	Lindeman, repainted
BO			G	$10,250	8/05	NEIN	Restored, round spokes front and rear
BO	1947		G	$15,500	7/05	SCIA	Lindeman, restored
BR	1937		F	$4,500	9/07	WCIL	Round spokes, cutoff front, teardrop drawbar, runs, has had some carburetor work - needs some more
BR	1941		G	$7,500	4/07	NEMO	All new rubber, hand start
BR	1945		G	$2,800	6/07	SCIL	Great all original, flat spoke rear wheels, electric start, dual fuel, open flywheel, PTO, double stick trans., hood 90%, fenders 75%, original paint
BR			G	$5,400	6/05	ECWI	Painted
BR			G	$5,500	7/05	SCIA	Full steel, restored
BR			G	$7,000	6/06	NEIN	Brass tag
BW			G	$35,000	7/06	WCIL	Running, restored, fenders, rare long frame, rare
BW			E	$35,000	7/06	NWIL	SN 57384, running, restored, fenders, rare long frame, very few built
BW	1939		E	$31,500	10/06	WCMN	Fenders, eagle beak, new paint, sharp
BW			G	$1,800	6/05	ECMN	WF
D			G	$7,500	3/08	WCOH	Restored
D	1944		G	$3,500	4/08	NENE	Styled, WF, hand crank start, PTO, restored
D	1947		G	$5,000	1/08	ECNE	Gas, SN 169513, styled, WF, 6-volt system, electric start, newer 16.9-30 rear tires and cast wheels, fenders, 1 hyd., rear hitch, runs, very straight
D	1947		E	$5,300	5/08	ECMN	Restored, styled, newer tires
D	1950		G	$3,700	4/08	NENE	Styled, WF, electric start, PTO, runs
D			F	$1,000	7/07	SESK	Canadian sale, antique tractor on rubber, SN 176434
D			F	$1,700	8/06	SCIL	Unstyled, older repaint, wheel extensions, water valve, ready to restore steel with lugs, engine loose, complete, hand start
D			G	$3,000	12/06	ECIN	Styled, repainted
D	1924		G	$57,500	4/07	NEMO	26" spoke flywheel, correct tractor, Van Zante hood, full steel, detailed restoration
D	1925		G	$40,000	4/07	NEMO	24" spoke flywheel, correct tractor, excellent restoration, full steel, ready for any show
D	1928		P	$1,700	9/07	WCIL	All steel, aero magneto, not running
D	1928		P	$2,100	9/07	WCIL	Full steel, extensions, not running, radiator poor
D	1928		G	$17,000	4/07	NEMO	Experimental John Deere D, SN X67564, full steel with extensions, early R2 magneto, great restoration, super-rare tractor, one of just a few known
D	1929		G	$3,100	7/07	WCSD	Parade-ready, steel wheels with rubber add-on, new block, pistons, rods, runs excellent

John Deere

Model	Year	Hours	Cond.	Price	Date	Area	Comments
D	1929		G	$5,600	11/07	WCIL	Running, restored, PTO
D	1930		E	$4,100	3/07	NWIL	All fuel, new sheet metal, brass carburetor, rubber lugs on steel, new front rubber
D	1935		F	$1,500	8/07	NWOH	Cutoffs, appears complete
D	1935		G	$3,200	8/07	SWSK	Canadian sale, 2WD, 40 PTO hp., 3 forward/2 reverse trans., running, needs restoration
D	1935		F	$4,600	9/07	WCIL	Older restoration, full steel, may need clutch work, came from Kansas
D	1936		F	$1,900	2/07	SCKS	WF, 14.2-28 steel rear wheels, flat spokes, engine free
D	1937		G	$2,500	11/07	WCIL	Older repaint
D	1937		G	$7,000	4/07	NEMO	Unstyled, very straight tractor, older repaint, PTO, full steel
D	1941		G	$2,000	6/07	SEID	Runs
D	1944		G	$4,250	6/07	SCIL	Styled, restored, dual fuel, hand start, new tires, runs good
D	1945		F	$1,800	2/07	SCKS	WF, engine free
D	1948		P	$1,550	9/07	NEND	Not running, electric start
D	1948		F	$1,900	2/07	SCKS	WF, steel wheels with lugs, has been rebuilt, engine free
D	1948		G	$4,000	8/07	WCMN	New 16.9-30 tires, electric start, restored, sharp, PTO
D	1949		G	$8,500	7/07	WCKS	Styled, totally restored
D	1950		G	$2,500	12/07	SCNE	Styled
D			G	$2,500	4/06	WCTX	Rear sand lugs, full steel
D			G	$2,600	1/06	NEIN	Unstyled, repainted
D			G	$4,750	7/06	NWIL	SN 50006, running, restored, fenders, belt pulley
D	1926		G	$7,000	9/06	ECMI	On steel wheels, decent unrestored tractor
D	1927		F	$1,200	9/06	ECMI	Unstyled, diesel, on rubber
D	1937		F	$1,300	3/06	NCOK	Did not run, motor free, restorable
D	1937		G	$2,250	2/06	WCIL	Restored and running
D	1937		F	$2,800	9/06	SCNE	Unstyled, running, new radiator
D	1937		G	$3,000	2/06	SCKS	Spoke front rims
D	1946		G	$3,500	9/06	ECMI	Styled, electric start
D	1946		G	$3,500	9/06	ECMI	Styled, diesel, electric start
D	1950		G	$4,000	10/06	SCSD	New rubber, motor OH, gas, original
D			G	$2,000	9/05	SCON	Canadian sale
D			G	$2,100	9/05	SCON	Canadian sale
D			G	$2,200	11/07	WCIL	Unstyled, WF, PTO
D			G	$3,150	6/05	ECWI	Hand start, painted, new rubber
D			G	$3,200	9/06	NEIN	Styled, turning brakes
D			G	$3,400	12/05	NCCO	
D			G	$3,600	6/05	ECWI	Styled, crankshaft, full steel
D			G	$3,750	7/05	SCIA	Unstyled, full steel, restored
D			G	$4,500	12/06	ECMN	Electric start, new paint
D			G	$4,900	6/05	ECWI	Unstyled, painted, rear round spokes
D			G	$17,000	7/05	SCIA	Spoker, 24", restored
D	1926		G	$6,000	6/05	SWOH	On steel, unstyled, accepted as the first Model D in Preble County
D	1933		G	$5,600	2/05	NWOH	Restored
D	1937		G	$8,600	9/05	SEMN	Styled, dual stacks, round front spokes
D	1941		G	$9,100	9/05	SEMN	Styled, rear weights, PowrTrol installed by John Deere dealer
D	1947		G	$2,600	4/05	NCCO	Completely restored, steel wheels
D	1950		G	$4,250	11/05	SCKY	
D			G	$4,000	3/05	ECMN	Restored

John Deere

Model	Year	Hours	Cond.	Price	Date	Area	Comments
G	1938		E	$7,000	5/08	WCIA	Unstyled, restored 2001, new tires all around, engine bored and doomed, this is a puller, very good compression
G	1941		E	$6,000	5/08	ECMN	Restored, NF, shutters, newer tires
G	1947		G	$3,800	5/08	ECMN	Restored
G	1947		E	$4,000	5/08	ECMN	Restored, WF, half-round fenders, newer tires
G	1947		G	$4,400	5/08	ECMN	Restored
G	1951		G	$5,000	4/08	NENE	NF, cast rear wheels, restored
G	1951		E	$7,100	2/08	SCMI	Gas, NF, restored
G	1951		G	$13,000	3/08	WCOH	Restored
G			F	$2,700	2/07	NCCO	
G			G	$3,200	11/07	SESD	
G			G	$3,300	8/06	SCIL	Really straight, all original late model styled G, fenders, lights, 3 pt., very straight, original paint is peeling, easy restoration
G			G	$3,400	11/07	SESD	
G			G	$3,700	9/07	ECMI	NF
G	1937		G	$12,500	4/07	NEMO	Unstyled, correct SN for a low radiator G, but has high radiator, older repaint, new tires all around, rears cut for pulling, strong runner
G	1938		G	$6,000	8/07	ECIA	Gas, row crop, SN 7290
G	1938		G	$6,500	9/07	WCIL	Single-rib front tire, cast rear, flat back, new rebuilt carburetor this week, runs
G	1948		F	$2,900	1/07	SWKS	Propane, WF, 3 pt., engine stuck
G	1948		G	$4,500	11/07	WCIL	Repaint, fenders, new rubber on front, M&W pistons, fresh major OH three months ago
G	1950		G	$4,500	9/07	SCMI	NF, gas, restored
G	1951		G	$2,400	11/07	WCIL	
G	1952		F	$2,250	9/07	WCIL	Aftermarket water pump, fair tin, runs, high compression head
G	1952		E	$2,500	10/07	NECO	Gas, double front, PowrTrol
G	1952		F	$3,200	9/07	WCIL	Rear cast water pump, fairly straight
G	1952		E	$3,750	3/07	NWIL	Gas, NF, new tires and paint, water pump, 2 fuel, Roll-o-Matic, standard PowrTrol
G	1953		F	$4,000	6/07	NWIL	LP, new tires, original, no 3 pt.
G			F	$1,400	5/06	SEND	PTO, PowrTrol, running condition
G			G	$2,650	1/06	WCTX	WF, new paint, 70% rubber
G			G	$2,900	7/06	NWIL	SN 7788, running, unrestored/original, PTO, belt pulley
G			G	$3,400	4/06	WCTX	Older restoration
G			G	$3,800	6/06	ECSD	Factory WF
G			G	$4,000	8/06	WCTX	Unstyled
G			G	$6,250	7/06	NWIL	SN 4341, running, restored, rear wheel weights
G	1938		G	$5,200	1/06	SCKS	New 13.6-36 rear tires, new 6.00-16 front tires, SN 5678
G	1948		G	$1,750	3/06	NCOK	Did not run, motor free, electric start, PTO
G	1949		G	$1,800	6/06	ECSD	New tires and paint
G	1952		G	$2,100	1/06	WCTX	NF, new rubber
G	1952		G	$4,500	10/06	SCMN	Factory, WF with original John Deere cloth umbrella, gas
G			P	$425	6/05	ECWI	For parts
G			G	$2,700	1/07	WCIL	NF, styled, PTO
G			G	$3,000	11/05	WCIL	Gas, single front
G			G	$3,100	11/05	WCIL	NF, deluxe seat, hyd.
G			G	$4,000	12/06	ECIL	NF
G			G	$7,400	6/05	ECWI	New tin, painted
G			G	$9,200	7/05	SCIA	Unstyled, tall radiator, restored

John Deere

Model	Year	Hours	Cond.	Price	Date	Area	Comments
G			G	$12,500	7/05	SCIA	Unstyled, low radiator, round spokes, restored
G	1938		G	$6,500	2/05	NWOH	Unstyled, restored, fenders
G	1939		G	$8,850	9/05	SEMN	Round rear spokes, unstyled, front weights
G	1951		G	$4,000	11/05	WCIL	NF, electric, belt pully
G	1952		G	$5,000	7/05	SCIA	Restored
G	1953		G	$6,250	11/05	SCKY	
GM			F	$2,500	9/07	WCIL	Electric start, flat back, rear cast, front pressed, runs, rebuilt carburetor, came from North Dakota
GM	1945		G	$3,500	9/06	ECMI	NF, electric start
GM			F	$2,500	7/05	SCIA	
GM			G	$3,500	6/05	ECWI	New rubber, painted
GM			G	$4,900	7/05	SCIA	Handstart, restored
GM			G	$11,000	7/05	SCIA	
GM	1945		G	$3,100	11/05	WCIL	#51 carb., new lights and gauges
GP	1931		G	$5,500	5/08	WCIA	Fully restored, new rings, pistons, bearings, carburetor reworked, rubber bands attached, lugs will go with tractor
GP			F	$1,700	8/07	SWIL	Unstyled, round spoke front wheels, flat spoke rear wheels, new front tires
GP			F	$2,400	8/07	SWIL	Multifuel unstyled, flat-top fenders, round spoke front wheels, flat spoke rear wheels
GP			G	$4,500	8/07	SWIL	Multifuel unstyled, flat-top fenders, round spoke front wheels, flat spoke rear wheels
GP			F	$5,000	9/07	WCIL	1930-32 Model, main case casting date February 1931, steel all around, lugs, complete, not currently running, kept in shed
GP	1928		G	$20,000	4/07	NEMO	Full steel, super nice restoration, correct magneto and carburetor.
GP	1929		G	$4,900	11/07	WCIL	Running, restored, fenders, belt pulley, OH, magneto-new fuel tank, new fenders, new hood, sandblasted and painted full steel
GP	1929		G	$6,250	6/07	SEID	New fenders (in box), mostly restored, runs good, on steel
GP	1931		G	$6,800	9/07	WCIA	On rubber, repainted
GP			P	$1,750	9/06	ECMI	For parts
GP			G	$2,000	4/06	WCTX	Power lift, cast rears
GP			G	$2,300	12/06	NWIL	SN 211797, running, unrestored/original, a very complete running original GP standard, runs good
GP			G	$6,500	4/06	WCTX	Older restoration
GP	1929		G	$2,200	8/06	ECNE	Cut-offs, runs
GP	1930		E	$20,000	10/06	WCMN	New tires, all restored, rare
GP	1931		G	$2,900	9/06	ECMI	Spoke wheels
GP	1931		G	$3,500	4/06	WCIA	Restored
GP	1938		G	$2,000	10/06	WCWI	
GP			G	$3,250	2/07	NENE	Steel wheels
GP			G	$3,300	11/05	WCIL	WF, not running, loose, cutoffs
GP			G	$6,600	6/05	NEND	Unstyled, spoke wheels, PTO
GP			G	$26,000	7/05	SCIA	Top steer, restored, 1 of 417 built
GP	1930		G	$3,300	11/05	WCIL	Hand start, fenders, belt pully
GP	1930		G	$4,600	2/05	NWOH	On hard rubber, restored
GP	1930		G	$7,900	9/05	SEMN	Factory round fronts and rears
GP	1931		G	$4,250	6/05	ECIL	Original tires
GP	1951		G	$3,300	12/05	NCCO	
GPWT			G	$5,500	7/06	WCIL	Loose, not running, front and rear cutoffs
GW			G	$5,000	5/08	ECMN	Row crop, WF, PTO

John Deere

Model	Year	Hours	Cond.	Price	Date	Area	Comments
H			G	$2,100	2/08	SCMI	Gas, NF
H	1939		E	$3,100	5/08	ECMN	Restored, NF, newer tires
H	1945		G	$3,000	4/08	NENE	NF, mounted cultivator
H			P	$700	8/05	WCIL	For parts
H			F	$2,700	6/07	NCIN	NF, restored, nice sheet metal, poor tires, nice paint, motor free
H			F	$3,250	9/07	WCIL	Cast fronts, hand start, older repaint, fair tin, runs
H	1939		F	$2,000	6/07	NCIN	NF, unrestored, rear tires and rims bad, motor turns
H	1939		G	$3,000	9/07	WCIA	
H	1939		G	$3,600	8/07	SWIL	New rear tires, older restoration, open flywheel
H	1940		G	$3,600	12/07	SEWY	Gas, single front
H	1940		G	$4,300	6/07	SCIL	Fenders, cast front wheels, hand start, good tires, PTO, double stick trans., dual fuel, starts and runs good, older restoration, ready for the show
H	1943		G	$4,250	4/07	NEMO	Hand start, connected tread rear tires, hyd., nice repaint
H	1944		G	$4,250	11/07	WCIL	Running, restored, PTO, lights, hyd. remote
H			F	$700	10/06	WCWI	
H			F	$800	4/06	WCWI	
H			F	$800	4/06	WCWI	
H			G	$3,900	10/06	NCWI	Restored, NF, gas, saw rig
H			E	$4,100	7/06	NWIL	SN 22664, running, restored, PTO, hyd. remote, belt pulley, very good
H			G	$4,400	4/06	ECNE	Fenders, restored to expo quality, new tires
H	1939		G	$3,300	10/06	SCSD	Hand start, gas, new rubber, mint
H	1940		G	$2,400	9/06	ECMI	Hand start, restored
H	1945		G	$2,250	10/06	SCMN	Electric start, hyd.
H	1946		G	$2,900	9/06	NCIA	9-32 tires
H			F	$1,250	11/07	NENE	PTO, drawbar, 9.5-32 tires
H			G	$1,900	1/05	SWMN	
H			G	$2,500	6/05	ECWI	
H			G	$3,250	6/05	ECMN	Row crop, tricycle, fenders, hand start
H			G	$3,500	9/06	NEIN	Restored, handstart, factory fenders
H	1940		G	$4,300	2/05	NCKS	Restored, new rubber, hand crank
H	1940		G	$4,750	11/05	SCKY	
H	1941		G	$2,600	7/05	SCIA	Restored
H	1941		G	$3,300	9/05	NENE	Electric start, hyd. pump and lift cyl., 9.5-32 tires
H	1941		G	$3,500	2/05	NWOH	Restored, new rubber
H	1945		E	$5,400	1/05	NEIA	
H	1947		G	$6,100	9/05	SEMN	Cast fronts, fenders, hyd., electric start, rear weights
HWH			G	$36,500	7/05	SCIA	Front weights, fenders, restored, 1 of 126 built
L			P	$800	6/07	NCIN	Rough
L	1937		G	$11,000	4/07	NEMO	Unstyled
L	1938		G	$2,800	8/07	WCMN	WF, new front tires
L	1940		P	$450	6/07	SCIL	For parts, rear wheels set up for duals
L	1940		F	$1,500	9/07	WCIA	Repainted
L			G	$3,200	8/06	NECO	
L			P	$2,500	8/05	NEIN	Frame and rear end
L			P	$2,900	8/05	NEIN	Styled, parts tractor
L			G	$3,100	6/05	ECWI	
L			G	$5,400	8/05	NEIN	Styled

John Deere

Model	Year	Hours	Cond.	Price	Date	Area	Comments
L			G	$5,700	8/05	NEIN	Styled
L			G	$6,000	8/05	NEIN	Styled, restored, belt pulley
L			G	$6,250	8/05	NEIN	Unstyled, restored, belt pulley
L			G	$8,500	9/06	NEIN	Styled, restored, belt pulley, lights, dual rear wheels
L			G	$8,700	8/05	NEIN	Unstyled, restored
L			G	$8,800	8/05	NEIN	Unstyled, repainted
L			G	$9,100	8/05	NEIN	Unstyled, restored, belt pulley
L			G	$9,500	11/05	SCKY	
L			G	$10,000	7/05	SCIA	Unstyled, belt pulley, restored
L			G	$10,200	8/05	NEIN	Unstyled, restored, mud lugs, 1-bottom plow
L	1937		G	$8,200	9/05	SEMN	Unstyled, belt pulley
L	1938		G	$1,950	3/05	WCIL	Ready to paint, gas
L	1939		G	$3,700	4/05	SCNE	Runs
L	1940		G	$7,750	9/05	SEMN	Styled, belt pulley, rear weights
LA			G	$5,000	3/08	WCOH	Restored
LA	1942		G	$4,000	2/08	SEAL	
LA	1941		G	$6,750	11/07	SWIN	Older repaint, electric start, new rear tires, cultivator, lights
LA	1941		G	$9,500	4/07	NEMO	New tires, belly sickle bar mower, good repaint
LA	1942		G	$7,250	11/07	WCIL	Running, restored, fenders, parade-ready, steering box not original
LA	1945		G	$4,000	9/07	WCIL	Older restoration, electric start, new rears, fronts very good, runs
LA			G	$1,550	3/06	NWOH	Gas, WF, restored
LA	1944		G	$3,900	10/06	SCMN	Restored original, belly pan, new rubber
LA	1946		E	$7,200	10/06	SCSK	Canadian sale, 2-cyl. gas, 3-rib front, 9×24 rear tires, fully restored, painted
LA	1946		E	$7,200	10/06	WCSK	Canadian sale, 2-cyl. gas, 3-rib front, 9×24 rear tires, fully restored, painted
LA			G	$2,400	2/05	NWOH	Restored, no serial plate
LA			G	$3,100	6/05	ECWI	New tires, painted
LA			G	$3,500	11/05	WCIL	WF, fenders
LA			G	$5,300	8/05	NEIN	Restored, lights, turf tires
LA	1945		G	$6,500	9/05	SEMN	Rear weights, belt pulley
LI	1941		G	$3,000	9/06	ECMI	Complete, restored
LI			G	$7,000	6/06	NEIN	Restored, factory blade, hyd., lights, weights
LI			G	$9,500	9/05	NEIN	Styled, restored, hyd., lights
LI	1946		G	$1,000	6/05	ECIL	34th from last produced, modified with add-on hyd. and homemade hood
M			G	$2,100	11/05	SEND	WF
M			E	$3,050	3/08	ECWI	Touch-O-Matic, 540 PTO, nice tin, collectible tractor
M			G	$2,800	10/07	NWWI	WF, side-mounted mower
M			G	$5,000	8/07	SWSK	Canadian sale, 2WD, 19.5 PTO hp,. 4 speed, 3 pt., PTO, pulley, 11.2/10-24 rubber
M	1947		F	$2,800	6/07	SCIL	Good original tractor, 2 pt., locking drawbar, has pulley, rear weights, front wheel hubcaps, field use for last 10 years
M	1948		G	$3,000	2/07	NENE	Woods belly mower
M	1948		G	$4,100	9/07	WCIA	Restored, nice
M			G	$1,850	1/06	SESD	Restored
M			G	$2,200	10/06	SENY	Unrestored
M			G	$3,500	3/08	WCMN	New paint
M	1947		E	$2,550	12/06	WCIL	Gas, new tires and paint
M	1948		G	$2,900	10/06	SCSD	Restored, gas, Touch-O-Matic

John Deere

Model	Year	Hours	Cond.	Price	Date	Area	Comments
M	1950		G	$4,000	10/06	SCMN	Restored original, new rubber
M			F	$1,600	3/05	WCIL	WF, rear wheel weights
M			G	$2,600	7/05	NEIA	Repainted
M			G	$3,200	11/05	NEIA	WF
M			G	$3,400	5/08	ECMN	Plow, WF, new paint
M			G	$3,500	6/05	ECIL	New paint, new tires all around, restored
M	1948		G	$6,500	9/05	SEMN	Fenders, belt pulley, PTO shield
MC			G	$3,250	6/05	ECWI	
MI			G	$7,500	4/08	NEIA	1 of 1,032
MT			G	$2,000	1/08	NEMO	
MT			F	$2,100	1/08	SESD	
MT			G	$4,600	3/08	WCOH	Restored
MT	1949		E	$6,250	5/08	ECMN	Restored, fenders, lights, rear wheel weights, 11.2-10-34 tires, new tires
MT			P	$700	6/07	SCIL	To restore
MT			G	$2,000	2/07	SCWA	
MT	1948		P	$1,150	6/07	SCIL	Engine stuck, tractor complete
MT	1949		F	$1,400	6/07	NCIN	Unrestored, original
MT	1949		F	$2,100	6/07	NCIN	Unrestored, original
MT	1950		F	$1,700	11/07	WCIL	Running, restored, fenders, PTO, lights, carburetor rebuilt, new oil, new tube rear tire, new ignition switch
MT	1950		G	$1,850	9/07	NWIL	NF, 3 pt., 540 PTO, dual Touch-O-Matic hyd. controls, rear belt pulley, gas
MT	1951		F	$1,700	6/07	SCIL	Belt pulley, original tractor, grill is rough, tin hood 95%, grill 50%, runs very good, 2 pt., locking drawbar, has radiator shutters, points, plugs, condensor
MT	1951		F	$2,200	2/07	SCKS	Original, 11.2-34 rear tires, 5:00-15 front tires
MT	1955		G	$1,800	9/07	ECMI	
MT			F	$1,300	12/05	WCWI	
MT			F	$1,450	3/06	NWOH	Gas, NF, did not run
MT			G	$1,700	9/06	NEIN	Repainted
MT			G	$2,750	1/06	SESD	Restored
MT	1950		G	$2,500	10/06	SCSD	New tires, lights, Touch-O-Matic, gas
MT	1951		G	$1,900	10/06	WCMN	New paint
MT	1951		G	$1,900	10/06	SCMN	Original, fenders
MT	1951		G	$2,500	11/06	SENE	NF, 10-34 rear tires, fenders, 540 PTO, shedded
MT			F	$1,200	7/05	SCIA	Runs good
MT			G	$1,500	11/05	WCIL	NF, restored, 3 pt.
MT			F	$1,900	7/05	SCIA	Plow
MT			G	$1,900	6/05	ECWI	Good rubber, lights
MT			G	$2,000	7/05	NEIA	Repainted
MT			G	$2,300	11/05	WCIL	NF, gas, electric start, 2 pt.
MT			G	$3,000	6/05	ECMN	Row crop, WF, gas, belt pully, new paint
MT	1951		F	$2,000	4/05	SCNE	WF, 11.2-34 tires, fenders, PTO, runs
MT	1951		G	$3,400	4/05	SCNE	WF, 10-34 tires, fenders, runs
MT	1952		G	$3,000	3/05	SEIA	
R	1949		E	$5,000	5/08	ECMN	Restored, diesel, standard, PTO, 2 hyd., newer tires

John Deere

Model	Year	Hours	Cond.	Price	Date	Area	Comments
R	1951		G	$5,000	5/08	WCIA	Main motor OH, mains, new bearings in crank, valves, tractor runs great, pony motor will be working by sale day, hyd., PTO, original rears, old repaint in 1980s
R			G	$2,200	6/06	SEID	Diesel, runs very good, pony motor strong
R			G	$3,200	11/07	SESD	Diesel
R	1951		F	$2,600	2/07	SCKS	WF, all original, never painted, straight tin, engine free, 18.4-34 tires
R	1951		G	$3,200	12/07	SEWY	Diesel, PTO, 1 hyd., pony start
R	1953		G	$3,750	9/07	WCIL	
R	1953		G	$4,750	6/07	SCIL	Nice older restoration, very straight, engine runs good, 1 hyd., rear tires matched, PTO, no PS
R	1953		E	$6,400	3/07	NWIL	Diesel, WF, 540 PTO, remote cylinder, new 18.4-34, new paint
R	1954		F	$4,000	9/07	WCIL	Diesel, runs, 1 rear weight, 1 hyd., PTO, fairly straight, just repainted, came from Lake City, KS, 15 years ago
R			F	$2,000	7/07	SEND	
R			G	$2,500	8/07	SESD	Original, diesel
R			G	$3,700	1/06	SESD	
R			G	$5,250	10/06	SWOH	Diesel, gas, pony motor, rear drawbar
R			G	$2,250	6/05	ECIL	Rear weights
R			G	$3,600	10/05	ECMN	Standard diesel, pony start
R			G	$4,200	6/05	ECWI	
R			G	$5,250	6/05	ECMN	Diesel, new paint, pony start
R			G	$5,250	1/07	WCIL	Diesel, restored, PTO, lights
R	1951		G	$4,000	12/05	NCCO	
R	1951		G	$4,400	3/05	SEIA	
R	1952		G	$2,950	4/05	NCCO	New pony engine, new front tires, PTO, remote, completely restored

Waterloo Boy

Model	Year	Hours	Cond.	Price	Date	Area	Comments
N	1918		E	$130,000	4/07	NEMO	Chain steer, Kenny Kass restoration, rubber belting over steel wheels, a really impeccable tractor
N	1920		E	$94,000	4/07	NEMO	Automotive steer, excellent restoration, everything is done, rubber tread over steel wheels, ready to take to any show
N/A			E	$65,000	7/06	NWIL	Restored, beautiful fresh restoration, this tractor runs and looks great

Ins and outs of auction action

Even if you don't have a dime in your pocket or room for one more tractor, a big auction with a good auctioneer is a grand show and cheap entertainment. For those of you new to the bidding game, I'd like to offer some suggestions for how you can come out still wearing your shirt, if not your overalls.

First, leave your credit cards at home and carry no more than $50 in your wallet. If you don't have this kind of self-discipline, just tell your wife where you're going, and she'll take care of the problem for you.

Once the bidding begins, especially on something you don't want like a slightly cracked chamber pot, pay attention to what's going on. If a fly lands on your nose, let it sit there until the auction is over. Otherwise, you're taking home a slightly cracked chamber pot.

Similarly, be sure before you go to . . . go. It doesn't take much full-bladder bouncing to catch an auctioneer's attention. I once ended up with a box of corset stays because of two cream sodas I drank on the way.

It's really important also to go early so you can walk around and see all the neat stuff you're not going to get. Jot down the items you're especially interested in with the top price you're willing to spend. This is important to smart bidding. That way, if you note that your top dollar for a Whiz Bang magneto is $17.52, you can be absolutely certain it's going to go for a final bid of $17.53.

You can plan all you like

OK, let's say you get a sale bill with a 1936 Allis Chalmers WC, unstyled, on steel, that is exactly the machine you've wanted for 10 years. So you go to that auction. Do not drag along your trailer unless you take special pleasure in dragging around an empty trailer. You are not going to get that Allis WC. It will go for somewhere around $27,500 since six of the previous owner's great-grandchildren will be there, all making sure none of the other grandchildren get it.

It is important to note that the WC you saw on another sale bill for an auction 6 miles the other side of town, in mint condition, will sell for $77. A buddy of yours will get it. And then never let you forget how close you came to a real bargain – if you hadn't gone to the wrong sale. But don't let this worry you. You'll always be at the wrong sale. Trust me on this.

So it might seem logical to ask a buddy to go to the other sale held the same day. That's almost like being in two places at the same time, right? Wrong! First of all, if your friend gets that tractor for iron price, he'll be surprised that you think it should be yours. He won't remember at all your conversation at the tavern at 11:30 the night before when he promised to do what he could for you. Secondly, it was pretty late, and you'd been having way too much fun, so you'll bid up to your limit, or maybe a couple hundred dollars above your limit, on the tractor at your sale only to find that you have been bidding against – your buddy. He thought you said he should be at this sale and that you'd be at the other one.

Weather, lunch, and timing

One of the true joys of farm auctions is the local church ladies' lunch counter where they serve things like homemade pie. I highly recommend you take advantage of this convenience. For one thing, you're going to get a great piece of pie. For another, you will save yourself a mess of money. Because the tractor you want will sell sometime while you're standing in line for the pie – count on it.

One thing you cannot count on is the weather. If it's good, there'll be a mob of bidders, all so cheerful about the good weather that they bid everything way high. If it's bad, buyers figure everything is a bargain because there are so few bidders and besides, if they've stood out here in the cold rain for four hours, they are not going home with an empty trailer, so they bid everything way high.

So go to a good farm auction soon. It's great entertainment. Just don't plan on coming home with that tractor.

– Roger Welsch

Massey Ferguson and Massey Harris

A realistic range is what you're looking for. Whether buying or selling an older tractor, what's needed is an accurate valuation of what it's worth. I'm here to help, folks. Take the Massey Ferguson 65 tractor, made from 1958 to 1964. What's a decent MF 65 worth? Start on the low end. See the 65 that sold for $450 in south-central California. Note this tractor was "for parts". On the high side, note the 65 that sold for $4,100 in south-central North Dakota in April 2008. The last four years, MF 65s have sold at auction for an average price of $2,258 to $2,500.

Now you know what it's worth.

Massey Ferguson

Model	Year	Hours	Cond.	Price	Date	Area	Comments
1085	1973		G	$3,000	2/06	WCKS	3 pt., PTO
1085	1976	3,697	G	$6,250	8/05	NCIA	Western style, cab, air, OH
1100		5,890	F	$6,200	3/07	SEMN	Duals, 3 pt.
1100			F	$1,600	3/06	SEPA	Sold as is, runs, WF, fair rubber
1100			G	$2,750	2/06	NWSD	Cab, 3 pt., Westendorf loader, scoop and grapple, one owner, diesel
1100			G	$3,900	4/06	NWSD	Diesel, cab, 3 pt., duals, Waldon 8' dozer
1100		7,466	G	$1,750	2/05	NEIN	Diesel, WF, no cab
1100			F	$2,300	12/05	NCCO	No cab
1100			G	$2,900	8/05	SEMN	
1100			G	$3,500	8/05	SEIA	Cab, 18.4-34
1100			G	$4,800	11/05	NCOH	New cab
1100	1968	5,720	G	$3,400	3/05	WCNE	Diesel, dual 325 quick-tach loader with grapple, 2 hyd., 18.4-38 rear tires, 11L-16 front tires, PTO
1100	1976		P	$1,175	5/05	NCKS	3 pt., PTO, really rough
1105	1974	4,135	G	$6,650	3/08	NCIL	Cab, air, 2 hyd., 3 pt., quick hitch, 540 PTO, diesel, 18.4-34 tires
1105	1975		G	$5,500	3/08	NEMO	18.4-38 tires, 246 loader, recent OH
1105			F	$4,500	2/07	WCKY	2 hyd.
1105		7,115	G	$4,500	3/07	SESK	Canadian sale, 2WD, 111 PTO hp., PTO, 2 hyd., 20.8-38 rear tires, 4-rib front tires
1105	1975		F	$2,500	8/07	SESD	Cab, 3 pt., diesel
1105		3,000	G	$5,400	11/06	NCKY	Diesel
1105	1976	6,783	P	$1,800	10/06	SWNE	Not running, cab, 3 pt., WF, diesel, 18.4-38 rear tires, 101 hp.
1105	1976	3,400	F	$5,000	9/06	NWIL	Cab, no weights, diesel
1105	1973	7,366	G	$4,300	8/05	NCIA	Open station, 2 hyd., 3 pt.
1130			P	$700	2/05	SCCA	Front weights, canopy, 3 pt., PTO, 2 hyd., 15.5-38
1130	1966	4,584	G	$4,600	8/05	NCOH	Diesel, H/L multi power, 2 hyd., 18.4-34 tires, four-season cab
1130	1973		G	$4,750	3/05	SWIN	Diesel, 2 hyd., full weights, 2-speed PTO, power-spread wheels, hub duals
1135	1974	3,066	G	$10,600	3/08	NCIL	18.4-38 clamp-on duals, diesel, 2 hyd., 3 pt., quick hitch, cab, air, 12 front weights
1135	1975	7,440	F	$2,000	12/07	WCMN	540 PTO, cab, heat, radio, 3 pt., 2 hyd., 18.4-34 tires
1135	1975	7,360	F	$5,500	10/06	SESK	Canadian sale, diesel, 2WD, cab, 2 hyd., 18.4-38 rubber, 7,360 hours showing
1135	1976	4,805	G	$6,400	2/06	WCNE	Diesel, 3 pt., cab
1135	1976	3,123	G	$8,300	6/06	NWMN	CAH, 3 pt., PTO, 2 hyd., front weights, 16.9-38 duals
1135	1973	6,025	G	$5,000	8/05	NCIA	Cab, 3 pt., 2 hyd.
1135	1977	4,740	G	$6,900	8/05	NCIA	Cab, air, OH, 3 pt., 2 hyd.
1150			F	$3,100	1/08	NEMO	Diesel
1150	1970		P	$600	6/07	SWNE	WF, 18.4-38 rear tires with duals, 11:00-16 front tires, 3 hyd., bad trans.
135			P	$3,300	4/08	SWOH	Rough, 1970 model year, diesel, hours unknown
135			G	$3,900	2/08	NWSC	Diesel, shuttle shift
135	1964		F	$2,600	5/08	SEIN	Tractor and loader
135		459	G	$5,400	6/07	ECIL	Gas, utility, PS, 3 pt., Massey Ferguson hyd. material and manure bkts., 13.6-28 tires
135	1962		G	$2,800	6/06	NECO	Gas, tach shows 3,385 hours, WF, 3 pt.
135	1969	5,753	G	$2,700	12/07	SEWY	Diesel, WF, 3 pt.
135	1969	3,577	G	$3,000	12/07	ECNE	Diesel, PTO, 13.6-28 tires

Massey Ferguson

Model	Year	Hours	Cond.	Price	Date	Area	Comments
135			G	$1,900	8/06	SETN	Gas
135			F	$3,000	8/06	NWIL	Woods finishing mower, gas
135			G	$3,100	8/06	NECO	Gas
135			G	$3,200	6/06	ECMN	Diesel
135		3,725	G	$3,700	6/06	NWIL	Gas
135	1969		E	$5,400	10/06	NECO	WF, 3 pt., gas, tach shows 1,688 hours
135			G	$2,950	10/05	WCTN	
135			G	$3,000	9/05	SCON	2WD, gas, loader, 3 pt.
135		2,327	G	$3,250	1/05	SWOH	Loader
135		3,145	G	$3,400	12/05	ECIN	Utility
135			G	$3,950	9/05	SCON	Canadian sale, loader, 3 pt.
135			G	$4,100	9/05	SCON	Canadian sale, gas, 3 pt.
135			G	$4,500	1/05	NWGA	3 pt., PTO
135			G	$4,650	3/05	NEKS	WF, 3 pt., PTO, low hours, new tires, gas
135		2,332	G	$5,050	1/05	SWOH	
135	1965		G	$4,000	10/05	WCIL	3 pt.
135	1967	5,450	G	$3,000	12/05	ECMN	Open station, 8 speed hi/lo, 3 pt., PTO, 13.6-28 rears
150			G	$2,750	1/07	WCIL	Utility, gas, WF
1505			P	$2,500	3/07	SESK	Canadian sale, 4WD, 185 engine hp., 3208 Cat, standard trans., 23.1-30 tires, 2 hyd., stored inside, SN 9C003753, new motor with 600 hours
165			F	$2,100	1/08	NEMO	
165	1974		P	$4,400	4/08	SWOH	Rough, diesel, hours unknown
165			P	$2,000	3/07	ECIL	Gas utility tractor, 3 pt., 540 PTO, 1 hyd., sells with Freeman 4000 loader, does not run, very rough
165			G	$3,600	11/07	WCIL	Gas, manual trans., hyd. brakes
165			G	$3,950	8/07	SESD	Diesel
165		6,180	G	$2,500	8/05	NEIN	Gas, WF
165			P	$2,700	2/05	NCIL	3 pt., 2 hyd., WF, rough
165			G	$3,100	11/05	WCIL	Perkins diesel, fenders
165			G	$3,250	12/05	SWMS	Diesel, 3 pt.
165		2,694	G	$3,350	1/05	SWOH	
165			F	$3,700	5/05	SEPA	Very clean, WF, 75% rubber, diesel
165			G	$3,750	9/05	SCON	Canadian sale, 2WD, 3 pt.
165	1973	3,205	G	$4,750	9/05	SWOH	WF, utility
175	1968	4,589	G	$6,000	4/08	SCMI	Diesel, 3 pt., 8 front weights, restored
175			F	$2,650	2/07	NEIN	Diesel, WF, steel wheels, 3 pt., forklift
175	1965	5,909	G	$3,900	3/07	SEWY	Diesel, WF, 3 pt.
175		3,314	F	$4,750	9/05	SEIA	Loader
180			F	$3,250	9/07	NWIL	Diesel
180	1969		G	$3,550	9/07	NEKS	Diesel, very good rubber
180			P	$1,300	10/06	SWNE	WF, diesel, 3 pt., not running, parts missing, 15.5-38 rear tires
180			P	$1,300	10/06	SWNE	WF, diesel, 3 pt., not running
180			F	$2,250	9/06	NWIL	Diesel, WF
180			G	$3,400	8/06	NENE	WF, Perkins 236 diesel engine, 15.5-38 tires
180	1966		G	$3,200	8/06	NENE	WF, Perkins 236 diesel, 13.6-38 rear tires
180	1966	6,034	G	$4,150	10/06	SWNE	WF, diesel, 64 hp., 15.5-38 rear tires, flotation front tires, 3 pt.
180	1969	3,485	G	$4,300	10/06	SWNE	WF, diesel, 64 hp., 3 pt., 15.5-38 rear tires, tri-rib front tires, Fassie hyd. valve, dual mounted 320 loader with 5' bkt. and grapple forks, runs well
180		2,624	F	$2,500	3/05	SCMI	Diesel, WF

Massey Ferguson

Model	Year	Hours	Cond.	Price	Date	Area	Comments
180			G	$2,600	9/05	SCON	Canadian sale, 2WD, diesel, 3 pt., hyd.
180			G	$5,300	1/05	SWOH	1 hour since OH
180	1966	3,599	F	$2,000	3/05	SEWY	No 3 pt., diesel
1805			F	$3,000	1/08	NWOH	4WD
1805			G	$4,000	1/08	NWOH	Cat 3208
1805		4,613	G	$8,000	6/07	NWSK	Canadian sale, 4WD, 3208 Cat, 192 PTO hp., 1,000 PTO, 3 hyd., 4,613 hours, 23.1-34 inside rubber
1805			F	$1,750	6/06	NCIA	4WD, PTO, air, 3208 Cat
1805	1975		P	$1,800	8/06	NENE	4WD, Cat 3208 diesel, 18.4-38, axle duals, front and rear 3 pt., not running
1805	1976	4,285	F	$2,800	10/06	NECO	Cab, bareback, 3 hyd., diesel
1805	1975	2,776	G	$4,500	2/05	WCIL	4WD, Cat 3208 motor, factory turbo, 4 hyd., air, cab, 3 pt., duals, 500 hours on rebuilt engine
20			G	$1,300	9/06	NWTN	Front loader, bale spear
204			G	$2,750	11/06	SCCA	Loader, 3 pt., PTO
204	1959		F	$2,500	10/06	NECO	Massey Ferguson loader, WF, 3 pt., gas, tach shows 3,424 hours
235	1975	1,329	G	$4,700	4/07	NCKY	Woods 5' rotary mower
235			G	$4,500	11/06	NCKY	
235	1976		F	$3,400	6/06	NECO	WF, 3 pt., gas
235			G	$3,800	6/05	ECPA	
245			G	$4,000	2/08	NWSC	
245		1,145	G	$4,400	8/06	SETN	
245			G	$4,500	9/05	SCON	Canadian sale, diesel, ROPS, 3 pt.
255	1976	3,635	G	$3,900	4/08	SCMI	Massey Ferguson 90 loader, 3 pt.
275	1976	3,285	G	$10,500	2/06	NWIL	Utility, ROPS, canopy, 3 pt., Bush Hog 4000 loader, 66" bkt.
275	1977	3,272	G	$6,500	12/06	WCIL	Utility, 540 PTO, 2 hyd.
275	1976	3,400	G	$8,000	10/05	WCIL	Massey Ferguson 236 hyd. loader, 3 pt., 2 hyd., roll bar
285	1977		G	$9,000	2/08	ECIL	Diesel, ROPS, 2 hyd., 540/1,000 PTO, 3 pt., 18.4-34 tires, sold with Great Bend 2524 quick-tach loader, 7' bkt.
30			F	$1,000	1/08	WCTN	
30	1951		G	$2,350	3/08	NCIL	3 pt., Woods RM59 5' rear-mount mower
30	1947		G	$1,200	11/07	WCIL	Running, unrestored/original, rear wheel weights, looks sharp runs good
30	1960		F	$1,700	9/07	NECO	Gas, tach shows 3,956 hours, loader, WF, 3 pt.
30			G	$1,050	10/06	WCMN	New paint
30			G	$5,000	1/06	ECIL	Industrial gas tractor, WF, 3 pt., weights, new 16.9-24 tires, ROPS, PS, attached hyd. quick-attach 80" bkt.
30	1953		F	$1,300	4/06	SEMI	Gas, WF, 3 pt., Wagoner loader with 3' and 6' hyd. bkts., front pump
30	1954		F	$3,700	12/06	WCMI	Gas, utility, Shawnee loader
30	1972	2,665	G	$3,200	1/06	NECO	WF, hydro, diesel, 32A loader, 3 pt.
35			G	$4,000	1/07	SCNE	Utility, diesel
35			G	$2,400	8/06	SCIL	Turf special, 4-cyl. gas, 3 pt., PTO
35			G	$4,000	8/06	SETN	3 cyl., diesel
35	1970		G	$4,600	4/06	WCMN	Canadian sale, diesel, hi/lo, 1 hyd.
35	1970		G	$4,600	4/06	SWMB	Canadian sale, 3 pt., 3-speed trans. with hi/lo, 1 hyd.
35			G	$1,500	9/05	SCON	Canadian sale, gas, 2WD, 3 pt.
35			G	$2,200	12/05	SWMS	Diesel, 3 pt.
35			G	$3,500	6/05	ECWI	Loader, restored
40			G	$2,850	9/05	SCON	Canadian sale, gas, loader, weight box

Massey Ferguson

Model	Year	Hours	Cond.	Price	Date	Area	Comments
40B			G	$6,750	3/05	ECNE	Grading tractor, Perkins 4 cyl., Massey Ferguson 34A loader, rear box blade, rear ripper
50			F	$1,800	2/07	ECMI	Gas, hi/lo range, 3 pt., PTO
50			G	$2,600	6/07	WCSD	Gas, 3 pt., PTO
50			G	$1,475	12/06	ECMO	
50			G	$1,600	11/05	NECO	Loader, box scraper
50	1958		G	$2,800	6/05	ECIL	3 pt., PS
50	1959		G	$3,100	4/05	SCNE	Gas, WF, 3,361 hours on tach, 12.4-38 tires, fenders, 3 pt., 2 hyd., runs
65			F	$1,000	2/08	WCOK	LP, 3 pt., hyd., PS
65			E	$2,000	3/08	NEKS	Gas, WF, OH, original paint, good rubber
65			G	$4,100	4/08	SCND	Utility, multipower, 3 pt., PTO, gas
65			F	$1,175	2/07	NEIN	Diesel, WF
65			F	$1,900	11/07	WCIL	
65		2,699	G	$3,000	2/07	NEIN	Gas, WF, 3 pt., 3-bottom plow
65			F	$3,000	8/06	NWIL	Gas, WF, loader
65	1959		G	$2,250	4/07	NECO	Gas, 3 pt., WF, tach shows 3,390 hours
65	1960		G	$2,250	4/07	NECO	Diesel, 3 pt., WF, tach shows 2,992 hours
65	1963	9,432	F	$3,000	5/07	SEWY	Ezee-On loader and 5' bkt., diesel
65			F	$2,100	5/06	SEWY	Wagner loader, 3 pt., gas
65			G	$4,600	11/06	NEOH	Gas, Freeman front-end loader
65	1958	4,890	G	$1,300	9/06	NEIN	WF, wiring needs work
65	1958	6,241	G	$1,600	10/06	NECO	Loader, WF, 3 pt., gas
65	1959		G	$1,800	8/06	NENE	WF, Perkins 203.5 diesel engine, 15.5-38 rear tires, excellent tin work
65			P	$450	2/05	SCCA	For parts
65			F	$1,250	11/05	NWIL	Gas, WF, 3 pt., loader, hand trip
65			G	$1,600	11/05	ECNE	3 pt.
65			F	$1,700	1/05	ECNE	WF
65			G	$4,200	2/05	ECNE	Loader, gas
65	1958		G	$4,350	2/05	NEIN	Gas, WF
85			P	$750	2/05	NCIN	Gas, rough, salvage
85			F	$1,400	2/05	SCMN	LP, WF, 3 pt.
85	1960	5,514	P	$925	9/05	NECO	WF, 3 pt., engine stuck, gas
88	1959		G	$2,600	3/06	NWKS	Gas, WF, 15-30 tires, PTO, dual 300 loader with 6' bkt. and bale forks
90	1963		G	$1,500	3/07	NCNE	WF, gas, 13.6-30 rear tires, 2 hyd., 540 PTO, rear fenders, SN CGM813563
90	1962		G	$4,000	3/05	NWMN	High crop, factory 3 pt., PTO, 2 hyd.
97	1963		G	$13,500	11/07	WCIL	Diesel, running, restored, fenders, PTO, deluxe seat, PS, new injector tips, new rear axle seats, new brake seals, rebuilt water pump
Super 90			F	$1,500	2/07	ECMI	Diesel, 3 pt., PTO, 1 hyd.
Super 90			G	$2,800	9/06	NWOH	3 pt., original
Super 90	1964	4,254	G	$2,250	9/06	NECO	Gas, WF, 3 pt., dual 340 loader
TO 20	1951		G	$750	9/06	NEIN	Blade, gas
TO 20	1950		G	$1,350	2/05	NEIN	Gas
TO 30			G	$2,700	8/07	SWOK	
TO 30	1954		G	$4,000	9/07	SEMN	4 speed, 3 pt., original rear tires, good paint
TO 30	1950		F	$1,400	7/06	NWIL	Gas, loader, new rear tires
TO 30	1955		G	$1,850	7/06	SEND	Utility, gas, 3 pt., PTO, good tires
TO 30			G	$2,000	3/07	SCNE	3 pt., runs, 11-38 tires
TO 35			G	$3,000	7/06	NWIL	Loader, cab, gas

Massey Harris

Model	Year	Hours	Cond.	Price	Date	Area	Comments
101			G	$3,250	5/08	ECNE	101 Senior Wheatland, 1 of 500
101	1940		G	$1,700	2/07	NCCO	101 Junior
101			P	$350	3/06	NWOH	101 Junior, gas, NF
101	1939		F	$1,500	8/06	NENE	Super standard, all four side panels good, grill halves and center cast are poor, tires poor, engine stuck
101	1944		P	$600	8/06	NENE	Senior standard, full steel with rear wheel extensions, engine bad
101	1944		P	$850	8/06	NENE	Senior, complete OH, tires fair
101	1944		F	$1,075	8/06	NENE	Junior row crop, good tin, missing rear side covers, tires bad, engine stuck
101			G	$975	8/05	WCIL	101 Junior, cleaned, painted, drawbar, gas, NF
101			G	$1,100	1/08	WCIL	Old restoration, big 6 cyl., NF
101			G	$1,850	11/05	NCOH	101 Junior
20	1946		F	$1,200	8/06	NENE	Row crop, tin and tires good, needs rear side panels, engine free
20	1947		P	$325	8/06	NENE	Row crop, for parts
20			G	$1,475	8/05	WCIL	New tires, cleaned, painted, drawbar, gas, NF
20	1946		G	$900	6/05	ECIL	Restored
22			G	$1,700	6/07	NCIN	Straight original, rear weights, good potential for restoration
22			G	$2,500	12/05	ECIN	
22			G	$1,050	1/06	WCIL	Cleaned, painted, drawbar, gas, NF
22			G	$2,400	11/05	ECNE	3 pt., 11.2/10-34 tires
22	1951		G	$1,150	8/05	NEIN	Gas, straight, no paint
30			G	$1,000	8/07	SESD	Standard
30			G	$1,000	8/07	SESD	Standard
30			G	$1,550	1/06	SESD	Standard
30	1948		G	$1,550	10/06	WCMN	Standard 30, new paint
30	1950		G	$1,800	10/06	SCMN	Row crop, restored original, very sharp
30			G	$1,050	1/06	WCIL	Clean, painted, drawbar, gas, NF
30			G	$1,100	1/08	WCIL	Cleaned, painted, drawbar, gas, NF
30			G	$1,650	11/05	NCOH	
33			F	$950	6/07	NCIN	Restored
33			G	$1,500	4/06	WCIA	Older repaint
33			G	$1,700	10/05	WCIL	WF, 3 pt., not painted, gas, drawbar
33			G	$2,000	11/05	WCIL	Cleaned, new tires, painted, drawbar, gas, NF
333			G	$3,000	6/06	ECSD	WF, live power
333		3,771	G	$2,450	11/05	NCOH	3 pt.
44			F	$1,200	4/08	NEIA	Special
44			F	$900	3/07	ECIL	NF, fenders
44			G	$900	7/07	SCSK	Canadian sale, PTO, 16.9-30 rubber
44			G	$1,100	7/07	ECIL	Special, NF, front wheel scrapers, lights, PTO, restored
44			G	$1,100	7/07	ECIL	NF, fenders, PTO, hyd. remote, restored
44			G	$1,950	7/07	NWOH	SN 764795MI, standard 4 cyl., show-ready
44			G	$2,050	8/07	SESD	LP gas, WF, PS, repainted
44			G	$2,600	7/07	SCSK	Canadian sale, cab, Robin front-end loader with 5' bkt.
44			G	$2,700	3/07	ECIL	NF, fenders, add-on 3 pt., 540 PTO., 1 hyd., restored
44	1948		F	$700	11/07	WCKS	Scoop
44	1948		F	$800	11/07	WCKS	3 pt., PTO, restorable
44	1950		G	$1,000	8/07	WCIL	Gas
44	1950		G	$1,000	8/07	ECNE	Gas, row crop, Farmhand loader
44	1951		G	$1,500	10/07	NECO	Gas, double front, Farmhand F11 loader, 4' dirt bkt.
44	1951		G	$1,500	10/07	NECO	Gas, double front, Farmhand F11 loader
44	1966		F	$700	8/07	ECIL	NF, one fender, John Deere 45 loader
44			F	$125	4/06	WCTX	

Massey Harris

Model	Year	Hours	Cond.	Price	Date	Area	Comments
44			G	$175	6/06	ECSD	Repainted
44			F	$400	1/08	ECNE	3 pt., needs steering wheel and tires, runs
44			P	$500	7/06	NWIL	Diesel, WF
44			F	$650	8/06	ECNE	12-38 rear tires, 1 hyd., 3 pt., fenders, Farmhand F10 loader with 3-tine grapple
44			F	$750	8/06	NWOH	Gas, NF
44			G	$850	10/06	SEWI	Straight tin
44			G	$1,400	10/06	SEWI	WF, rear weights
44			G	$1,600	8/07	WCMN	44 Special, standard, new OH, new paint
44			G	$1,800	4/06	WCTX	Restored, new tires
44	1944		F	$700	2/06	WCSD	Runs
44	1946		F	$800	1/06	SESD	Dual loader
44	1948		F	$850	10/06	NECO	Gas, Farmhand F11 loader with dirt bkt., double front
44	1948		G	$1,350	9/06	ECMI	NF, hydraulic
44	1952		F	$1,000	5/06	SEWY	Gas, WF, Farmhand F11 loader, 5' bkt.
44	1954		G	$2,000	2/06	WCKS	3 pt., PTO
44			G	$900	11/05	WCIL	WF
44			G	$1,000	11/07	WCIL	Clean, painted, drawbar, gas, NF
44			G	$1,050	1/06	WCIL	6 cyl., gas, cleaned, painted, NF, drawbar
44			G	$1,050	1/06	WCIL	Motor OH, not painted, gas, NF
44			G	$1,050	1/06	WCIL	OH, gas, NF, cleaned, painted, drawbar
44			G	$2,400	11/05	NCOH	44 Special
44-6			G	$1,300	11/05	WCIL	6 cyl., older repaint, good runner
444			G	$1,600	6/05	ECMN	Gas, WF
444			G	$2,400	7/05	SESD	3 pt., PS
444	1956		G	$1,900	7/06	NWOH	Original, WF, 3 pt.
444	1957		G	$4,250	11/05	WCIL	NF, gas, spinout wheels, factory 3 pt.
50		4,080	G	$1,800	3/08	SWOH	WF, 12.4-28 tires
50			G	$3,750	10/07	NWWI	WF, high arch, paint
50			G	$2,150	8/05	NCOH	
55			E	$5,700	5/08	ECMN	Standard, diesel, restored, new tires
55	1949		F	$500	11/07	WCKS	
55	1950		G	$4,000	11/07	WCIL	Gas, restored
55			G	$1,800	10/06	WCMN	Duals, new paint
55			G	$2,000	10/06	SEWI	Rusty
55	1948		G	$800	9/06	NECO	Wheatland, new paint, gas
555			G	$5,750	9/07	ECIN	Repainted
555			G	$3,750	6/06	ECSD	
555	1957		G	$3,500	3/05	WCNE	Diesel, 4 speed, PS, 15-34 rear tires, 7.50-18 front tires, PTO
81			F	$1,000	5/08	ECNE	4 cyl., older paint
81			F	$900	6/05	ECWI	
81			F	$1,025	8/05	WCIL	Not painted, cleaned, gas, NF, drawbar
81	1946		G	$1,000	9/05	NENE	PTO, drawbar, 12.4-28 tires
Challenger			F	$4,000	9/07	WCIA	Original, cultivator
Mustang			G	$2,050	3/06	NWOH	Gas, 3 pt., WF, restored
Mustang	1953		G	$3,750	9/07	WCIL	WF, 3 pt., original rear tires
N/A			G	$11,000	2/05	ECNE	4WD, SN 3022907, 6 cyl., older restoration
Pony			G	$1,600	2/06	WCIL	Running, old restoration/repaint, fenders, PTO, lights, sound tractor, looks and mows great
Pony			G	$3,200	3/05	WCIL	
Pony			G	$4,000	12/06	ECIL	Light utility, fenders, rear hyd. lift, PGA21895, nice restoration
Pony			G	$1,600	2/06	WCIL	WF, repaint, 3 pt.
Pony			F	$2,000	8/05	WCIL	WF, OH
Pony			G	$2,400	9/05	NENE	N62 Continental engine, WF, live hyd., 9.5-24 tires
Pony			G	$3,900	7/05	SCIA	Restored
Super 101			G	$4,025	8/05	WCIL	Full tin, cleaned, painted, gas, NF

YOU KNOW YOU LOVE OLD TRACTORS...

... WHEN YOUR WALLET HAS PICTURES OF YOUR TRACTOR
INSTEAD OF YOUR FAMILY.

Minneapolis-Moline, Minneapolis, and Moline

A piece of history. That's what tractors made in the 1920s are today, very cool pieces of history. Imagine what it was like in the roaring late '20s, just before the Great Depression hit. That's what I think about when I see tractors like the three Minneapolis models sold at auction September 22, 2007, in northeast Illinois. A 1926 Minneapolis 17-30 for $17,000, a 1928 model 39-57 for $24,000, and the topper, a 1926 Model 30-50 (one of only nine built) for $40,000.

Pretty cool bits of history indeed.

Minneapolis-Moline

Model	Year	Hours	Cond.	Price	Date	Area	Comments
17-30	1927		G	$9,500	9/07	WCIA	Repainted
20-35			G	$6,500	8/07	SWIL	Twin City
21-32			F	$2,400	8/07	NWOH	Twin City, full steel, straight
21-32	1929		G	$4,750	9/07	WCIA	Twin City, on full steel
27-44	1926		G	$24,000	9/07	NEIL	Twin City, major OH, complete restoration, canopy added
4 Star	1959		G	$2,200	5/08	NEMO	13.6-38 new rear tires and water pump, 4-cyl. gas engine, 3 pt., PS
4 Star			G	$1,400	11/05	WCIL	Running, unrestored/original, fenders, PTO, lights, hyd. remote, deluxe seat, 3 pt., PS, TA
445	1956		F	$1,700	5/08	NEMO	4-cyl. gas engine, WF, PS, 1 hyd., 3 pt., runs good, 14.9-38 rear tires
445			F	$2,200	11/07	WCIL	Street sweeper
445	1958		G	$2,500	8/07	ECNE	LP, 3 pt., NF
445			G	$1,350	10/06	WCMN	Gas
5 Star	1957		G	$2,100	8/07	NWOH	3 pt., fenders, straight
5 Star	1957		G	$2,200	8/07	NWOH	3 pt., fenders, straight
5 Star	1957		G	$7,400	6/07	NCIN	Propane, WF, restored, 3 pt.
5 Star			G	$2,750	1/06	SESD	LP gas, standard, 1957 model
5 Star			G	$3,000	11/05	WCIL	Western air cleaner, Wheatland LP
602			F	$2,325	3/07	NWIL	LP gas, NF
602	1968		F	$2,400	1/06	SCNE	LP, WF, 1 hyd., 18.4-34, front weights
670	1965		G	$3,000	11/07	WCIL	Running, unrestored/original, fenders, PTO, lights, hyd. remote, TA, new brakes, rebuilt generator in 2006
670			P	$1,400	4/05	NCOH	Salvage, gas, 3 pt., fenders, weights
670			F	$1,650	7/05	SCMN	WF, 3 pt., needs restoring
706			F	$2,600	11/05	WCIL	FWA, LP, PTO
BF			F	$900	3/05	NWIL	Gas
BF			G	$1,500	11/05	WCIL	Running, old restoration/repaint, runs good, looks good
BF			G	$3,100	11/05	WCIL	WF, gas, rear wheel weights, fenders, PTO, lights, belt pully, all original
BF			G	$4,000	9/06	NEIN	Repainted, 3 pt., belly mower
BG			G	$4,600	8/07	NWOH	1 row, SN 57901104, older restoration, new front tires, 3 pt.
D			G	$1,500	8/07	NWOH	Corn sheller
FTA			F	$1,500	8/07	SWIL	Twin City, flat-spoke front steel wheels, rear rubber tires
G			F	$1,000	8/07	NWOH	LP
G			G	$1,700	11/07	SESD	Standard, gas, WF
G			G	$1,900	6/07	NCIN	Original
G			G	$2,600	8/05	WCIL	Running, unrestored/original, fenders, PTO, lights, belt pulley
G	1952		F	$800	3/06	NWKS	403-A gas engine, 18.4-34 tires, fenders, restorable
G			G	$1,050	9/05	SCON	Canadian sale
G			G	$5,200	11/05	WCIL	Standard, LP, rear wheel weights, PS, cab
G1000	1966		F	$3,750	5/08	NEMO	Cab, 6-cyl. LP engine, White 1840 loader, 18.4-34 rear tires, dual PTO, 2 hyd., 3 pt.
G1000	1966		G	$5,300	11/07	WCIL	Wheatland, running, unrestored/original, fenders, lights, hyd. remote, PS
G1000			G	$2,600	7/05	SESD	Vista standard, LP gas, 1967 model
G1000			F	$3,300	3/06	NWIL	LP
G1000			G	$1,900	4/05	NCOK	LP, Wheatland model, cab, water cooler
G1000			G	$5,100	3/05	WCKS	LP, 3 pt., excellent tires, PTO, 2 hyd., bad seat
G1355	1974	3,691	G	$7,000	2/06	WCNE	Diesel, 3 pt., PTO, White cab, air, duals
G350		3,174	E	$8,000	3/06	NCND	Utility, 3-cyl. diesel, open station, 3 pt., PTO, 13.6-28 rear tires
G705	1963		F	$2,100	8/07	NWOH	Diesel, factory cab, duals, 1 hyd., PTO, fair original condition

Minneapolis-Moline

Model	Year	Hours	Cond.	Price	Date	Area	Comments
G707	1965		F	$2,900	8/07	NWOH	Standard, 2 hyd., PTO, decent fenders and tin
G707	1965	5,935	G	$3,750	11/07	WCIL	Running, unrestored/original, rear wheel weights, fenders, PTO, lights, hyd. remote, PS, very nice, clean original, low production tractor, new 23.1-30 tires
G707			G	$6,750	3/07	ECMN	New paint, WF, standard diesel
G750			G	$7,250	8/07	NWOH	Early 1970s tractor, gas, powershift rears, front weights, 3 pt.
G900			F	$3,000	11/07	SWNE	Cab, 3 pt., 2 hyd., runs
G900	1968		G	$4,600	11/07	WCIL	Running, unrestored/original, rear wheel weights, fenders, PTO, lights, hyd. remote, deluxe seat, TA
G900	1970		G	$5,200	8/06	NCOH	WF, 2 hyd., 34" duals
G955	1973	2,826	G	$5,100	8/06	NCOH	WF, cab, 2 hyd., over/under hyd. shift, duals
GB	1949		G	$2,100	5/08	NEMO	Propane, good paint, runs good, 14-34 tires, needs tires
GB			F	$800	11/07	WCIL	Original, fairly complete, not running
GB	1955		G	$2,200	6/07	NCIN	LP gas, original, good lights, rear weights, runs
GB	1958		G	$1,300	11/07	WCIL	Running, old restoration/repaint, rear wheel weights, fenders, PTO, hyd. remote, has a custom flat deck with LP tank relocated in the rear like a Vista model
GB			G	$2,000	11/07	SESD	LP gas, standard, 1956 model
GB	1956		F	$2,300	5/05	WCNE	Been in storage for 25 years, diesel
GTA			G	$3,300	11/05	WCIL	Wheatland, gas, electric, 4 cyl.
GTB	1958		F	$2,000	5/08	NEMO	Round fenders, PS, 18.4-34 tires, widened rear rims
GTB			P	$450	2/07	NCCO	Project
GTB			F	$1,850	8/06	NWIL	Gas
GVI			G	$1,500	11/05	WCIL	LP
GVI	1959		G	$2,500	6/07	NCIN	Propane, original, Western air cleaner, rear weights, fairly straight, runs
GVI	1959		G	$4,000	8/07	NWOH	Diesel, 1 hyd., no PTO, decent fenders and tin
GVI	1960		G	$1,700	6/07	NCIN	LP, original
Jet Star	1959		F	$2,100	5/08	NEMO	4-cyl. gas engine, WF, 3 pt., rebuilt PTO, good 13.6-28 rear tires
Jet Star			F	$1,400	1/08	SESD	Loader
Jet Star 2	1963	3,900	G	$4,950	4/06	WCMN	WF, full hyd. loader
Jet Star 3	1965		F	$1,800	11/07	WCIL	Loader
Jet Star 3			G	$2,400	9/06	SCMN	Gas, full hyd. loader, 2 bkts., 3 pt., 14.9-28 spin-out rubber
Jet Star 3	1964		G	$3,250	9/06	NEIN	Gas, WF
Jet Star 3			G	$1,200	11/07	WCIL	WF, gas
Jet Star 3			G	$3,100	11/05	WCIL	WF, gas, electric
JT			F	$1,300	8/07	NWOH	Cutoffs, power lift, fairly straight
JT			F	$1,500	8/07	NWOH	Cutoffs, appears complete
KT	1930		F	$1,300	5/08	NEMO	Twin City, 14-24 rear tires, 4-cyl. gas engine, runs
KTA			F	$1,250	8/07	NWOH	Factory rubber, older restoration
KTA			G	$2,500	6/07	NCIN	Factory rubber, original
M			F	$1,900	3/06	NWOH	Utility-M, SN 05400089, military tractor, with air compressor, dual industrial rear wheels, fenders, restorable
M			G	$700	2/08	ECIL	Rear and front weights, power adj.

Minneapolis-Moline

Model	Year	Hours	Cond.	Price	Date	Area	Comments
M5	1960		F	$2,200	5/08	NEMO	WF, 15.5-38 rear tires, 4-cyl. gas engine, PS, 3 pt.
M5			G	$2,400	11/07	WCIL	
M5	1960		G	$3,200	11/07	SESD	3 pt., WF, repainted
M5			F	$800	11/05	ECNE	Gas, WF, 15.5-38, hyd., PTO, 3 pt., fenders
M5	1961		G	$1,200	11/05	WCIL	PTO, 3 pt.
M602	1964		G	$3,200	8/07	SWIL	Diesel, 3 pt., running
M670			F	$3,600	3/08	NWKS	Gas, Hi-Master Great Bend loader and grapple, no low TA
M670	1966		G	$3,500	5/08	NEMO	Rebuilt torque, 15.5-38 rear tires, good paint, 4-cyl. gas engine, WF
M670	1967	6,613	F	$2,650	2/06	WCNE	LP, 3 pt., PTO, Minneaoplis-Moline cab
M670	1970	3,500	G	$3,700	2/06	WCNE	LP, 3 pt., PTO, Minneaoplis-Moline cab
M670			G	$1,800	11/05	WCIL	New seat, LP
MTA	1938		P	$650	5/08	NEMO	Twin City, NF, 12-36 synthetic rear tires, motor locked up
MTA	1938		P	$700	5/08	NEMO	Twin City, NF, rear steel wheels, runs, needs rear-end
MTA			G	$2,100	9/07	WCIA	Twin City, NF, older restoration
MTA			G	$3,500	8/07	NWOH	Factory rubber, power lift, older repaint
N/A			P	$700	12/07	WCNE	Locked, gas, all there
N/A			F	$750	6/07	SWMN	Model Mule Uni tractor, two-row picker, gas
N/A			G	$3,150	3/07	SCTX	Eagle loader landscape tractor SN 40010242 powered by diesel engine, PTO, 3 pt.
N/A			G	$900	11/06	SCCA	Gas, PTO
N/A	1936		F	$550	8/06	NCIA	Twin City, JT, Z motor, skeleton wheels
R			P	$375	8/07	NWOH	Belt pulley, original condition, can't read SN
R			F	$1,600	6/07	NCIN	NF, right lights, engine free, unrestored
R			G	$1,900	6/07	NCIN	NF, new tires, nice sheet metal
R			F	$2,000	1/05	NWOH	Austin road machinery street sweeper on Minneapolis-Moline R tractor
R			G	$2,600	6/07	NCIN	Unrestored, original, runs
R	1949		F	$1,000	8/07	NWOH	Waterloo decals
R	1950		F	$1,500	8/07	NWOH	Lights, fenders, 1 new rear tire, fairly straight
R			G	$1,100	10/06	WCMN	New paint
R			F	$900	1/05	SWOH	
RTI			F	$900	6/06	NWSD	Gas, runs, industrial loader and basket
RTU			G	$3,000	6/07	NCIN	Red nose, 13.6-38 rears, front weight bracket and wheelie bars, ready for pulling, older restoration
RTU	1949		G	$2,600	9/07	WCIA	Older repaint
U	1950		F	$800	5/08	NEMO	4-cyl. gas engine, 15.5-38 rear tires
U			F	$575	3/07	SEND	PTO, hyd.
U			G	$2,000	3/06	NCCO	Restored and tractor puller
U	1950		G	$1,550	8/07	WCIL	Gas
U	1951		P	$500	9/07	WCIL	Rear tires and rims off and sitting behind tractor
U	1951		F	$1,500	8/07	NWOH	Belt pulley, fenders, old repaint fairly straight
U	1951		G	$1,550	8/07	WCIL	Gas
U	1952		G	$2,300	6/07	NCIN	Standard, unrestored, gas, runs
U			G	$550	9/06	NWMN	NF, converted 45'×10" auger with 8" takeout
U			G	$600	4/06	WCTX	
U			F	$1,300	3/06	NWIL	Gas, NF, fenders
U			G	$1,800	9/06	NEIN	LP, original
U	1950		F	$950	8/05	NWIL	Gas, NF
U302	1965		G	$3,250	8/07	NWOH	WF, 2× front weights, rear weights, 3 pt., straight original
UB	1954		F	$900	8/07	SESD	
UB	1954		G	$1,900	8/07	WCIL	Gas, hand clutch
UB	1955		F	$1,000	8/07	NWOH	
UB	1955		F	$2,100	8/07	NWOH	Diesel

Minneapolis-Moline

Model	Year	Hours	Cond.	Price	Date	Area	Comments
UB	1955		G	$2,800	7/07	ECIL	Pulling tractor, running on propane, NF, new front and rear rubber, restored
UB			F	$3,500	4/06	WCTX	
UB	1955		F	$3,000	2/06	NWIA	NF, 2 pt., clamshell fenders
UB	1955		G	$2,550	6/05	SEND	WF, 1 hyd., PTO, gas, 13-38 tires
UT			G	$1,225	1/06	SESD	LP gas, 1950 model
UTE	1951		G	$1,900	8/07	WCIL	Running, unrestored/original, fenders, PTO, lights, hyd. remote, low production tractor, very clean original
UTE			G	$1,200	11/07	WCIL	WF, live PTO
UTS			F	$400	11/07	WCIL	
UTS			F	$700	11/07	WCIL	Running, unrestored/original, rear wheel weights, fenders, belt pulley, belts are new, generator replaced in 2007, new battery
UTS			F	$1,000	6/07	NCIN	Runs
UTS			G	$1,800	6/07	NCIN	Old repaint, new tires, Western air cleaner
UTS	1955		G	$1,400	6/07	NCIN	Belt pulley, orginal, good 18.4-30 rear tires
UTS	1952		G	$1,100	11/05	WCIL	LP, lights, hyd.
UTU			F	$900	6/07	NCIN	NF, LP, unrestored, needs full restoration
UTU			F	$1,300	6/07	NCIN	Unrestored, original
UTU			G	$2,200	8/07	NWOH	Red nose, good tin
UTU			F	$300	4/06	WCTX	
UTU			G	$1,400	10/06	SEWI	
Z			P	$200	8/07	NWOH	For parts
Z			P	$500	8/07	NWOH	
Z			P	$750	8/07	NWOH	
Z			F	$1,100	6/07	NCIN	Unrestored, original, hood rusted through
Z			F	$1,100	6/07	NCIN	NF, rear wheel weights, unrestored
Z			G	$1,900	6/07	NCIN	NF, runs
Z			G	$500	10/06	SEWI	
Z			F	$1,000	4/06	WCTX	
Z			G	$2,100	11/07	WCIL	Hand start, mag rebuilt
ZAN	1951		F	$800	5/08	NEMO	Single front wheel, 4-cyl. engine, engine locked up, 11.2-38 tires (good), needs paint
ZAU			P	$250	2/08	NEMO	NF, as is, barn find
ZAU	1949		P	$165	5/08	NEMO	For parts, no engine
ZAU	1949		F	$900	8/07	NWOH	No fenders, fairly straight
ZB			P	$525	1/07	SCNE	WF, for parts
ZB			P	$800	2/07	NCCO	Project
ZB			F	$1,500	8/07	NWOH	Greasy but straight
ZB			G	$800	2/08	WCIL	NF
ZB			G	$1,800	6/05	ECMN	
ZTS	1938		G	$2,500	5/08	NEMO	4-cyl. gas engine, 12.4-36 rear tires
ZTU	1944		P	$125	5/08	NEMO	For parts, no engine, has fenders and trans.
ZTU			P	$300	9/07	ECIN	Needs work
ZTU			P	$600	8/07	NWOH	Red nose, fenders, incomplete
ZTU			F	$1,200	6/07	NCIN	Red nose, fenders, older repaint
ZTU	1947		F	$900	8/07	NWOH	Red nose, fenders
ZTU			F	$975	6/05	WCMN	NF, wide fenders
ZTU	1940		G	$850	11/05	WCIL	NF, missing side shield

Minneapolis

Model	Year	Hours	Cond.	Price	Date	Area	Comments
17-30	1926		G	$17,000	9/07	NEIL	Restored, crossmotor 4-cyl. engine
30-50	1926		G	$40,000	9/07	NEIL	9 built
39-57	1928		G	$24,000	9/07	NEIL	12 built, restored

Moline

Model	Year	Hours	Cond.	Price	Date	Area	Comments
Universal	1919		G	$10,000	9/07	WCIA	Motor OH, repainted

YOU KNOW YOU LOVE OLD TRACTORS...

...WHEN YOU CAN SPOT AN OLD TRACTOR UNDER A TANGLE OF WEEDS... ONE HALF MILE OFF THE ROAD... DRIVING DOWN THE INTERSTATE.

Oliver and Hart-Parr

Smile for the camera. About five years ago, Dave Mowitz, machinery editor for *Successful Farming* magazine and editor of the *Ageless Iron Almanac* mentioned he needed some new pictures of me to run with my "Machinery Pete" columns. So the photographer and I set out to find machinery to shoot next to. We wound up in a southeastern Minnesota collector museum featuring 20 to 30 beautiful old Oliver/Hart-Parr tractors. I've been a big fan ever since.

Apparently others are, too. Oliver 1850s, made from 1964 to 1969, have appreciated 79.3% in value from 2002 to 2007, up to an average sale price of $4,686.

Oliver

Model	Year	Hours	Cond.	Price	Date	Area	Comments
1250			G	$2,200	1/08	NEMO	Gas
1265			G	$3,500	11/05	WCIL	Running, unrestored/original, fenders, PTO, lights, PS
1265			G	$2,700	10/06	NCWI	Diesel, high clearance, orchard
1355			G	$2,500	9/06	NWTN	Front loader, bale spear
1365		3,812	G	$4,000	2/07	NEIN	Diesel, WF, 1510 loader and 6' bkt., 3 pt., 1 hyd.
1365			G	$4,500	10/05	WCMT	
1450			G	$1,100	1/08	WCIL	Diesel
1550		4,751	G	$4,100	3/07	NWIA	One owner, actual hours, loader
1550	1966		G	$6,500	11/07	WCIL	Utility, running, unrestored/original, fenders, PTO, lights, hyd. remote, PS
1550	1968		G	$2,600	12/07	NWOH	
1555	1975		G	$7,000	11/07	WCIL	Running, restored, fenders, PTO, lights, hyd. remote, deluxe seat, new paint and decals
1555		3,000	G	$4,150	6/06	ECSD	WF
1555			G	$6,000	3/07	WCIL	High crop, diesel
1555	1974		G	$3,950	3/05	SEND	Livestock special, WF, gas, hydra power drive, PTO, Farmhand F11 loader, fenders, new tires
1600		3,000	G	$4,000	1/08	NEMO	Gas, tach changed approx. 2,000 hours ago
1600			G	$2,500	3/06	NCIL	Gas, WF, 2-speed hyd. power drive
1600		4,500	F	$4,400	1/06	SCMI	Gas, WF, 3 pt., Oliver 1610 front end loader
1600			G	$2,250	2/05	NCOH	
1600	1963	2,500	G	$4,100	4/05	NCOH	Diesel, NF, hydra power, 3 pt., 1 hyd.
1650			G	$3,750	1/08	NWOH	
1650	1969	7,312	G	$4,000	2/08	SWMN	Diesel, WF, PTO, 3 pt.
1650			F	$2,600	10/07	NEKS	Diesel, cab
1650			P	$3,550	8/07	SCMN	As is, antifreeze in oil, diesel engine
1650	1965	5,590	G	$4,700	8/07	NCIA	Gas, factory WF, 3 pt., 16.9-34
1650	1965	2,700	E	$6,000	2/07	WCIL	Gas, WF, 2 hyd., 540 PTO, 16.9-34 & 6.5-16.5SL tires, 2,700 hours, professionally restored and extra sharp
1650	1966	8,279	G	$3,600	2/07	NEIL	Gas, 15.5-38 tires
1650	1968		G	$3,700	1/07	NEOH	Gas, WF, 15.5-38 tires
1650			F	$2,500	8/06	SCMI	WF, gas, 3 pt., OH in 2004
1650		3,426	G	$3,000	5/06	SCIL	
1650		6,486	G	$3,000	3/06	NEIA	One family owned, row crop, gas, OH around 3,500 hours, like-new 16.9-34 tires
1650			G	$3,000	8/06	SEMN	Good rubber
1650	1965	4,671	G	$4,000	2/06	WCIL	WF, 3 pt., fenders, 2 hyd., gas, 16.9-34
1650	1967		G	$3,850	8/06	WCIL	
1650			P	$1,800	10/05	ECIL	Row crop tractor, gas, rough shape, 3 pt., 15.5-38, hyd. front brush blade
1650			G	$2,500	1/05	SWMN	Schwartz loader
1650			G	$2,550	10/05	ECMN	Diesel, cab, 3 pt.
1650			G	$2,900	1/05	ECIA	Older, gas, NF, 3 pt., fenders
1650			G	$3,500	11/05	WCIL	NF, diesel, PS, deluxe seat
1650			G	$4,800	11/05	NCOH	
1650			G	$6,250	1/07	WCIL	High crop, front spokes, fenders
1650	1967	7,750	F	$3,450	10/05	NCOH	Gas, needed rear tires
1655			G	$5,400	8/06	NEIA	Gas, 6,000 hours on OH, 3 pt., 2 hyd., NF, 16.9-34 tires
1655			G	$5,500	7/05	NEIA	Diesel, 900 hours on OH, 3 pt., 2 hyd., 16.9-34 tires, WF, Year-A-Round cab
1655			G	$5,600	8/07	SCMN	Cab, gas
1655	1969	4,823	F	$4,100	5/07	SEWY	Diesel, WF, cab, 3 pt.
1655	1971	5,456	G	$5,100	2/07	NEIL	Gas, 16.9-34 tires
1655		3,942	G	$4,000	10/06	SEWI	Diesel, cab, front weights
1655			F	$5,500	2/06	WCIL	
1655	1971		G	$7,000	2/06	NCOH	Diesel, WF, 16.9-34 rubber, 2 hyd., over/under trans., SN 229075

Oliver

Model	Year	Hours	Cond.	Price	Date	Area	Comments
1655	1973		G	$6,500	4/06	NWOH	Gas, cab very clean
1655			G	$3,900	11/05	WCIL	Deluxe seat, wheel weights, fenders
1655			F	$4,000	7/05	SCMN	Cab, 500 hours on OH, diesel
1655			G	$5,600	8/07	SCMN	
1655			G	$6,100	6/05	NCIL	5,400 hours, WF, 3 pt., good tin
1655	1972		G	$4,200	12/05	ECNE	Diesel, 2 hyd., over/under hyd. shift trans., 16.9-34, 540 PTO
1655	1972		G	$7,900	9/05	NCND	Gas, row crop, 540 PTO, dual 320 loader, very clean
1750			F	$2,700	8/07	NEIA	Gas
1750			G	$3,200	11/07	SESD	WF, 3 pt., dual fender fuel tank, Westendorf WL42 loader and grapple
1750	1969	5,338	G	$3,900	9/07	SCMI	1 hyd., 3 pt.
1750			F	$2,100	11/07	WCIL	WF, gas
1750			F	$3,500	8/05	NWIA	WF, cab, 3 pt., Koyker K5 loader
1750		3,216	G	$4,500	7/05	SCMI	Gas, WF, Hi-Master 900 loader
1750	1967		G	$2,675	2/05	NEIN	Diesel, WF, tach shows 1,461 hours
1755	1974	3,065	G	$7,100	12/07	SEIA	Hiniker cab, 3 pt., 2 hyd., one owner
1755	1974		G	$14,000	11/07	WCIL	Running, old restoration/repaint, rear wheel weights, fenders, PTO, lights, hyd. remote, front weight, deluxe seat, 3 pt., PS, 5.9 Cummins diesel conversion, hd clutch
1755		9,244	F	$3,300	11/05	SCMI	Diesel, WF
1755	1972		G	$3,800	12/05	NWIL	Clamp duals, WF, 2 hyd., short axles, diesel
18-27	1935		G	$1,800	4/06	WCIA	Older restoration
18-28			G	$1,900	10/06	SEWI	Cutoffs, variable-speed governor
18-28			G	$1,900	10/06	SEWI	Cutoffs, variable-speed governor
1800			F	$3,000	11/07	SESD	Gas, standard
1800			G	$6,750	11/07	WCIL	Running, restored, beautifully restored cab with all Oliver parts
1800	1963		G	$10,000	11/07	WCIL	B Series, running, unrestored/original, FWA, tin perfect, hydra power works great, original paint
1800	1964		G	$4,200	7/07	ECIL	Checkered board, NF, rear wheel weights, front weights, PTO, hyd., 3 pt., PS, lights, nice restoration
1800			G	$1,550	6/06	ECSD	Wheatland, MFWD, new rubber
1800	1963		E	$6,500	6/06	NECO	WF, no cab, dual loader, gas, tach shows 839 hours
1800			F	$1,300	8/07	WCIL	WF, gas, fenders, PTO, lights
1800			F	$2,100	8/05	NWIA	3 pt.
1800			F	$2,900	4/05	SEND	3 pt., diesel, cab
1800			G	$3,250	11/05	SEND	Gas, 2 hyd., 3 pt., 540 PTO, fenders, band duals
1800			G	$5,200	11/05	WCIL	Std., diesel, wheel weights, PTO shift, front weights, Western tractor
1800	1964		G	$3,900	3/05	ECNE	3 pt., 2 hyd., 540 PTO, diesel, WF, rear fenders, 18.4-34, clamp-on duals, rear weights
1850		5,203	G	$4,300	4/08	ECMN	Diesel, hydra power drive, new 18.4-38 hub duals, 3 pt., PTO, 1 hyd.
1850		4,700	G	$6,000	1/08	NWOH	Great Bend loader
1850	1967		G	$4,600	5/08	ECMN	Diesel, WF, 3 pt.
1850	1969		G	$3,250	1/08	ECMI	No cab, 3 pt., PTO, diesel
1850			F	$2,100	11/07	SEND	Cab, 3 pt., gas
1850			F	$3,500	11/07	SESD	3 pt., WF
1850		2,472	G	$5,750	2/07	WCIL	Diesel, Year-A-Round cab, WF, hyd. shift, 18.4-34 and 10.00-16 tires, 2 hyd., 540 PTO, tach showing 2,472 hours, straight, SN is covered up by the cab and cannot be read
1850	1966	5,820	G	$5,850	2/07	WCIL	Diesel, WF, hydra-power drive, fenders, 18.4-34 and 11L-15 tires, 2 hyd., 540 PTO

Oliver

Model	Year	Hours	Cond.	Price	Date	Area	Comments
1850	1966		G	$7,000	1/07	NEOH	Diesel, Year-A-Round cab, 3 pt., 2 hyd., 18.4-34 rears
1850	1968		P	$600	4/07	NECO	Diesel, 3 pt., WF, cab, needs work
1850	1968		G	$8,000	1/07	NEOH	Diesel, open station, 18.4-38 rears, 3 pt., 2 hyd.
1850			G	$2,200	7/06	NWIL	Gas
1850	1964		G	$4,000	11/06	NCKS	Hydra power, open station, 3 pt., 2 hyd.
1850	1965		F	$1,750	9/06	NEIN	Diesel, noisy leaking valves
1850	1965		G	$3,800	11/06	ECND	Diesel, Year-A-Round cab, Koyker fast-attach loader
1850			F	$2,750	7/05	ECND	Wheatland, XL cab, 2 hyd., PTO, Farmhand F11 loader and grapple
1850	1967	6,900	G	$3,950	2/05	NEIN	Diesel, OH
1850	1967		G	$4,400	9/05	SEIN	
1850	1968		G	$3,400	2/05	NENE	Diesel, 18.4-38 tires, 2 hyd., WF
1855			P	$1,000	4/08	ECMN	For parts with rear end, WF, tires, radiator and fenders
1855	1973	6,099	G	$3,900	4/08	ECMN	Diesel, WF, 2 hyd., 3 pt., PTO, 20.8-38 tires
1855			F	$3,250	8/07	SESD	WF, 3 pt., 1971 model
1855	1973	3,200	G	$5,400	12/07	SEIA	Hiniker cab, 3 pt., 2 hyd., one owner
1855	1972		G	$4,900	2/06	NCIL	Diesel, 2 hyd.
1855	1973	3,400	G	$5,300	3/06	NCIL	WF, 2 hyd., 3 speed, gas, over/under
1900			F	$2,400	3/08	NWTN	Diesel, needs repair
1900			G	$3,800	8/07	SESD	GM engine
1900			G	$3,500	11/07	ECSD	Detroit diesel
1950			G	$3,500	11/05	WCIL	GM diesel, running, unrestored/original, PS, good tin
1950			G	$5,000	2/06	NWSD	Industrial tractor, 4WD, yellow, underground cable plow
1950T			G	$7,750	11/05	WCIL	Running, restored, very nice paint, has over/under that works like it should, 3 pt., about 500 hours on a new Waukesha diesel engine, runs perfect
1955	1971	5,304	G	$4,300	4/08	ECMN	Diesel, 3 pt., PTO, 2 hyd., 18.4-38 duals
1955		4,587	G	$8,250	3/05	ECMI	Diesel, cab, heat, axle duals, dual PTO, stone box and front weights, power beyond
2255			G	$7,100	9/06	NEIN	4WD
2255			G	$6,600	3/05	WCIL	Cab, weights, 3150 engine, diesel
2655			G	$6,750	2/05	NCIN	LP, 3 pt., PTO, 800-cubic-inch engine
55			G	$1,600	2/06	WCIL	Ind WF, gas, lights, fenders, restored
55			G	$2,800	11/05	WCIL	
550	1964	1,532	G	$5,200	5/08	ECIL	WF, utility, gas, 3 pt., PS, attached Oliver 59A hyd. bkt. loader, original one-owner tractor
550			G	$0	2/07	NEIA	No sale at $3,300
550			F	$2,500	9/07	WCIL	Rear weights, drawbar
550		1,600	G	$5,300	1/07	WCIL	Utility, WF
550	1959		G	$2,500	11/07	WCIL	Running, unrestored/original, fenders, PTO, 3 pt.
550	1959		G	$4,000	2/07	SWIL	
550	1959		G	$5,750	9/07	WCIA	WF
550	1965		F	$2,000	10/07	WCIL	Did not run, small auction but still five bidders from farther than 50 miles away on this tractor
550			G	$3,000	3/08	SCIL	PS
550	1958		G	$3,200	12/06	SCNE	Gas, 3 pt.
555			G	$3,900	3/05	NWIL	Loader, blade, gas
60			P	$475	8/07	SESD	For parts
60			F	$1,900	8/07	SESD	

Oliver

Model	Year	Hours	Cond.	Price	Date	Area	Comments
60			G	$3,750	12/06	WCIL	Running, restored, fenders, lights, belt pulley, this is a frame-up restoration with original synthetic rear tires, everything works including the lights
60	1941		F	$850	8/07	NWOH	Right side curtains, straight tin, fenders, lights, rear tires fair, front tires poor
60	1942		G	$7,000	9/07	NEIL	Standard
60	1943		E	$3,800	3/07	NWIL	Gas, power lift, PTO, new 9.5-32 tires, new paint
60			G	$925	12/06	NWIL	Fenders
60			G	$1,200	6/05	WCMN	Row crop, original, curtains, 4 speed
60			F	$1,700	4/06	WCTX	
60			G	$2,100	3/08	NCIA	Row crop, mid- to late-1940s, 9.5-32 new rubber
60	1941		G	$3,300	10/06	SCSD	Row crop, fenders, NF, gas, older restoration
60			G	$1,100	1/08	WCIL	NF
60			G	$1,300	11/05	WCIL	Restored
60			G	$1,300	11/05	SCKY	
60			G	$1,600	6/05	ECMN	Gas, saw rig
60			G	$4,300	11/05	WCIL	NF, gas, side dress fertilizer
66			G	$3,600	6/07	NCIN	NF, missing side curtains, fenders, restored, runs
66	1949		F	$1,400	6/07	NCIN	Missing one side curtain, runs
66	1950		G	$2,300	7/07	NCIA	Row crop, gas, good tin, 12 volt, fenders
66	1951		G	$7,000	11/07	SWIN	Standard, restored, nice, inside rear weights, belt pulley
66	1951		G	$7,250	11/07	WCIL	Industrial, running, restored, rear wheel weights, fenders, lights, no PTO, flat trans. cover, all new tires
66	1953		G	$1,900	8/07	WCIL	Gas, hyd.
66			G	$3,600	5/06	SCIL	Restored
66	1956		P	$1,250	7/06	ECND	NF, stuck, not running
66			G	$825	3/05	SCMI	New paint, row crop
66			G	$825	3/05	ECMI	Row crop, new paint
66			G	$1,800	11/05	WCIL	NF, gas
66			F	$2,000	7/07	SEND	Belly mower
66			G	$3,000	11/05	WCIL	Row crop, NF, gas, restored
66			G	$3,250	11/05	WCIL	Std., gas
66			G	$21,000	7/06	WCIL	Orchard, gas
66 RC	1949		G	$3,250	11/07	WCIL	Running, old restoration/repaint, rear wheel weights, fenders, lights, belt pulley, new tires 2003, gauges all work, 8-volt battery, side shields all there
660	1962		G	$6,550	8/07	SCMN	One owner, gas, WF
660			G	$3,450	11/05	ECMN	Gas, older loader
70			P	$275	2/08	NCOH	Original shape, runs, as is
70	1937		G	$3,200	2/07	ECNE	Full steel, side shields, original
70			G	$2,100	6/07	NCIN	Belt pulley, fenders, flat lens headlights, old restoration, runs
70	1939		F	$1,600	6/07	NCIN	Two flat lens lights, fenders with extensions (fair-poor), runs
70	1946		F	$500	11/07	WCIL	Row crop, running, unrestored/original
70	1947		E	$21,000	11/07	WCIL	Row crop, running, restored, PTO, lights, deluxe seat, belt pulley, Raby cab, very low-hour tractor, cardboard still under Monroe upgrade seat
70	1948		F	$1,300	6/07	NCIN	Restored, missing both side curtains, flat lens headlights (full set), runs
70			G	$1,400	10/06	SEWI	
70	1930		G	$1,250	8/06	NCIA	Skeleton wheels, NF
70			G	$650	11/05	WCIL	Original

Oliver

Model	Year	Hours	Cond.	Price	Date	Area	Comments
70			F	$800	4/05	SCMI	NF, gas
70			G	$2,600	8/05	WCIL	Styled rowcrop, original side panels, converted to 12 volt, NF, 6-speed trans., new spark plugs and wires
70	1946		G	$1,100	3/05	SEND	NF, older restoration
77			G	$2,550	3/08	SEIL	Row crop
77			F	$1,000	2/07	NEIN	Gas, NF, row crop
77			G	$1,250	1/08	SESD	NF
77			G	$11,000	1/08	WCIL	Orchard, older restoration
77	1948		F	$1,900	6/07	NCIN	WF, fenders, headlights, fair tin and tires, runs
77	1949		F	$800	1/07	NEOH	Not running, gas, good tin, fenders, side screens, restorable
77	1949		G	$4,000	11/07	SWIN	Standard, restored, new tires, belt pulley, hydroelectric
77	1950		G	$4,000	2/07	NEMO	NF, original sheet metal
77	1951		F	$1,800	6/07	NCIN	Missing one side curtain, belt pulley, restored, runs
77	1952		G	$2,850	8/07	WCIL	Gas
77	1953		F	$800	8/07	NWOH	Old repaint, fenders and lights
77			G	$1,750	10/06	SEWI	Belt pully
77			F	$2,000	1/05	NWOH	NF, gas, row crop
77	1949		F	$1,700	9/06	ECMI	Row crop, WF, all original
77	1951		G	$5,000	10/06	SCSD	Row crop, gas, WF, fenders, lights super nice
77	1955		F	$950	7/06	ECND	WF, not running
77			G	$900	11/05	WCIL	Std., gas, 12 volt, toolbox radiator, center brake link, original gauges
77			G	$1,000	11/07	WCIL	WF
77			G	$1,300	11/05	WCIL	Gas
77			G	$1,400	11/05	WCIL	Row crop, gas, coil battery and plugs
77			F	$1,400	11/07	WCIL	Gas, new rear tires, good tin
77			G	$1,750	3/05	SEND	NF, hyd., PTO, older restoration, new 14.9-38 tires
77			G	$2,000	11/05	WCIL	Standard, diesel
77	1949		F	$1,300	4/05	SCNE	Row crop, PTO, gas, fenders, 13-38 rear tires, lights, PS, runs
77	1950		F	$425	11/05	ECNE	Gas
77	1953		F	$1,100	4/05	SCNE	Row crop, single front wheel, runs
77 RC			G	$1,900	11/05	WCIL	Row crop, diesel, running, old restoration/repaint, fenders, lights, all it needs is paint, does have sidepanels, all new tires
77 RC	1951		G	$2,250	7/07	ECIL	NF, new front wheels, 70% rear, fenders, lights, PTO, side curtains, restored
770			F	$1,000	1/08	NEMO	Gas
770		4,568	G	$2,700	1/08	NWIL	NF, rear weights
770	1965		E	$4,250	5/08	ECMN	Restored, gas, NF, square fenders, newer 13.5-38 tires
770			F	$1,500	7/07	SESK	Canadian sale, antique tractor, 2 hyd., belt pulley, gas motor, loader with bkt.
770			F	$1,800	1/07	SEKS	NF, Duncan loader, gas
770		3,400	G	$2,500	8/07	SESD	WF
770			F	$2,600	12/07	SCMN	Gas, late model, Schwartz loader
770			F	$3,500	11/07	NENE	NF
770	1963		G	$15,000	11/07	WCIL	Running, restored, orchard fenders
770	1965		E	$13,000	11/07	WCIL	LP, orchard, SN 153-613-732, running, restored, fenders, lights, very rare, low production tractor, excellent first class restoration
770			P	$625	3/06	SWOH	Diesel, row crop, bad head, for parts or restoration
770			G	$1,625	9/06	SCMN	Gas, fenders, rock box
770			G	$2,400	3/06	NEIA	Second owner, row crop, gas
770	1962		G	$2,350	1/06	NWIL	Row crop, new paint, gas

Oliver

Model	Year	Hours	Cond.	Price	Date	Area	Comments
770	1964		F	$1,700	9/06	NECO	WF, 3 pt., PS, gas, tach shows 3,385 hours
770	1966		F	$2,900	9/06	NECO	Gas, Farmhand F11 loader and grapple, WF, 3 pt., tach shows 1,620 hours
770			G	$1,400	11/05	WCIL	Orchard, diesel, fenders
770			G	$1,700	10/05	WCIL	Single front, lights, 3 pt.
770			F	$2,100	4/05	SCMI	WF, gas
770			G	$2,500	11/05	WCIL	Motor needs work, orchard
770			E	$3,400	3/05	SCIL	Factory 3 pt., original good paint, NF, 1962 model, original paint
770		3,370	G	$3,900	1/05	SWMN	WF
770			G	$5,000	11/05	WCIL	High-boy sprayer, WF, high crop, LP
770	1958		E	$2,300	3/05	SCIL	PS, 2 hyd., NF, original paint, straight, original paint
80			F	$850	8/07	SESD	
80			G	$1,500	11/05	WCIL	Gas, NF, front steel
88			G	$3,000	11/05	WCIL	Puller, Chevy V-8 small block
88	1951		G	$1,700	6/07	NCIN	Row crop, gas, WF, nice tractor, clean, engine free, old repaint, runs
88	1951		G	$26,000	11/07	WCIL	Orchard, running, restored
88			G	$2,700	10/06	SEWI	Old style, has had Fleetline Series sheet metal installed
88	1948		F	$450	3/06	SWNE	Row crop, 6 cyl., gas, NF, 13.6-38
88	1949		F	$1,000	5/06	SEWY	Gas, double front, live PTO
88	1951		F	$500	3/06	SWNE	Row crop, diesel, WF, 13.6-38, front fenders, pulley
88	1953		G	$1,550	8/06	NCIA	Gas, NF, PS, side panels, fenders
88	1956		P	$1,100	7/06	ECND	WF, diesel
88			P	$550	3/05	SEND	WF, for parts
88			G	$900	11/05	WCIL	NF, gas, unrestored
88			G	$1,500	11/05	WCIL	Brass throttle, WF, gas, PTO, hyd.
88			F	$1,600	8/05	WCMN	NF, row crop 88, side tins, PTO fenders, lights
88			F	$1,700	3/07	SCNE	Row crop, Farmhand F11 loader, gas, 13.6-38 tires, 8' bkt., quick-tach, runs
88			G	$1,800	3/08	NCIL	Row crop, repainted, good tin
88			G	$1,800	11/05	WCIL	Diesel
88			G	$2,000	11/05	WCIL	Standard, gas
88			G	$2,900	11/05	WCIL	Row crop, NF, new wrist pins and bushings
88			G	$7,800	11/05	WCIL	Old style, gas, OH, fenders, PTO, lights
88	1949		F	$1,700	4/05	SCNE	Row crop, 6-cyl. gas engine, WF, 13.6-38 tires, fenders, runs
88	1953		F	$1,100	12/05	NWIL	Standard WF, side shields, good rubber, runs
88 RC	1952		G	$2,500	7/07	ECIL	NF, fenders, PTO, side curtains, restored
88 RC	1953	3,200	G	$3,400	6/05	ECIL	
880			G	$2,400	7/05	SESD	WF, PS, gas
880		4,845	F	$2,500	3/08	NWWI	LP, WF, fenders
880			F	$3,250	12/06	NWWI	LP, WF, fenders, rebuilt engine, loader
880	1958		G	$3,000	5/08	ECMN	Diesel, WF, fenders, 16.9-34 tires
880	1960		F	$1,400	4/08	NEIA	Diesel
880			G	$2,300	1/08	SESD	
880			G	$3,400	5/08	ECMN	Gas, WF, standard hitch
880			G	$3,750	12/06	WCIL	Standard, older repaint, fender ext.
880	1958		G	$5,100	7/07	ECIL	Pulling tractor, rear wheel weights, fenders, front and mid-tractor weight brackets, nice
880	1958		G	$5,500	11/07	WCIL	Pulling tractor, restored, fenders, lights, this is a proven winner, very strong tractor, makes the earth shake when it's running
880	1959		G	$1,900	12/07	SEIA	LP, NF
880	1963		G	$2,500	8/07	ECNE	Diesel
880			F	$1,700	9/05	NWIA	
880			F	$3,300	6/05	NCIL	9,913 hours, WF, gas, good tin
880	1959		G	$2,450	9/05	SEIN	

Oliver

Model	Year	Hours	Cond.	Price	Date	Area	Comments
950			G	$6,750	1/08	SESD	Wheatland, diesel
99			F	$3,600	9/07	ECIN	Repainted
99			G	$4,000	8/07	SESD	Diesel
99			P	$525	4/06	WCSD	For parts
99				$4,800	11/05	WCIL	Diesel, rear wheel weights, lights, Western
990			G	$19,000	7/06	WCIL	GM, running, restored, fenders, lights, fresh restoration of a great tractor, really nice body shop restoration, great runner, started with a good original tractor
N/A			G	$4,500	1/08	ECNE	LXK3 steel track tractor
Super 44	1957		G	$15,500	11/07	WCIL	Running, restored, rear wheel weights, fenders, PTO, front wheel weights, lights, hyd. remote, 3 pt.
Super 44			G	$7,600	9/06	NEIN	Restored
Super 44			G	$10,000	8/06	SCIL	
Super 55			F	$1,900	11/07	WCIL	Trip loader
Super 55			G	$2,100	8/06	SCIL	Oliver, hyd. bkt.
Super 55			G	$2,800	2/06	SWNE	4 cyl., PTO, rear weights, one owner
Super 55			F	$3,400	3/05	NCIL	Loader, backhoe
Super 55	1954		G	$2,150	9/05	SEIN	Loader
Super 66			G	$5,000	11/05	WCIL	
Super 66			G	$7,300	5/06	SCIL	Diesel, belly mower
Super 77			G	$3,600	3/08	SEMN	Diesel, WF
Super 77			F	$1,600	6/07	NCIN	Gas, NF, good rubber, bent grill, fenders, missing front side panels, runs
Super 77	1956		G	$1,950	1/07	ECIL	Row crop, hyd.
Super 77			F	$2,000	1/05	NWOH	WF, gas
Super 77			G	$2,400	8/06	SCIL	Oliver loader, hyd. bkt.
Super 77			G	$3,900	8/06	SEMN	Diesel, 3 pt., New Idea loader
Super 88			F	$2,500	3/06	NCOH	Row crop, not running
Super 88			G	$3,000	11/05	WCIL	Diesel, running, old restoration/repaint, PTO, hyd. remote, PS, runs perfect, drives great
Super 88			P	$375	9/06	NEIN	For parts, diesel
Super 88			G	$500	5/06	SCIL	
Super 88			F	$850	8/06	SEMN	
Super 88			F	$1,250	8/05	WCMN	WF, good tin, motor and clutch weak
Super 88			F	$1,750	12/05	NWIL	Diesel, NF, good rubber, runs
Super 88			E	$3,100	3/05	SCIL	PS, 2 hyd., new paint, NF, 1957 model, diesel
Super 99	1955		G	$21,000	9/07	NEIL	3-71 GM diesel engine, hyd., PTO
Super 99			G	$8,200	11/05	WCIL	Unrestored

Hart-Parr

Model	Year	Hours	Cond.	Price	Date	Area	Comments
10-20			G	$30,000	10/06	SEWI	
10-20			G	$6,300	6/05	ECWI	
18-36			G	$5,250	6/07	SEID	Mechanically restored, new fenders
18-36			G	$11,500	9/07	ECIN	Rubber lugs
18-36		5,000	G	$1,929	9/07	WCIA	Older restoration
18-36			F	$3,000	4/05	NCCO	Steel wheels, not stuck and complete, extra radiator, restorable, complete
22-40	1926		G	$25,000	9/07	NEIL	Easy starting, very sweet running tractor
28-44	1934		F	$1,600	9/06	ECMI	Rubber tires
28-50			G	$8,000	11/07	WCIL	Crossmotor, running, restored, excellent older tractor, nice rubber treads for parade use, new valves, rebabbited bearings, rebuilt lubricator, new fenders and fuel tank, new oil feed lines, professional restoration
70			G	$2,200	4/06	WCIA	Oliver
N/A			F	$3,100	6/07	NCIN	Single-wheel row crop, rear steel, round spoke front, belt pulley, older repaint, manifold damaged

Lesser-Known Classics

A 1918 Huber 35-70 (big prairie-type) sold for $250,000. A 1918 Rumeley 30-60 OilPull Model E for $165,000. What fun to find these gems on the auction market. They sure don't show up every day. Probably why there is such an incredible level of interest in these rare tractors when they do appear for sale.

Oh, the 1918 Rumeley 30-60 that sold for $165,000? According to the auction sale bill, it sold for $4,300 brand-new 90 years ago!

Lesser-Known Classics

Atlas Topp Stewart

Model	Year	Hours	Cond.	Price	Date	Area	Comments
N/A	1919		G	$45,000	9/07	NEIL	4WD, full restoration, bought by township of Caspian, MI, in 1919 to build roads, sold to farmer who owned until 1985, used for plowing

Aultman Taylor

Model	Year	Hours	Cond.	Price	Date	Area	Comments
30-60	1918		G	$95,000	9/07	NEIL	90" drive wheels, full canopy, older restoration

Avery

Model	Year	Hours	Cond.	Price	Date	Area	Comments
12-25	1920		G	$52,500	9/07	NEIL	Recently restored, 7,500 lbs., 2-cyl. opposed engine
20-35	1923		G	$67,500	9/07	NEIL	Original paint, starts, runs great
25-50	1923		G	$67,500	9/07	NEIL	Restored, very rare, canopy
25-50	1923		G	$70,000	9/07	NEIL	
45-65	1923		G	$150,000	9/07	NEIL	Meticulous restoration in last two years
5-10	1917		G	$62,500	9/07	NEIL	Cost $295 new, 4 cyl.-engine, total restoration
8-16	1917		G	$70,000	9/07	NEIL	4,900 lbs., complete restoration, show tractor
A			G	$1,900	7/07	ECIL	Single front wheel, fenders, lights, PTO
A			G	$2,500	11/05	WCIL	BF Avery A, SN 16A745, running, restored, fenders, PTO, lights, belt pulley
A			G	$3,000	11/05	WCIL	BF Avery A, SN 5A188, running, restored, fenders, PTO, belt pulley
A	1947		G	$4,500	11/07	WCIL	Running, restored, fenders, PTO, lights, belt pulley, award-winning show tractor
BF				$1,400	11/05	WCIL	WF, Hercules, restored, fenders, lights, new manifold/paint, rebuilt trans., new decals
BF				$1,900	6/05	ECWI	
N/A				$725	9/07	NWIL	Gas
N/A				$2,000	11/05	WCIL	WF, gas, rear wheel weights

Baker

Model	Year	Hours	Cond.	Price	Date	Area	Comments
25-50	1926		G	$30,000	9/07	NEIL	Only known factory blue tractor, restored

Cletrac

Model	Year	Hours	Cond.	Price	Date	Area	Comments
BDH	1947		E	$3,900	10/07	NEKS	Dozer blade, electric start
HG42			F	$900	6/07	SEID	Crawler, tracks in fair condition, restorable
N/A			P	$500	4/05	NCOH	Dozer, parts only
N/A			F	$1,500	9/05	NENE	Oliver Cletrac crawler tractor

Co-Op

Model	Year	Hours	Cond.	Price	Date	Area	Comments
E2			G	$3,600	8/05	SEND	SN 2045
E3			G	$1,400	11/07	SEND	WF, row crop, PTO, 12.4-38 rear tires
E3			G	$1,550	9/07	NCND	NF, gas, like-new 12.4-38 tires, 540 live PTO, wheel weights
E3			G	$3,750	6/07	NCIN	1953-54
E3			F	$950	3/06	NWOH	Gas, NF, restored
E3	1949		P	$600	9/06	ECMI	Does not run, NF
E3			G	$800	8/05	WCMN	Very good 12.4-38 tires, good rubber and tin

Lesser-Known Classics
Co-Op continued

E3			G	$1,600	8/05	SEND	
E3			G	$1,900	11/05	WCIL	NF, live PTO
E3	1942		G	$3,200	11/05	WCIL	WF, gas, restored

David Bradley

Model	Year	Hours	Cond.	Price	Date	Area	Comments
N/A			F	$1,000	9/06	ECMI	Hi-trac, belly mower, all original, restored, SN 917

David Brown

Model	Year	Hours	Cond.	Price	Date	Area	Comments
200			G	$2,500	11/05	WCIL	Select-o-Matic, running, restored, fenders, PTO, lights, hyd. remote, deluxe seat, 3 pt., PS, beautiful professional repaint recently, diesel runs great but slobbers a bit
1200	1970		F	$1,400	8/07	NWIL	Diesel, WF, Select-o-Matic trans.
1210			G	$4,000	12/05	SCIA	Westendorf WL-21 loader
780			G	$2,400	4/07	SCKS	Diesel, PTO, 3 pt., 1 hyd.
880			G	$3,000	8/06	SETN	Front loader
885		3,034	G	$2,400	8/05	NEIN	WF, diesel
990			F	$2,900	2/07	NEIN	Diesel, 3 pt., 1 hyd., loader
990		2,959	G	$1,800	3/06	SEIA	3 pt., 2,959 hours
990		3,200	G	$4,500	11/06	NEOH	Diesel, good rubber
990			G	$5,900	8/06	NCMI	Diesel, hyd. loader and material bkt., 1 hyd., 16.9-30 tires
995			G	$4,900	3/07	WCIL	Case loader, diesel, ROPS
995		4,638	G	$2,100	9/06	NEIN	Diesel, works good
995		2,706	G	$6,600	7/05	ECND	3 pt., 540/1,000 PTO, Allied loader, grapple

Deutz

Model	Year	Hours	Cond.	Price	Date	Area	Comments
100-06	1976		G	$3,400	2/05	NEIN	CAH, diesel
5506	1972	2,569	F	$3,500	8/07	ECIL	Made in Germany, WF, weights, 3 pt.

Eagle

Model	Year	Hours	Cond.	Price	Date	Area	Comments
12-22	1920		G	$45,000	9/07	NEIL	

Emerson B.

Model	Year	Hours	Cond.	Price	Date	Area	Comments
12-20	1920		G	$55,000	9/07	NEIL	Emerson Brantingham, older restoration, rebuilt radiator, magneto, and carburetor

Gibson

Model	Year	Hours	Cond.	Price	Date	Area	Comments
N/A			G	$1,100	11/05	SCKY	Blade and plow

Greyhound

Model	Year	Hours	Cond.	Price	Date	Area	Comments
N/A	1929		G	$29,000	9/07	WCIA	On steel, roof canopy, rare, one of a dozen known

Huber

Model	Year	Hours	Cond.	Price	Date	Area	Comments
25-50	1927		G	$24,000	9/07	NEIL	Drop hood version, only 7 built
35-70	1918		G	$250,000	9/07	NEIL	Only three of this size known, 8' rear wheels, 5' front wheels, big prairie tractor
LC			G	$5,500	8/07	NWOH	Straight
Light 4	1917		G	$65,000	9/07	NEIL	Restored, crossmotor 4-cyl. engine

Lesser-Known Classics

Kubota

Model	Year	Hours	Cond.	Price	Date	Area	Comments
L260	1975		G	$1,600	6/06	NECO	2WD, 3 pt., PTO, diesel

Lauson

Model	Year	Hours	Cond.	Price	Date	Area	Comments
25-45	1928		G	$40,000	9/07	NEIL	6-cyl. engine

Rumeley

Model	Year	Hours	Cond.	Price	Date	Area	Comments
12-20			G	$35,000	9/07	ECIN	OilPull, K heavyweight
16-30			G	$31,000	9/07	ECIN	OilPull, H heavyweight
20-30			G	$21,000	9/07	ECIN	OilPull, W lightweight
20-35			G	$24,000	9/07	ECIN	OilPull, M lightweight
20-40			G	$39,000	9/07	ECIN	OilPull, G heavyweight
25-40			G	$24,000	9/07	ECIN	OilPull, X lightweight
30-50			G	$28,000	9/07	ECIN	OilPull, Y lightweight
30-50			G	$30,000	9/07	ECIN	OilPull, Y lightweight
30-60			G	$50,000	9/07	ECIN	OilPull, S lightweight
30-60	1918		G	$165,000	9/07	NEIL	OilPull, Model E, cost $4,300 brand-new
6A			P	$2,600	9/07	ECIN	For parts
M	1926		G	$80,000	9/07	NEIL	Professional restoration

Satoh

Model	Year	Hours	Cond.	Price	Date	Area	Comments
S650G	1975		P	$350	4/07	NECO	Gas, 3 pt., WF, PTO, tach shows 1,381 hours
S650G	1975		G	$1,350	4/07	NECO	Gas, 3 pt., WF, PTO, tach shows 740 hours

Sawyer Massey

Model	Year	Hours	Cond.	Price	Date	Area	Comments
11-22	1918		G	$40,000	9/07	NEIL	Older restoration

Sears

Model	Year	Hours	Cond.	Price	Date	Area	Comments
Economy	1936		G	$3,000	9/07	WCIA	Repainted

Shephard

Model	Year	Hours	Cond.	Price	Date	Area	Comments
SD3			G	$7,500	11/07	WCIL	Diesel, SN 6E9921, running, old restoration/repaint, belt pulley, engine tuned up, new paint, new 13.6-38 tires, WF, equipped with belt pulley, runs good
SD4			E	$8,500	11/07	WCIL	Diesel, running, restored, engine restored, new paint, new 14.9-38 tires, runs good

Lesser-Known Classics

Steiger

Model	Year	Hours	Cond.	Price	Date	Area	Comments
Bearcat	1976	5,650	F	$10,350	3/05	SCIL	Series III, 855 Cummins engine, 3 pt., 3 hyd., ST 220
Bearcat II	1974	5,433	G	$6,200	6/05	SEND	4WD, Cat 3208 engine, 10 speed, 3 hyd., 20.8-34 band duals (20%)
Bearcat II	1975	3,346	G	$7,000	8/05	NCIA	4WD, 3208 Cat, air, 3 hyd., heavy axle and trans., 18.4-38 duals
Cougar	1971		F	$3,400	3/05	ECND	4WD, 3306 Cat, 18.4-34 duals (70% rubber), 2,980 hours on OH
Panther II	1974	5,000	G	$12,750	7/07	NWIL	4WD, 855 Cummins diesel, 30 hours on OH, 3 pt., 24.5-32 duals
Panther II	1975		G	$9,500	8/06	WCIA	4WD, 310-hp. Cummins, 3 pt., quick hitch, 3 hyd., 24.5-32 duals
PT 270	1977		F	$5,200	1/05	SWOH	4WD, 500 hours since it had a top end OH
ST 220	1976	12,205	G	$7,500	4/06	NWMN	Bearcat, 4WD, Series III, 855 Cummins, 10-speed manual, 4 hyd., 20.8-34 duals (85%), extensive reconditioning
ST 220	1976	12,205	G	$7,500	4/06	NWMN	Bearcat, 4WD, Series III, 855 Cummins, 10-speed manual, 4 hyd., 20.8-34 duals (85%), extensive reconditioning
ST 220	1977	3,160	G	$15,500	7/06	NWIL	4WD, Bearcat III, bareback, 3 hyd., full set of duals, 200 hours on complete new Cummins diesel engine
ST 270	1978		G	$15,000	3/08	NEMO	4WD, 20 speed, 270 hp., 3306 Cat, bareback
ST 310	1974	10,000	G	$9,500	3/06	SEND	Panther II, 4WD, 855 Cummins, 10 speed, 3 hyd., 23.1-34 duals (85%)
Tiger II	1974	5,740	G	$4,750	5/06	SEWY	Turbo Tiger II, 4WD, 3 pt., 12' dozer
Wildcat III	1976	4,152	G	$7,500	12/07	NWIN	4WD, 3208 Cat, 10 speed, 3 pt., quick hitch, 3 hyd., duals

Tillsoil

Model	Year	Hours	Cond.	Price	Date	Area	Comments
18-30	1920		G	$125,000	9/07	NEIL	1 of 5 known, 3 in museum in Canada, frame-up restoration with engine rebuild

Titan

Model	Year	Hours	Cond.	Price	Date	Area	Comments
10-20	1921		G	$22,000	9/07	NEIL	Early gas tractor

Versatile

Model	Year	Hours	Cond.	Price	Date	Area	Comments
150	1978	6,523	F	$6,000	5/07	SEWY	4WD, loader and bkt., rear 3 pt., diesel
300	1973		G	$9,000	3/07	ECND	Hydro mech. 4WD, V-6 Cummins, 3 hyd., hyd. toplink, 3 pt., 1,000 PTO, new Goodyear 16-9-38 DT710 tires, air ride seat, cab, heat, new injectors, starter, extensive reconditioning
500	1977	4,669	G	$13,500	3/08	NCMI	4WD, 3 pt., PTO, 3 hyd., 18.4-38 axle duals

Lesser-Known Classics

Versatile

Model	Year	Hours	Cond.	Price	Date	Area	Comments
500	1977	4,574	G	$11,000	3/05	SCMI	4WD, 3 pt., cab, air, PTO, 3 hyd., 18.4-38, duals
700	1973		F	$3,250	7/06	SEND	4WD, 3 hyd., 18.4-38 tires, four tires brand-new
750	1976	9,412	G	$6,500	3/06	NWMN	4WD, 3 hyd., 20.8-38 duals (70%)
750	1978	7,800	F	$5,000	9/05	ECND	4WD, new turbo, air ride seat, 18.4-38 (75%), clean
800	1973	8,100	G	$6,600	3/07	SEND	4WD, 12 speed, turbo, 4 hyd., 18.4-38 duals (60%)
800	1977		F	$8,000	3/07	WCKS	4WD
800	1974		G	$4,200	7/06	ECND	4WD, CAH, duals
800	1975	6,543	G	$6,000	9/06	NWMN	4WD, CAH, 2 hyd., recent engine major, 30.5-32, 50% rubber
800	1976	9,134	G	$5,000	7/06	NWMN	Series II, 4WD, 12 speed, 3 hyd., return flow, 18.4-38 duals, 30% rubber
800	1976		G	$6,500	12/05	SEND	4WD, Series II, 3 hyd., 1,900 hours on in-frame OH, 200 hours on rebuilt pump, set up for air seeder, 18.4-38, duals, 85% rubber
825	1977	7,520	G	$6,000	3/05	WCMN	4WD, 855 Cummins, 4 hyd., 3 pt., major OH, 18" rims, tube type
850	1976		G	$9,000	3/08	ECND	Series II, 4WD, CAH, 12-speed gear drive, 3 hyd., 20.8-38 duals
850	1974	4,300	G	$7,900	11/06	WCMN	4WD, 30% rubber
850	1976	9,106	F	$3,900	7/06	ECND	4WD, Series II, CAH, 2 hyd., duals, 3,000 hours on OH
850	1976		G	$7,700	6/06	SEND	4WD, Series II, 3 hyd., 20.8-38 duals, R134A conversion, 4,000 hours on OH at 310 hp.
850	1976	8,669	G	$11,500	7/06	NWMN	Series II, 4WD, 12-speed gear, 3 hyd., 24.5-32 duals, 80% rubber, 400 hours on OH
850	1974		P	$4,100	3/05	ECND	4WD, Series I, 3 hyd., 1,500 hours on OH
850	1974	5,500	F	$4,200	6/05	SEND	4WD, Cummins, 20.8-38 duals
850	1976	8,000	F	$5,900	7/05	ECND	4WD, Series II, 12 speed, 4 hyd., air drill hyd., duals, 1,500 hours on major OH
850	1977	7,885	G	$7,250	7/05	ECND	Series II, 3 hyd., 20.8-38 duals, 1,000 hours on new rod and main bearings, planetary 7-axle bearings
855	1978	9,300	G	$7,500	7/07	SEND	4WD, 4 hyd., 20.8-38 tires (60%)
875	1978	8,561	G	$9,750	4/08	ECND	4WD, CAH, 12-speed gear, 4 hyd., 20.8-38 triples
875	1978	4,400	F	$11,000	7/05	ECND	4WD, 4 hyd., duals
875	1978	10,000	F	$17,000	3/05	SCKS	4WD, 4 hyd., 2 years on OH, big cam NTA 360 Cummins, good tires, interior fair
900	1976	13,225	G	$5,300	7/06	NWMN	Series II, 4WD, 12 speed, 3 hyd., 20.8-38 duals, 30% rubber
950	1977		F	$6,500	8/05	WCMN	4WD, new clutch past 2 years, new water pump past 2 years

Finish by wiring right

By Dave Mowitz

As tempting as it may be, don't scrimp or scrape by rewiring a tractor to complete its restoration.

Compared to the cost of engine parts or a paint job, rewiring expense is nominal.

Old cloth- or plastic-covered wiring, although it may look sound on the surface, may be long past its prime for transferring electricity efficiently. And wiring terminals and connections that are corroded, rusted, or have disappeared entirely not only compromise engine operation, but also pose a fire hazard.

Selecting wire

The best advice is to go with entirely new wiring. The good news is there are a number of firms that offer a wide selection of cloth wiring to match the original equipment on your vintage tractor.

Selecting a wire size depends on a wide variety of conditions such as the voltage and amperage expected, where it is used (transferring power from the battery to the starter vs. from a magneto to a spark plug), and the length of its run. Size difference reflects the fact that as electricity moves through a wire it builds up resistance. This resistance, in turn, creates heat.

Now, the longer the wire and the greater the load it is carrying, the greater the resistance. Using too small a diameter wire in certain situations can cause enough heat to be generated that it will make a wire's covering melt.

Refer to the manual

When it doubt, check your tractor's service manual for the proper wire size recommendations. If no recommendations are provided, then refer to the wiring chart in an automotive guide. Many public libraries carry Chilton Guides, which are often a great source of information.

And then there is my favorite source of restoration information – other collectors. Get online and go to a Web site that caters to the brand of tractor you are working on. I'll guarantee you a whiz-bang mechanic at that site will respond to your discussion group plea for advice.

Installing wire

After buying the proper size wire, take a bit more care to install it properly. Use new terminal ends that are properly crimped or soldered to assure a solid connection. Thoroughly clean original terminal ends to reduce resistance. And tighten locking washers, when used, so that they also make solid contact. I like to use terminal boots or a covering on all connections that are exposed to weather.

Other installation rules to follow:
- Avoid making sharp turns in wiring runs.
- Keep wiring away from high heat surfaces, such as a manifold, or from any areas where it might be exposed to oil or fuel.
- Always use a rubber grommet to protect wires passing through sheet metal.
- Run multiple wires in a harness or loom.
- Secure wiring using wire clamps or clips.

When sizing wiring to a particular job, bear in mind that its carrying capacity must match the total load for a particular circuit. The greater the load it has to carry and the longer the path it must run, the greater the resistance in the wire will be.

This resistance can be reduced with a larger diameter wire.

Cars

Absolutely, I include auction sale prices on cars in the data I compile. Not just late model ones either, but cars of every vintage, going all the way back to Ford Model T's. I just love running across cars like the 1912 Ford Model T sold for $10,000 on a September 2007 auction in west-central Iowa. Check out the "Specs" column – this car was used a "pie wagon" back in the day.

In my humble opinion, I think the world would be a better place if we had pie wagons today. As long as they all were stocked with my favorite, banana cream.

Cars

Make	Model	Year	Cond.	Price	Date	Area	Comments
All American	N/A	1918	F	$1,750	9/07	WCIA	Less box
Buick	55	1923	G	$26,000	9/07	NEIL	Series 230 Model 55, sport touring, $1,675 just spent, older restoration
Buick	Century	1976	F	$50	10/06	SWNE	Custom sedan, 4 door, 350 V-8 gas engine, 129K miles, red
Buick	N/A	1927	G	$6,000	9/07	WCIA	4-door standard sedan
Buick	N/A	1926	G	$7,600	4/05	NENE	2-door sedan, restored, green with black top, 22,283 miles
Buick	Series 57	1933	G	$36,000	9/07	NEIL	V-shape grill, complete restoration 7 years ago
Buick	Skylark	1961	F	$900	2/07	ECNE	2 door; 215-cubic-inch, aluminum block V-8; auto., partially restored
Cadillac	Eldorado	1975	F	$125	2/05	ECIA	2 door, auto., PS, power brakes, stored
Cadillac	Eldorado	1977	G	$2,600	4/05	NCIA	Barritz, 82,400 miles, looked good for age
Chevy	Baby Grand	1917	G	$9,500	9/07	WCIA	Touring
Chevy	Belair	1967	P	$110	6/07	NENE	For parts, no title
Chevy	Belair	1955	F	$425	4/06	NWSD	Pink
Chevy	Belair	1957	G	$8,750	7/05	ECND	4-door sedan, V-8, 3 speed, AM radio, 50,347 total actual miles, restored
Chevy	Blazer	1978	F	$550	1/06	WCCA	4WD
Chevy	Camaro	1968	G	$20,000	3/05	ECCO	Super Sport, build 350-cubic-inch engine, 750 hp. with 250-hp. nitrous shot, OT, 67K miles on body
Chevy	Camaro	1974	E	$4,500	11/05	NWPA	LT, 2 door, one owner, stored for over 20 years
Chevy	Deluxe	1951	E	$14,000	11/06	NCKY	17K actual miles, original interior, gangster whitewalls, second owner
Chevy	El Camino	1973	G	$2,400	2/08	NCCO	350, auto., air, new paint
Chevy	El Camino	1977	G	$3,900	2/07	NWIA	50K miles, restored, red
Chevy	Impala	1964	G	$1,800	2/08	NWSC	Impala station wagon
Chevy	Impala	1963	F	$825	6/06	NCIA	4-door station wagon, auto., air, good body
Chevy	Impala	1972	G	$1,100	11/06	SCNE	Hardtop, 75,087 miles
Chevy	Impala	1972	G	$1,350	12/06	WCIL	Custom care, 350 motor, 64,441 miles
Chevy	Impala	1973	F	$300	4/06	NWSD	4 door
Chevy	Monte Carlo	1973	G	$5,000	2/06	NWKS	Monte Carlo, 350 V-8 engine, air, AM/FM/cassette, 113,940 miles, shedded
Chevy	Nova	1975	F	$350	2/05	WCIL	
Cord Beverly	N/A	1937	G	$65,000	9/07	WCIA	4-door sedan, older restoration
Dodge	Charger	1977	F	$500	11/07	SWNE	2 door, 132,581 miles on odometer
Dodge	N/A	1925	G	$4,800	5/07	ECND	4-door sedan, running order
Dodge	N/A	1938	G	$3,000	10/06	WCSK	Canadian sale, 4 door
Douglas	N/A	1917	F	$2,500	9/07	WCIA	Less box
Edsel	N/A		P	$90	1/07	SWKS	2 door, hardtop, needs work
Essex	K6	1932	G	$25,000	9/07	WCIA	Terraplane Roadster, rumble seat, parade car
Ford	A	1930	G	$14,750	8/07	SWOK	

Cars

Make	Model	Year	Cond.	Price	Date	Area	Comments
Ford	Coupe	1938	G	$14,000	3/07	ECNE	Deluxe 4-door coupe, V-8 gas engine, suicide doors, 6:00-16 tires, 6,400 actual miles
Ford	Customline	1953	G	$7,000	9/07	NEIL	Restored 6 years ago, V-8, 3 speed
Ford	Fairlane	1963	G	$1,400	2/08	NWSC	6 cyl., auto.
Ford	Fairlane	1956	E	$16,250	8/06	SCMN	292 Thunderbird engine, 3-speed on column, 3" white wall tires, fender skirts
Ford	Galaxy 500	1962	G	$2,750	2/08	NWLA	Shows 97,478 miles
Ford	Galaxy 500	1963	G	$1,700	4/08	NCND	4-door car, 289 V-8 automatic, stored inside, straight, clean
Ford	Galaxy 500	1967	G	$1,900	1/08	WCIL	390 V-8, 55,200 miles, air, PS, 4 door, very clean
Ford	Galaxy 500	1960	F	$140	6/06	NCIA	4 door, hardtop
Ford	Galaxy 500	1966	G	$5,500	4/06	ECND	Convertible, 2 door, 352 engine, 165K miles, all original, family-owned car
Ford	Model A	1928	G	$30,000	9/07	NEIL	Rare Ford Model A has a rare after-market snowmobile attachment, everything still works just as it did with the original owner in northern Minnesota when he needed it to navigate the snowy backroads, exceptionally hard to find in this condition, runs great
Ford	Model T	1915	G	$4,200	3/08	NCTN	Great old original 1915 Ford Model T Coupe with flatbed
Ford	Model T	1912	G	$10,000	9/07	WCIA	Pie wagon
Ford	Model T	1919	G	$7,500	9/07	WCIA	Touring
Ford	Model T	1920	G	$19,000	9/07	NEIL	Roadster, recent updates to this little car include a new interior, new tires and electric start, a great original car that is ready to go
Ford	Model T	1921	G	$5,750	9/07	WCIA	
Ford	Model T	1924	G	$10,500	9/07	NEIL	Dump bed, open cab and early dump bed make it a little more unusual than the average Model T, runs good
Ford	Model T	1926	G	$7,300	9/05	NENE	Wire wheels, runs
Ford	Mustang	1966	F	$300	6/06	NCIA	Coupe, 6 cyl., 3 speed, pony interior
Ford	Mustang		G	$1,250	6/05	SWCA	4 cyl., gas, auto., PS, PB, air, cassette, AM/FM, convertible
Ford	N/A	1959	G	$27,500	2/07	ECNE	4 door, partially restored, ran well
Ford	Thunderbird	1955	G	$22,500	9/07	WCIA	Both tops, beautiful, a real looker
Ford	Thunderbird	1967	F	$1,500	11/07	SWNE	2 door, stored in for years, some body damage
Ford	Thunderbird	192	G	$1,200	12/05	SEID	2 door, air, cruise, tilt
Ford	Victoria	1956	G	$14,500	9/07	NEIL	2-year-old restoration, 4 door, 292 V-8 engine, fender skirts
Jeep	N/A	1967	G	$1,700	4/07	SCOK	Jeep, 1¼-ton feed truck
Jeep	N/A	1967	G	$2,100	3/07	WCNE	4×4 Jeep truck, Torino in-line 6-cyl. gas engine, 4-speed manual trans., lube truck setup, 9:00-20 tires
Jeep	N/A	1967	G	$4,200	4/07	SCOK	Jeep, 1¼ ton, 6-cyl. engine
Jeep	N/A	1968	G	$3,000	4/07	SCOK	Jeep, 1¼ ton, flatbed, 350 V-8 engine

Cars

Make	Model	Year	Cond.	Price	Date	Area	Comments
Jeep	N/A	1968	G	$3,100	4/07	SCOK	Jeep, 1¼ ton, 350 V-8 engine
Jeep	N/A	1953	G	$6,100	8/06	SETN	Gas, Willys, 4-cyl. waterproof engine
Jeep	N/A	1960	F	$500	9/06	NENE	4×4, L226 gas engine, 3-speed manual trans., 97,250 miles
Jeep	N/A	1965	P	$400	6/06	NWIL	Snowplow, not running
Kaiser	N/A	1967	F	$900	3/07	WCNE	Army Jeep, 4×4, Torino in-line 6-cyl. gas engine, 4-speed manual trans., lube truck setup, (2) oil barrels with pumps and valves, used for oil and waste oil
Koehler	N/A	1916	F	$3,400	9/07	WCIA	
Mercury	N/A	1957	G	$3,250	3/08	ECNE	Montclair Turnpike Cruiser, 4 door, push-button auto., 79,055 miles
Mercury	N/A	1965	G	$5,250	12/07	SWNE	Comet Caliente power convertible, lavender paint, original interior and drivetrain
Mercury	Park Lane	1966	G	$6,250	9/07	WCIA	Convertible
Oldsmobile	98	1959	E	$45,000	9/07	NEIL	Convertible, owned by two bachelor brothers in PA, driven to church on Sundays, all original, power seats and windows, 40K miles
Oldsmobile	Delta 88	1972	F	$700	1/06	WCCA	
Plymouth	Belveder	1962	F	$325	9/06	NENE	Sedan, 4 door, 318 8-cylinder gas engine
Plymouth	Fury III	1967	F	$200	5/07	WCSD	Low miles
Plymouth	N/A	1950	E	$4,000	1/08	ECNE	4-door sedan, flathead 6-cyl. gas engine, 7,430 actual miles on new engine (Arizona car), sun visor, original white wall tires, 3-speed, AM radio, always shedded
Pontiac	Catalina	1974	P	$130	1/07	SWKS	2-door sedan
Pontiac	Executive	1967	F	$2,800	2/07	ECNE	2-door hardtop, 400, V-8, auto., air, 114K actual miles, leather seats, excellent inside, rust inside, rust on fender wells
Pontiac	Grand Prix	1970	G	$4,000	2/08	NWSC	Model J, all original, ready to paint
Pontiac	Grandville	1974	F	$1,800	3/08	ECNE	Grandville convertible, 400 gas engine, automatic, leather interior, tilt steering, power windows, power locks, AM/FM, 102,647 miles
Pontiac	N/A	1929	F	$7,500	11/07	WCIL	Sedan, not running/loose, old restoration/repaint, good solid Kansas car, black with wire wheels, 6 cyl.
Republic	N/A	1917	F	$3,250	9/07	WCIA	Less box
Studebaker	N/A	1949	F	$800	5/08	ECMN	Land Cruiser, 4 door, 44K miles, unrestored, does not run
Volkswagen	Beetle	1972	G	$3,400	2/08	NWSC	Low miles, light green
Volkswagen	Bug	1960	G	$1,250	10/06	WCSK	Canadian sale
Volkswagen	N/A	1968	G	$1,500	11/05	ECNE	Dune buggy, #1131, 2 door, blue, has title
Willys	CJ5	1955	G	$2,700	11/06	SCNE	Jeep, 4×4, Overland
Willys	Jeep		F	$1,000	3/08	SWKY	4×4, runs good, no title
Willys	Jeep	1954	G	$3,000	3/07	NWIL	2,514 miles on rebuild, gas, 4×4, soft top, 4-cyl. engine
Willys	Jeep	1962	F	$800	9/07	WCIL	Tall hood, new points and plugs, needs fuel pump

Trucks

Used farm equipment values zoomed higher beginning in late November 2007, pushed along by the commodity price surge. Also forcing used values higher were the price increases and limited availability of new equipment. One segment of the used market where I really saw this trend play out involved used grain trucks and semis.

Wow, did values rise rapidly for nice-condition older trucks. How about $18,000 for a 1975 Chevy C65 tandem-axle grain truck with 40,398 miles and 18-foot steel box sold January 2008 in south-central Kansas? Makes a guy wish he'd have bought four or five similar trucks a couple years ago. Sell them now and make a pretty penny.

Trucks

Make	Model	Year	Cond.	Price	Date	Area	Comments
Chevy	10	1972	P	$300	2/06	NWSD	Custom 10, standard
Chevy	10	1975	G	$1,550	2/06	WCIL	Scottsdale
Chevy	10	1976	G	$950	12/06	NEMO	Scottsdale, 4WD, flatbed, dually, fifth-wheel hitch, 350, V-8
Chevy	10	1977	P	$325	4/05	SEND	4×4, 350, auto
Chevy	20		G	$650	2/07	SEFL	Service truck, gas, 8' mechanic's body, rear air hose reel
Chevy	20	1968	F	$350	6/07	NENE	2WD, 6-cyl. gas engine, 4 speed
Chevy	20	1976	F	$500	3/07	NWIL	Scottsdale, 4×4, 156K miles
Chevy	20	1967	G	$950	9/06	NCIA	6 cyl., 4 speed, decent body
Chevy	20	1972	G	$1,900	3/06	NCIA	Cheyenne 350, auto, 86,410 miles
Chevy	20	1966	F	$500	8/05	NCIA	V-8, 4 speed
Chevy	2500	1975	F	$550	11/05	SEND	292, 6 cyl., 4 speed
Chevy	30	1978	G	$1,400	1/06	NWIL	2WD, 8' flatbed, 454, V-8, 74,465 miles
Chevy	3100	1953	G	$19,000	1/05	NENE	28K miles, one owner, 3 speed on column, 6 cyl.
Chevy	3100	1956	G	$8,750	3/05	ECCO	4WD, auto, shaved handles
Chevy	3600	1951	F	$600	10/06	SWNE	10' wood box, hoist, 6-cyl. gas engine, 4-speed manual trans., 7.00-18 tires, 129K miles
Chevy	3800	1957	G	$1,100	8/05	NCIA	9' step side box, 6 cyl., 4 speed, originally from California
Chevy	40	1967	G	$3,000	3/08	WCIL	40 Series grain, 12' bed, 4 speed, 8.25-20 tires, 283 engine, 53,262 miles
Chevy	4100	1953	G	$4,300	2/05	ECNE	Obeco 10' wood box with hoist, 46,346 actual miles, 6-cyl. gas engine, 4 speed, shedded
Chevy	4400		F	$200	10/06	SWNE	Box, hoist
Chevy	4400	1949	F	$150	10/06	SWNE	13.5' wood box, hoist, 6-cyl. gas engine, 4 speed
Chevy	50		G	$3,200	2/05	SCMN	366 engine, 4×2 speed, Omaha standard 16' box & hoist, 9:00-20 tires
Chevy	50	1963	F	$350	2/07	NCNE	Gas, 5 speed, 14' steel box
Chevy	50	1967	F	$500	8/07	SWOH	V-6, 5×2 speed, 16' grain bed, no hoist, 104K miles
Chevy	50	1967	F	$900	5/07	WCSD	4×2 speed, Farmhand 450 power manure spreader box
Chevy	50	1967	G	$1,950	8/07	ECNE	18' box, tag axle, 6 cyl., 292 motor, 88,300 miles
Chevy	50	1968	F	$750	10/07	NEKS	13' wooden bed with racks, no hoist
Chevy	50	1968	G	$5,000	12/07	ECIL	Grain, 16' Knapheide box, cargo doors & grain door, twin hyd. cylinders, 396 engine board to 400, 4×2 speed, dead tag axle, Hendrickson suspension, 102,966 miles
Chevy	50	1967	G	$3,600	3/06	WCSD	Clean, metal floor in wood 18' combo box, hoist, 4×2 speed, rebuilt motor, 4,000 miles on OH
Chevy	50	1970	G	$3,600	11/06	SCNE	Grain truck, V-8, 4×2 speed, 16' steel side box, 45K actual miles
Chevy	5700	1950	G	$1,600	4/05	NCKS	6 cyl., 4×2 speed, 250-bu. gravity box, rollover tarp, hyd. auger
Chevy	60	1962	G	$1,600	1/08	WCKS	Viking, one owner, Knapheide 16' steel box, V-8 gas, 31,227 miles
Chevy	60	1966	G	$1,700	2/08	NWSC	Dump truck
Chevy	60	1960	P	$225	3/07	SCNE	13.5' box, doesn't run
Chevy	60	1966	G	$1,950	9/07	NCIA	16' Omaha standard box, hoist, 6 cyl., 4×2 speed, 64,571 miles
Chevy	60	1968	G	$11,750	8/07	ECNE	Omaha standard 20' box, tag axle, 5×2 speed, V-8, like-new 8.25×20 tires, nice and clean, 32,730 miles

Trucks

Make	Model	Year	Cond.	Price	Date	Area	Comments
Chevy	60	1970	F	$2,300	2/07	ECNE	Cabover, 366 engine, 4×2 speed, tag axle, hoist, 20' wood box with 48" sides
Chevy	60	1970	G	$4,500	3/07	NEMI	Tilt cab dump truck, box, hoist
Chevy	60	1960	E	$1,450	11/06	NEKS	Dump, almost new tires
Chevy	60	1960	G	$2,600	3/06	NCCO	13.5' box, side hoist, 4×2 speed, one family owned, 35,080 actual miles
Chevy	60	1962	G	$950	12/06	SCNE	327 engine, 4×2 speed, 16' box and hoist
Chevy	60	1966	G	$4,600	3/06	NCCO	15.5' box, 4×2 speed, electric opener, one-family owner, 47,400 actual miles
Chevy	60	1967	G	$2,500	2/06	NECO	20' box, 427 engine, 5×3 speed
Chevy	60	1976	F	$2,000	4/06	SEND	350, 4×2 speed, 5-yard gravel box
Chevy	60	1964	P	$500	2/05	WCIL	Does not run, 14' steel box, tag axle, 5×2 speed, needs engine work
Chevy	60	1964	F	$750	11/05	NEND	Single-axle fuel truck, 6 cyl., 4×2 speed, 4 compartments, hose and pump
Chevy	60	1966	G	$1,600	12/05	WCIN	Single-axle grain, 13' wooden bed with steel floor, scissor hoist, 6 cyl., 4 speed, 45,815 miles
Chevy	60	1966	G	$1,850	1/05	SCNE	Grain truck, 87,500 miles, 4×2 speed, 15.5' box, hoist, 350, V-8
Chevy	6100	1955	F	$500	12/05	ECNE	Single axle, 4×2 speed, Bradford 240-316 gravity box, 8.25-20 tires
Chevy	6100	1957	G	$1,200	3/05	WCMN	Single axle, Parker 2000 gravity box and hyd. brush auger
Chevy	6400	1957	G	$1,150	6/07	ECND	Tag tandem, V-8, 4×2 speed, 17' steel box, headlift hoist
Chevy	6400	1949	G	$700	9/06	NWMN	Single axle, 6 cyl., 4 speed, 14' steel box, hoist
Chevy	6400	1955	F	$350	7/06	NEND	Single axle, 6 cyl., 4×2 speed, 15.5' Knapheide box, hoist, roll tarp, 77K miles
Chevy	6400	1955	F	$400	9/06	NENE	First-edition grain, single axle, combination stock rack, single-cyl. hoist, 235 gas engine, 4×2 speed, 8.25-20 tires
Chevy	6400	1957	G	$525	11/05	NWMN	Single axle, 4×2 speed, 13' Omaha standard box, hoist
Chevy	6500	1955	F	$800	12/07	SCMT	Rear hoist and box
Chevy	6500		F	$600	2/05	SCCA	Single-axle dump, V-8, 16' box
Chevy	70	1961	G	$1,150	11/05	SEND	Tilt cab tag axle, 409 engine, 5×2 speed, PS, 19' wood box, hoist
Chevy	80		P	$200	1/07	SEKS	Truck tractor, fifth-wheel plate
Chevy	80	1974	F	$1,350	3/07	NWIL	10' Henderson 10' spreader box, 403 V-6, 5×2 speed, gas
Chevy	Bison	1977	G	$1,600	7/05	ECND	9 speed, 290 Cummins
Chevy	C10	1962	E	$13,000	9/07	NEIL	Recent restoration that included a new bed and a lot of extras, it has great chrome, V-8, 4 speed, air, PS, lots of pizazz
Chevy	C10	1970	P	$175	11/07	ECNE	Rebuilt V-8 gas engine, 3 speed (on tree) parts truck
Chevy	C10	1970	F	$900	11/07	NEND	½ ton, 350, 4 speed, 2WD
Chevy	C10	1970	G	$4,250	3/07	WCSK	Canada sale, ½ ton, 350 V-8, auto, rally wheels, 152,205 miles, good for restoration
Chevy	C20	1972	G	$1,100	1/08	SCKS	350 V-8 engine, auto
Chevy	C20	1974	F	$700	4/07	SCKS	Runs good, dually, flatbed
Chevy	C20	1976	F	$1,600	2/06	NWMN	350 engine, 4 speed, 8' flatbed
Chevy	C25	1964	G	$5,600	8/06	ECNE	6-cylinder gas engine, 4 speed, 97,928 miles, one owner
Chevy	C30	1968	G	$2,600	11/07	ECKS	1 ton, 9.5' grain box with stock racks & hoist, 16,845 miles, lost title, will sell with bill of sale and seller will help file for lost title, truck has been in the family its whole life

Trucks

Make	Model	Year	Cond.	Price	Date	Area	Comments
Chevy	C30	1970	F	$600	4/07	ECND	1-ton dually, 350 engine, auto, service body, 2 service tanks with electric pumps
Chevy	C30	1971	G	$2,700	4/07	NEMO	Dually, 4 speed, 350 cu.in., 3-ton Sudenga feed/seed, spring ride bed
Chevy	C30		G	$1,100	4/06	WCSD	Red pickup, 1 ton, dually, tool boxes, Omaha standard flat box
Chevy	C30	1969	G	$2,000	2/06	NCOH	1 ton, 4 speed, 10' bed & scissors hoist, only 32,430 miles
Chevy	C30	1970	G	$1,125	4/06	WCSD	Scottsdale, 1 ton, flatbed, used for service truck, blue
Chevy	C30	1963	G	$600	4/05	SENE	Gas, 8' box, hoist
Chevy	C30	1970	G	$1,150	6/05	NEND	1-ton dually, 350 V-8, 4 speed, flatbed, 400-gal. fuel tank and 12-volt pump
Chevy	C30	1973	G	$1,400	1/05	NWGA	
Chevy	C50	1968	F	$800	3/08	NWKS	15' bed, hoist
Chevy	C50	1969	G	$2,500	2/08	WCIL	350, 4×2 speed, 13' bed
Chevy	C50	1969	G	$2,500	3/08	ECND	Single axle, 350, 4×2 speed, 15' box, hoist, 42,479 miles
Chevy	C50	1970	G	$5,100	4/08	NWMN	Single axle, 366, 4×2 speed, 16' Westgo box, hoist, 61K miles
Chevy	C50		P	$450	2/07	ECNE	84K miles, 350 motor, 4×2 speed, 13' box, frost plugs blown to engine frozen
Chevy	C50		G	$2,800	6/07	SCMN	Single axle, 350 V-8, 4 speed, 47,304 miles showing, 15' steel box & hoist, 9R22.5 tires
Chevy	C50	1968	F	$1,900	11/07	ECKS	Speedstar 71 drilling unit, less than 100 hours on deck OH, runs
Chevy	C50	1968	G	$3,600	9/07	NCND	14' steel box, hoist with Shur-Lok roll tarp, 8.25 20 tires (good), inline 6-cyl. gas engine, 4×2 speed, 80,766 miles
Chevy	C50	1969	P	$300	2/07	SWKS	Tradewinds steel box, 366 engine, 4×2 speed, 48,296 miles
Chevy	C50	1969	F	$1,050	2/07	SWKS	366 engine, V-8, 4×2 speed, 49,350 miles, AM/FM radio
Chevy	C50	1970	P	$300	3/07	SCNE	Midwest 15.5' box and hoist
Chevy	C50	1970	G	$3,000	1/07	NCNE	Tri-axle, 5×2 speed, 366 engine
Chevy	C50	1970	F	$3,500	7/07	NCSK	Canadian sale, single-axle grain, 350 V-8, 4×2 speed, 14' steel box, 9:00×20 rubber, 50,223 miles showing
Chevy	C50	1970	G	$9,500	1/07	NCNE	Tri-axle, 5×2 speed, 366 engine
Chevy	C50	1971	G	$2,100	3/07	WCIL	13.5' box, 110K miles, gas
Chevy	C50	1971	G	$4,100	2/07	NENE	350 engine, 4×2 speed, steel side box, 42,321 miles
Chevy	C50	1972	F	$800	2/07	ECNE	Single axle, 16' box, 48" sides, no brakes
Chevy	C50	1972	G	$2,500	9/07	ECND	Single axle, 350, 4×2 speed, stake box, hoist, grip rubber, recent repaint, 100K miles
Chevy	C50	1962	P	$200	8/06	ECNE	Tandem axle, steel flatbed, hoist, stretched frame, runs, needs work
Chevy	C50	1963	F	$425	3/06	ECNE	Grain truck, 14' wood box, 95K miles, 350 V-8, 4 speed
Chevy	C50	1967	F	$450	2/06	NCIL	4 speed, 6-cyl. engine
Chevy	C50	1967	P	$550	12/06	WCMN	Box, hoist
Chevy	C50	1967	G	$3,250	6/06	NWKS	366, 4×2 speed, aux. fuel tank, 15' Omaha B&H, 5,000 miles on new motor
Chevy	C50	1967	G	$4,500	2/06	NWKS	Single-axle grain truck, V-8 gas engine, 4×2 speed, 16' steel box, 19,500-lb. GVW, 9.00-20 tires, 50,700 miles
Chevy	C50	1968	G	$1,700	8/06	WCIL	13' bed, 6 cyl., 4 speed, 93,300 miles
Chevy	C50	1968	G	$2,600	2/06	ECIL	4×2 speed, Knapheide grain bed hoist, 52,510 miles

Trucks

Make	Model	Year	Cond.	Price	Date	Area	Comments
Chevy	C50	1968	G	$3,200	3/06	NWMN	Tag tandem, 327, 4×2 speed, 16' Knapheide box, hoist, 70K miles
Chevy	C50	1969	G	$2,700	8/06	WCIL	13' bed, 350 V-8, 4 speed, 74,500 miles
Chevy	C50	1970	G	$11,250	10/06	NCKS	Winch truck, clean, 350, V-8, 4×2 speed
Chevy	C50	1971	F	$1,700	4/06	NCKS	16' steel bed
Chevy	C50	1971	G	$2,650	3/06	NEKS	Bed and hoist
Chevy	C50	1971	G	$4,000	3/06	SWKY	Grain bed and dump
Chevy	C50	1971	G	$4,000	10/06	ECSK	Canadian sale, C50 single-axle grain truck with 350 V-8, 4×2 speed, 14'×8' steel box with steel floor, 9.00×20 rear rubber 8.25-20 front tires, pintle hitch, 43,244 miles showing, shedded
Chevy	C50	1972	P	$400	4/06	NEND	Single axle, 2 ton, 15' box, hoist
Chevy	C50	1972	G	$2,300	4/06	SEND	Fuel truck, 350, V-8, 4×2 speed, 5 compartment, 1,500 gal.
Chevy	C50	1972	G	$2,600	12/06	ECMO	Grain truck, 16' bed
Chevy	C50	1972	G	$6,200	4/06	WCMT	350 engine, 4×2 speed, 8,000-lb. front springs, 9.00-20 tires, 16' box, hoist, stock rack, wooden floor, plumbed for a drill fill, only 45K miles, nice shape
Chevy	C50	1964	P	$500	2/05	WCIL	13' farm box, twin hoist, 6-cyl. engine, 4×2 speed
Chevy	C50	1968	G	$1,200	12/05	SWMS	Single axle, trash body
Chevy	C50	1968	G	$4,900	3/05	ECMI	Tandem-axle grain truck, 16' wood box, twin post hoist, 4×2 speed, V-8 gas engine, 45,310 miles
Chevy	C50	1969	F	$800	9/05	SEIA	Grain truck, 90,646 miles
Chevy	C50	1969	F	$950	3/05	WCMN	Single axle, 14' wood box scissor hoist, 4×2 speed, 350, V-8
Chevy	C50	1969	G	$3,200	7/05	ECND	Single axle, 350, 4×2 speed, 14.5' steel box, hoist, 63,079 miles
Chevy	C50	1969	E	$4,000	1/05	SENE	Gas, 18' box, hoist
Chevy	C50	1970	G	$1,400	3/05	SEND	Single axle, 350, 4×2 speed, 14' steel box
Chevy	C50	1970	G	$2,000	1/05	NENE	16' wood combination box and hoist, 6 cyl., 5×2 speed, 77,529 miles, shedded, not used in 4 years
Chevy	C50	1970	G	$3,500	11/05	ECWA	350 gas engine, 4×2 speed, PS, 9:00-20 rubber (good); 16' bed (diamond plate) with twin-cylinder hoist (Anthony lift; both cyl. have been rebuilt), 48" grain racks
Chevy	C50	1971	G	$1,150	11/05	SEND	Single axle, small block 400 engine, 5×2 speed, wood box
Chevy	C50	1971	F	$2,000	4/05	NWOK	Feed, 6 cyl., BJM C900B mixer bed
Chevy	C50	1971	G	$2,200	6/05	NWMN	Tag tandem, plumbed, 350, V-8, 4×2 speed, 20' steel box, hoist and roll tarp
Chevy	C50	1972	P	$300	3/05	SEIA	13' silage box
Chevy	C50	1972	F	$1,050	1/05	NCPA	350 engine, 4 speed, 18' bed with 4' high removable side racks, wooden bottom, new tires on rear
Chevy	C50	1972	G	$3,200	1/05	NWIL	350 V-8, 4×2 speed, 14.5' steel grain box, 62,728 miles
Chevy	C50	1972	G	$3,400	1/05	NWIL	350 V-8, 4×2 speed, 14.5' steel grain box, 42,828 miles
Chevy	C50	1972	G	$3,650	1/05	NWIL	350 V-8, 4×2 speed, 14.5' steel grain box, 42,332 miles
Chevy	C50	1975	G	$3,750	7/05	NWMN	Single axle, 350, V-8, 4×2 speed, Kiefer 16' box, hoist, roll tarp, 62,177 miles
Chevy	C50	1976	G	$3,400	11/05	NECO	Dump truck, 1 axle, gas, 4×2 speed
Chevy	C60	1958	G	$1,150	1/08	SCKS	Viking single-axle grain truck, Knapheide 13' steel box with wood floor

Trucks

Make	Model	Year	Cond.	Price	Date	Area	Comments
Chevy	C60	1960	G	$9,250	3/08	NWMN	Tag axle, V-8, 4×2 speed, 2,000 H&S galvanized silage 18' box, plastic floor, 29K miles
Chevy	C60	1965	G	$3,700	1/08	SCKS	Single-axle grain truck, 16' Knapheide box with wood floor, 4×2 speed
Chevy	C60	1968	F	$3,000	3/08	ECND	Tag tandem, 427 gas, 5×2 speed, 20' Frontier box, roll tarp
Chevy	C60	1969	G	$3,500	4/08	NEND	Tag tandem, dead tag, 366, 5×2 speed, 18' French combination box, twin-post hoist, 59,126 miles
Chevy	C60	1969	E	$5,800	1/08	ECIL	4×2 speed, 366, 13.5' Midwest bed, 49,334 miles
Chevy	C60	1973	G	$2,900	2/08	WCIL	104K miles, 4×2 speed, 350, 13.5' bed
Chevy	C60	1973	G	$8,000	3/08	NEMO	Grain truck, Parkhurst 14' bed, 350 engine, 4 speed, 9.00-20 tires
Chevy	C60	1973	E	$8,700	1/08	ECIL	5×2 speed, 366, side fuel tanks, 14.5' Midwest bed, 68,217 miles
Chevy	C60	1973	G	$11,500	3/08	WCKS	Tandem axle, 366 motor, 5×2 speed, 56,490 miles, 20' bed, hoist, rollover tarp
Chevy	C60	1974	G	$2,250	2/08	SEIA	350, V-8, 4×2 speed, 16' farm box, hoist, 131,900 miles
Chevy	C60	1974	G	$3,000	4/08	SCND	Single axle, 350, 4×2 speed, staked 16' box, hoist, tarp, silage sides
Chevy	C60	1975	G	$4,200	3/08	SCKS	16' metal bed and hoist, 825×20 tires, 4×2 speed, 32,820 miles
Chevy	C60	1975	G	$5,000	4/08	ECMO	350 engine, 4×2 speed, PS, single axle, duals, Doyle spreader (lime bed) with double fan
Chevy	C60	1975	G	$9,500	4/08	NCND	Grain truck, 366 engine, 5×2 speed, roll tarp
Chevy	C60	1976	G	$5,200	1/08	WCIL	Grain truck, 47,800 miles, 4×2 speed, 13' bed, hd hitch
Chevy	C60	1976	G	$6,000	1/08	WCIL	Custom deluxe, V-8, 4×2 speed, PS, 15' (like-new) grain bed with 52" nonvented sides, twin hoist, wood floor, 110,086 miles
Chevy	C60	1977	F	$1,400	3/08	WCMN	Single axle, V-8 gas engine, 4×2 speed, 15' box, hosit, 56K miles, 8.25-20 tires
Chevy	C60		F	$1,300	3/07	NWMO	327 motor, 9:00-20 rubber, 46K miles, 4×2 speed, Omaha standard 15.5' wooden box, dual hoist, equipped with electric over hyd. planter box auger system, one owner, shedded
Chevy	C60		G	$2,100	6/07	NCSK	Canadian sale
Chevy	C60		G	$10,500	8/07	SCMN	108,558 miles, 366 engine, 5×2 speed, Scott 16' box, 50" sides, cargo doors, grain door, 900×20 tires
Chevy	C60	1958	F	$500	1/07	SWKS	Single-axle grain truck, 14' steel box, 4×2 speed, white cab with red box
Chevy	C60	1960	P	$300	2/07	SWKS	Winch truck, 350 gas engine, needs new brakes
Chevy	C60	1962	G	$1,000	3/07	SCNE	4×2 speed, 16' box with hoist, new paint
Chevy	C60	1964	F	$750	1/07	SWKS	Single-axle grain truck, steel box, 4×2 speed, 173,360 miles, green cab with red box
Chevy	C60	1964	G	$2,600	9/07	NCND	16' steel box and hoist with Shur-Lok roll tarp, (8). 25 20 traction tires, inline 6-cyl. gas engine, 4×2 speed
Chevy	C60	1965	F	$500	3/07	ECND	Single axle with pusher, V-8, 5×2 speed, shop built 16' combo box, hoist, 9:00×20 tires

Trucks

Make	Model	Year	Cond.	Price	Date	Area	Comments
Chevy	C60	1965	F	$700	1/07	SWKS	Single-axle grain truck, 14' steel box, hoist, 4×2 speed, blue cab/white box
Chevy	C60	1965	G	$1,500	11/07	NWMN	Tag tandem, V-8, 4×2 speed, 16' box, hoist and roll tarp
Chevy	C60	1968	P	$800	3/07	SEMN	Single-axle farm truck with 16' box and hoist, V-8, 4 speed, very rough shape
Chevy	C60	1969	G	$3,250	11/07	SEID	Nurse truck, 427 gas, 5×4 speed, tandem axle, (2) 1,300-gal. poly tanks, (2) 35-gal. poly chemical hoppers, Honda 5.5-hp. pump
Chevy	C60	1969	G	$3,250	11/07	WCID	427 gas, 5×4 speed, tandem axle, with (2) 1,300-gal. poly tanks, (2) 35-gal. poly chemical hoppers, Honda 5.5-hp. pump
Chevy	C60	1971	F	$400	12/07	WCMN	Grain truck, 5 speed, needs work
Chevy	C60	1971	F	$2,400	4/07	NWMN	Tag tandem, V-8, 5×3 speed, 19' box with hoist and roll tarp
Chevy	C60	1972	F	$1,500	3/07	ECND	Tag tandem, 366, 5×2 speed, air up tag with shop built 18' combo box, hoist, plumbed for drill fill, 10:00×20 rears, 9:00-20 fronts
Chevy	C60	1973	F	$1,500	7/07	WCMN	Single axle, 6 cyl., 4×2 speed, 15' box, hoist, new paint
Chevy	C60	1973	G	$2,000	7/07	NWIL	Gas, 4×2 speed, 2 ton, 14' grain box, hoist, 350 V-8 engine, runs good
Chevy	C60	1973	G	$3,600	1/07	SWNE	Single-axle grain truck, 16' steel box, hoist, flip-up stock racks, roll tarp, 350 gas engine, 4×2 speed, 9:00-20 tires
Chevy	C60	1974	P	$550	12/07	WCMN	Single axle, steel box, hoist
Chevy	C60	1974	F	$1,200	9/07	ECND	Single axle, 350, V-8, 4×2 speed, 15' box, hoist, roll tarp, 84K miles
Chevy	C60	1974	G	$2,900	10/07	NEND	Single axle, 366, 16' Knapheide box, 4×2 speed hoist, roll tarp, 42,397 miles
Chevy	C60	1974	G	$4,000	9/07	ECND	Single axle, 366, 4×2 speed, Frontier box, rebuilt floor and poly, roll tarp, 62,853 miles
Chevy	C60	1974	G	$5,100	9/07	NEND	Grain truck, 366 big block engine, 4×2 speed, 34,800 miles
Chevy	C60	1974	F	$5,700	3/07	SCSK	Canadian sale, single-axle grain, 350 V-8, 4×2 speed, 14' steel box with inland roll tarp, 9.00×20 rubber, aux. fuel tank, drill fill plumbing
Chevy	C60	1974	G	$9,500	12/07	NEMO	Grain truck, 4×2 speed, 350 V-8, 15' bed, 53" sides, Shur-Lok tarp, 73,515 miles
Chevy	C60	1975	F	$1,500	2/07	WCIL	15' bed, lots of miles
Chevy	C60	1975	G	$2,800	8/07	WCIL	New engine
Chevy	C60	1975	F	$4,200	4/07	NWND	Tag tandem, 366, V-8, 5×2 speed, lift tag, 18' box, hoist and roll tarp
Chevy	C60	1975	G	$7,250	2/07	NENE	350 engine, 4×2 speed, steel side box, 37,146 miles
Chevy	C60	1975	G	$9,100	9/07	WCIA	Tag axle, 17' box, 4×2 speed, 366 engine
Chevy	C60	1976	G	$5,000	3/07	ECNE	Single-axle grain truck, 16' wood box, 360 gas engine, 5 speed, 9:00-20 tires, 88K miles
Chevy	C60	1976	G	$5,400	9/07	NCND	16' Rugby steel box & hoist with tip tops and Shur-Lok roll tarp, SSR Pump Co. swing-out end gate, 9:00 20s, 350 V-8 engine, 4×2 speed, 35,610 actual miles
Chevy	C60	1978	G	$3,000	1/07	WCTN	Flatbed dump truck, 4×2 speed, 36,735 miles

Trucks

Make	Model	Year	Cond.	Price	Date	Area	Comments
Chevy	C60	1978	G	$3,800	4/07	NEAB	Canadian sale, single-axle grain truck, 4×2 speed, 350 V-8, 9.00×20 tires, 15-16' wooden box
Chevy	C60	1978	G	$4,100	4/07	SEND	Single axle, 4×2 speed, 18.5' Frontier box, roll tarp, 47K miles, one owner
Chevy	C60	1978	F	$7,750	4/07	SCSK	Candian sale, deluxe single-axle grain, 350 V-8, 5×2 speed, 9.00×20, 16' steel box with steel floor
Chevy	C60		P	$200	11/05	ECNE	Viking single-axle grain
Chevy	C60		F	$500	1/06	WCCA	Nurse truck, V-8
Chevy	C60		G	$800	9/06	NEIN	Grain box, gas
Chevy	C60		G	$1,750	9/06	WCTX	Diesel, single axle
Chevy	C60		F	$2,200	3/06	NWPA	16' wooden dump bed
Chevy	C60		G	$6,500	4/06	WCSD	20' bed, double hoist, Shur-Lok roll tarp, hyd. brakes
Chevy	C60	1960	G	$900	12/06	ECNE	Single-axle grain truck, 13' wood box, fold-down stock rack, 283 propane engine, 4×2 speed, 8.25-20 tires
Chevy	C60	1962	G	$150	1/06	NEMO	Grain truck
Chevy	C60	1962	P	$275	3/06	WCMN	Tag axle truck, 18' flatbed, hoist, V-8, 5×2 speed
Chevy	C60	1963	G	$2,000	11/06	NCKS	4×2 speed, V-8 engine, 16' bed steel sides, wood floor, hoist
Chevy	C60	1964	F	$150	1/06	NEMO	Grain truck
Chevy	C60	1964	P	$350	4/06	NEND	Single axle, 2 ton, 14' box, hoist
Chevy	C60	1964	F	$800	3/06	ECNE	Grain truck, 283, 8-cyl. gas engine, 4×2 speed, hyd. brakes, single axle, Kory 220 steel gravity box
Chevy	C60	1964	G	$2,100	3/06	NCIN	Grain truck, 15' Midwest bed and hoist, V-8, 4×2 speed
Chevy	C60	1966	E	$3,000	8/06	ECIL	Tandem, 327, V-8, 4×2 speed, 18' Schien bed, 69,484 miles
Chevy	C60	1967	G	$1,700	9/06	NEIN	Flatbed, gas, 5 speed, 13K miles, used to be a fire truck
Chevy	C60	1969	G	$4,000	3/06	NCCO	16' box, 4×2 speed, second owner, electric side opener, 49,950 actual miles
Chevy	C60	1970	F	$1,750	11/06	NWMN	Twin screw, 427, 5×4 speed, Westgo 18' box, hoist, twin fuel tanks, 69,895 miles
Chevy	C60	1970	G	$2,900	10/06	NCND	Single axle, 366, 5×2 speed, 16' steel box, hoist, roll tarp
Chevy	C60	1970	E	$5,750	8/06	ECIL	366, V-8 engine, 5×2 speed, 16' Knapheide bed, 57,050 miles
Chevy	C60	1971	G	$700	2/06	WCME	Custom-built 24' wood bulk body, 427 gas
Chevy	C60	1972	G	$2,600	4/06	WCSD	Tandem axle, 5×2 speed, 22' steel box and hoist
Chevy	C60	1973	G	$1,850	9/06	SEIA	Grain, twin-cyl. hoist
Chevy	C60	1973	G	$2,500	8/06	NENE	Grain, Hobbs 16' steel box, 350 gas engine, 4×2 speed, good tires
Chevy	C60	1973	G	$2,700	12/06	NWIL	350 V-8, 4×2 speed, 13.5' farm box, roll tarp, twin hoist, 83K miles with 20K miles on new GM engine, gas
Chevy	C60	1973	G	$2,900	12/06	NENE	Single-axle grain, 16' box, twin-cyl. hoist, 4×2 speed
Chevy	C60	1973	G	$4,250	6/06	SCMN	Single axle, V-8, 4×2 speed, 12' steel flatbed with 1,000-gal. and 300-gal. poly tanks
Chevy	C60	1973	G	$8,000	4/06	SWSD	Omaha 16' standard combo box and tarp, 43K actual miles, one owner, well cared for, very clean, stored inside
Chevy	C60	1973	E	$9,000	3/06	NCOK	47,910 miles, Mabar 16' steel bed and hoist, V-8, 4×2 speed

Trucks

Make	Model	Year	Cond.	Price	Date	Area	Comments
Chevy	C60	1973	G	$9,600	4/06	WCSD	20' Scott steel grain box and steel ear corn rack, hoist, roll tarp, tandem axle, twin-cyl. hoist, tag axle, 5×2 speed, 366 gas, 90K miles
Chevy	C60	1974	G	$900	2/06	WCME	22' flatbed, 350 gas, 5×2 speed
Chevy	C60	1974	G	$3,300	3/06	ECNE	Grain, 18' wood box with Mid Equipment twin-cyl. hoist, 350 gas engine, 5×2 speed, 51,120 miles
Chevy	C60	1974	G	$3,650	1/06	NECO	16' steel box
Chevy	C60	1974	F	$3,800	6/06	SEND	Lift tag tandem, 366, 5×2 speed, 20' Frontier box, roll tarp, plumbed for drill fill
Chevy	C60	1974	G	$6,100	3/06	NWKS	Single-axle grain, 24' steel box, roll tarp, hd twin-cylinder hoist, 366 V-8 gas engine, 5×2 speed, rear tag axle, 21,000 GVW, 9:00-20 tires, 25,021 miles
Chevy	C60	1975	G	$1,300	3/06	WCKS	32,871 miles, 16' bed and lift, 5×2 speed, single axle, propane or gas, runs good
Chevy	C60	1975	G	$1,300	3/06	SCIN	16' grain bed, hoist, 366 engine, 5×2 speed
Chevy	C60	1975	F	$3,500	4/06	NESD	16' steel box, wood floor, double acting hoist, 4×2 speed
Chevy	C60	1975	G	$7,200	9/06	NWMN	Tag tandem, V-8, 4×2 speed, 16' steel box, hoist, 42,150 miles
Chevy	C60	1975	G	$13,000	6/06	NWIL	Tandem axle, twin screw truck, 5×4 speed, 427 engine, 2,000 miles on engine
Chevy	C60	1976	G	$2,600	12/06	NWIL	350 V-8, 4×2 speed, 13.5' farm box, twin hoist, 8.25-20 tires, one owner, 88,875 miles, gas
Chevy	C60	1976	G	$2,700	5/06	SEND	Single axle, 350, V-8, 4×2 speed, 15' Omaha standard box, hoist, roll tarp, 51,289 miles
Chevy	C60	1978	G	$2,900	3/06	NEKS	Custom single axle, 16' wood box, Harsh single-cyl. hoist, 350, 4×2 speed, 95,092 miles
Chevy	C60		G	$5,500	8/07	NWOH	Single-axle grain, V-8, 4×2 speed, Knapheide 15' body, 27,219 miles, waxed
Chevy	C60	1960	F	$525	7/05	ECND	Single axle, 6 cyl., 4/5, 8,000 miles on new engine, 51,046 miles
Chevy	C60	1964	G	$1,700	11/05	ECNE	16.5' box and hoist, V-8, 4×2 speed
Chevy	C60	1966	G	$3,400	11/05	ECKS	2-ton truck, 5×2 speed, 9.00 tires, 16' steel bed and hoist, 54" sides
Chevy	C60	1966	G	$3,400	1/05	NWIL	327 V-8, 4×2 speed, 14.5' steel grain box and hoist, 47,346 miles
Chevy	C60	1967	G	$2,600	7/05	NCMN	V-8, 5×4 speed, 18' Knapheide box, hoist
Chevy	C60	1969	G	$3,100	11/05	ECWA	366 gas engine, 5×2 speed, 10:00-20 rubber, 86K total miles, 16' flatbed, hoist, new wood deck, 300-bu. grain box
Chevy	C60	1969	G	$4,000	12/05	WCMN	Tandem axle, V-8, 5×2 speed, 20' wood box and hoist, 136,690 miles
Chevy	C60	1970	F	$700	2/05	WCOH	1,200-gal. ss tank and spray booms
Chevy	C60	1970	G	$4,400	9/05	WCMN	427, V-8, 5×2 speed, 20' Schwartz wood floor box, air tag/air brakes, 11-22.5
Chevy	C60	1971	G	$1,000	7/05	ECND	Engine rebuilt 2002, new hyd. pump, roll tarp
Chevy	C60	1972	P	$450	12/05	NWIL	V-8, 4×2 speed, 14' grain box and hoist
Chevy	C60	1972	G	$5,600	6/05	NEND	Single axle, 366 V-8, Allison auto, 2-speed rear end, 16' Knapheide steel box with hoist, 39K miles
Chevy	C60	1972	G	$7,750	12/05	ECMN	Twin screw, 427, 5/4 speed, 18' Knapheide box, hoist, 91,485 miles

Trucks

Make	Model	Year	Cond.	Price	Date	Area	Comments
Chevy	C60	1972	G	$12,750	12/05	SCMN	Twin screw, 20' Crysteel box, hoist, 427 engine, 5×4 speed, air brakes, 10-20 rear rubber, roll tarp
Chevy	C60	1972	G	$12,750	12/05	SCMN	Twin screw, 20' Crysteel box, hoist, 427 engine, 5×4 speed, air brakes, 10-20 rear rubber, roll tarp
Chevy	C60	1973	P	$3,600	1/05	NCKS	16' steel bed with hoist, 5×2 speed, 350, V-8, rough
Chevy	C60	1973	G	$4,250	3/05	NWMN	Lift tag tandem with steel box and hoist, new roll tarp, 350 engine, 5×2 speed
Chevy	C60	1974	P	$300	6/05	ECNE	Dump truck
Chevy	C60	1974	F	$2,300	6/05	SEND	Single axle, V-8, 4×2 speed, 15' box, hoist, flip-down stock rack, roll tarp, 138K miles
Chevy	C60	1974	G	$2,450	9/05	WCIL	13' Knapheide box, dual hoists, 350 V-8, 2 ton, 4×2 speed
Chevy	C60	1974	F	$2,500	2/05	WCIL	13.5' farm box with new floor, hoist, 350 V-8, 4×2 speed, 147K miles, needs bushings
Chevy	C60	1974	G	$2,800	4/05	SEMI	Grain box, cattle rack, 16' hoist, 68K miles
Chevy	C60	1974	G	$3,600	2/05	WCIL	45,200 miles, 13' bed
Chevy	C60	1974	F	$3,600	5/05	NCKS	16' bed and hoist, V-8, 4×2 speed
Chevy	C60	1974	F	$3,950	2/05	NECO	Jacobs 16' steel box, power up & down rear hoist, new 350 V-8, always shedded
Chevy	C60	1974	G	$4,950	1/05	NWIL	350 V-8, 5×2 speed, 14.5' steel grain box and hoist, tarp, 75,683 miles
Chevy	C60	1974	G	$6,650	11/05	ECNE	4×2 speed, V-8, 17.5' OS box, 74,450 miles
Chevy	C60	1975	F	$1,700	8/05	NWIA	Single axle, 16' wood box
Chevy	C60	1975	F	$2,700	3/05	WCKS	16' steel bed, hoist, 82,215 miles, stock rack, 5×2 speed
Chevy	C60	1975	G	$2,800	11/05	SCIA	366 motor, 5×2 speed, steel grain box, 86K miles
Chevy	C60	1975	G	$3,500	11/05	NEIL	
Chevy	C60	1975	G	$7,200	1/05	SWOH	4×2 speed, 53K miles, 10' dump bed
Chevy	C60	1976	F	$3,200	3/05	SCKS	V-8, 5×2 speed, 18' Knapheide steel bed and sides, showing 75,117 miles, poor brakes
Chevy	C60	1976	G	$3,400	2/05	WCOK	Bobtail grain, Westfield drill fill auger on back
Chevy	C60	1977	G	$6,000	8/05	ECIL	16' Midwest bed, 4×2 speed, 78K miles, one owner, 9×20 tires
Chevy	C60	1977	G	$6,500	12/05	WCIL	350, V-8, 4×2 speed, 14' Schien bed, 48,500 miles
Chevy	C65		G	$9,250	3/08	SEND	Twin screw, 427, 5×4 speed, 20' Frontier box, roll tarp, 52,864 miles, one owner
Chevy	C65	1974	G	$5,300	3/08	NCIL	44,937 miles, 427, V-8 engine, 5×2 speed, 14' grain box, roll tarp
Chevy	C65	1974	G	$5,750	4/08	NEND	Twin screw, 427, 5×4 speed, 20' French combination box, twin post hoist
Chevy	C65	1974	G	$11,100	1/08	WCKS	366 motor, 5×2 speed, 59,766 miles, 16' bed & hoist, rollover tarp
Chevy	C65	1974	G	$12,000	4/08	ECND	Tag axle, 366, 5×2 speed, 18' box, hoist, roll tarp
Chevy	C65	1975	G	$13,000	3/08	WCMN	Twin screw, 366 engine, 5×4 speed, 70,069 miles, 20' steel Frontier box and hoist
Chevy	C65	1975	G	$15,000	3/08	ECND	Twin screw, 427, 20' Bert's staked box, hoist, roll tarp, Allison auto, odometer shows 71,468 miles

Trucks

Make	Model	Year	Cond.	Price	Date	Area	Comments
Chevy	C65	1975	G	$18,000	1/08	SCKS	Tandem-axle grain, 18' steel box with 52" sides, roll tarp, 427 gas engine, 5×2 speed, 9:00-20 tires, 40,398 miles
Chevy	C65	1976	G	$7,250	3/08	ECNE	Custom deluxe boom bucket, 366 V-8 gas, 52,673 miles, Allison auto AT540, 10:00-22.5 tires, budd rims, Pitman Pace Mate service body, & Pitman Hot Stick Model HS-50HAL 4-hyd. bucket boom
Chevy	C65	1977	G	$5,800	3/08	NWKY	366 V-8, red, 15' metal bed, hoist
Chevy	C65	1977	G	$9,600	1/08	SWKY	Tandem, roll tarp
Chevy	C65	1978	F	$1,600	3/08	WCMN	Tandem truck, 19' grain box
Chevy	C65	1978	F	$1,800	2/08	SEMO	10 wheeler, 22' metal bed, hoist, V-8, 79K miles
Chevy	C65	1978	G	$6,300	1/08	SWKY	Grain bed, dump
Chevy	C65	1978	G	$8,200	2/08	SEIA	366 V-8, 5×2 speed, 15' farm box, 91K miles
Chevy	C65	1978	E	$9,250	1/08	WCIL	366, V-8, 5×2 speed, Scott 13.5' farm box, 50" steel sides and cargo doors, 60-gal. fuel tank, red exterior, 114,261 miles
Chevy	C65		F	$4,500	4/07	NWMN	Twin-screw, tandem, 427 V-8, 5×4 speed, 20' box, hoist, roll tarp, 10:00-20 rubber
Chevy	C65	1972	G	$6,900	1/07	WCIL	18' grain box, tandem axle, 75,200 miles, 427 gas engine
Chevy	C65	1973	G	$2,100	4/07	ECND	Tag axle, tandem truck, 366 engine, 5×2 speed, 18' box, French combination box, roll tarp
Chevy	C65	1973	G	$6,600	11/07	SESD	18' steel box, rebuilt 366 engine, 4×2 speed, cheater axle, new battery, good tires
Chevy	C65	1973	G	$7,750	1/07	NCKY	Dump, 14' steel grain box
Chevy	C65	1973	G	$8,000	2/07	SWKS	427, 2,625-gal. water tank with (2) 5.5-hp. 2" Honda pumps, side board and ends
Chevy	C65	1973	E	$11,200	4/07	SCKS	Twin screw, air brakes, 427 gas, Shur-Lok rollover tarp, shedded
Chevy	C65	1974	F	$550	3/07	WCIL	250-bu. center dump bed, hyd. auger, runs on LP
Chevy	C65	1974	P	$1,400	2/07	ECNE	366 engine, 4×2 speed, 18' box, Heil hoist 98K miles, silage end gate
Chevy	C65	1974	F	$1,500	12/07	WCMN	Twin screw, steel box
Chevy	C65	1974	G	$11,000	6/07	ECND	Lift tag tandem, 427 V-8, 5×2 speed, 19' box, hoist, roll tarp, 9:00×20 tires, HEI ignition, 3,000 miles on OH, one owner, 46,860 actual miles
Chevy	C65	1975	F	$3,000	4/07	NWMN	Twin screw tandem, V-8, 5×4 speed, air brakes, 18' Lode King 2-compartment tender box
Chevy	C65	1975	F	$3,750	1/07	NWIN	V-8, 5×2 speed, 16' bed and hoist with dummy axle
Chevy	C65	1975	E	$7,750	2/07	WCIL	15' Knapheide grain bed with 52" sides, 427 engine, 5×2 speed, color bright red with white tip tops on the bed, only 71K miles, very sharp
Chevy	C65	1975	G	$8,000	2/07	SWKS	Grain truck, 20' box, Shur-Lok tarp, 427 gas engine
Chevy	C65	1975	G	$10,250	10/07	NEND	Twin screw, 427, 5×4 speed, 19' 1980 strong box, hoist, roll tarp
Chevy	C65	1975	G	$13,500	1/07	NCNE	366 engine, 5×2 speed, air
Chevy	C65	1975	G	$13,500	1/07	SCNE	Triple axle, lift tag, 5×2 speed, 366 engine, Knapheide 20' steel comb box with tip tops, roll tarp, air

Trucks

Make	Model	Year	Cond.	Price	Date	Area	Comments
Chevy	C65	1975	G	$15,400	2/07	SCNE	427 engine, 5×2 speed, hyd. tag, 20' steel side high side box, new engine in 1999, near-new radial tires, 72K miles, extra sharp
Chevy	C65	1976	F	$3,900	6/07	ECND	Tag tandem (lifting push axle), 427 V-8, 5×2 speed, 16' box with hoist and roll tarp, low miles on engine rebuild and new clutch
Chevy	C65	1976	G	$8,500	11/07	ECND	Tag tandem, 366, 5×2 speed, 20' Frontier box, hoist, roll tarp, 10:00-20 rubber
Chevy	C65	1976	G	$10,250	10/07	NEND	Twin screw, 427, 19' Knapheide stake box, 5×4 speed, hoist, roll tarp, 3-piece end gate, Firestone radials, 74,193 miles
Chevy	C65	1976	G	$13,750	10/07	NEND	Twin screw, 427, 5×4 speed, 19' 1980 strong box, hoist, roll tarp, beet equipped
Chevy	C65	1977	F	$950	3/07	SWNE	Grain, box, hoist, sold on this auction but located elsewhere
Chevy	C65	1977	G	$4,250	12/07	WCIL	14' box
Chevy	C65	1977	G	$5,700	4/07	NEMI	Custom deluxe, lift axle, 15' drop-side box with hoist, 5 speed with od
Chevy	C65	1977	G	$15,000	2/07	SCKS	Tandem-axle grain, 20' steel box, 52" sides, Harsh twin-cylinder hoist, 366 gas engine, 5×2 speed, dual fuel tanks, new brakes, 9:00-20 tires, 61,470 miles
Chevy	C65	1978	G	$5,500	2/07	SWKS	Tandem-axle grain, 22' steel box, 427, V-8, 5×2 speed, AM/FM radio, 9:00-20 tires
Chevy	C65	1978	G	$6,000	12/07	SCMN	Grain
Chevy	C65	1978	G	$6,700	11/07	SWNE	18' steel box, hoist
Chevy	C65		G	$1,000	11/06	SECA	Flatbed
Chevy	C65		G	$14,000	11/06	SCMN	Tag
Chevy	C65	1973	F	$700	7/06	WCMN	Cab and chassis, single axle, 427 gas, 5×2 speed
Chevy	C65	1973	G	$7,200	3/06	SWMN	Grain, twin screw, 427, 5×4 speed, V-8, 18' steel box and hoist, 12K miles on major OH in 2000
Chevy	C65	1973	G	$13,000	3/06	ECND	Twin screw, 427, 13 speed, 20' aluminum box, twin-post hoist, roll tarp, pintle hitch, twin fuel tanks, 98,211 miles
Chevy	C65	1974	F	$1,250	4/06	NWMN	Single-axle fuel truck, 366, 5×2 speed, 5 compartments, 2 hose reels
Chevy	C65	1974	F	$4,100	3/06	NCCO	Peterson 18' steel box, tag axle, end dump, 5×2 speed, 129,900 miles
Chevy	C65	1974	G	$5,900	3/06	NCIA	427, 4×2 speed, 20' Knapheide steel box, hoist, fixed tag, PS
Chevy	C65	1974	G	$7,250	1/06	NEMO	Grain, 69K miles
Chevy	C65	1974	G	$9,250	2/06	NWMN	427, twin screw, auto, 100k+ miles, Berts 19.5' box, hoist, 3 piece end gate, roll tarp
Chevy	C65	1974	G	$10,000	4/06	NWMN	Twin screw, 427, auto, Kiefer 19' box, hoist, roll tarp, air brakes, 100K miles
Chevy	C65	1974	G	$10,500	4/06	ECND	Lift tag tandem, 366, 5×2 speed, twin fuel tanks, Westgo 19' box, twin-post hoist, roll tarp
Chevy	C65	1974	G	$12,000	1/06	ECIL	Grain, 10 wheeler, air tag axle, cargo doors with 18' bed and 64" sides, 5×2 speed, 9/20 rubber, roll tarp
Chevy	C65	1975	G	$3,300	8/06	NCIA	Tandem truck, 366, V-8, 5×2 speed, near new tires, 18' box and hoist
Chevy	C65	1975	F	$3,600	11/06	NWMN	Tag tandem, 427, 5×2 speed, 15' Kiefer box, hoist, roll tarp, air up tag, pintle hitch, air brakes

Trucks

Make	Model	Year	Cond.	Price	Date	Area	Comments
Chevy	C65	1975	G	$7,800	12/06	WCMI	18' Midwest rack and hoist, 10 speed, 51K miles
Chevy	C65	1975	G	$8,000	3/06	ECND	Twin screw, 427, Allison 5 speed, auto, Midland box, combination end gate, hoist, 104K miles
Chevy	C65	1975	G	$8,400	4/06	NEND	Hyd. lift tag tandem, 427, 5×2 speed, 20' Crysteel box, hoist, roll tarp, 42K miles on new engine
Chevy	C65	1975	G	$10,000	2/06	NWKS	Grain truck, 69,158 miles, 5×2 speed, V-8, 22' steel box with hoist, roll tarp
Chevy	C65	1975	E	$13,000	8/06	WCKS	Tandem truck, 366 motor, 5×2 speed, 38,986 miles
Chevy	C65	1976	G	$1,600	2/06	WCME	14' dump body, 366 gas, 13 speed
Chevy	C65	1976	G	$3,000	7/06	NEND	Single axle, 366, 5×2 speed, 16' Rugby box, hoist, roll tarp, 65,800 miles
Chevy	C65	1976	G	$3,000	12/06	SEIA	16' grain box
Chevy	C65	1976	G	$4,000	10/06	NCND	Twin-screw truck with Logan Live bottom box, 5×4 speed, 427, 28K actual miles
Chevy	C65	1976	G	$6,300	11/06	SCNE	5×2 speed, 366, V-8, Midwest 18' steel box, 61,674 miles
Chevy	C65	1976	G	$6,500	1/06	NWIL	14' farm box, hoist, 366V-8, 5×2 speed, like-new tires, 50,150 miles
Chevy	C65	1976	G	$7,000	2/06	WCKS	Tandem truck, V-8 motor, 5×2 speed, 22' bed and hoist, rollover tarp
Chevy	C65	1976	G	$9,000	10/06	NCND	Single axle, 5×2 speed, 16' steel box, hoist, 3-piece end gate, plumbed for drill fill, 10:00×20 rubber
Chevy	C65	1977	F	$1,500	2/06	WCIL	Tag axle, 81K miles, 366 motor, 5×2 speed, 21' bed with hoist
Chevy	C65	1977	G	$2,200	8/06	SCMN	Tandem-axle twin screw, 427 gas, grain box and hoist
Chevy	C65	1977	F	$2,750	6/06	NWMN	Twin-screw tandem, 427, V-8, 5×2 speed, air shift, low differential, 19' Frontier beet equipped box, hoist and roll tarp
Chevy	C65	1977	F	$3,500	4/06	NWKS	366, V-8, 5×2 speed, 9:00-20 tires, 15'×42" Hillsboro box, hoist, rollover tarp
Chevy	C65	1977	G	$4,200	10/06	NCND	Twin screw, 15' gravel box, 427, 5×4 speed, 32K actual miles
Chevy	C65	1977	G	$7,250	1/06	NWIL	15' farm box, hoist, 366V-8, 5×2 speed, like-new tires, 71,040 miles, 3,663 miles on rebuilt engine
Chevy	C65	1977	G	$15,000	3/06	NWMN	Twin screw, 427, 5×4 speed, 19' Polar box, twin-post hoist, roll tarp, 71,751 miles
Chevy	C65	1978	G	$1,200	9/06	NEIN	Rollback, gas, 5×2 speed
Chevy	C65	1978	F	$2,800	8/06	WCIA	350 gas, 4×2 speed, cheater axle, loaded box, 450-bu. corn
Chevy	C65	1978	G	$5,000	8/06	SCMN	Gas, 360 engine
Chevy	C65	1978	G	$12,000	2/06	ECIL	Tandem grain, 20' bed, 5×2 speed, air tag and cargo doors, 56,950 miles
Chevy	C65		G	$3,100	8/05	SEMN	
Chevy	C65		E	$7,000	12/05	ECIL	14' Midwest bed and hoist, 366 engine, 39K miles, new 900/20 tires, 5×2 speed
Chevy	C65	1961	F	$750	6/05	ECNE	6-cylinder gas engine, 4 speed, PTO, Progress 1,200-gal. fuel tank
Chevy	C65	1973	P	$350	2/05	SWNE	Single-axle dump truck
Chevy	C65	1973	G	$2,250	1/05	NWGA	350, 5 speed, 10' box dump
Chevy	C65	1973	G	$4,500	3/05	SWIN	5×4 speed, 427 gas engine, Hendrickson suspension, air brakes, tandem, roll tarp, metal floor, grain sides

Trucks

Make	Model	Year	Cond.	Price	Date	Area	Comments
Chevy	C65	1973	G	$4,900	1/05	ECIL	Tandem axle, 466 engine, 15' aluminum bed, hoist, gravel gate, 5×2 speed, air brakes
Chevy	C65	1973	G	$10,100	12/05	SCMN	18' steel box, hoist, twin screw, 427 gas, PS, 5×4 speed, Syntex roll tarp
Chevy	C65	1974	G	$4,100	12/05	NCKS	Grain truck, steel box, bought new
Chevy	C65	1974	G	$4,500	6/05	NWMN	Tag tandem, 366, V-8, 5×2 speed, 16' steel box, hoist, roll tarp
Chevy	C65	1974	F	$4,900	11/05	NWMN	Twin screw, 427, 5×4 speed, 19' steel box, tip tops, 3-piece end gate, 52,800 miles
Chevy	C65	1974	G	$5,400	7/05	ECND	Automatic, 20' Buffalo box, center-mount hoist, 40K miles
Chevy	C65	1974	G	$6,500	3/05	NEMT	16' Parkhurst steel combination box, 48" sides, rear hoist, 366 V-8, 5×2 speed, 9.00×20s, 61K miles
Chevy	C65	1974	G	$7,000	6/05	ECNE	
Chevy	C65	1975	F	$500	2/05	NCIN	Grain, 16' grain box
Chevy	C65	1975	G	$6,000	3/05	ECND	Single axle, 366, 5×2 speed, 16' box, hoist, roll tarp, 67,100 miles
Chevy	C65	1975	G	$7,000	7/05	NCMN	Twin screw, 427 V-8, 19' Knapheide box and hoist
Chevy	C65	1976	F	$1,850	2/05	NWKY	10 wheeler, 5×2 speed, 18' metal bed, twin hoist
Chevy	C65	1976	G	$3,500	8/05	NWIA	Tag axle, 366, V-8 engine, 21' box, Shur-Lok roll tarp
Chevy	C65	1976	G	$5,800	2/05	WCIL	15' farm box, hoist, 366 V-8, 5×2 speed, 38,165 miles
Chevy	C65	1976	F	$7,100	8/05	NWIL	95,409 miles, V-8, 15' grain box, 5×2 speed
Chevy	C65	1976	G	$8,250	2/05	SCMN	427 engine, Allison auto, twin screw, 6K miles on new motor, 76K total miles
Chevy	C65	1976	G	$13,750	8/05	NCIA	Twin screw, 366, V-8, 5×4 speed, PS, hoist, 20' Scott steel box, cargo doors, roll tarp, 101,486 miles
Chevy	C65	1976	G	$14,000	2/05	NWMN	Scottsdale tandem axle, 427 gas, 5×4 speed, 18' Buffalo box, 3-piece end gate, roll tarp, twin fuel tanks
Chevy	C65	1976	G	$16,000	2/05	WCIA	91K miles, hyd. tag, 18' wood box, roll tarp, 366 engine, 5×2 speed, new clutch
Chevy	C65	1977	F	$500	9/05	SEIA	Dump truck
Chevy	C65	1977	G	$10,000	1/05	SCNE	Allison 534 auto, grain truck, 20' steel box, twin screw
Chevy	C65	1977	G	$10,100	6/05	NENE	Custom deluxe, Omaha standard 20' wood box with Harsh hoist, hyd. rear tag, 44,400 miles, 366, V-8, 5×2 speed
Chevy	C65	1978	G	$20,000	3/05	NEMT	20' Midland steel stakeless box, 60" sides, roll tarp, rear hoist, 427 V-8, 5×4 speed, twin screw, 9.00×20s, 59K miles
Chevy	C6500	1974	G	$5,400	9/06	NWOH	427 engine, 5×2 speed, 16' Omaha grain bed and hoist, air brakes, 71,770 miles
Chevy	C70	1974	G	$13,200	1/08	SCKS	Tandem-axle grain truck, 20' steel box, 52" sides, roll tarp, 366 gas engine, 5×2 speed, dual 50-gal. fuel tanks, 9:00-20 tires, 99K miles
Chevy	C70	1976	G	$7,500	1/08	WCIL	Knapheide 15' grain box, 5×2 speed, 366 engine, 9:00-20 tires, ladder, good tarp, only 33,840 miles
Chevy	C70	1973	F	$1,000	11/07	SEID	671 Detroit, Allison 653 trans., tandem axle
Chevy	C70	1973	G	$1,000	11/07	WCID	671 Detroit, Allison 653 trans., tandem axle

Trucks

Make	Model	Year	Cond.	Price	Date	Area	Comments
Chevy	C70		G	$3,000	12/05	ECWA	Bucket, 8.2L diesel, 2-speed rear axle
Chevy	C70		G	$3,300	2/05	WCIA	
Chevy	C70		F	$3,500	9/05	SEIA	Hoist, 86,005 miles
Chevy	C70		E	$11,750	11/05	NWOH	Single-axle grain truck, V-8, 4×2 speed, Omaha 14' body, 13,967 miles, waxed
Chevy	C80	1971	F	$1,250	3/06	ECIN	Gas, tandem axle
Chevy	C90	1970	G	$5,000	12/07	WCMN	Single axle, air tag axle, 18' box, roll tarp, 372,124 miles, 250 Cummins, 9 speed, approx. 5,000 miles on OH
Chevy	Cheyenne	1976	F	$600	4/06	ECND	½ ton, 400, auto, 4WD, 7' flatbed
Chevy	Cheyenne	1973	G	$400	6/05	SEND	Automatic, 2WD, super 10,350 engine
Chevy	Cheyenne	1973	G	$1,700	3/07	SEND	Cheyenne 20, 454, auto, 103,925 miles
Chevy	Cheyenne		F	$200	5/06	SCCA	Diesel, auto
Chevy	Custom	1977	F	$975	11/07	WCTN	111,137 miles
Chevy	Custom	1978	G	$2,300	7/07	NCND	Custom deluxe ¾-ton dually service pickup, custom-made flatbed, V-8, 4-speed manual, 400-gal. service tank with 12-volt pump
Chevy	Custom	1976	F	$350	5/06	SEND	Custom deluxe pickup, V-8, 4 speed
Chevy	Custom 10	1978	F	$800	2/05	NEMI	4×4, 350 engine, auto, 8' box, dark blue
Chevy	Custom 20	1974	F	$370	3/07	SEND	Utility box, 4 speed, crew cab
Chevy	Custom 20	1976	P	$150	1/07	SWKS	20 custom deluxe ¾-ton pickup, auto, white, needs engine work
Chevy	Custom 20	1971	G	$3,400	8/06	NCIA	4×4, 350 V-8 rebuilt 20K miles ago, auto, pretty good body
Chevy	Custom 20	1973	E	$7,750	3/06	NCOK	4WD, new bale bed, V-8, 4 speed, ¾ ton, new paint
Chevy	Custom 30	1972	G	$2,000	2/07	ECIL	1 ton, 350 engine, 9' steel bed with hoist, runs good
Chevy	Custom 30		G	$1,100	5/06	SWCA	Crew cab, V-8, auto
Chevy	Deluxe 30	1976	G	$6,300	12/07	ECIL	1 ton, V-8, 4 speed, PTO hoist, 9' grain bed with stock racks, hd springs
Chevy	K20	1978	F	$800	2/05	NWIL	157,473 miles, 4×4
Chevy	K60	1966	F	$1,400	3/05	SWIN	Dump truck, 5×2 speed, heavy bed
Chevy	N/A	1963	F	$1,000	3/08	NEKS	2 ton, 6 cyl., bed, hoist
Chevy	N/A	1965	G	$1,650	4/08	ECND	Tag tandem, 1,400-gal. stainless tank
Chevy	N/A	1969	G	$4,200	3/08	SCKS	Grain truck, V-8, 4×2 speed, Omaha standard 16' bed & hoist, plumbed for hyd. drill fill auger
Chevy	N/A	1970	F	$1,450	3/08	ECND	Cabover tag tandem, 427, 20' wood box, roll tarp
Chevy	N/A	1977	G	$9,600	1/08	SWKY	Dump
Chevy	N/A		P	$600	7/07	WCMN	1 ton, flatbed, 350 gas, runs good
Chevy	N/A		G	$1,000	1/07	NENC	
Chevy	N/A		G	$14,500	11/07	WCIL	½-ton old pickup, running, old restoration/repaint, good older restoration, sharp looker, nice style pickup, rust-free Kansas truck
Chevy	N/A	1948	G	$1,200	6/07	SCSK	Canadian sale, ton truck, 235 6 cyl., 4×2 speed, wood box and hoist
Chevy	N/A	1949	F	$200	3/07	SECO	2 ton
Chevy	N/A	1956	F	$350	2/07	SCNE	13' box and hoist
Chevy	N/A	1957	G	$1,750	8/07	SWOK	2 ton, lift
Chevy	N/A	1963	F	$1,300	6/07	SCND	Single axle, 292 6 cyl., 4×2 speed, 13.5' Knapheide box, hoist, 8:25-20 tires
Chevy	N/A	1964	G	$3,000	8/07	SWOK	2 ton, lift, 37,459 miles, shedded
Chevy	N/A	1965	G	$2,750	2/07	NCCO	Bed
Chevy	N/A	1966	G	$1,000	3/07	SECO	2 ton, Oswalt feed box

Trucks

Make	Model	Year	Cond.	Price	Date	Area	Comments
Chevy	N/A	1966	F	$1,000	2/07	ECNE	Single axle, 366 engine, 4×2 speed, 16' box with 42" sides
Chevy	N/A	1966	G	$4,400	2/07	ECNE	Short box pickup, 3 speed on the column, V-6
Chevy	N/A	1968	G	$6,000	7/07	NCND	2 ton, tag tandem, 5×2 speed, 18' steel box and hoist, new roll tarp, 366 engine, 84K miles
Chevy	N/A	1968	G	$7,750	12/07	SCNE	½-ton step-side pickup, 400-cubic-inch big block engine, auto, 15K miles since complete rebuild and customization
Chevy	N/A	1969	G	$5,250	4/07	NEND	Twin-screw tandem, 366 V-8, 5×4 speed, 20' box with hoist, beet equipped
Chevy	N/A	1970	G	$2,700	2/07	NCCO	Bed
Chevy	N/A	1970	G	$4,400	2/07	NCNE	Dump
Chevy	N/A	1972	G	$1,100	11/06	SCNE	Cabover, V-6 Detroit, 5×2 speed, tender box, 30 yard or 1,500-gal. capacity, tag axle
Chevy	N/A	1972	G	$10,500	7/07	WCKS	Tandem, 20' bed and hoist, 5×2 speed
Chevy	N/A	1973	P	$225	2/07	NENE	305, auto, utility boxes, does not run
Chevy	N/A	1974	F	$400	2/07	ECIL	1-ton service
Chevy	N/A	1974	F	$800	6/07	ECND	¾ ton, V-8, auto, sprayer pickup
Chevy	N/A	1975	G	$12,000	10/07	NEND	Twin screw, 427, 5×4 speed, 19.5' Frontier box, hoist, roll tarp
Chevy	N/A	1977	F	$1,250	6/07	ECND	1-ton dually, crew cab, 4×4, rebuilt 350 V-8, 4 speed, air suspension, grain hitch, dual service tanks with pump
Chevy	N/A		F	$200	10/06	SWNE	Early 1940s Chevy cab and chassis
Chevy	N/A		G	$1,400	6/06	NEIN	4×4 pickup, gas
Chevy	N/A		G	$5,800	5/06	SWCA	Extended cab, turbo diesel, auto
Chevy	N/A	1945	F	$150	7/06	NWOH	Straight, 16' flatbed, not running, no title
Chevy	N/A	1946	F	$170	5/06	SEND	1½-ton truck
Chevy	N/A	1946	G	$925	5/06	SEND	¾-ton pickup, 6 cyl., ran good
Chevy	N/A	1949	F	$350	11/06	SENE	1½-ton truck, 13.5' fold-down box, Harsh twin-cyl. hoist
Chevy	N/A	1958	F	$650	4/06	NWMN	Single axle, box, hoist
Chevy	N/A	1962	F	$150	8/06	NCIA	6 cyl., 1,600-gal. banded tank, 5-hp. pump, hoses, chemductor, 3" quick fill
Chevy	N/A	1964	F	$275	3/06	NENE	½-ton pickup, 3 speed on column, 6 cyl.
Chevy	N/A	1964	F	$800	2/06	WCKS	292 motor, 4 speed, 15.5' bed and hoist, 12-volt drill fill auger
Chevy	N/A	1964	F	$900	6/06	SEND	½ ton, 6 cyl., 3 speed, front hitch
Chevy	N/A	1964	G	$1,700	2/06	NWKS	Fleetside pickup, red
Chevy	N/A	1965	F	$450	4/06	NEND	¾ ton, 6 cyl., 4 speed, 4WD, 65' sprayer, 500-gal. tank
Chevy	N/A	1966	G	$1,250	9/06	NEND	Tag tandem V-8, 5×2 speed, 16' box, hoist
Chevy	N/A	1969	F	$550	4/06	SEND	1-ton dually, V-8, 4 speed
Chevy	N/A	1969	G	$1,700	9/06	ECMN	2½ ton, 5 speed, 24' Schwartz bed, tilt, wench, 2-speed rear end
Chevy	N/A	1973	G	$7,200	2/06	WCKS	Tandem truck, 366 motor, 5×2 speed, 20' B&H, row tarp
Chevy	N/A	1974	F	$250	8/06	NCNE	1-ton dually, flatbed
Chevy	N/A	1974	F	$1,500	6/06	NWMN	Tandem-axle, twin screw, 318 Detroit, 13 speed, box and hoist
Chevy	N/A	1974	G	$9,250	2/06	SCKS	4 speed, 13.5' steel box, roll tarp, V-8, 58,779 miles
Chevy	N/A	1974	G	$10,200	8/06	NCIA	Tandem truck, 427, Allison auto, good rubber
Chevy	N/A	1975	F	$800	3/06	SEND	½ ton, 400, auto, 4WD, single owner

Trucks

Make	Model	Year	Cond.	Price	Date	Area	Comments
Chevy	N/A	1975	G	$14,000	4/06	SEND	Twin-screw tandem, 427 engine, 5×4 speed, 20' steel box, hoist, roll tarp
Chevy	N/A	1976	F	$350	11/06	SCNE	½ ton, auto, 350, V-8
Chevy	N/A	1976	F	$600	9/06	NENE	Custom deluxe, 4×4, 350 gas engine, auto, hd rear bumper, 135K miles, light blue
Chevy	N/A	1978	F	$350	3/06	NWMN	¾ ton, 4×4, V-8, auto, no title
Chevy	N/A	1978	F	$535	8/06	SWWI	4×4 stepside ½ ton
Chevy	N/A	1978	G	$1,750	3/06	SCIN	1 ton, dump truck
Chevy	N/A		F	$250	4/05	SCMI	¾ ton
Chevy	N/A		F	$350	2/05	NEIN	Grain box
Chevy	N/A		F	$550	9/05	NEIN	Gravel, dump
Chevy	N/A		G	$1,300	12/05	ECIN	1 ton
Chevy	N/A		G	$1,900	2/07	NEIN	400K miles, 4WD, duals
Chevy	N/A		G	$2,250	6/05	SWCA	1,500-gal. water tank, V-8, 5 speed
Chevy	N/A		G	$2,800	12/05	ECIN	Single-axle dump truck
Chevy	N/A		G	$3,400	3/06	ECIN	Single-axle dump truck
Chevy	N/A	1923	G	$5,000	4/05	NCIA	Repainted, new box and bed
Chevy	N/A	1941	G	$1,500	2/05	NWKS	Shedded, restorable
Chevy	N/A	1946	G	$1,050	8/05	NCIA	Extra early 6-cyl. Chevy engine
Chevy	N/A	1947	F	$1,350	8/05	NEIA	Pickup, rusty but runs
Chevy	N/A	1959	G	$900	3/05	SEND	2 ton, single axle, 6 cyl., 4×2 speed, box, hoist, low miles
Chevy	N/A	1966	F	$1,500	2/05	NECO	47,543 miles, Freeman 15' beet box, rear hoist, 5×2 speed, 327 engine
Chevy	N/A	1968	F	$1,800	2/05	NWOH	Cabover, 16' grain bed and hoist, 366, 5×2 speed
Chevy	N/A	1969	G	$1,900	9/05	NENE	46K miles, 6-cyl. engine, 3-speed on the column, long box
Chevy	N/A	1971	G	$4,000	1/05	NEMO	Tandem, Scott 18' bed, cargo doors, roll tarp
Chevy	N/A	1972	G	$3,300	7/05	NCIA	V-8, 4×2 speed, single axle, 14' box, cargo doors
Chevy	N/A	1972	G	$3,800	1/05	SCMN	15' steel box
Chevy	N/A	1974	P	$3,500	3/05	NEND	Tandem, 366, V-8, 5×2 speed, 20' Frontier box, hoist, roll tarp
Chevy	N/A	1974	G	$5,000	1/05	SCMN	15' steel box
Chevy	N/A	1975	G	$1,900	7/05	NCMN	4WD, 400 gas, auto, flatbed
Chevy	N/A	1976	F	$700	3/05	ECND	Regular cab, 400 V-8, 4WD
Chevy	N/A	1976	F	$925	9/05	SEIA	
Chevy	N/A	1976	F	$3,200	9/05	SEIA	1 ton, 70,300 miles, hoist
Chevy	N/A	1977	P	$500	3/05	WCMN	¾ ton, 4WD, rebuilt trans., flatbed, fifth-wheel ball
Chevy	N/A	1977	P	$2,900	12/05	SEND	Twin screw, 366, 5×2 speed, 20' Knapheide box, used last week for hauling, brakes might be weak
Chevy	N/A	1977	G	$5,200	2/05	NWKS	Feed truck, 28 Oswalt box, 4×2 speed, scales
Chevy	Scottsdale 10	1977	G	$1,050	3/06	NEND	Scottsdale 10, ½ ton, 350, 4 speed, 4WD, lockouts
Chevy	Scottsdale 20		F	$750	2/05	SCCA	
Chevy	Scottsdale 30	1975	G	$3,000	10/06	WCSK	Canadian sale, 30 1-ton dually grain, 350 V-8, 4 speed, 12' steel box and hoist, steel floor, 37,490 original miles
Chevy	Scottsdale	1978	G	$1,000	3/08	ECND	½ ton, auto
Chevy	Scottsdale	1975	G	$3,000	10/06	SCSK	Canadian sale, 30 1-ton dually grain truck, 350 V-8, 4 speed, 12' steel box and hoist, steel floor, 37,490 original miles

Trucks

Make	Model	Year	Cond.	Price	Date	Area	Comments
Chevy	Silverado		F	$700	12/07	WCMN	4×4, spray, MT3000 spray controller, 40' boom, 450-gal. poly tank, 5.5-hp. Honda, foam markers, double nozzle
Chevy	Silverado	1975	F	$300	1/07	SWKS	Silverado 20, 3×3 trailering special pickup
Chevy	Silverado	1978	G	$1,150	6/07	ECND	½ ton, 350, V-8, 4 speed, 4WD, lockouts
Chevy	Silverado		G	$1,300	3/05	WCCA	V-8, gas, auto, dual wheels, brush guard
Chevy	Silverado	1977	F	$600	3/05	ECCO	4WD, short box
Chevy	Viking	1958	F	$300	7/07	NWOH	Grain, 16' grain bed
Chevy	Viking	1968	G	$2,000	6/07	SCSK	Canadian sale, cab over, 4 speed, 292 6 cyl., 12' box and hoist, 30,778 miles showing
Chevy	Viking	1961	F	$300	8/06	ECNE	Grain, M&W gravity wagon, 350 gas engine, 4×2 speed, runs
Chevy	Viking	1961	F	$600	3/06	NWKS	Grain, 16' steel box, 327 V-8 gas, 4×2 speed, 8.25-20 tires, 48,656 miles, one owner
Chevy	Viking 60	1959	F	$550	1/07	SWKS	Viking 60 boom
Denby	N/A	1918	F	$2,600	9/07	WCIA	Less box
Diamond Reo	N/A	1974	G	$4,500	3/05	NEMT	Water, 3,000-gal. tank, 3" pump, 400 Cummins, 15 speed, Q100 rears, tandem twin screw
Diamond T	662	1958	G	$6,500	5/08	ECNE	Dump, IHC 6-cyl. gas engine
Diamond T	922	1965	G	$25,000	5/08	ECNE	Cummins 250, 6 cyl., diesel, 5 speed, 3-speed rear end, custom-built 22' roll back bed, repainted
Diamond T	U	1922	G	$50,000	9/07	NEIL	2½-ton stake truck, professionally restored, show winner
Dodge	100	1973	F	$400	11/06	NCKS	Power wagon, ½ ton, 4×4, 4 speed, V-8, high mileage
Dodge	150		G	$900	8/06	SWCA	Ram charger
Dodge	150	1978	P	$240	2/05	WCIL	As is, power wagon, 4×4, V-8, auto
Dodge	400	1971	G	$3,100	10/06	SESK	Canadian sale, grain, steel box, hoist
Dodge	500	1969	G	$1,100	3/08	WCMN	Grain, 8-cyl. engine, 4×2 speed, 1 hyd., 16' steel and wood box and hoist, 55,580 miles
Dodge	500	1969	F	$750	3/07	WCMI	Single-axle, 16' Knapheide metal grain box, twin post hoist
Dodge	500	1976	G	$4,750	5/07	NWSK	Canadian sale, single-axle grain truck with 318 V-8, 4 speed, 14' steel Western Indiana box and hoist, roll tarp, drill fill plumbing, 7.50-20 rubber, 43K miles
Dodge	500		F	$1,500	11/06	NEND	Single axle, wood box and hoist
Dodge	500	1963	F	$600	4/05	SEND	15.5' steel box, 4×2 speed, V-8 engine
Dodge	500	1969	F	$400	11/05	SCMI	54K miles, 15' grain box, no hoist
Dodge	600	1974	G	$1,400	4/08	ECND	Single axle, 318, V-8, 4×2 speed, 14' Omaha standard box, hoist, roll tarp
Dodge	600	1972	F	$450	11/06	SCCA	Cab, chassis, 8 speed
Dodge	600	1976	G	$450	11/05	NCCA	2 axle
Dodge	800	1975	G	$6,600	4/08	ECND	Twin screw, 417, V-8, 5×4 speed, 22' Knapheide box, hoist, roll tarp
Dodge	800		P	$200	5/06	SWCA	Cab and chassis, gas engine
Dodge	800		P	$500	11/06	SWCA	3-5 yard box
Dodge	800	1970	G	$3,000	3/06	NWKS	Grain, 22' steel box, roll tarp, twin-cylinder hoist, V-8 gas engine, RT-160 10 speed, twin screw, 50,000-lb. GVW, air brakes, 10:00-20 tires
Dodge	800	1964	F	$900	3/05	NWMN	Twin screw, 15.5' steel box and hoist
Dodge	800	1978	G	$5,600	12/05	SCND	Tandem axle, 20' Scott box and hoist, roll tarp
Dodge	D100	1977	P	$425	7/05	ECND	½ ton, short box, 318, V-8, auto, 4WD
Dodge	D250	1975	F	$750	6/05	NWMN	¾ ton, 4×4, 360 V-8, auto, air, cruise

Trucks

Make	Model	Year	Cond.	Price	Date	Area	Comments
Dodge	D600	1974	G	$4,400	7/07	SEND	Single axle, 318, 4×2 speed, 15.5' Omaha box, hoist
Dodge	D600	1975	G	$3,000	4/06	SEND	318 engine, 2 speed, 15' Wilrich box, hoist, roll tarp, 42,800 miles
Dodge	D600	1976	G	$4,500	4/06	SEND	360 engine, 2 speed, 15' Wilrich box, hoist, roll tarp, 37,600 miles
Dodge	F120		F	$200	11/07	NEND	Single axle, 6 cyl., standard trans., 13' wood box
Dodge	LC	1936	G	$20,000	9/07	NEIL	Restored, brakes recently OH
Dodge	N/A	1949	G	$5,100	5/08	ECNE	B-1-D-126 1-ton express, 6-cyl. flathead, 4 speed, 88,144 miles
Dodge	N/A	1969	F	$675	4/08	ECND	Single axle, 6 cyl., 4×2 speed, 14' Omaha standard box, hoist
Dodge	N/A	1945	E	$27,500	9/07	NEIL	2-ton gasoline tanker, bright restoration, starts and runs well
Dodge	N/A	1953	G	$1,100	9/07	NCND	2-ton single-axle grain truck, 15' wood box and hoist
Dodge	N/A	1964	F	$900	6/06	SEND	½ ton, 6 cyl., 4 speed, 36K actual miles
Dodge	N/A	1967	P	$450	10/07	NEKS	Not running, rough, big water tank
Dodge	N/A	1971	F	$200	2/07	SCWA	
Dodge	N/A	1947	F	$175	3/06	NWKS	12' steel floor box, hoist, restorable
Dodge	N/A	1947	G	$600	6/06	ECND	
Dodge	N/A	1948	F	$550	3/06	ECND	2½-ton single axle, 5×2 speed, 14' box, hoist
Dodge	N/A	1961	P	$250	6/06	NWMN	½-ton pickup
Dodge	N/A	1966	P	$175	8/06	NENE	Gravel, steel box, no front axle
Dodge	N/A	1966	G	$1,500	5/06	WCKS	4 speed, V-8 motor
Dodge	N/A	1972	G	$3,500	3/06	ECND	Twin screw, 20' Frontier box, twin post hoist, roll tarp, 361,V-8, 5×3 speed
Dodge	N/A	1959	P	$350	4/05	NCOK	13' bed & hoist, 5×2 speed
Dodge	N/A	1962	G	$1,750	4/05	NCOK	4×2 speed, 15' steel bed, hoist
Dodge	N/A	1963	P	$1,150	4/05	NCOH	1 ton, 9' hoist bed, 66,589 miles
Dodge	N/A	1968	G	$400	3/05	NECO	Single axle, fifth wheel, 318 engine, 4×2 speed
Dodge	N/A	1976	F	$1,000	9/05	SEIA	Flatbed, hoist
Dodge	N/A	1977	G	$3,900	3/05	WCMN	1½ ton, fiberglass service body, V-8, 4 speed, 350-gal. fuel tank, Fill-Rite 110 pump, Katolight Model 850
Dodge	N/A	1978	F	$800	7/05	ECND	½ ton, 2WD, short box, 225 6 cyl., 4 speed OD, 2WD
Dodge	Power W.	1946	G	$4,700	1/07	SWKS	Power wagon truck, 4×4, PTO driven winch
Duplex	N/A	1916	G	$30,000	9/07	NEIL	4WD, engine and drivetrain rebuilt
Fisher	N/A	1927	F	$2,750	9/07	WCIA	Less box
Ford	100	1965	F	$550	6/06	NEIN	
Ford	100	1966	F	$700	4/06	SEND	V-8, 4 speed
Ford	500	1968	P	$375	2/07	SEIA	Box and hoist, good rubber, needs engine work
Ford	500	1969	G	$3,600	7/07	NCSK	Canadian sale, 330 V-8, 4×2 speed, 12' steel box, Michael's roll tarp, 8.25×20 rubber
Ford	500	1972	G	$5,000	8/07	SWSK	Canadian sale, single-axle grain, 330 V-8, 4×2 speed, 15' steel box and hoist, 47,556 miles
Ford	500		G	$4,600	9/06	SCNE	Single axle, John Deere 714A chuck wagon feed box, 8-cyl. Ford 330 engine
Ford	500	1968	F	$775	2/05	NEMI	Single-axle grain, 14' wood rack and hoist, 390 gas engine, 4 speed, 98K miles
Ford	600	1967	G	$2,250	2/08	WCMN	5×2 speed, 20' box and hoist, roll tarp
Ford	600		G	$800	9/06	NEIN	Gas, 10 speed, single axle, 16' box and hoist
Ford	600	1969	F	$600	3/07	NCNE	15.5' beet box, side hoist, 4×2 speed

Trucks

Make	Model	Year	Cond.	Price	Date	Area	Comments
Ford	600	1969	G	$1,550	10/07	NEKS	Obeco 15.5' steel combination bed, hoist
Ford	600	1975	F	$3,750	4/07	SCSK	Canadian sale, single-axle grain truck, 330 V-8, 4×2 speed, 14' steel box, hoist, roll tarp, 8.25×20 front tires, 9.00×20 rear tires, 87,435 miles
Ford	600	1968	G	$3,200	2/06	WCKS	4×2 speed, V-8 motor, 13.5' bed and hoist, rollover tarp, 12-volt drill fill auger
Ford	600	1969	G	$1,100	11/06	SCIL	Louisville, V-8, 14' bed, 107K miles
Ford	600	1969	G	$2,800	8/06	ECMI	Single-axle grain, 13' metal grain box, hyd. hoist, 98K miles, housed and sharp
Ford	600	1972	G	$2,300	8/06	NCIA	330, V-8, 5×2 speed, PS, new tires, 14' wood box and hoist, cheater axle
Ford	600	1974	G	$2,200	8/06	NCIA	361, V-8, 5×2 speed, PS, 16' steel box and hoist, new tires, cheater axle, 60K miles
Ford	600	1975	G	$3,000	3/06	NENE	18' steel box, hoist, 5×2 speed
Ford	600	1978	E	$7,000	3/06	ECMI	Cabover grain, new 16' box, 330, V-8, 4 speed, 77K miles, cherry condition
Ford	600		F	$900	12/05	WCNJ	Roll off
Ford	600	1969	G	$2,500	9/05	WCMN	Single-axle gravel, 4-yard box, 10' single-cyl. hoist, 127K miles, 330, V-8, 4×2 speed
Ford	700	1971	F	$1,900	3/07	SEND	Louisville tag tandem, V-8, 5×2 speed, 19' box and hoist
Ford	700	1974	G	$3,100	2/07	SEIA	Cabover, 18' steel Knapheide combo grain and livestock box, hoist, cheater axle, 5×2 speed, low miles on 390 engine
Ford	700		G	$1,600	5/06	SWCA	Cab and chassis, Cat 3208 diesel, 5 speed
Ford	700	1978	F	$1,200	4/06	SEND	Tilt cab, V-8, 5×2 speed, 24' enclosed van body
Ford	700		G	$1,800	8/05	SCMI	Flatbed
Ford	700	1965	G	$2,200	11/05	NWPA	V-8, 5 speed, 14' steel dump
Ford	700	1970	F	$950	4/05	NEIN	47,027 miles, 14' grain dump box
Ford	7000	1976	F	$650	8/07	ECMO	Straight, Cat diesel engine, 5×2 speed, hyd. brakes, knuckle boom crane on bed, posthole digger attachment
Ford	7000		F	$1,000	10/06	ECMO	Flatbed
Ford	7000	1972	G	$1,500	9/06	NEMO	Dump, gas, Anthony 12' rock bed
Ford	7000		F	$800	3/05	NCIL	1160 Cat, 10 speed, 15' box
Ford	7000	1973	G	$1,350	3/05	NCIL	3208 Cat, 10 speed, 16' box
Ford	750	1964	G	$1,200	3/08	SEND	Single axle, 4×2 speed, 14' Midwest box
Ford	750	1968	P	$800	7/06	SEND	Tandem, 2½ ton, 18.5' box
Ford	750	1975	G	$2,000	3/06	ECND	Tilt cab, single axle, 361, V-8, 5×2 speed, 18' box, hoist, roll tarp
Ford	750		G	$3,500	11/05	NECO	
Ford	750	1972	G	$3,200	11/05	NCCA	2-axle dump
Ford	750	1978	F	$850	3/05	SCMI	16' box, hoist, single axle, 391 gas engine, 2-speed axle
Ford	800	1974	P	$225	3/08	WCMN	Single-axle gravel, 10' box, V-8 gas, auto, in grove
Ford	800		F	$2,800	12/07	WCMN	Utility, WF, 5 speed
Ford	800	1972	G	$3,500	3/07	NCNE	Feed, Allison auto, gas engine, BJM 12', mixer feeder, scales, working daily
Ford	800	1974	G	$5,400	6/07	SWMN	Twin screw, V-8, 5×3 speed, 20' steel box and hoist, 87,441 miles, one owner
Ford	800	1974	G	$6,500	2/07	NWKS	Tandem-axle grain, 24' box, roll tarp, 5" hyd. drive auger, 391 V-8 gas engine, 5×2 speed, tag axle, 9:00-20 tires, 46,736 miles
Ford	800	1975	H	$3,000	1/07	SCIL	Gas, 5 speed, 1,600-gal. liquid nurse tank, belt drive pump
Ford	800	1976	G	$4,200	3/07	NCOH	16' Omaha bed, 5×2 speed, 391 engine, V-8, motor noise

Trucks

Make	Model	Year	Cond.	Price	Date	Area	Comments
Ford	800		G	$3,750	11/06	SWCA	Dump, 5-6 yard box, diesel, 5 speed
Ford	800	1971	G	$7,250	7/06	NWMN	Louisville twin screw, 477, 5×4 speed, 19' Buffalo box, hoist, roll tarp, 91,359 miles, twin tanks, beet equipped
Ford	800		F	$6,250	12/05	SEMN	18' box and hoist
Ford	800	1968	F	$850	6/05	SEND	Cabover tag tandem, air lift, V-8, 5×2 speed, Frontier box, hoist
Ford	800	1972	F	$1,700	12/05	ECMI	59,227 miles, 20' box and hoist, tandem axle, 391 engine, 5 speed, gas
Ford	800	1976	P	$450	8/05	SEMI	16' potato bed, DNR, parts only
Ford	800	1978	G	$4,000	3/05	SWIN	5.5 hp. 2" pump, ss fertilizer tank, chemical inductor, mounted on 1978 Ford 800 truck with 5×2 speed, 361 engine
Ford	800	1978	G	$12,000	3/05	SWIN	3,350-gal. liquid manure tank, hyd. pump and spray nozzle mounted on 1978 Ford 800 twin-screw truck with Cat 3208 engine, 5×4 speed, Hendrickson suspension, hd frame, good tires
Ford	8000	1976	G	$9,000	2/08	ECKS	Tandem-axle grain, Hibo 18' steel box, Harsh hoist, roll tarp, silage side extensions, silage end gate, Cat 3208 diesel engine, 13-speed trans., 222" wheelbase, 44,800-lb. GVW, spring ride suspension, 10:00-20 tires, 93,049 miles
Ford	8000		F	$7,700	5/07	ECMN	Dump, 713,278 miles, tri-axle, J-Craft box, heavy front, good rubber
Ford	8000	1975	G	$7,800	12/07	SCNE	Tandem, twin screw, 3208 Cat, 4-speed Browning, Allison auto, BJM 20' spreader box and slop gate
Ford	8000		G	$700	11/06	SECA	Water, Cat 3208, 5 speed
Ford	8000	1970	G	$2,750	1/06	WCCA	3 axle
Ford	8000		G	$1,600	8/05	NEIN	Diesel, no bed
Ford	8000		G	$2,500	12/05	ECIN	Single axle, diesel
Ford	8000		G	$3,000	2/05	SCCA	Single axle, Cat 3208, 13 speed
Ford	8000		G	$3,200	1/06	WCOH	Dump bed grain, 14,044 miles
Ford	8000	1970	G	$12,000	3/05	ECND	Louisville twin screw, 3208 Cat, 10 speed, 20' Buffalo box, hoist, roll tarp
Ford	8000	1972	G	$3,500	3/05	ECID	Tandem axle, twin screw, 1450 Cat, auto, Logan 20' self-unloading bed, elec./hyd., PTO, 24" belt
Ford	8000	1977	G	$3,000	3/05	ECMN	Dump truck, tandem axle
Ford	8000	1977	G	$5,000	12/05	WCWA	16' flatbed
Ford	8000	1977	G	$5,500	6/05	ECMN	Boom
Ford	8000	1977	G	$6,500	11/05	NECO	Water, 4,000-gal. capacity, 350 Cummins, 10 speed
Ford	8000	1977	G	$10,000	2/05	WCWI	18' aluminum box, hoist, roll tarp, 5×2 speed, 60K miles
Ford	8000	1977	G	$10,750	3/05	SEMN	Twin screw, 3208 Cat, diesel, 19'9" steel box, 85% rubber
Ford	850	1969	G	$3,000	3/08	WCMI	Live tandem straight, 12-ton Porta box with hyd. unloading auger, 10-speed Roadranger, 3 speed aux. big V-8
Ford	850	1968	P	$1,150	2/05	WCWI	18' box, hoist, for parts, did not run
Ford	880	1973	G	$10,000	3/07	SEND	Louisville twin screw, V-8, 5×4 speed, 20' Buffalo box, roll tarp, 10:00×20 rubber, rebuilt 7/95 at 61,476 miles, now 67,972 miles
Ford	880	1973	G	$13,000	3/07	SEND	Louisville twin screw, V-8, 5×4 speed, 20' Buffalo box, roll tarp, 9:00×20 rubber, majored in 1996 at 56,444 miles, now 63,735 miles

Trucks

Make	Model	Year	Cond.	Price	Date	Area	Comments
Ford	880	1974	G	$6,200	3/07	ECND	Twin screw tandem, 534 V-8, 10 speed, air brakes, 19' steel box, hoist, 3-piece swing-out end gate, roll tarp, 2,000 miles on OH
Ford	880	1974	G	$9,500	3/07	SEND	Louisville twin screw, V-8, 5×4 speed, 20' Buffalo box, 9:00×20 rubber, majored in 1996 at 62,668 miles, now 67,612 miles
Ford	880	1975	F	$9,500	4/07	WCSK	Canadian sale, tandem-axle grain, 475 V-8, 5×4 speed, lux 18'×8.5'×4.5' steel box, Michael's roll tarp, 11.00×20 front tires, 10.00×20 rear rubber, 89,789 miles showing
Ford	880	1978	F	$750	2/07	SCWA	Gas, auto, 20' potato bed
Ford	880	1978	F	$750	2/07	SCWA	Gas, auto, 20' potato bed
Ford	880	1978	F	$750	2/07	SCWA	Gas, auto, 20' potato bed
Ford	880	1978	F	$750	2/07	SCWA	Gas, auto, 20' potato bed
Ford	880	1978	F	$1,200	2/07	SCWA	Gas, auto, 20' potato bed
Ford	880	1978	F	$1,250	2/07	SCWA	Gas, auto, 20' potato bed
Ford	880	1976	F	$5,000	4/06	ECND	Twin screw, 475, 13 speed, air brakes, 20' Cancade box, hoist, roll tarp, 3-piece end gate
Ford	880		G	$2,500	6/05	ECWI	Gas, 8 speed, 13' dump box
Ford	880	1974	G	$5,100	6/05	SEND	Twin screw, 534, 5×4 speed, 18.5' Frontier steel box, tip tops, 2-cyl. hoist, 72,700 miles
Ford	880	1974	G	$5,600	2/05	WCIN	10 wheeler with 477, 5×3 speed, air brakes, 18' Scott bed, 72" sides, cargo doors, hoist
Ford	900	1976	G	$2,300	3/07	NCNE	Gas engine, 5×4 speed, hd winch bed on roll back, runs good
Ford	900		P	$550	10/06	SWOH	Dump, tandem axle, no engine or trans., camelback rear suspension
Ford	900	1977	G	$4,000	11/06	SCCA	Cat 3208 diesel, aluminum box
Ford	900		G	$4,250	2/05	NEIN	Grain box, diesel
Ford	900	1970	G	$3,700	3/05	NWMN	Louisville twin-screw tandem with 20' steel box and hoist with roll tarp, 534, V-8, Roadranger trans.
Ford	900	1976	G	$1,750	8/05	NCMI	Fire truck, gas, pumper unit, 500-gal. holding tank, ladder
Ford	9000	1971	G	$18,000	2/08	NCOH	Live tandem-axle, 318 Detroit diesel, 10 speed, 18' grain bed with twin-cyl. hoist, roll tarp, new rear brakes and drums, air brakes and seat
Ford	9000	1975	G	$5,700	3/08	WCMI	Tandem axle, day cab tractor, 290 Cummins, wet kit, PS
Ford	9000	1978	G	$16,000	3/08	SCID	Single axle, Detroit diesel, 5×2 speed, 10.00-20 rubber, 4,000 gal., Berkeley PTO pump, Haldex controls, front, rear & side spray
Ford	9000		G	$5,500	4/06	WCSD	Tandem axle, recent OH on Detroit engine, 13 speed, 20' steel box with hoist
Ford	9000		G	$7,250	3/07	NWIA	Semi tractor
Ford	9000	1972	F	$10,000	3/07	NWIL	358,815 miles, 5 speed, Cummins engine
Ford	9000	1973	G	$2,500	2/07	SCNE	350 Detroit, turbo, Roadranger 13 speed, US 20' truck crane, trolley boom, twin screw
Ford	9000	1974	F	$2,000	7/07	WCMN	Tandem axle, 6 cyl., Detroit 10 speed, 14' gravel
Ford	9000	1974	G	$2,450	8/07	SWMN	Day cab, Ford diesel engine, 13 speed, tandem axle, full screw, spring ride suspension, fifth-wheel plate, dual stacks, dual fuel tanks, 11R22.5 spokes

Trucks

Make	Model	Year	Cond.	Price	Date	Area	Comments
Ford	9000	1975	F	$900	3/07	WCNE	Single-axle, 671 Detroit diesel engine, 10-speed trans., hoist, 2,000-gal. tank, 5.5-hp. gas motor with pump, 10:00-20 rear tires, 15-22.5 front tires
Ford	9000	1976	G	$3,750	6/07	WCMO	Dump, Cummins diesel engine, 10-speed trans., full screw, tandem axle, AM/FM radio, 14' dump box, 10:00-20 tires, approximately 687K miles
Ford	9000	1976	G	$4,750	2/07	SCWA	Dump
Ford	9000	1977	G	$900	8/07	SWMN	Cab & chassis, Detroit 8V-92 diesel engine, 10 speed trans., tandem axle, full screw, third air tag
Ford	9000	1978	G	$11,250	3/07	SEND	Louisville twin screw, third axle lift, dual trailing, 270-hp. Cummins, 9 speed, 24' box, hoist, roll tarp, 3-piece end gate, rear pintle hitch, air, 10:00×20 rubber, 358,290 miles
Ford	9000	1970	G	$6,800	3/06	SWMN	Grain, air brakes, 3406 Cat engine, 5×4 speed, 20' steel box with wood floor, twin screw
Ford	9000	1972	F	$550	10/06	SEMN	Semi
Ford	9000	1973	G	$1,200	3/06	SCIN	Cummins engine, 10 speed
Ford	9000	1973	G	$12,000	7/06	NWMN	Twin screw, air up/down pusher, Cummins 250, 9 speed, 293,100 miles
Ford	9000	1975	F	$3,000	8/06	SEMN	Dump truck, Detroit, 13 speed
Ford	9000	1976	G	$3,000	6/06	SEID	8V71 Detroit, 5×4 speed, 20' self-unloading bed, elec./hyd., PTO
Ford	9000	1976	G	$4,800	4/06	SEND	Twin screw, Detroit, 13 speed, 10-yard gravel box
Ford	9000	1978	G	$2,500	12/06	ECIN	Dump truck, 290 Cummins, 9 speed, tandem axle
Ford	9000	1978	G	$5,700	2/06	NCIL	Custom cab, 290 Cummins, 10 speed, wet kit, 313,740 miles
Ford	9000	1974	G	$11,000	3/05	SEND	Louisville twin screw, 8V72 Detroit, 9 speed, Henderson suspension, Scott 20' all steel box and hoist, plumbed for drill fill
Ford	9000	1974	G	$16,500	2/05	NECO	398K miles, 350 Cummins, 20' slip box with full hyd. end gate, silage racks
Ford	9000	1975	G	$11,000	3/05	NCMT	19' box, new headlift hoist, 318 Detroit, 13 speed, twin screw, 10.00×20s
Ford	9000	1976	G	$13,000	3/05	WCMN	Twin screw, 318 Detroit diesel, 12,000 miles on major, 10 speed, 22' box and hoist, roll tarp
Ford	9000	1978	G	$14,000	11/05	NWMN	Twin screw, 290 Cummins diesel, 13 speed, third-axle air lift, single tire pusher, 20' Knapheide steel box, Shur-Lok roll tarp, soft end caps
Ford	A	1928	F	$2,200	8/07	SESK	Canadian sale, for restoration, stored inside
Ford	A	1931	G	$15,750	1/07	ECKS	Pickup, fully restored
Ford	A	1929	G	$2,000	8/05	SEND	Stake box
Ford	C600		G	$4,000	8/07	NENE	Cabover, 18' box and hoist
Ford	C600	1958	G	$650	3/07	WCIL	Cabover, gas
Ford	C600	1960	G	$1,200	3/07	NEND	Cabover, 15.5' steel box, hoist, V-8, 5×2 speed
Ford	C600	1972	G	$2,700	12/07	SCNE	16' Melroe aluminum box, hd 330 engine, 5×2 speed, 900 tires, PS
Ford	C600	1955	F	$350	3/06	SEMN	Good hoist
Ford	C600	1972	G	$3,700	11/06	SENE	2-ton grain, 48" sides on box, Crysteel hoist
Ford	C750	1966	F	$850	6/07	ECND	Tilt cab, 5×2 speed, 15' steel box, roll tarp

Trucks

Make	Model	Year	Cond.	Price	Date	Area	Comments
Ford	C750	1961	G	$1,600	11/06	SENE	15.5' box, heel hoist, fold-down racks
Ford	C800	1975	G	$4,400	2/05	NCIL	5×2 speed, air tag, 20' grain body, hoist, roll tarp
Ford	C800	1977	F	$3,000	3/05	NWMN	Twin-screw tandem, 20' Midwest box, hoist with roll tarp, 534, V-8, 5×2 trans.
Ford	C850	1969	G	$4,250	4/07	NWOH	Water, SN C85HUF50982, 477 gas, 5×2 speed, PS, air brakes, water tank, 11R22.5 tires
Ford	F	1950	P	$300	11/06	SENE	F1, for parts, restorable
Ford	F100	1957	F	$300	11/07	WCKS	4 speed
Ford	F100	1973	G	$1,600	11/07	SCNE	2WD, auto
Ford	F100	1974	G	$575	3/07	SEND	460, V-8, auto, 2WD
Ford	F100	1955	G	$9,300	11/06	NCKY	New red paint, oak bed, new bumpers, original body, 3-speed auto in floor, 352 V-8
Ford	F100	1972	P	$50	11/06	SCSD	2WD, 302 auto, newer trans, engine bad
Ford	F100	1975	P	$75	11/06	SCSD	2WD, extended cab, 8' box, 390 engine, auto, 30K miles on OH
Ford	F100	1974	P	$850	3/05	WCKS	½ ton, 4×4, Dudrey wire roller, V-8, auto
Ford	F100	1978	G	$1,575	3/05	NCAL	156K miles, auto, brown
Ford	F150	1977	P	$450	3/08	ECND	400 engine, 4WD
Ford	F150	1978	P	$350	3/08	ECND	Regular cab, 300, 6 cyl., 4 speed, 4WD
Ford	F150	1978	F	$375	3/08	ECND	Standard cab, 8' box, 400 gas, 4 speed, 4WD, 169,508 miles
Ford	F150		P	$250	2/07	SEFL	Gas
Ford	F150	1976	F	$500	3/07	SEND	460, automatic, 2WD
Ford	F150	1976	F	$550	3/07	SEND	460, V-8, auto, topper
Ford	F150	1978	F	$600	2/07	NCNE	4WD, gas, auto
Ford	F150	1978	P	$200	10/06	SWOH	Gas, 5 speed
Ford	F150	1978	F	$500	8/06	NCIA	Custom 4×4, topper, rebuilt 400 engine
Ford	F150	1978	F	$700	3/06	ECND	Ranger, V-8, auto, 4WD, air, 20K miles on OH and front end
Ford	F150	1978	G	$1,100	4/05	NCOK	4×4, flatbed, hyd. bale splice, V-8, 4 speed
Ford	F150 XLT	1978	G	$950	9/07	NEIA	4×4, V-8, auto, topper, 70K miles
Ford	F150 XLT	1978	G	$1,000	11/07	SCNE	4×4, auto
Ford	F250	1976	G	$1,400	1/08	NEMO	16' Knapheide grain bed
Ford	F250	1976	G	$4,500	2/08	ECMT	Agro III fertilizer box 7', Dickey-john controls, radar, 4WD, 429 V-8, 22/24" row spacing
Ford	F250	1978	F	$800	1/08	SCKS	Custom flatbed pickup
Ford	F250	1964	F	$275	3/07	ECND	6 cyl., 4 speed
Ford	F250	1974	G	$900	7/07	SEMN	
Ford	F250	1975	F	$400	2/07	SWKS	Pickup, 4×4
Ford	F250	1977	F	$400	12/07	ECNE	Super cab, 2WD, 460, documented repairs
Ford	F250	1977	F	$950	3/07	SECO	4×4, flatbed, engine rebuilt
Ford	F250	1978	G	$1,600	11/07	ECND	Regular cab, 460, V-8, auto, 2WD, 110,884 miles
Ford	F250	1970	F	$400	7/06	SEND	¾ ton, Custom Camper Special, 93,800 actual miles
Ford	F250	1973	P	$850	7/06	SEND	460, 4WD, blown motor, gas
Ford	F250	1975	G	$650	4/06	NWMN	Custom, 390, V-8, auto, 2WD, 122,440 miles
Ford	F250	1976	G	$1,500	7/06	NWIL	4×4, big engine, gas
Ford	F250	1978	F	$850	4/06	NWMN	4WD, 110-gal. tank and pump
Ford	F250	1954	P	$120	4/05	NCOK	2WD, flatbed, rebuild project
Ford	F250	1969	G	$800	3/05	WCKS	89K miles, ¾ ton, 4×4, V-8, 4 speed
Ford	F250	1975	F	$1,000	3/05	ECND	Highboy, 390, 4 speed, 4WD
Ford	F250	1977	F	$800	9/05	NWOR	Contractor's rack
Ford	F250	1978	G	$3,950	8/05	NCIA	4×4, V-8, auto, 59K miles
Ford	F350	1969	F	$900	3/08	NWKY	White, metal service bed

Trucks

Make	Model	Year	Cond.	Price	Date	Area	Comments
Ford	F350		F	$500	8/07	SWIL	Older, dually, flatbed
Ford	F350	1974	G	$4,000	10/07	NWMN	Drill truck, 390, 4 speed, 4WD, core drill for gravel testing
Ford	F350	1976	F	$200	4/07	SCKS	No title, ran, 2WD
Ford	F350	1978	F	$100	1/07	SECO	Cab & chassis
Ford	F350	1969	G	$1,400	3/06	NCND	Dually, 360 V-8, 4 speed, 2WD, 8' steel service bed, engine drive air compressor, PTO fuel pump, 220-gal. tank, 80-gal. air tank
Ford	F350	1973	F	$600	4/06	NWKS	Dually, V-8, 4 speed, fifth-wheel plate
Ford	F350	1975	F	$1,500	10/06	SESK	Canadian sale, duals, 4 speed, 360 V-8, steel flat deck
Ford	F350	1978	G	$2,600	12/06	SEND	1-ton flatbed, 400 V-8, 4 speed, 12' flatbed, 107,700 miles, yellow
Ford	F350	1978	G	$3,100	3/06	NWMN	Tonner dually, 351 V-8, 4 speed, 2WD, service bed to include (2) 110-gal. field service units, 7' side toolboxes, 66,423 miles
Ford	F350	1967	F	$1,300	3/05	ECND	1 ton dually, 352, V-8, 4 speed, 58K original miles, 8', storage compartment
Ford	F350	1972	G	$2,050	2/05	SWIN	1 ton, steel floor, hoist, V-8, 4 speed, 98K miles
Ford	F350	1974	F	$850	12/05	SCND	Flatbed, 1-ton service truck
Ford	F350	1977	G	$1,050	2/05	NCIL	Flatbed, duals, gooseneck hook up
Ford	F350	1977	G	$2,350	2/05	NCIL	Service
Ford	F4	1951	G	$3,050	2/07	WCIL	1-ton grain, 47K original miles, owner spent $1,200 on it before sale getting it to run perfectly
Ford	F4	1948	F	$150	9/06	SCNE	Single-axle grain, 10' wooden box, hoist, runs
Ford	F500	1953	F	$500	1/08	WCKS	Boom, 8-cylinder flathead gas engine, 4 speed, blue
Ford	F500	1971	G	$4,250	2/08	SCKS	Single-axle grain, 13.5' steel box with 40" sides, Harsh hoist, 330 V-8, gas, 4 speed, 27,229 miles
Ford	F500	1972	F	$350	8/07	ECMO	Gas engine, 4 speed, single axle, hyd. brakes, short-side dump bed
Ford	F500	1975	F	$4,600	7/07	SWSK	Canadian sale, single-axle grain, V-8, 4×2 speed, Westeel 13.5' steel box, Michael's roll tarp, 8.25-20 tires, 41,842 miles showing
Ford	F500		F	$350	10/06	SWOH	Single axle, gas, 4 speed, 8' steel dump box
Ford	F500	1963	G	$1,550	8/06	NEKS	57K miles, V-8, 5 speed
Ford	F500	1969	F	$1,300	3/06	ECND	Single-axle service truck, 331, V-8, 5 speed, 10' service body
Ford	F500	1963	G	$600	3/05	NECO	13' box, hoist, 292 engine, 4×2 speed
Ford	F550		G	$8,400	10/06	SWOH	Utility, 4WD, single axle, Triton V-10 gas, auto
Ford	F6	1948	G	$550	7/06	SEND	Flatbed, V-8, grain box, hoist
Ford	F6	1950	G	$2,500	6/05	ECNE	Flatbed
Ford	F600		F	$1,900	2/08	SWMN	Single axle, 4×2 speed, 15' steel box & hoist
Ford	F600	1966	F	$700	3/08	NWOH	Grain, 12' hyd. grain body, 63K miles
Ford	F600	1968	G	$3,200	4/08	NCND	2-ton grain, steel box, hoist, 4×2 speed, 330 V-8 engine
Ford	F600	1969	G	$2,050	2/08	NEKS	330 V-8, 4×2 speed, 15.5' bed, hoist, steel floor, blue cab
Ford	F600	1969	G	$5,200	1/08	SCKS	Single-axle grain, Midwest 16' steel box with 40" sides, steel floor, 4×2 speed, 55,999 miles

Trucks

Make	Model	Year	Cond.	Price	Date	Area	Comments
Ford	F600	1971	G	$950	1/08	WCKS	Cab & chassis, 330 V-8 gas engine, 4-2 speed trans., 12,557 actual miles, bought new, used as feed truck
Ford	F600	1973	G	$5,400	1/08	ECNE	Grain, 330 V-8 gas engine, 4×2 speed, hyd. brakes, PS, micro brake lock, Omaha standard 15.5' body with hd hoist, 8.25-20 tires
Ford	F600	1974	G	$2,100	2/08	ECIL	Grain, 13.5' bed, 52" sides, 54K miles
Ford	F600	1975	G	$2,000	3/08	WCMN	16' box and hoist, Westfield brush auger, Shur-Lok roll tarp
Ford	F600	1976	G	$2,100	3/08	WCNE	Bobtail, 5×2 speed, mounted Haybuster H1100 grinder, Detroit D power unit
Ford	F600	1976	G	$4,000	3/08	SEIL	Grain, 351, 4×2 speed, 9/20, 16' bed
Ford	F600	1976	G	$5,300	1/08	ECNE	Grain, 46,536 actual miles, 330 V-8 gas engine, 4×2 speed, hyd. brakes, PS, Obeco combination 13.5' body with hd hoist, 8.25-20 tires, always shedded
Ford	F600	1958	F	$400	3/07	ECND	Single-axle, V-8, 4×2 speed, 14' steel box, hoist
Ford	F600	1960	F	$250	3/07	SCNE	Steel-sided box
Ford	F600	1962	F	$600	3/07	SCID	Bobtail, 15.5' box with rear hoist
Ford	F600	1964	F	$450	3/07	SEWY	4×2 speed, 330 V-8, 14' Jacobs bed, 8.25×20 tires, runs good
Ford	F600	1965	F	$1,100	2/07	NEND	Tag tandem, V-8, 4×2 speed, 18' Knapheide box, hoist
Ford	F600	1966	P	$500	11/07	ECND	Single axle, V-8, 4×2 speed, 16' wood box, hoist
Ford	F600	1966	G	$2,300	12/07	SCNE	68K miles, hd 330 V-8, 4×2 speed, 16' giant combination box, 8.25-20 tires
Ford	F600	1967	G	$4,250	11/07	SEID	Fuel, gas, 5×2 speed, 1,500-gal. 5-compartment tank, dual pumps & hose reels, 25K miles
Ford	F600	1967	G	$4,250	11/07	WCID	Gas, 5×2 speed, 1,500-gal. 5-compartment tank, dual pumps and hose
Ford	F600	1968	F	$2,900	4/07	ECSK	Canadian sale, single-axle grain, 330 engine, 5XSP, 14' Trailrite steel box with hoist, roll tarp
Ford	F600	1968	G	$3,500	11/07	ECND	Single axle, 330, V-8, 4×2 speed, 14' Knapheide box, hoist, roll tarp, 96,279 actual miles
Ford	F600	1969	G	$2,000	3/07	SEWY	New 330 V-8 motor and clutch, 4×2 speed, 15' box, 42" sides, rear hoist, 8.25×20 tires
Ford	F600	1969	G	$3,100	9/07	NCND	Gas engine, equipped with Flying L triple-axle implement flatbed, all-steel semitrailer with fold-down ramps
Ford	F600	1970	F	$1,550	2/07	NWIL	80,200 miles, V-8 engine, 4×2 speed, 13' metal grain box, hoist, lots of rust
Ford	F600	1971	G	$2,700	4/07	NEMO	Grain, 15' bed, steel floor, 370 engine, 4×2 speed, Shur-Lok rollover tarp
Ford	F600	1971	G	$2,800	3/07	ECND	Tag tandem, 391 V-8, 5×2 speed, 18' steel box, hoist, 3-piece end gate, roll tarp
Ford	F600	1971	G	$3,700	2/07	NEMO	60" sides, 4×2 speed, twin-cyl. hoist, 361 engine
Ford	F600	1972	F	$1,600	9/07	ECND	Single axle, V-8, 4×2 speed, 16' box, hoist, roll tarp, 8:25-20 tires, 84K miles
Ford	F600	1973	G	$1,600	2/07	NEMO	Grain, 16' bed, 330 engine, 4×2 speed, tarp, 50,657 one-owner miles
Ford	F600	1973	G	$2,750	3/07	NENE	Single-axle grain, Obeco 16' wooden box, twin-cylinder hoist, 330 V-8 gas engine, 4×2 speed, 9:00-20 tires, 46,197 miles, shedded

Trucks

Make	Model	Year	Cond.	Price	Date	Area	Comments
Ford	F600	1973	G	$2,750	12/07	NEMO	Grain, 15' bed
Ford	F600	1973	G	$3,000	2/07	NENE	330 engine, 4×2 speed, 1,250-gal. 5-compartment fuel tank with pump
Ford	F600	1974	F	$1,600	12/07	ECNE	Single-axle grain, spring cheater axle, 18' box, Omaha standard hoist, 60K miles
Ford	F600	1974	G	$2,400	12/07	ECNE	Single-axle grain, 16' box, Omaha standard hoist, 84,843 miles
Ford	F600	1974	G	$2,400	11/07	WCMN	Single-axle grain, 15' steel box, hoist, 64,813 miles, V-8, 4×2 speed, one owner
Ford	F600	1974	G	$3,000	1/07	SCNE	4×2 speed, 16' steel box and hoist
Ford	F600	1974	G	$3,500	11/07	ECND	Single axle, 330, V-8, 4×2 speed, hd axle, 16' Knapheide box, hoist, roll tarp, 96,279 miles
Ford	F600	1974	G	$3,800	1/07	NWIN	PS, 15' bed and hoist, 45K miles
Ford	F600	1974	G	$4,000	12/07	WCMN	Grain, single axle, budd wheels, 70,014 miles, 4×2 speed, 330 Ford V-8 gas engine, 14' box
Ford	F600	1975	G	$1,050	9/07	WCIA	Flatbed dump, 390 V-8 gas engine, 4×2 speed, 16,000-lb. GVW, 158" wheelbase, 79,930 miles, AM radio, heat, 8.25-20 tires, 14' steel flatbed body with scholastic hd scissor hoist, headache rack, 12" wood sides
Ford	F600	1975	G	$1,400	6/07	SWMN	Single axle, V-8, 4×2 speed, 18' steel box, 9:00-20 tires, 6,761 miles
Ford	F600	1975	G	$1,950	2/07	WCNE	Less than 2,000 miles since complete OH, 4×2 speed, 16' Midwest steel box, 56K actual miles
Ford	F600	1975	G	$3,000	6/07	SWMN	Tandem axle, 18' box and hoist, V-8 361, 4×2 speed, 10,833 miles, 9:00-20 tires
Ford	F600	1975	G	$3,200	12/07	SWIN	Grain, 72K miles, 330XD gas engine, 4×2 speed, single-drive axle with single wheel dead axle, 14' bed, steel sides with extensions, new steel floor, rear swing doors with grain chute, roll tarp
Ford	F600	1975	G	$4,600	9/07	NWOH	Grain, V-8 engine, 5 speed, 49,950 miles
Ford	F600	1975	E	$10,000	8/07	NCSK	Canadian sale, custom cab single-axle grain, 391 V-8, 5×2 speed, 900×20 rubber, 15' steel box and hoist, Michael's electric side roll tarp, 41,620 miles showing
Ford	F600	1976	F	$850	4/07	NCNE	Harsh 12' mixer feeder, scales
Ford	F600	1976	G	$2,900	4/07	SEAB	Canadian sale, single-axle flat deck, 351 V-8, 5×2 speed, 16'×8' steel frame deck with wood floor, 850-gal. poly water tank with valves, 42,362 miles
Ford	F600	1976	F	$3,600	4/07	ECSK	Canadian sale, single-axle grain, 360 V-8 engine, 4×2 speed, 15' Westeel steel box with hoist, roll tarp, plumbed for drill fill, 9:00×20 tires
Ford	F600	1976	G	$6,300	2/07	SEIL	15' grain bed, hoist, 4×2 speed, PS, 9/20, 45,951 total miles
Ford	F600	1977	G	$5,300	12/07	SWNE	Cab and chassis, 360, 5 speed, 4×4, high and low range, front winch
Ford	F600	1977	G	$6,500	4/07	ECSK	Canadian sale, single-axle grain, 361 V-8, 5×2 speed, Westeel 16' box and hoist, roll tarp
Ford	F600	1977	G	$7,700	4/07	NCSK	Canadian sale, single-axle grain, 361 V-8, 5×2 speed, Westeel 14'×8.5' box with steel floor, roll tarp, 9.00-20 front and rear rubber (newer), plumbed for hyd., 95,617 kilometers

Trucks

Make	Model	Year	Cond.	Price	Date	Area	Comments
Ford	F600	1978	E	$4,400	12/07	WCIL	15' Knaphiede bed, twin hoist, 361 engine, 4×2 speed, 42K miles, 9.00-20 rear tires
Ford	F600	1978	G	$13,000	4/07	ECSK	Canadian sale, single-axle grain, 361 V-8, 5×2 speed, 14'×8'×42" box & hoist, 10.00×20 tires, 47,700 original kilometers
Ford	F600		G	$800	7/06	WCCA	Tank, V-8, 5×2 speed, 1,000-gal. tank, pump
Ford	F600		F	$1,800	5/06	SWCA	7.0L, 5×2 speed
Ford	F600		G	$2,750	5/06	SWCA	Dump, 7-8 yard box, gas engine, 5 speed
Ford	F600		G	$2,750	5/06	SWCA	Pressure washing truck, 12' bed, 4 speed, 500-gal. water tank, Landa 3000 steam pressure washer, 7-amp generator, toolboxes
Ford	F600	1953	F	$600	4/06	NWMN	Single axle, 50th Anniversary flathead V-8, 4×2 speed, wood 13' box, hoist
Ford	F600	1955	F	$400	4/06	NWMN	Single axle, 292, V-8, 4×2 speed, wood 13' box, hoist
Ford	F600	1959	F	$600	11/06	SENE	Grain, 16' Obeco combination wood box with 42" sides, twin-cyl. hoist, 292 V-6 gas engine, 5×2 speed
Ford	F600	1960	G	$3,000	4/06	WCMT	V-8 engine, 4×2 speed, 8.25-20 tires 14' box and hoist, 69,300 miles
Ford	F600	1961	F	$550	3/06	NWKS	Custom grain, 14' steel box, 6-cyl. gas engine, 4×2 speed, 8.25-20 tires, 112,164 miles
Ford	F600	1963	F	$100	6/06	WCMN	Single axle, 14' wood box, new engine
Ford	F600	1963	F	$450	7/06	NWMN	Single axle, flatbed
Ford	F600	1965	F	$800	4/06	NWMN	Single axle, 330, V-8, 4×2 speed, Knapheide 14' box, hoist, roll tarp
Ford	F600	1966	F	$1,000	7/06	NEND	Single axle, 331, 4×2 speed, 15' Knapheide box, hoist, 40,900 miles
Ford	F600	1966	G	$1,400	8/06	NEKS	95K miles, V-8, 4×2 speed
Ford	F600	1968	G	$1,300	3/06	NWMN	Single axle, V-8, 4×2 speed, 14' box with hoist
Ford	F600	1968	G	$2,000	2/06	NWIL	13.5' farm box, 330 V-8, 75K miles
Ford	F600	1969	G	$1,100	1/06	WCIL	Grain, V-8, 4×2 speed, 15' Knapheide bed, 95K miles
Ford	F600	1969	G	$1,250	3/06	NWKS	Grain 16' steel box, hoist, V-6 gas, 4×2 speed, 9:00-20 tires, under 70K miles
Ford	F600	1969	F	$5,000	8/06	SCMI	44,038 miles, 12' bed and hoist, 330 V-8 gas engine, 5 speed, single axle
Ford	F600	1970	G	$600	3/06	ECIN	Dump
Ford	F600	1972	F	$1,600	8/06	SCMI	63,353 miles, 12' bed & hoist, gas engine, 5 speed, single axle
Ford	F600	1972	G	$2,600	3/06	NEND	Tilt cab tag tandem, 361, V-8, 5×2 speed, 18' Frontier box, hoist, roll tarp, low miles on OH
Ford	F600	1972	G	$3,400	3/06	NEND	Single axle, 331, V-8, 4×2 speed, 15.5' Knapheide box, hoist, roll tarp, low miles on OH, 9:00×20 tires
Ford	F600	1972	G	$3,700	1/06	SCNE	Grain, 34,500 miles, single axle, 4×2 speed, Schwartz box with combination sides, 8.25-20 tires, set up for hyd. grain fill auger
Ford	F600	1973	G	$575	4/06	NCOK	V-8, 4 speed, flatbed bobtail, rebuilt motor 9K miles ago
Ford	F600	1973	G	$575	4/06	NCOK	V-8, 4 speed, flatbed bobtail, rebuilt motor 9K miles ago
Ford	F600	1973	G	$1,500	1/06	ECNE	Grain, 15.5' Obeco box, 5×2 speed, V-8 gas engine, PTO, cylinder hoist

Trucks

Make	Model	Year	Cond.	Price	Date	Area	Comments
Ford	F600	1973	G	$5,500	12/06	SCNE	16' box and hoist, 25K miles, one owner, very nice
Ford	F600	1974	F	$1,900	4/06	WCMB	Canadian sale, 2 ton
Ford	F600	1974	G	$2,500	4/06	NEND	Single axle, V-8, 4×2 speed, 15' steel box, with hoist, quick-tach roll tarp, 30,683 miles, 9:00-20 rubber
Ford	F600	1974	G	$3,750	3/06	SCIN	Contractor bed and hoist
Ford	F600	1974	E	$8,400	3/06	NCOK	108,494 miles, Mabar 18' bed & hoist, V-8, 4×2 speed, new trans.
Ford	F600	1975	F	$1,100	5/06	NENE	Wrecker truck, single axle, 4 speed, 5.9L V-8 gas, duals, Holmes Model 500 wrecker
Ford	F600	1975	G	$3,000	4/06	SEND	V-8, 2 speed, 15.5' Western box and hoist, 25K miles
Ford	F600	1976	G	$8,000	8/06	SCMN	21,075 miles, 16' Scott box and hoist
Ford	F600	1978	G	$1,800	4/06	WCSD	14' flatbed, 2 ton, 4×2 speed, 370 engine
Ford	F600		P	$150	6/05	ECWI	Service, gas, 5×2 speed, single axle
Ford	F600		P	$400	12/05	ECIN	Tandem-axle dump
Ford	F600		F	$750	3/05	NCAL	Dump
Ford	F600		G	$2,000	6/05	SWCA	Service, V-8, 5 speed
Ford	F600	1956	G	$900	1/05	NEMO	Grain
Ford	F600	1959	F	$325	6/05	NWMN	Single axle, 292, V-8, 13' steel box and hoist
Ford	F600	1960	G	$700	3/05	NECO	13' box, hoist, 291 engine, 4×2 speed, single axle
Ford	F600	1961	F	$800	9/05	WCMN	Single axle, 4×2 speed, 292, V-8, 13.5' box, 100K miles
Ford	F600	1963	F	$600	11/05	ECNE	13' wood box, 8.25-20 tires, 6-cyl. gas engine
Ford	F600	1964	G	$925	1/05	NENE	16' steel flatbed
Ford	F600	1966	G	$2,000	3/05	NWMN	Tag tandem, 17.5' steel box and hoist
Ford	F600	1967	F	$900	3/07	WCNE	Grain, 16' steel box, roll tarp, scissor hoist, 4×2 speed, 9:00-20 tires, engine compartment fire damage
Ford	F600	1969	G	$1,700	12/05	SCND	Single axle, box, hoist, roll tarp
Ford	F600	1971	F	$1,600	7/05	ECND	Single axle, 302, 4×2 speed, Frontier 14' box, hoist, roll tarp
Ford	F600	1971	E	$2,900	11/05	NCIN	15.5' Delphi bed and hoist, 28K miles, 4×2 speed, sharp
Ford	F600	1971	G	$3,500	2/05	SCKS	79,656 miles, 4×2 speed, V-8, Knapheide 16' grain bed
Ford	F600	1972	F	$525	11/05	NCIN	13' bed and hoist
Ford	F600	1973	G	$500	12/05	WCWA	Flatbed
Ford	F600	1973	G	$3,500	9/05	ECMN	Roadside sprayer
Ford	F600	1973	G	$4,100	3/05	NECO	20' box, hoist, 361 engine, 5×2 speed, tag, low miles on OH
Ford	F600	1974	G	$2,050	12/05	NCIL	Grain, 390 engine, 4 speed, 62,395 miles, Knapheide 15' box
Ford	F600	1974	G	$2,600	2/05	ECIL	Grain, 51,600 miles, 14' bed, 4×2 speed
Ford	F600	1974	G	$2,800	3/05	WCNE	Grain, 16' Omaha standard steel box, 360 gas engine, 4×2 speed, 9:00-20 tires, 50,290 miles
Ford	F600	1974	G	$3,200	2/05	WCIL	2 ton, 361 hd V-8 engine, 5×2 speed, dual hoists, stock racks, 16' bed, good rubber, gas
Ford	F600	1975	G	$1,400	11/05	NECO	Dump, 1 axle, gas, 4×2 speed
Ford	F600	1975	F	$1,700	3/05	NECO	16' steel box, single axle, split shift, 361 motor
Ford	F600	1975	E	$5,600	1/05	SENE	24,695 miles, 13' box, hoist
Ford	F600	1975	G	$6,750	8/05	WCIA	42K miles, 14' Obeco box

Trucks

Make	Model	Year	Cond.	Price	Date	Area	Comments
Ford	F600	1976	F	$2,100	12/05	NWIL	58,412 miles, 361, V-8, 4×2 speed, 15' Midwest grain box and hoist, roll tarp
Ford	F600	1976	G	$2,600	10/05	NEND	Single axle, 370 V-8, 4×2 speed, 16' box, hoist, roll tarp, plumbed for drill fill, 9:00×20 tires, 61,143 actual miles
Ford	F600	1977	F	$600	7/05	ECND	Single axle, 392, V-8, 5×2 speed, 14' flatbed, hoist
Ford	F650		G	$1,700	4/06	WCSD	16' box and hoist
Ford	F700	1965	G	$2,500	1/08	NWOH	10,700 miles
Ford	F700	1967	G	$2,850	1/08	SCKS	Single-axle grain, Midwest 16' steel box with 40" sides, 361 gas, 4×2 speed, 104,125 miles
Ford	F700	1973	G	$4,400	1/08	WCKS	Single-axle grain, Midwest 16' box, Ford 361, V-8 gas, 5×3 speed, 42,579 miles
Ford	F700	1978	F	$500	3/08	WCMN	Single axle, 5×2 speed, V-8
Ford	F700	1969	G	$6,750	1/07	SCNE	391 V-8, 20' steel box, roll tarp, cargo doors, 9:00 tires, 5×2 speed, drag tag
Ford	F700	1974	P	$100	1/07	WCIL	Cab and chassis only, 391, 5×2 speed, body rough
Ford	F700	1975	G	$10,000	2/07	ECNE	Tandem axle, 22' steel box, twin-cyl. hoist, 391 gas, electric over hyd. tag axle, dual fuel tanks, micro brake
Ford	F700	1977	G	$3,500	7/07	SEND	Single-axle pumper truck, 389 V-8, 5 speed, 10,726 miles
Ford	F700		G	$12,500	11/06	SCMN	Tag, 20' Crysteel bed, hoist, 85K miles
Ford	F700	1968	G	$2,700	4/06	NEND	Tilt cab, 5 speed, 330, V-8, Westeel Roscoe 16' box, hoist, roll tarp, 3-piece end gate
Ford	F700	1972	F	$2,600	4/06	SEMI	Single axle, gas, 330-cubic-inch engine, 5 speed, 16' Omaha box, hoist
Ford	F700	1974	F	$1,400	12/06	ECMO	Dump, 61,437 miles
Ford	F700	1975	G	$4,250	2/06	NWIL	15' farm box, new floor, 361, V-8, 5×2 speed, 64,650 miles
Ford	F700	1975	G	$4,600	11/06	SCNE	4×2 speed, 39,715 miles, 16' giant combination box and hoist
Ford	F700	1975	G	$6,200	3/06	ECND	Single axle, 370, V-8, 5×2 speed, Westgo 16.5' box, hoist, roll tarp, 46,300 miles, one owner
Ford	F700	1976	G	$3,250	9/06	NEIN	Septic waste truck, 1,750-gal. waste tank, 250-gal. fresh water tank, new clutch, single axle, 5×2 speed
Ford	F700	1976	G	$8,700	8/06	SCMN	20,493 miles, 16' Scott box and hoist
Ford	F700	1977	G	$5,500	2/06	ECIL	361 engine, 4×2 speed, 14.5' Midwest grain bed with hoist, roll tarp, 50,310 miles
Ford	F700	1978	G	$800	3/06	ECIN	Gas, single axle
Ford	F700	1978	F	$5,100	2/06	NCCO	Harsh 354 box, scales, 5×2 speed
Ford	F700		P	$400	12/05	SWMS	Single axle
Ford	F700		P	$1,150	2/05	WCOK	BJM mixer feeder, truck poor, mixer fair
Ford	F700		G	$2,600	3/05	NCAL	Dump
Ford	F700	1967	G	$2,700	2/05	WCIL	2 ton, 361 hd V-8 engine, 5×2 speed, dual hoists, good rubber, 16' bed
Ford	F700	1973	G	$9,700	4/05	SCKS	Custom 16.5' bed, Harsh twin-cyl. scissor hoist, 5×2 speed, 361 motor, 57,916 miles, one owner, good tires
Ford	F700	1978	G	$800	3/05	NCAL	Dump
Ford	F7000	1978	G	$15,500	3/07	SEND	Cat 318 diesel, 5×2 speed, service body complete with 800-gal. fuel tank, retractable hose reel and meter
Ford	F7000	1971	F	$1,000	3/06	ECNE	Cabover cab and chassis, Cat 3208 diesel, 5×2 speed, fifth-wheel plate, 24,000 GVW, recent OH, 173K miles, 295/75R22.5 tires

Trucks

Make	Model	Year	Cond.	Price	Date	Area	Comments
Ford	F750	1974	G	$5,000	3/08	WCKS	391 motor, 5×2 speed, 22' bed and hoist
Ford	F750	1975	G	$6,500	3/08	WCMN	Single axle, rear tag, 389 Ford gas engine, 5×2 speed, 20' Frontier grain box, roll tarp, aux. hyd. for drill fill, 61,545 miles
Ford	F750	1971	F	$900	12/07	WCMN	Grain, tandem tag, 18' box with hoist, 2 speed
Ford	F750	1975	G	$4,100	1/07	SCNE	391 V-8, 5×2 speed, 20' steel side box, 10:00 tires, air brakes, drag tag
Ford	F750	1973	F	$3,300	9/06	NWIL	V-8 engine, 5×2 speed, 14' metal grain box with cargo doors and hoist
Ford	F750	1974	F	$2,750	7/06	NWMN	Tag tandem, box, hoist
Ford	F750	1974	G	$7,000	7/06	NWMN	Twin screw, 390, 5×3 speed, 19.5' Buffalo box, hoist, roll tarp, 46,800 miles
Ford	F750	1975	G	$7,250	7/06	NWMN	Twin screw, 390, V-8, 5×3 speed, 20' Buffalo box, hoist, roll tarp, 52,941 miles
Ford	F750	1977	F	$900	11/06	SCIL	361 V-8, 15' bed, 44,800 miles
Ford	F750	1959	F	$2,000	3/05	NCAL	Fire truck
Ford	F750	1974	G	$5,700	2/05	SCKS	94,998 miles, cracked windshield, fender dent, 5×2 speed, V-8, Mabar 18' grain bed
Ford	F750	1975	G	$1,000	12/05	SWMS	Single axle, spread body
Ford	F800	1969	F	$1,500	4/08	NCND	Tandem, twin-screw grain, steel box, hoist, Shur-Lok roll tarp, 5×3 speed, 391 V-8 engine
Ford	F800		P	$400	2/07	SEFL	Water truck, 2,000 gal.
Ford	F800		G	$4,950	4/07	ECMA	Dump, 7.8L diesel, auto, 5-7 yard body
Ford	F800	1978	G	$16,000	2/07	SENE	Custom tandem-axle grain, Obeco 20' steel box with 54" sides, twin-cyl. hoist, (3) end gates, 7.8L gas engine, Eaton Fuller 13 speed, air brakes, dual fuel tanks
Ford	F800		G	$13,000	11/06	SWCA	Dump truck, 5-7 steel box, diesel, 7 speed
Ford	F800	1974	F	$600	12/06	NEMN	
Ford	F800	1974	G	$5,200	8/06	ECNE	Single-axle grain, 22' steel box with 58" sides, twin-cyl. hoist, roll tarp, 460 gas engine, 5×2 speed, air tag axle, 10:00-20 tires
Ford	F800	1978	G	$4,100	12/06	SEIA	Bulk feed,10-ton Krause feed body
Ford	F800	1978	G	$10,000	3/06	NEKS	Custom cab, single-axle grain truck, 59,022 miles, 18' steel box, 389 gas engine, 5×2 speed, dual fuel tanks, 10:00-20 tires
Ford	F800		G	$1,500	12/05	ECWA	Bucket, diesel, 5×2 speed, Telsta bucket
Ford	F800		G	$10,000	11/05	SWCA	Fuel lube, 3 axle, diesel, auto
Ford	F800	1972	G	$8,100	12/05	NCCO	18' end dump, 392, rollover tarp and beet gate, twin screw
Ford	F800	1977	F	$1,200	12/05	WCIN	Single-axle grain, 391, V-8, 17' wooden bed with steel floor, 98K miles
Ford	F800	1978	E	$15,600	3/05	SEMN	Gas grain, 5×4 speed, twin screw, 19.5' Crysteel box and hoist, saddle tanks, all new rubber with only 19,200 actual one-owner miles, very clean, DOT'd
Ford	F880	1975	G	$29,500	3/08	SCKS	Diesel, tandem axle, twin screw, 22' bed, twin-cyl. hoist, auto, 10:00×20 tires, 34,995 miles
Ford	F900	1972	G	$4,000	8/07	NENE	534 engine, 10 speed, 20' box and hoist
Ford	F9000	1971	G	$7,500	5/06	ECNE	Tandem-axle dump, Peabody 10-12 yard steel dump box, Cummins diesel engine, 13 speed, 54,000-lb. GVW, AM/FM radio, 10:00-20 recapped tires
Ford	L700	1972	G	$1,050	4/07	NCND	Grain truck
Ford	L750	1970	G	$2,200	7/06	ECND	Single axle, V-8, 4×2 speed,16' Westgo box, hoist

Trucks

Make	Model	Year	Cond.	Price	Date	Area	Comments
Ford	L750	1974	G	$6,500	3/05	SEND	Tag tandem, V-8, 5×2 speed, Westgo 19' diamond box, hoist, roll tarp, 56,990 miles
Ford	L8000	1970	F	$600	12/05	WCNJ	Dump truck, diesel, 10' steel body, single axle
Ford	L8000	1978	G	$4,500	1/05	NWGA	Cat, air locking differential, limited boom, air compressor
Ford	L880	1974	F	$5,200	3/06	NWMN	Tag tandem, 475, V-8, 5×2 speed, 19' box, hoist and roll tarp
Ford	L9000		F	$1,750	10/06	SWOH	Tandem axle, V-8 engine, two-stick 10 speed, 14' steel dump box
Ford	L9000		G	$3,500	1/06	WCPA	Fuel/lube, tandem-axle fuel truck, 4,000-gal. fuel tank
Ford	L9000	1977	F	$3,000	11/06	WCMN	Twin screw, 318 Detroit, 13 speed, 16' SteelCraft box, new LTL grill
Ford	L9000	1978	G	$4,500	6/06	SEID	3208 Cat, 9 speed, 12' dump bed
Ford	L9000		G	$5,500	6/05	SWCA	Cab, single axle, Detroit diesel
Ford	LN600	1974	G	$8,250	1/08	SCKS	Single-axle grain, 16' box, 330 gas, 4×2 speed, 36,546 miles
Ford	LN600	1970	G	$3,600	12/07	WCKS	V-8 motor, 4,000 miles on OH, 4×2 speed, 16' bed and hoist
Ford	LN600	1973	G	$4,300	1/07	SEKS	Single axle, 16' box with steel floor, 4×2 speed, 52,215 miles
Ford	LN600	1974	G	$3,700	2/07	WCNE	Newer engine, 5×2 speed, 16' steel box, 1' steel top extension, extra hyd.
Ford	LN6000	1972	F	$1,100	7/05	ECND	Cat, 5×2 speed, 14' flatbed
Ford	LN700	1978	P	$400	12/07	SWNE	John Deere 716A feed box
Ford	LN700	1970	F	$1,300	3/06	NWIL	V-8 engine, 5×2 speed, 13.5' grain box, hoist, miles unknown
Ford	LN700	1974	G	$1,800	4/06	SCKS	16' bed and lift, new 361 engine, 5×2 speed
Ford	LN700	1975	F	$850	3/06	NWKS	16' steel box, 5×2 speed
Ford	LN700	1975	G	$2,100	4/06	WCKS	Tip tops, Allison, 20,000 on OH
Ford	LN700	1971	G	$4,850	1/05	SCNE	110K miles, 5×2 speed, Tradewinds 16' steel box/hoist, 50" sides, 361, V-8, single axle
Ford	LN700	1977	G	$7,000	2/05	WCIL	Grain, 53,609 miles, cheater axle, 14' box, Shur-Lok roll tarp, 361 V-8 engine, 5×2 speed, new clutch, dual gas tanks
Ford	LN7000	1972	G	$16,000	1/08	WCKS	5×2 speed, Oswalt 420 mixer box, scales
Ford	LN7000	1975	F	$1,300	6/07	ECND	Tag tandem, Cat 3208, 5×2 speed, 18' bag/bulk feed body with floor and discharge auger
Ford	LN7000	1974	G	$9,000	3/05	WCKS	Feed, 350 Harsh mixer bed & scales, $16K spent on reconditioning, feed box in excellent shape, truck fair
Ford	LN750	1973	G	$1,200	2/07	ECIL	Grain, 391 engine, Knapheide 14' bed, 78,300 miles
Ford	LN800	1978	G	$3,500	11/06	SCSD	Grain, 534 gas, 5×2 speed, tandem, 20' box with roll tarp
Ford	LN800	1978	G	$9,000	9/05	NECO	50K miles, 18' beet box, Harsh 2-way side and rear hoist, silage racks for side dump
Ford	LN8000	1972	G	$9,000	1/07	ECNE	Grain, Cat 1160 diesel engine, 5×4 speed, air brakes, 50-gal. step fuel tank, 54,000-lb. GVW, twin screw, Knapheide 19' all-steel stakeless grain body with Harsh hd. 2-cylinder scissor hoist (inside frame), 3-door rear cargo doors, Shur-Lok roll tarp
Ford	LN8000	1977	F	$5,300	5/06	SEND	Twin-screw dump truck, 3208 Cat, 13 speed, 15' gravel box, hoist

Trucks

Make	Model	Year	Cond.	Price	Date	Area	Comments
Ford	LN8000	1978	G	$10,000	3/05	NEND	Tag tandem, 5/2, 18' Knapheide box with 60" sides, double acting Knaphoist, roll tarp, plumbed
Ford	LN880	1976	G	$10,000	4/06	NWMN	Twin screw, 477, V-8, Allison auto, Buffalo 20' box, hoist, roll tarp, 3-piece end gate, 92,486 miles, 20K miles on trans. OH, 9:00-20 rears, 11×22.5 steering
Ford	LN880	1973	G	$6,000	6/05	ECNE	Dump, Caterpillar 1160 diesel engine, 7 speed, 44,800-lb. GVW, air brakes, PS, twin screw, 10:00-20 tires, (1) 50-gal. step fuel tank, flip hood, 2000 Bibeau Model MO 15' steel dump body with hyd. hoist
Ford	LN900	1973	P	$900	3/07	NWMN	Twin screw, V-8, 5×3 speed, 20' bed with hoist
Ford	LN9000	1976	G	$10,750	2/07	ECNE	LN9000SE, tandem-axle grain, Brehmen 22' box, 534 gas engine, 5×2 speed, air tag axle, 85,700 miles
Ford	LN9000	1977	G	$2,100	3/06	NECO	Cab and chassis, Cummins SC 299 engine, new 9-speed trans.
Ford	LN9000	1978	F	$14,200	2/06	NCCO	20' Peterson steel box, Detroit diesel, 13 speed, rebuilt engine
Ford	LNT800	1978	G	$9,250	6/07	ECND	Twin-screw tandem, 534 V-8, 5 speed with air shift, 2 speed, 20' box with hoist and roll tarp, recent manufacturing engine and new clutch
Ford	LNT9000	1978	F	$1,000	8/07	SWMN	855 Cummins diesel engine, manual trans., tandem axle, full screw, Willmar 16-ton three-compartment ss dry tender box, rear auger, all hyd., rollover tarp
Ford	Louisville	1977	P	$1,050	3/08	SEND	6V92 Detroit, 13 speed, for parts
Ford	LT9000		G	$2,500	10/06	SWOH	Tandem axle, day cab, Cummins diesel, road grader trans., aluminum fenders
Ford	LTS9000	1977	G	$11,500	3/05	ECMN	3406 Cat, tandem axle, Braco 130 log loader, grapple attach, hd flatbed
Ford	N/A	1973	G	$8,000	4/12	WCMN	Dump, Cat 3208, Allison auto, 3-speed backup, tandem axle
Ford	N/A	1959	G	$1,900	3/08	SEKS	2-ton grain, 14' wood bed and sides with hoist and new rear brakes, 73K miles
Ford	N/A	1963	F	$510	3/08	NWKS	16' box, hoist
Ford	N/A	1970	F	$300	3/08	ECSD	½ ton, auto, 360, 2WD
Ford	N/A	1970	G	$5,800	3/08	ECIL	1 ton, 4 speed, 360 engine, grain box with hoist, 26,252 miles
Ford	N/A	1971	F	$300	3/08	ECSD	¾-ton pickup, auto, 260, 2WD
Ford	N/A	1972	G	$5,300	3/08	ECIL	1 ton, 4 speed, 390 engine, grain box with hoist, 31,716 miles
Ford	N/A	1974	G	$3,800	3/08	NCTN	Grain
Ford	N/A	1975	F	$700	3/08	WCMN	¾ ton, 4×4, Ford 360 gas engine
Ford	N/A	1978	G	$3,100	1/08	SWKY	Lift axle tandem
Ford	N/A		P	$300	5/07	WCSD	Tandem axle, 20' box, hoist, not running
Ford	N/A	1922	G	$2,300	9/07	WCIA	1 ton, white
Ford	N/A	1946	G	$500	2/07	ECIL	2 ton, flatbed, hoist
Ford	N/A	1950	F	$775	4/07	NWIA	Antique F2, original
Ford	N/A	1968	F	$500	2/07	ECIL	Grain, 16' Midwest bed and hoist
Ford	N/A	1972	G	$1,100	3/07	NCIA	Grain, 14' wood box & floor, 330-cubic-inch engine, 4×2 speed, 20,000 GVW
Ford	N/A		F	$800	9/06	NEIN	Semi tractor, smokes on start, diesel
Ford	N/A	1947	E	$1,300	3/06	NCOK	52,611 miles, straight 6 cyl., 4 speed, 14' bed and hoist, one-way hoist
Ford	N/A	1962	F	$500	2/06	WCKS	292, V-8 motor, 4 speed, 15.5' bed and hoist
Ford	N/A	1970	F	$300	6/06	NWMN	½ ton, 4×4, service body, 300, 6 cyl.
Ford	N/A	1976	F	$600	6/06	NWMN	Box and hoist

Trucks

Make	Model	Year	Cond.	Price	Date	Area	Comments
Ford	N/A	1977	F	$100	7/06	WCOH	Crew cab, 93,101 miles
Ford	N/A		P	$300	8/05	NCIA	Econoline pickup, for parts or to restore
Ford	N/A		F	$425	8/05	NEIN	400 bu.
Ford	N/A	1952	F	$550	5/05	NCKS	13.5' bed, hoist
Ford	N/A	1974	F	$1,650	2/05	WCOH	250 bu., 5×2 speed, dual hoist, 16' grain bed, 32,435 miles
Ford	N/A	1975	P	$1,700	3/05	NECO	Feed truck, 82K miles, 5 speed, 390 motor, Mohrlang 15' mixer box, good flighting, poor skin
Ford	N/A	1975	G	$3,200	1/05	SCMN	Fuel truck
Ford	N/A	1977	G	$6,250	9/05	ECMN	Water truck
Ford	N/A	1978	F	$700	3/05	WCMN	1-ton dually 8×12 flatbed, ag hitch
Ford	N600	1969	E	$4,750	4/05	NCOK	V-8, 4×2 speed, 16' Mabar bed & hoist, 36" sides, Poorboy rollover tarp, in family since new!
Ford	N700	1975	F	$1,250	7/07	SEND	Fuel truck, 5 compartment, rebuilt engine
Ford	Ranger	1978	P	$450	7/05	NEIA	300 6-cyl. engine, 81K miles
Freightliner	N/A	1971	P	$500	2/07	SCNE	Cabover, 335 Cummins diesel, 13 speed, tandem axle, full screw, air slide fifth-wheel plate, 11R22.5 tires
Freightliner	N/A	1970	F	$700	4/06	ECND	Cabover, 250 Cummins, 13 speed, spring ride
Freightliner	N/A	1972	P	$175	11/06	NWKS	For parts, cabover
Freightliner	N/A	1978	F	$1,000	3/06	NCNE	Cabover, Cat 3406 diesel engine (knocks), 13 speed, rebuilt clutch, air ride, aluminum headache rack, sliding fifth wheel, dual fuel tanks, air, 11R24.5 tires
GMC	1500	1972	G	$2,100	6/07	SCSK	Canadian sale, 350 auto, 3 speed, 113,038 miles showing, approx. 15,000 miles on new engine
GMC	2500	1970	F	$400	3/06	NWKS	1-ton dually service truck, V-6 gas engine, auto, steel utility box, 105,156 miles
GMC	2500	1974	P	$250	3/06	NWKS	Super custom 2500, ¾ ton, 4×4, auto, towbar, fuel tank, one owner, trans. weak
GMC	350	1954	P	$900	3/07	WCSD	13.5' box and hoist, for salvage
GMC	370	1958	F	$1,250	4/07	NWMN	Short nose, tag axle tandem, 389 V-8, 5×2 speed, 17.5' box with hoist
GMC	4000	1966	G	$1,000	3/07	WCIL	12' box, V-6, gas
GMC	4000	1966	E	$1,900	11/06	NEKS	67K miles, dump, almost new tires
GMC	4000		F	$1,100	3/05	SEND	Single axle, V-8, 4×2 speed, poly tank, mix cone and pump
GMC	5000	1960	F	$575	3/08	ECNE	Single-axle truck tractor
GMC	5500		F	$200	1/06	WCCA	Flatbed
GMC	5500	1969	P	$1,600	7/06	SEND	Single axle, 15' steel box with hoist, 4-speed split trans.
GMC	5500	1970	G	$1,000	6/06	SEID	V-6, 5×2 speed, 18' flatbed, single axle
GMC	5500	1972	G	$2,200	1/06	SCNE	Grain, 88,500 actual miles, Omaha standard 13.5' box and hoist, 47.5" sides, box extensions, 4×2 speed, V-8, good 9.00-20 tires
GMC	5500	1976	F	$500	9/06	WCND	Twin screw, V-8, auto, 20' French combination box, electric
GMC	5500	1968	G	$800	12/05	SEMN	Grain
GMC	5500	1970	G	$1,850	12/05	NCOH	14' New Leader lime bed, cheater axle, 40K miles, 350 engine
GMC	5500	1971	F	$1,700	4/05	NCOK	Single rear axle, 15.5' steel bed with 40" sides, rollover tarp, 4×2 speed
GMC	5500	1972	F	$1,850	3/05	NECO	20' box, 5×2 speed, tag axle, 350 motor
GMC	5500	1972	G	$4,300	2/05	ECIL	Grain, 62,300 miles

Trucks

Make	Model	Year	Cond.	Price	Date	Area	Comments
GMC	600	1974	F	$1,100	3/08	NWKS	Single-axle grain, Knapheide 16' steel box, single-cylinder hoist, tarp, 366 gas engine, 4×2 speed, 9:00-20 tires, 47,503 miles
GMC	6000	1973	F	$1,900	1/07	NWIN	Knapheide 15' bed and hoist
GMC	6000	1974	G	$5,200	7/07	SEMN	Single-axle grain, 350 V-8, 4×2 speed, 16' wooden box and hoist, 56,338 miles
GMC	6000	1976	G	$2,750	2/07	NWKS	Sierra, single-axle grain, 16' steel box with 40" sides, twin-cyl. hoist, 366 V-8 gas engine, 4×2 speed, 9:00-20 rear tires, 8.25-20 front tires, 36,303 miles
GMC	6000	1973	F	$800	11/06	NCOH	Spray, 1,000-gal. ss tank and booms
GMC	6000	1973	G	$4,000	3/06	SWNE	350, V-8, 4×2 speed, 43,453 miles, 18' steel box
GMC	6000	1975	G	$7,000	10/06	ECSK	Canadian sale, Sierra single-axle grain, 350 V-8, 4×2 speed, 15'×8'×42" steel box with steel floor, Shur-Lok roll tarp, 9.00×20 tires, saddle tank, drill fill plumbing, 78,550 miles, shedded
GMC	6000	1976	G	$3,400	4/06	ECND	Sierra Grande, 350, V-8, 4×2 speed, 16' Scott box, hoist
GMC	6000	1977	F	$2,700	2/06	SWNE	Cab and chassis, V-8, 5×2 speed, single axle
GMC	6000	1977	G	$8,100	3/06	SWKY	V-8, grain bed and dump
GMC	6000		F	$430	3/05	SCMI	Gas, single axle, 14' flatbed
GMC	6000	1973	F	$1,375	9/05	SEIN	15' Omaha bed, hoist
GMC	6000	1974	G	$5,350	1/05	WCIL	14' farm box, 350 V-8 engine, 5×2 speed, 42,112 miles
GMC	6000	1975	G	$2,100	7/05	WCMN	Grain, 4×2 speed, 38,482 miles, recent OH, single axle, hyd. plumbed, 16' steel box, roll tarp
GMC	6000	1975	P	$2,600	8/05	WCMN	Grain, 16' steel grain box, 31,500 actual miles
GMC	6000	1977	G	$7,500	12/05	ECIL	Grain, 14' Schien box, 350, V-8, 5×2 speed
GMC	6000	1978	G	$2,950	4/05	SENE	82,250 miles, gas, 16' box, hoist
GMC	6500	1973	G	$14,500	3/08	WCMN	Tandem tag axle, 366, V-8, 5×2 speed, 18' steel box, 58K actual miles
GMC	6500	1975	F	$8,750	4/08	NWMN	Twin screw with third axle pusher, 427, 5×4 speed, 20' strong box, hoist, roll tarp, 3-piece end gate, beet equipped, 122K miles
GMC	6500	1977	G	$15,500	3/08	NWKS	Tandem-axle grain, 20' steel box, 366, V-8, 5×2 speed
GMC	6500	1978	G	$6,200	2/08	WCIL	427, 5×2 speed, 15' bed, tilt hood, air brakes, 63K miles
GMC	6500		F	$5,400	6/07	SCSK	Canadian sale, grain, 366 V-8, 5×2 speed, 9.00×20 front tires, 10.00×20 rear tires, 8.5'×16' box and hoist, Univision roll tarp, 63,500 miles
GMC	6500		G	$17,000	11/07	NEND	Sierra Grande tandem-axle twin screw, 427, 5×4 speed, 20' steel box, Shur-Lok roll tarp
GMC	6500	1971	G	$6,100	1/07	WCIL	Tandem axle, 18' grain box, 108K miles, 427 gas engine
GMC	6500	1973	G	$10,750	1/07	SENE	5×2 speed, 20' steel
GMC	6500	1975	G	$3,500	12/07	ECNE	Single-axle grain, 18' wood box, twin-cyl. hoist, bale rack, 366 V-8 gas, 5×2 speed
GMC	6500	1975	G	$4,250	4/07	SCKS	Big wheel, air brakes, 427 gas, Allison auto, liquid fertilizer spreader
GMC	6500	1975	G	$7,800	12/07	SCNE	366 engine, 5×2 speed, steel side box, near-new hoist, 63,100 miles

Trucks

Make	Model	Year	Cond.	Price	Date	Area	Comments
GMC	6500	1975	G	$8,250	4/07	NEND	Twin-screw tandem, 427 V-8, 5×4 speed, 20' box with hoist and roll tarp, beet equipped
GMC	6500	1976	G	$10,000	11/07	SEID	Tender truck, V-8, 5×3 speed, PTO, wet kit, 18' Agri-Box seed and fertilizer hoppers
GMC	6500	1976	G	$10,000	11/07	WCID	V-8, 5×3 speed, PTO, wet kit, 18' Agri-Box seed, fertilizer hoppers
GMC	6500	1976	G	$11,500	4/07	ECSK	Canadian sale, High Sierra single-axle grain, 366 V-8, 5×2 speed, 10.00×20 rubber, 15' steel box with steel floor, Michael's roll tarp, 21,212 original miles
GMC	6500	1977	G	$9,000	3/07	SCSK	Canadian sale, single-axle grain, 366 V-8, 5×2 speed, 15'×8.5'×42" Univision steel box, roll tarp
GMC	6500		F	$1,200	4/06	NWSD	Dump
GMC	6500	1967	G	$7,000	10/06	ECSK	Canadian sale, single-axle grain, 366 V-8, 5 speed, 16'×8.5'×42" steel box, 900×20 rubber, drill fill plumbing, step tank, 58,620 kilometers
GMC	6500	1970	G	$4,500	6/06	ECND	Single axle, 366, 5×2 speed, new engine 3 years ago, 15.5' Omaha box, hoist, roll tarp, 158,887 miles
GMC	6500	1972	P	$200	12/06	NEMO	Grain truck, 5×2 speed, decent tires
GMC	6500	1973	G	$2,700	2/06	WCME	24' Sylvin bulk body
GMC	6500	1974	G	$4,300	2/06	ECKS	
GMC	6500	1975	P	$1,600	10/06	SESK	Canadian sale, single-axle grain, 5×2 speed, 366 V-8, 11R22.5 rear tires, steel box, hoist, Challenger side roll tarp
GMC	6500	1975	G	$4,000	2/06	NCOH	Gas, 4×2 speed, 14' grain bed, three-stage twin-cyl. hoist
GMC	6500	1976	F	$300	2/06	WCME	24' custom, built cover top steel bulk body, 366 gas, 5×2 speed
GMC	6500	1976	G	$1,400	2/06	WCME	20' Dylvin bulk body
GMC	6500	1977	F	$2,550	1/06	WCCA	Fire truck
GMC	6500	1977	G	$14,500	4/06	ECND	Lift tag trandem 427, auto, 20' Buffalo box, twin-post hoist, 3-piece end gate, roll tarp, twin fuel tanks
GMC	6500	1977	G	$16,000	12/06	SCNE	Sierra Grande, 366 engine, 5×2 speed, 20' steel side box, hyd. tag axle, 67,156 one-owner miles
GMC	6500	1977	G	$17,500	3/06	ECNE	Grain, 427, 13 speed, 20' steel box, roll tarp, 39,600 miles
GMC	6500	1978	G	$7,000	11/06	SCNE	5×2 speed, 366, V-8, Midwest 18' steel box and hoist, 58,164 miles
GMC	6500	1978	G	$9,000	12/06	SCNE	427 engine, 5×2 speed, 20' steel side box, air brakes, air tag axle, cargo doors, roll tarp, 10:00×20" tires, 140K miles
GMC	6500		F	$800	9/05	NWOR	V-8 gas, steel flatbed
GMC	6500		G	$15,000	5/06	SWCA	Bobtail dump, 5-6 yard, Caterpillar 3126 diesel, Allison trans.
GMC	6500		G	$15,000	5/06	SWCA	Bobtail dump, 5-6 yard box, Cat 3126 diesel
GMC	6500	1974	F	$5,500	2/05	NWOH	16' grain bed and hoist, 366, 5×2 speed, roll tarp
GMC	6500	1974	G	$10,200	3/05	WCMN	Twinscrew, 366, V-8, 5×4 speed, 20' J-Craft box and hoist, roll tarp, new motor 2003
GMC	6500	1975	E	$9,500	1/05	ECNE	2 ton, 427 V-8, 5×2 speed, 22' box with twin-cylinder hoist, stock rack
GMC	6500	1977	G	$10,700	1/05	WCIL	15' bed, 48K miles

Trucks

Make	Model	Year	Cond.	Price	Date	Area	Comments
GMC	6500	1978	G	$4,750	11/05	ECWA	366 gas engine, 5×2 speed, air brakes, 9:00-20 rubber, pintle hitch,16' bed, Allied power up/down hoist, 40" grain racks
GMC	7000		F	$650	4/08	ECMN	V-8 fuel, 5×2 speed, 1,200-gal. 2 pumps and 2 hose reels
GMC	7000		G	$2,500	5/06	SWCA	Flatbed dump, 18' bed, headache rack, V-8, 5×2 speed
GMC	7000		G	$4,500	12/05	ECWA	5-yard dump, Detroit 8.2L, 5×2 speed
GMC	7000		G	$5,500	6/05	SWCA	Equipment carrier, 20' bed diesel
GMC	7500	1967	G	$1,550	1/07	ECIL	Grain, tandem axle, twin screw, 18' bed, 5×3 speed
GMC	7500		G	$5,000	12/05	WCCA	Water, Detroit 8.2L, 10 speed, PTO pump
GMC	7500	1967	F	$1,700	11/05	ECNE	20' wood box
GMC	9500	1956	G	$900	7/07	NCSK	Canadian sale, single-axle grain, 6 cyl., 4 speed, 12' steel box with wood floor
GMC	9500	1970	G	$4,800	1/07	ECIL	Grain, tandem axle, twin screw, 18' bed, 10 speed
GMC	9500	1973	F	$1,100	3/07	WCNE	Flatbed, Detroit 671 diesel engine, 10 speed, hoist, 1,500-gal. poly tank with transfer pump, 300-gal. fresh water tank, 10:00-20 tires
GMC	9500	1970	G	$2,500	3/06	NCIL	Detroit 6-71, 13 speed, tandem axle with 14' spreader box, hyd. drive
GMC	9500	1971	G	$8,000	3/06	NCND	Tag tandem, 671 Detroit, 10 speed, Lodeline 20' box, tip tops, roll tarp, hoist, 3-piece end gate, 10:00-20 rubber all around
GMC	9500	1974		$0	12/05	SEND	Cabover twin screw, 318 Detroit, 13 speed, 674K miles
GMC	9500	1974	F	$2,500	12/05	ECMI	20' box, hoist, tandem axle, 318 Detroit, 10 speed
GMC	Brigadier	1977	G	$12,500	7/06	NWMN	Twin screw, 6V92 Detroit, 10 speed, PS, Frontier 20' box, hoist, roll tarp
GMC	Brigadier	1978	G	$3,500	4/06	ECND	739,649 miles, 400-gal. ss water tank, Honda water pump, air operated discharge valves
GMC	Brigadier	1978	G	$3,500	4/06	ECND	
GMC	C50	1971	F	$1,000	3/06	ECNE	Grain, 18' wood box, hoist, 327 gas engine, 4×2 speed, needs carburetor work
GMC	C60		G	$1,675	12/05	NCIL	350 engine, 4 speed, fertilizer spreader truck
GMC	C6500	1977	G	$3,800	12/06	WCIL	16' box, new tarp, gas
GMC	General	1978	G	$12,500	2/07	NENE	Tandem, Detroit 6V92 engine, air lift third axle, Scott 20' box and hoist, Allison trans., good rubber
GMC	General		G	$6,000	5/06	SWCA	Dump, 12-15 yard box, Detroit diesel, 9 speed
GMC	General	1978	F	$650	8/05	SEMI	5 Star General, 21' potato bed, 350 Cummins, 13 speed
GMC	K2500	1975	G	$1,400	3/06	SWNE	¾-ton pickup, auto, 4×4, V-8, flatbed with headache rack
GMC	N/A	1940	F	$2,200	5/08	ECMN	Open air fire truck, pumper with lights & sirens, 10,249 actual miles, no brakes
GMC	N/A	1973	G	$1,500	3/08	ECSD	1-ton, flatbed, 450-gal. tank and pump
GMC	N/A	1973	G	$5,600	1/08	ECMI	Tandem, hoist
GMC	N/A	1973	G	$8,250	2/08	ECMT	Conventional, 671 Detroit, 13 speed, Tradewind box 20'×96"×60", Shur-Lok roll tarp, twin screw, 10.00×20 tires
GMC	N/A	1977	G	$3,500	1/08	ECMI	General semi tractor, conventional cab, 318 Detroit, 13 speed, twin screw, wet kit, 392K miles
GMC	N/A	1978	G	$4,000	1/08	SWKY	Dump

Trucks

Make	Model	Year	Cond.	Price	Date	Area	Comments
GMC	N/A		G	$1,600	6/07	WCMO	Road Boss, Cummins diesel engine, 9 speed, full screw, tandem axle, 12,000 front axle, air brakes, Texoma guard post driver bed, self-contained 4-cylinder gas on turntable
GMC	N/A	1939	G	$40,000	9/07	NEIL	AFR Series cabover, SN 752339
GMC	N/A	1949	F	$650	9/07	NEND	¾ ton
GMC	N/A	1966	P	$225	3/07	SCNE	Flatbed, V-6
GMC	N/A	1966	P	$275	2/07	NENE	Flatbed, V-8, does not run
GMC	N/A	1970	P	$750	3/07	WCSD	18' box for salvage
GMC	N/A	1971	F	$250	7/07	WCSD	2WD, ½ ton, oil tank
GMC	N/A	1972	G	$6,500	2/07	NENE	Single-axle cab and chassis, 360 engine, 18' wood box, roll tarp, 5×2 speed, 45,729 miles
GMC	N/A	1973	F	$900	3/07	NEMI	Detroit 318 engine, day cab, wetlines
GMC	N/A	1973	G	$3,800	4/07	NEAB	Canadian sale, 350 V-8, 4×2 speed, VM 15' wooden box, hoist, 9.00×20 tires
GMC	N/A	1974	G	$1,250	1/07	WCNY	Rollback
GMC	N/A	1974	G	$2,900	3/07	SECO	2 ton, bed, hoist
GMC	N/A	1974	G	$6,100	11/07	WCKS	5,000 miles on rebuilt motor, 671 Detroit motor, 13 speed, 20' bed, hoist, twin screw, rollover tarp
GMC	N/A	1978	G	$3,600	2/07	NENE	Tandem straight, 20' box, new Scott twin-cyl. hoist
GMC	N/A	1952	G	$2,200	3/06	NEND	6×6 Army, GM 292 inline 6 cyl., Allison 4 speed, Pleasure Products 2,500-gal. tank
GMC	N/A	1956	G	$3,000	8/06	NEKS	½ ton, 4-speed manual, 65,746 miles, one owner
GMC	N/A	1960	F	$700	3/06	SEND	Single axle, V-6, 4×2 speed, 16' box, hoist, 8:25×30 tires
GMC	N/A	1962	F	$1,500	1/06	NCCO	18', side hoist, tag axle
GMC	N/A	1966	P	$325	3/06	WCMI	Live tandem truck chassis, Detroit 671 diesel, 3 speed aux.
GMC	N/A	1967	F	$2,500	7/06	NWOH	Tandem grain, V-6 engine, double stick trans., Omaha bed/hoist, good bed
GMC	N/A	1968	F	$300	9/06	ECNE	14' steel box and hoist
GMC	N/A	1969	G	$4,100	11/06	WCMN	½ ton, small block V-8, air, PS, tinted glass, power brakes, auto, push button radio, no-glare mirror, original unrestored except paint, 130K miles
GMC	N/A	1972	G	$3,400	2/06	WCIL	Tandem grain, 20' bed, tag tandem, 5×2 speed, 125K miles
GMC	N/A	1973	G	$1,000	11/06	SWIA	¾-ton pickup, auto, flatbed with hd hitch and pull hitch
GMC	N/A	1973	F	$1,900	5/06	NENE	Truck & frame, 671 Detroit diesel, 13-speed Roadranger, heavy front axle
GMC	N/A	1977	F	$600	3/06	SWOH	½ ton, 127K miles, auto
GMC	N/A	1952	P	$200	4/05	NCOK	Not running, wood bed, #6100
GMC	N/A	1952	G	$3,750	12/05	WCWA	Boom, 6×6, Pitman hyd. crane
GMC	N/A	1953	F	$425	6/05	NEND	½ ton
GMC	N/A	1963	F	$550	6/05	NEND	Stepside, ¾ ton, V-6
GMC	N/A	1967	F	$1,200	9/05	SEIA	1 ton, dump
GMC	N/A	1969	G	$1,950	1/05	SENE	2 ton, box, hoist
GMC	N/A	1969	G	$15,000	10/05	SWSD	Farm truck, tag axle, 366, V-8, 20' Plains box with twin cyl., 3 stage, SRT roll top tarp
GMC	N/A	1971	G	$3,750	11/05	NCCA	2 axle
GMC	N/A	1971	G	$3,750	11/05	ECWA	321 V-6 gas engine, 5×2 speed, 9:00-20 rubber, pintle hitch, 16' bed, Allied power up and down hoist, 40' grain racks
GMC	N/A	1972	F	$900	11/05	NWIL	1 ton, grain box, good tires, 100K miles
GMC	N/A	1972	G	$7,750	3/05	ECID	Service

Trucks

Make	Model	Year	Cond.	Price	Date	Area	Comments
GMC	N/A	1973	G	$2,800	2/05	NWKS	16' bed and hoist, 5×2 speed, roll tarp
GMC	N/A	1976	G	$3,100	9/05	NCIA	366, V-8, 4×2 speed, Obeco 14' steel box with hoist, roll tarp, 109K miles
GMC	N/A	1978	F	$1,050	7/05	ECND	Single axle, V-8, 5×2 speed, 16' flatbed, fold-down sides
GMC	N/A	1978	G	$8,100	12/05	ECIL	Grain, 14' Schien box, 366, V-8, 5×2 speed, 57,027 miles
GMC	Sierra	1976	G	$2,000	11/07	NEND	Sierra Grande 350, auto, 4WD, 12K miles on engine
GMC	Sierra	1975	G	$5,800	2/06	WCIL	Grain 6000 V-8 engine, 4×2 speed, 13' bed, 78,172 miles
GMC	Sierra	1977	G	$1,000	3/06	NCWI	V-8, 4 speed, duals, flatbed with hoist
GMC	Sierra	1975	G	$15,500	7/05	ECND	Twin screw, 427, Browning 5×4 speed, 20' Knapheide
GMC	Sierra 15	1976	P	$160	3/06	SWOH	Rough
GMC	Sierra 35	1975	F	$1,100	5/05	SWIN	1 ton, hoist, grain bed, 350, 4 speed
Graham Bros.	N/A	1927	G	$31,000	9/07	NEIL	1-ton express truck, restored, 26K miles
Hendrickson	N/A	1974	G	$2,000	6/05	ECWI	Cat diesel, 13 speed
IHC	1100	1966	G	$3,500	10/06	SCSK	Canadian sale, 1 ton, 4×4, 264 6 cyl., 4 speed, good steel deck, fifth-wheel hitch, 51,550 miles
IHC	1100	1966	G	$3,500	10/06	SCSK	Canadian sale
IHC	1200	1968	P	$250	8/07	NWOH	Pickup, 2WD, V-8, 4 speed, 67K miles, rusty
IHC	1300	1967	G	$1,300	10/06	SCSK	Canadian sale, 1 ton, 4 speed, 7.00-17 light rubber, wooden box and hoist, 46,433 miles
IHC	1500	1966	G	$3,400	4/07	NCSK	Canadian sale, single-axle grain, 6 cyl., 4×2 speed, 12'×8' wood box, hoist, 7.50×20 tires
IHC	160	1959	G	$450	1/07	ECIL	Grain bed, 44,450 miles
IHC	1600	1967	G	$1,500	3/08	ECSD	Silage, new clutch, V-8, 4×2 speed, Schweiger's deluxe box
IHC	1600	1968	G	$3,750	1/08	NWOH	Loadstar, grain, hoist
IHC	1600	1970	F	$850	3/08	NCMI	Loadstar single-axle grain, hyd. wood grain box, V-8 gas
IHC	1600	1975	G	$4,750	3/08	WCIL	Grain, 13.5' Knapheide bed, 4×2 speed, 9.00-20 tires, 23,043 miles
IHC	1600	1977	E	$7,000	2/08	NCIN	Loadstar grain, 15' bed and hoist, fold-down stock racks, V-8, 4×2 speed, 35,400 miles
IHC	1600	1978	G	$3,350	3/08	NEKS	Loadstar, 2 ton, 18' bed, hoist, 5×2 speed
IHC	1600	1978	G	$8,100	3/08	ECND	Loadstar single axle, 345, 4×2 speed, 16' Omaha box, hoist, roll tarp, 28,739 miles, one owner
IHC	1600			$0	2/07	WCIL	15' bed
IHC	1600	1963	F	$650	4/07	ECND	Single axle, 304 engine, 4×2 speed, Westgo 15' box, hoist, roll tarp
IHC	1600	1966	F	$1,250	7/07	SEND	Loadstar, box, hoist
IHC	1600	1966	F	$1,700	12/07	ECNE	Tanker fire truck, 4 speed, 345-cyl. engine, 1,250-gal. Luverne tanker body
IHC	1600	1967	G	$900	3/07	WCIL	Good box, 35K miles on OH, gas
IHC	1600	1968	F	$800	12/07	ECNE	4×2 speed, 1,000-gal. poly liquid fertilizer tank, 5-hp. 2" Banjo pump, 3" fill valve, AM/FM, 100,370 miles
IHC	1600	1968	G	$3,100	8/07	WCSD	Loadstar, Omaha standard wood combination box, hoist
IHC	1600	1970	P	$325	8/07	WCMN	Loadstar, single axle, 5×2 speed, V-8, box and hoist
IHC	1600	1970	G	$1,200	3/07	WCNE	Fuel, 4×4, 345 gas engine, 4 speed, 1,500-gal. tank with pump, 8.25-20 tires

Trucks

Make	Model	Year	Cond.	Price	Date	Area	Comments
IHC	1600	1970	G	$3,700	7/07	SEMN	Loadstar, single-axle truck, 345, V-8, 5×2 speed, Obeco 16' steel box and hoist, like-new rubber
IHC	1600	1973	P	$500	11/07	ECKS	Loadstar 1600 grain, 13.5' box, has engine and trans. but not installed in truck, will sell with truck, hood and pieces included but ultimately will sell as is
IHC	1600	1973	F	$800	8/07	ECIL	Loadstar grain, 13.5' box, twin cyl., 144,848 miles
IHC	1600	1973	F	$1,600	3/07	SCNE	Loadstar 1600 truck, V-8., 4×2 speed, 15.5' box and hoist
IHC	1600	1974	F	$2,500	12/07	WCMN	Loadstar single axle, 15' steel box and hoist, 23,300 actual miles
IHC	1600	1975	G	$2,100	2/07	WCIL	15' bed, good paint
IHC	1600	1975	G	$3,500	3/07	SCNE	Loadstar, 35,735 miles, 5×2 speed, Midwest steel 16' combination box with hoist
IHC	1600	1976	G	$4,300	6/07	ECND	Loadstar, single axle, 345, V-8, 4×2 speed, 16' Westgo box, hoist, roll tarp, 20,000 actual miles
IHC	1600		F	$800	4/06	WCWI	Loadstar, hoist, gear box
IHC	1600		G	$1,000	11/07	WCIL	Loadstar, V-8 engine, 4×2 speed, 81,700 miles
IHC	1600	1966	F	$500	3/06	SEMN	Loadstar, good motor, trans, rear end with 20' box and hoist
IHC	1600	1966	G	$1,100	6/06	NWMN	Single axle, 15' steel box hoist, low mileage
IHC	1600	1967	F	$4,000	10/06	ECSK	Canadian sale, Loadstar single-axle grain, 304 V-8, Michael's roll tarp, CIM 14'×8.5'×48" steel box, 9.00×20 tires, 97K miles
IHC	1600	1968	F	$200	8/06	ECNE	Loadstar grain, Schwartz 14' wood box, twin-cyl. hoist, 345 gas engine, 8.25-20 tires
IHC	1600	1968	F	$650	2/06	SEIA	Loadstar, 4×2 speed, 302 engine, 13.5' grain bed
IHC	1600	1969	F	$1,800	12/06	ECIL	Loadstar, grain bed, shedded, needs clutch work
IHC	1600	1969	G	$2,600	11/06	NCOH	Loadstar, 304 gas, 5×2 speed
IHC	1600	1972	F	$500	12/06	WCMI	Loadstar, 12' box, hoist, 54K+ miles, gas, single axle
IHC	1600	1972	G	$4,500	3/06	NENE	Grain, 19' steel box, 48,980 miles, twin-cyl. hoist
IHC	1600	1973	G	$1,600	4/06	NEND	Loadstar single axle, 345, V-8, 4×2 speed, 15' Knapheide box, hoist, roll tarp, 18,491 miles, plumbed for drill fill
IHC	1600	1973	G	$2,300	4/06	WCSD	Loadstar, Schwartz comb. box, single axle, 4×2 speed
IHC	1600	1973	G	$2,300	7/06	NEND	Loadstar, single axle, 345, 4×2 speed, Knapheide 15' box, hoist, roll tarp, plumbed for drill fill, 77K miles
IHC	1600	1973	G	$2,500	2/06	ECIL	Grain, 13.5' bed, 5×2 speed, 93,164 miles
IHC	1600	1973	G	$3,600	2/06	NWSD	Loadstar, single axle, 16' steel comb. box, plumbed for hyd. drill fill
IHC	1600	1974	F	$800	3/06	SEMN	Loadstar, good 16' box and hoist, truck not running
IHC	1600	1974	G	$1,800	7/06	NEND	Loadstar single axle, 345, 4×2 speed, 15' Knapheide box, hoist, roll tarp, 58K miles
IHC	1600	1974	G	$2,150	2/06	NCIN	Grain, 16' bed and hoist, single axle, PS, V-8, 5×2 speed
IHC	1600	1974	G	$4,000	3/06	NCND	Loadstar single axle, 345, V-8, 4×2 speed, 15' Knapheide box, hoist, 50,945 miles
IHC	1600	1974	G	$7,000	2/06	ECKS	16' bed and hoist

Trucks

Make	Model	Year	Cond.	Price	Date	Area	Comments
IHC	1600	1975	G	$6,300	2/06	SCMI	Loadstar, straight, gas, 5 speed, single axle, 60K miles, Kilbros 390 box, tarp
IHC	1600	1976	G	$2,100	3/06	SENE	16' box, hoist, 78,100 miles, gas
IHC	1600	1962	G	$2,000	4/05	NWOK	Loadstar, 2 ton, 18' Tradewind B&H, 5×2 speed, Westfield hyd. drill fill auger
IHC	1600	1963	F	$700	4/05	SEND	Loadstar, 15.5' steel box and floor, 4×2 speed
IHC	1600	1963	G	$1,200	3/05	NESD	16' Flasco wood, twin-cyl. hoist, 304 engine, 4×2 speed, plumbed
IHC	1600	1964	F	$600	6/05	NEND	Loadstar, 13.5' steel box, hoist
IHC	1600	1966	G	$1,500	6/07	SCNE	72,954 miles, 4×2 speed, 15.5' box and hoist, kept inside
IHC	1600	1968	P	$450	6/05	SEND	Loadstar single axle, V-8, 4×2 speed, 15' box, hoist, plumbed, bad brakes
IHC	1600	1969	P	$850	4/05	NCOH	Loadstar, 12' hoist bed, V-8
IHC	1600	1970	F	$900	2/05	WCIN	10 wheeler, 18' bed, hoist
IHC	1600	1971	F	$550	8/05	SWOH	Loadstar, 16' Omaha bed, 5×2 speed, 127K miles, no hoist
IHC	1600	1974	G	$3,000	2/05	SCMN	Loadstar, 1,000-gal. poly tank
IHC	1600	1974	G	$5,100	9/05	NCND	Loadstar, 16' steel box with 48" sides and roll tarp, 4×2 speed, 56,856 miles
IHC	1600	1975	G	$2,250	9/05	NWMO	Loadstar, 2 ton, 16' Knapheide box and hoist
IHC	1600	1975	G	$2,700	3/05	NESD	16' Omaha steel box, hoist, 345 engine, 4×2 speed, silage liner
IHC	1600	1975	G	$6,100	12/05	WCIL	Loadstar, 345, V-8 engine, 4×2 speed, 14' Schien bed, 40,100 miles
IHC	1600	1975		$7,000	10/05	NCND	Loadstar, 16' steel box and floor with steel fold-down stock rack, low mileage
IHC	1600	1976	G	$2,900	1/05	ECNE	Loadstar, 20K miles
IHC	1600	1976	G	$3,500	3/05	SWNE	Loadstar, 345, 4×2 speed, PS, 18,116 miles, 15.5' steel bed, hoist
IHC	1610	1973	F	$3,600	4/07	SESK	Canadian sale, Cargostar cab over truck with 345 V-8 motor, 5×2 speed, rebuilt hd trans., 10:00×20 rubber, box and hoist, round bale rack
IHC	1610A	1961	F	$2,100	12/05	NCOH	Cabover, 16' hoist bed, V-8, 22,375 miles, no brakes
IHC	170	1956	F	$475	4/08	ECMN	96,400 miles, 9' landscape box
IHC	1700	1972	F	$400	4/08	SCND	Loadstar, single-axle gravel truck, 4×2 speed, old state truck
IHC	1700	1974	G	$5,400	2/08	NCOH	Loadstar, 16' Midwest bed with excellent wood floor, twin-cyl. hoist
IHC	1700	1975	F	$2,500	2/08	ECKS	Grain, Knapheide metal box, Harsh hoist, 404 gas engine with hole in block, 5-2 speed trans., 187" wheelbase, 9:00-20 tires, 64,700 actual miles, bought new
IHC	1700	1976	G	$3,400	1/08	SEIL	1,600-gal. ss tank and pump
IHC	1700	1977	G	$4,000	3/08	ECNE	Loadstar, 404 gas engine, 5×2 speed, PS, hyd. brakes, 9:00-20 tires, spoke wheels, Omaha standard all steel 16'×52" sides
IHC	1700	1977	G	$8,000	2/08	NEKS	404 V-8, 4×2 speed, 16' bed, twin-cyl. scissors hoist, steel floor, roll tarp, blue cab
IHC	1700		F	$3,500	12/07	WCMN	Loadstar, single axle, 390 IH gas, 100,159 original miles (300 miles on new engine), 5 speed with splitter, rebuilt carb., new radiator, new trans., new 9.00-20 rear tires, new exhaust
IHC	1700	1966	G	$1,950	11/07	WCKS	Tandem truck, 340 motor, 5×2 speed, 20' bed, hoist, rollover tarp

Trucks

Make	Model	Year	Cond.	Price	Date	Area	Comments
IHC	1700	1972	P	$1,500	2/07	NEND	Loadstar tag tandem, 392, V-8, 5×2 speed, 18' Frontier box, hoist, roll tarp
IHC	1700	1974	G	$1,200	4/07	ECSK	Canadian sale, Loadstar tandem-axle grain truck, 5×2 speed, V-8, Westeel steel box with wood floor
IHC	1700	1975	G	$7,750	1/07	SCNE	Loadstar, 18' steel box, roll tarp, 5×2 speed, 10:00 tires, drag tag
IHC	1700	1976	G	$6,500	3/07	WCSK	Canada sale, Loadstar single-axle grain, 345 V-8, 5×2 speed, 10:00-20 tires, Westeel 8.5'×15' steel box, hoist, Michael's roll tarp, 60K miles, one owner, shedded
IHC	1700	1977	G	$8,000	7/07	SESK	Canadian sale, Loadstar single-axle grain, 404 V-8, 5×2 speed, Univision 15' steel box, hoist, 175" WB, 51,548 kilometers
IHC	1700	1968	F	$1,000	7/06	SEND	Tandem, Westfield 240-bu. seed tender
IHC	1700	1972	F	$1,350	2/06	NCIN	16' bed, hoist, single axle, PS, V-8, 5×2 speed
IHC	1700	1974	F	$450	10/06	SWOH	Dump truck
IHC	1700	1976	G	$2,900	3/06	NECO	Loadstar, 20' bed, swing end gate, rollover tarp, 404 gas engine, 5×2 speed, hyd. tag
IHC	1700		F	$1,700	2/05	SCCA	Single-axle dump
IHC	1700		F	$1,800	6/05	WCMN	Loadstar, single axle, V-8, 5 speed, splitter, 15' steel box, hyd. silage and grain gate
IHC	1700		F	$2,500	12/05	SEMN	Loadstar
IHC	1700	1965	F	$700	2/05	WCOH	Cabover, 5×2 speed, dual hoist, 20' grain and livestock bed, 60,513 miles
IHC	1700	1970	G	$2,400	6/05	SEND	Loadstar, single axle, 345, 5×2 speed, 16' Omaha standard steel box, tip tops, 123,808 miles
IHC	1700	1970	G	$6,800	7/05	ECND	Tag tandem, gas, 5×2 speed, 18' grain box, 24,970 miles
IHC	1700	1972	F	$1,375	3/05	SCMI	Loadstar, single axle, gas, Kilbros 385 box and auger
IHC	1700	1972	G	$2,900	5/05	NWAB	Canadian sale, grain, single axle, 16' wood box
IHC	1700	1973	F	$750	3/05	ECND	Loadstar single axle, V-8, 5 speed, 16' flatbed, hoist
IHC	1700	1975	G	$2,000	4/05	NWIL	Loadstar grain, 2 speed, 345, hyd. lift,13' box, 46,690 miles
IHC	1700	1976	G	$700	3/05	NECO	18' box, hoist, 392 engine, 5×2 speed, tag
IHC	1700	1977	G	$1,800	2/05	SCMN	
IHC	1710	1974	G	$2,150	3/06	SEMN	Cargo star single-axle truck, V-8, 4×2 speed, 16' box and hoist
IHC	1710A	1972	G	$7,300	1/05	WCIL	16' bed, 55K miles
IHC	1710B	1975	G	$2,000	5/05	SCMN	Cargostar, water truck, gas, auto, right-hand steer
IHC	1750		G	$4,500	11/06	SWCA	Dump, 5-6 yard box, IH 466DT diesel
IHC	1750	1978	G	$0	1/05	SCNE	No sale at $9,600, grain, 96K miles, 5×2 speed, 20' box, Shur-Lok tarp
IHC	1800		G	$5,700	2/08	SWMN	Twin screw, V-8, 5×4 speed, 18' steel box and hoist, full end gate, drill fill
IHC	1800	1970	F	$4,400	2/08	NCOH	18' bed, 5×2 speed, V-8, cheater and hoist
IHC	1800	1974	G	$11,500	3/08	ECND	Loadstar, twin screw, V-8, 5×3 speed, Westgo diamond box, hoist, roll tarp, 44,194 miles
IHC	1800	1977	F	$3,500	4/08	NWMN	Twin screw, 466, 5×3 speed, 18.5' Frontier box, hoist, roll tarp, beet equipped, plumbed for drill fill, twin fuel tanks, 103,391 miles

Trucks

Make	Model	Year	Cond.	Price	Date	Area	Comments
IHC	1800	1977	G	$19,500	2/08	NCOH	Loadstar, live tandem axle, 5×4 speed, 18' bed with center gate door and roll tarp, twin-cyl. hoist, dual tanks
IHC	1800		G	$2,250	2/07	SCWA	Boom
IHC	1800	1964	G	$5,000	9/07	NCND	Twin-screw tandem grain, swing-out gate, 345 V-8, 15 speed, Rugby 20' steel box & hoist with jiffy roll tarp
IHC	1800	1969	G	$3,000	12/07	WCMN	Loadstar, twin screw, steel box, hoist, tarp, DOT'd
IHC	1800	1970	F	$2,400	3/07	WCIL	Tandem twin screw, 20' bed, hoist, freight doors, gas, runs good
IHC	1800	1974	F	$3,750	4/07	NWMN	Twin-screw tandem, V-8, 5×3 speed, 19' box with hoist and roll tarp
IHC	1800	1974	G	$6,500	11/07	NWMN	Twin-screw tandem, V-8, 5×3 speed, flatbed, 3,000-gal. Pleasure Products fiberglass tank, 2" gas engine pump, PTO drive chemical keg air pump, 10:00×20 rubber
IHC	1800	1975	F	$600	2/07	SCNE	Diesel, 5 speed, 2-speed axle, tandem axle, full screw, air brakes, 18' box & hoist, hoist needs repair, 11R22.5 tires
IHC	1800	1975	F	$2,500	4/07	ECND	Twin-screw tandem truck, 466 engine, 5×3 speed, 18' steel box, hoist, roll tarp
IHC	1800	1975	F	$5,000	4/07	ECND	15K miles on new engine, twin-screw tandem, 466 engine, 5×3 speed, Frontier 18' box, hoist, roll tarp
IHC	1800	1976	G	$2,100	3/07	WCMI	Loadstar single-axle grain truck, 14' Midwest metal grain box, V-8 gas, PS
IHC	1800	1976	F	$5,750	9/07	ECND	Loadstar twin screw, 446, 5×3 speed, Westgo diamond box, hoist, roll tarp
IHC	1800	1964	F	$500	11/06	SCSD	Loadstar tandem dump truck
IHC	1800	1966	F	$300	3/06	WCMI	Live tandem flatbed truck, V-8 gas, 5×3 speed
IHC	1800	1971	G	$7,200	4/06	NEND	Loadstar twin screw, 392, V-8, 5×2 speed, Frontier, 20' box, hoist, roll tarp, plumbed for drill fill
IHC	1800	1972	G	$1,800	8/06	NCOH	Loadstar, 12', 350 bu., hoist bed, 5×2 speed, 132K miles
IHC	1800	1973	G	$2,500	2/06	ECIL	Loadstar, 5×2 speed, 10/20, 14' grain bed
IHC	1800	1973	G	$4,500	7/06	NWMN	Loadstar twin-screw tandem, new engine in 2005, box and hoist
IHC	1800	1974	F	$3,000	9/06	NEND	Twin-screw tandem, 5×3 speed, 18' box hoist, roll tarp
IHC	1800	1975	P	$800	9/06	NEND	Twin-screw tandem, V-8 Allison auto, 18' box, hoist, roll tarp
IHC	1800	1975	F	$2,500	9/06	NEND	Twin-screw tandem, V-8, 5×4, 18' steel box, hoist
IHC	1800	1975	G	$7,250	12/06	SEND	Twin-screw truck, V-8 gas engine, Allison auto, 20' Frontier box, hoist, roll tarp
IHC	1800		F	$500	12/05	SWMS	Diesel, tandem axle
IHC	1800		F	$2,600	6/05	WCMN	Loadstar, twinscrew, 5×4 speed, 20' steel box and hoist
IHC	1800	1973	F	$3,600	6/05	SEND	Loadstar, 392, V-8, 5×2 speed, 19' Frontier box, hoist, roll tarp, 54,192 miles

Trucks

Make	Model	Year	Cond.	Price	Date	Area	Comments
IHC	1800	1973	G	$5,100	6/05	SEND	Twin screw, 392, 5×4 speed, 18' Omaha standard steel box, double tip tops, plumbed for drill fill, 114,664 miles, rebuilt engine fall of 2004
IHC	1800	1974	G	$3,000	3/05	NWMN	Twin screw tandem, 9.5' steel box and hoist, roll tarp, V-8, 5×3 speed
IHC	1800	1974	G	$6,500	3/05	NCMT	20' box, hoist, 392 V-8, auto, twin screw, 10.00×20's
IHC	1800	1975	G	$2,000	3/06	ECND	French 19' 3-in-1 box, center-mount hoist, 40K miles, trans. OH
IHC	1800	1975	G	$2,750	2/05	SCMN	1,000-gal. poly tank
IHC	1800	1975	G	$3,500	7/05	ECND	French 19' 3-in-1 box center-mount hoist, 40K miles, trans. OH
IHC	1800	1976	G	$500	7/05	NCMN	Twin screw, V-8, 5×4 speed
IHC	1800	1976	G	$5,000	7/05	NCMN	Twin screw, V-8, 19' Westgo box
IHC	1800	1976	G	$9,000	3/05	NWMN	Twin-screw tandem, V-8, Allison auto, 19' strong box, scissor hoist and roll tarp, 93,475 miles
IHC	1800	1977	G	$9,800	3/05	WCMN	Loadstar, twin screw, 446, 5×4 speed, new clutch, 20' box and hoist, roll tarp, DOT'd
IHC	1800	1978	F	$275	8/05	SEMI	Loadstar, 16' flatbed, tandem axle, gas
IHC	1800	1978	G	$4,000	3/05	NEKS	Loadstar, bucket, Pitman hotstick
IHC	1810		F	$1,900	2/08	SCIL	Loadmaster tandem axle with 25' log grapple and 16' bed, 404 gas with fresh OH
IHC	1850	1969	F	$1,800	7/06	ECND	Twin screw, V-8, 5×3 speed, 20' Frontier box, hoist
IHC	1850	1976	F	$4,000	9/06	WCND	Twin-screw tender, 466 diesel, 13 speed, direct PTO, New Leader 16-ton twin compartment
IHC	190	1967	F	$800	3/07	WCIL	Road tractor, 549 gas motor, collectible
IHC	1900	1968	G	$6,250	3/08	WCMI	Fleetstar grain,18' metal grain box with center post hoist, dead tandem air lift, big 6 cyl., 5×2 speed
IHC	1910	1973	G	$7,500	6/06	NWMN	Twin-screw tandem, 478, V-8, 5×3 speed, 20' Knapheide box with hoist, roll tarp
IHC	1910A	1972	F	$2,200	3/07	NWMN	Cabover tandem, 549, V-8, auto, 20' box with hoist
IHC	2000	1971	G	$1,900	3/07	WCNE	Tandem-axle grain, 20', steel box with wood floor, 52" sides, dual rear cargo doors, (2) corn gates, Detroit 671 diesel engine, 10 speed, 10:00-20 tires, needs work
IHC	2000	1973	G	$6,500	4/07	SESK	Canadian sale, tandem-axle grain, 6V71 Detroit, 4+4 trans., 18' long × 64" tall wooden box, Nordic hoist
IHC	2000	1965	F	$1,650	3/05	NWMN	Twin-screw tandem gravel
IHC	2000	1966	G	$4,100	11/05	ECNE	Fleetstar, twin screw, Detroit DWL5 diesel engine, Fuller 9-speed Roadranger, air brakes, Knapheide 18' steel box, twin-cyl. hoist
IHC	2010		G	$2,750	5/06	SWCA	Fleetstar, diesel, conventional cab, tandem axle, diesel
IHC	2010A	1969	G	$8,500	12/05	SCMN	Twin screw, 538 engine, 5×2 speed, 19' steel box & hoist, roll tarp, air brakes, hookups for pup

Trucks

Make	Model	Year	Cond.	Price	Date	Area	Comments
IHC	2050	1975	G	$3,900	8/07	WCMN	Twin screw, 5×4 speed, air seat, brakes, 16' Frontier box, hoist, roll tarp
IHC	2050	1975	G	$3,400	8/06	SWCA	
IHC	2070	1975	F	$6,750	6/07	SEND	Tandem, 850 Cummins, 13 speed, 20' Frontier unibody box, roll tarp
IHC	2275	1978	F	$1,500	1/05	SWOH	450,930 miles, 13 speed, 350 Cummins, twin screw
IHC	2500	1978	P	$950	12/07	WCMN	Twin-screw semi, Cummins, 13 speed, needs work
IHC	2575	1978	G	$8,000	2/08	NCOH	Tandem-axle semi tractor, 300 Cummins engine, 13 speed, wet line, air ride
IHC	2575	1977	G	$6,750	11/06	SCCA	Dump, Cummins, 9 speed
IHC	3450	1969	G	$2,500	10/07	NWWI	Hauler, 26' roll back and tilt bed, new brakes
IHC	4070	1974	P	$850	11/07	NEMO	Transtar cabover semi tractor, sleeper, 13-speed Fuller trans., 11R22.5 radial tires
IHC	4070	1977	G	$9,600	12/07	ECNE	Transtar, tandem axle, cabover grain truck, 20' steel box, rear cargo doors, Cummins diesel engine, 9 speed, red/silver
IHC	4070		P	$700	3/08	WCMN	Cabover, 6V71 Detroit, 13 speed
IHC	4070		F	$2,100	3/06	NWIL	Transtar, diesel, 50K miles on OH, cabover, good tires
IHC	4070	1971		$0	1/06	NCCO	20' bed, Cummins diesel, 13 speed, Harsh rear and side hoist, side board closer
IHC	4070	1972	G	$4,000	11/06	ECND	Cabover tri-axle, front lift duals, 318, 13 speed, steel 24' box, hoist and extension, 279,518 miles
IHC	4070		F	$4,250	9/05	NWOR	Transtar, cabover, 20' steel bed with hoist, rollover tarps
IHC	4070A	1969	G	$7,500	4/05	NENE	Transtar grain, 318 Detroit, twin screw, differential lock, 13 speed, Harsh triple-stage hoist, 20' wooden box with steel floor, rear air tag axle, roll tarp
IHC	4070A	1971	F	$600	4/05	SCMI	Cabover semi tractor, transtar, Cummins engine, 354K miles
IHC	4070B	1974	F	$4,000	1/07	NWIN	Cabover semi tractor
IHC	4070B	1978	F	$1,500	12/07	NWIA	Cummins diesel, 13 speed (rebuilt), tandem axle, full screw, air tag third axle, 240" wheelbase, Lann 22' aluminum grain box, triple rear doors, hoist, pintal hitch, 11R24.5 tires
IHC	4200	1973	G	$4,000	6/06	SEID	Logan self-unloading bed
IHC	4200	1974	F	$1,000	6/06	NWMN	Transtar, 318 Detroit and 4×4 trans., twin-screw tractor
IHC	4200	1974	G	$7,250	6/06	SEID	Detroit, 13 speed, 2001 Spudnik 22' self-unloading bed, electric/hyd., PTO, pintle hitch, missing left rear axle, drum, tires and wheels
IHC	4200	1974	G	$1,250	7/05	NCMN	Twin screw
IHC	4200	1974	G	$1,250	7/05	NCMN	Twin screw
IHC	4200	1974	G	$2,100	7/05	NCMN	Twin screw
IHC	4200	1974	G	$2,200	7/05	NCMN	Twin screw
IHC	4200	1974	G	$3,100	7/05	NCMN	Twin screw
IHC	4200	1978	G	$4,100	7/05	ECND	700K miles, 13 speed, dual-tandem axle pusher axle in front

Trucks

Make	Model	Year	Cond.	Price	Date	Area	Comments
IHC	4300	1975	G	$4,000	3/08	NWKY	Transtar, road tractor, 350 Cummins, 13 speed
IHC	4300	1978	G	$11,750	2/08	SWMN	Day cab, wet kit, 219K miles, 350 Cummins, double frame, 13 speed
IHC	4300	1975	F	$4,000	3/07	WCIL	Transtar, 13 speed, good tires
IHC	4300	1978	E	$16,200	4/07	SWIA	Transtar, 350-hp. Cummins, 9 speed, differential rebuilt last fall, Obeco 18' steel grain box, rollover tarp, always shedded
IHC	4300	1972	G	$6,750	12/06	ECNE	Transtar tractor, day cab, Cummins 400 diesel engine, 13 speed, 200" wheelbase, jake brake, dual 80-gal. fuel tanks, air ride seat
IHC	4300	1972	G	$7,300	12/06	ECNE	Transtar tractor, Cummins 350 diesel engine, 13 speed, 206" wheelbase, jake brake, dual 80-gal. fuel tanks, single sleeper, air ride seat
IHC	4300	1973	G	$7,000	12/06	ECNE	Transtar tractor, Cummins 350 diesel engine, 13 speed, 240" wheelbase, pusher tag axle, jake brake, air ride suspension, 40,000-lb. rear axles, dual 108-gal. fuel tanks
IHC	4300	1973	G	$9,000	3/07	ECND	Twin screw, third axle pusher, air up/spring down, 855 Cummins, 13 speed, 22' flatbed, spring suspension, (2) 1,500-gal. poly tanks, chemical mix, Honda 5-hp., 2" pump, 2" hose reel
IHC	4300	1974	G	$4,500	6/06	SEID	NTC 350 Cummins, 5×4 speed, fifth wheel
IHC	4300	1975	g	$5,000	6/06	SEID	NTC 350 Cummins, 9 speed, Logan 22' self-unloading bed, electric/hyd., 24" belt pintle hitch, missing rear differential
IHC	4300	1975	G	$7,300	11/06	NWKS	20' steel box, twin screw
IHC	4300	1977	G	$1,500	6/06	SEID	Cummins, 13 speed, 1993 Logan 22' self-unloading bed
IHC	4300	1977	G	$3,100	3/06	ECIN	Tandem axle
IHC	4300	1977	g	$4,000	6/06	SEID	350 Cummins, 13 speed, 1996 Spudnik self-unloading bed, electric/hyd., PTO, 30" belt, engine has a knock
IHC	4300	1977	G	$7,000	6/06	SEID	NTC 400 Cummins, 9 speed, 13' dump box
IHC	4300	1977	g	$9,000	6/06	SEID	NTC 350 Cummins, 13 speed, 1993 Logan 22' self-unloading bed, electric/hyd., 24" belt
IHC	4300	1978	g	$9,000	6/06	SEID	Detroit, 13 speed, 1993 Logan 22' self-unloading bed, elect/hyd, 24" belt
IHC	4370	1974	P	$1,900	4/06	WCSD	Dump truck tandem axle with pusher and 16' Fruehauf dump box, 400 Cat engine, 13 speed, not running, missing some tires and rims
IHC	5070	1975	G	$900	6/05	ECPA	Dump truck
IHC	A	1932	G	$7,000	8/06	SCIL	1½ ton
IHC	A160	1959	G	$800	10/05	NCND	Wood box, hoist
IHC	B160	1960	F	$150	1/07	SWNE	Single-axle grain truck, 14' wood box
IHC	B160	1960	F	$100	8/06	ECNE	Single-axle grain truck, 13' wood box
IHC	B160	1962	G	$1,500	2/06	NWIL	13.5' farm box, 345 V-8 engine
IHC	BC160	1959	F	$100	1/08	WCKS	13.5' box, no sides, 6-cylinder engine
IHC	BC160	1960	G	$1,400	4/05	NWOK	Steel bed, hoist, 4×2 speed, runs great!
IHC	C	1933	G	$8,250	8/06	SCIL	Stake truck

Trucks

Make	Model	Year	Cond.	Price	Date	Area	Comments
IHC	K	1947	G	$300	8/06	SCIL	Fire truck
IHC	K5	1949	G	$100	8/06	SCIL	Needs restoring
IHC	K5	1949	G	$125	8/06	SCIL	Needs restoring
IHC	KB5	1937	G	$1,600	8/06	ECNE	Cab, chassis, 37,319 miles
IHC	KB5	1948	F	$400	3/06	ECND	Single-axle flatbed, 1,000-gal. poly tank and cone
IHC	KB6	1947	G	$1,650	8/06	ECNE	Fire truck, 6,193 miles, runs good
IHC	KB6	1948	F	$250	8/06	SCIL	
IHC	KB6	1946	G	$700	6/05	SEND	Single axle, 4×2 speed, wood box, hoist, 88K miles
IHC	KB6	1948	G	$1,300	4/05	NWOK	Wood grain bed, grain sides, stock rack
IHC	L170	1950	G	$3,600	8/06	SCIL	Fire truck
IHC	Loadstar	1974	G	$6,100	4/07	SESK	Canadian sale, single-axle grain truck, 392 V-8, 5×2 speed, 15' steel box
IHC	Loadstar		F	$600	1/06	SCMI	Gas, dump truck, 5-yard box, single axle
IHC	Loadstar		G	$2,250	6/05	SWCA	Dump truck, 3-5 yard box
IHC	Loadstar		G	$3,250	11/06	SWCA	Dump truck, 3-5 yard box
IHC	Loadstar	1975	G	$6,300	2/06	SCMI	Loadstar seed truck, 390 Killbros box/seed jet, Shur-Lok tarp, new springs, front and rear breakes in 2005, 65,375 miles
IHC	Loadstar		F	$2,000	1/06	WCCA	Nurse truck, 1,500-gal. poly tank, Honda 5.5-hp. motor
IHC	Loadstar	1976	G	$4,400	8/05	ECNE	Grain truck, 15.5' box, hoist, 48,400 miles
IHC	Loadstar	1976	G	$5,000	3/05	WCMN	Single axle, 5×2 speed, rear end, 1600 gas V-8, Farm Star 16' steel box, roll tarp
IHC	N/A	1961	F	$700	3/08	SEND	Single axle, flatbed, always shedded
IHC	N/A	1917	G	$3,500	9/07	WCIA	1½-ton truck
IHC	N/A	1921	G	$28,000	9/07	NEIL	Nicely restored truck, complete with brass headlights and polished brass hardware, fitted with oak stake bed and new hard rubber tires, starts and runs great
IHC	N/A	1961	G	$950	2/07	ECNE	V-8 motor, 8.25-20 rubber, 16' wood box, hoist
IHC	N/A	1972	F	$500	9/07	WCIA	Pickup, yellow
IHC	N/A	1975	G	$4,250	11/07	SEID	Dump truck, IHC army truck, 6×6, 6 cyl., gas, 5 speed, front winch
IHC	N/A	1976	G	$5,000	3/07	WCNE	International Transtar tandem-axle grain truck, 22' steel box with 52" sides, twin-cylinder hoist, rear cargo doors, Cummins 903 diesel engine, 10 speed, 115-gal. fuel tank, 11R22.5 tires, no PS
IHC	N/A	1978	G	$600	6/07	WCMO	Gas engine, 4 speed, single axle, lime spreader box, dual spinners, 11R22.5 tires
IHC	N/A	1978	G	$2,900	4/07	NWND	Cargostar single axle, V-8, 5×2 speed, 18' flatbed, 10:00-20 tires
IHC	N/A		F	$200	1/06	WCCA	V-8, 4 speed
IHC	N/A		F	$500	9/05	NEIN	Semi tractor, diesel, Cat 1160 engine
IHC	N/A		F	$900	7/06	WCPA	Water truck, gas, 5×2 speed, 2,500-gal. stainless tank
IHC	N/A		F	$2,500	7/06	WCCA	Cummins 6 cyl., 9 speed
IHC	N/A		G	$4,000	11/06	SCCA	Dump truck
IHC	N/A	1921	G	$19,000	8/06	SCIL	Stake bed, hard rubber tires
IHC	N/A	1926	G	$6,250	8/06	SCIL	Flat bed, open cab
IHC	N/A	1928	E	$10,000	8/06	SCIL	2-ton flat bed, stake sides
IHC	N/A	1936	G	$3,500	8/06	SCIL	Dump, needs restoration
IHC	N/A	1950	G	$39,500	8/06	SCIL	KB-14, recently restored, red paint, black fenders, new tires, Cummins C165 diesel, 5 speed

Trucks

Make	Model	Year	Cond.	Price	Date	Area	Comments
IHC	N/A	1961	G	$500	8/06	SCIL	Emeryville truck
IHC	N/A	1964	F	$550	2/06	NWSD	Loadstar, box and hoist
IHC	N/A	1971	G	$1,800	12/06	SCMT	Bale retriever bed
IHC	N/A	1975	F	$650	3/06	NCNE	Cabover, 2,000-gal. diesel barrel
IHC	N/A	1975	g	$1,600	12/06	ECMN	Paystar, 15' gravel box, 13 speed
IHC	N/A	1975	G	$3,100	3/06	WCMI	Transtar live tandem grain truck, air lift axle, 20' metal grain box and twin post hoist, Detroit 318 diesel, 13 speed
IHC	N/A	1976	F	$4,000	5/06	ECNE	Transtar Eagle truck tractor, 503K miles, Cat 3406 diesel, 13 speed, 44,800-lb. GVW rating, dual fuel tanks, 11R24.5 tires, aluminum front rims
IHC	N/A	1978	F	$950	3/06	NEMI	Transtar road tractor, 350 Cummins, 13 speed
IHC	N/A	1978	F	$2,000	3/06	NCKS	Semi
IHC	N/A		F	$150	12/05	WCWA	
IHC	N/A		G	$700	11/05	NCCA	3 axle
IHC	N/A		F	$800	11/05	ECWA	Boom truck, gas engine, 4×2 speed, hydraulic controls
IHC	N/A	1926	G	$2,650	3/05	WCNE	8.5' Omaha standard wood box, 32×6 tires
IHC	N/A	1951	G	$1,475	3/05	WCIL	Dump truck, former fire truck, 18K miles
IHC	N/A	1962	P	$340	4/05	SEND	Fleetstar, 671 Detroit engine
IHC	N/A	1963	P	$130	4/05	SEND	Single axle, wood box
IHC	N/A	1969	F	$950	1/05	SEMI	Twin screw, 15 speed, twin hoist
IHC	N/A	1970	F	$800	9/05	SEIA	Grain truck
IHC	N/A	1973	F	$1,000	9/05	SCON	Dump truck
IHC	N/A	1973	G	$4,500	3/05	SWIN	Tri-axle grain truck, 5×3 speed, Cat 3208 engine, Hendrickson suspension, air brakes, 20' wood floor, metal grain sides, tarp
IHC	N/A	1973	G	$6,000	3/05	ECMI	Cabover tandem-axle grain truck, 18' steel grain box and hoist, V-8, 5×2 speed, 24,084 miles
IHC	N/A	1973	G	$7,500	12/05	WCMN	Twin screw, 318 Detroit diesel, 13 speed, 22' poly lined steel box and hoist, 18,000-lb. axles, roll tarp
IHC	N/A	1976	G	$3,600	12/05	WCIL	Transtar, cabover, 13-speed Eaton Fuller trans., budd wheels, wet kit, dual fuel tanks, 350 Cummins engine, PS, air
IHC	N/A	1977	G	$2,800	6/05	NCIA	Twin screw truck, figerglass tilt cab, 22' steel box and hoist, 5×3 speed
IHC	N/A	1978	G	$7,000	3/05	ECMN	Dump truck, tri-axle, tandem axle, 14' box
IHC	N/A	1978	G	$8,250	11/05	NECO	Reel carrier, Cummins diesel, 10 speed
IHC	R160	1953	G	$1,000	9/07	SCMI	Gas, single axle, 10' box and hoist, 63,955 miles
IHC	S1955	1977	G	$4,000	2/07	SCWA	Dump truck
IHC	S2200	1977	G	$12,750	11/06	NEND	Twin screw, Cummins power, 13 speed, 21' Polar box, roll tarp
IHC	S2500	1978	G	$7,500	2/06	NWKS	Semi truck, 180" wheelbase, 13 speed, Cummins 350
IHC	Scout		G	$5,000	8/06	SCIL	Red
IHC	Scout	1961	G	$5,000	8/06	SCIL	2WD, new metallic green paint, low miles
IHC	Scout	1963	G	$5,000	8/06	SCIL	4WD, yellow
IHC	Scout		G	$1,250	9/05	SCON	Canadian sale
IHC	Transtar	1974	F	$750	6/05	SEND	318 Detroit, 13 speed, 10:00-20 rubber
IHC	Transtar 2	1976	F	$5,000	2/06	NCCO	Semi, Cummins 350, 13 speed
IHC	Transtar 2	1978	G	$6,500	3/06	NWKS	Grain truck, service body, 350 Cummins, 20' Scott steel box, cargo doors, 10 speed, Harsh hoist, twin screw, tag axle, rollover tarp

Trucks

Make	Model	Year	Cond.	Price	Date	Area	Comments
Kenworth	K100	1976	F	$3,500	12/07	ECNE	Truck tractor, sleeper
Kenworth	K100		P	$600	11/06	NEND	Cabover, Cummins, 9 speed, spring ride
Kenworth	N/A	1978	G	$7,500	3/08	SCND	Conventional, 400 Cummins, Fuller 13 speed, jake brake, PS, new rears, single sleeper, air ride
Kenworth	N/A	1969	F	$1,600	12/07	SWNE	Dump truck, Cummins, 5×3 speed, twin screw, 13' box, 275/80R24.5 rubber, anti-freeze in fuel, may need fuel pump
Kenworth	N/A	1970	G	$2,300	3/07	SEND	Cabover, 855 Cummins, 13 speed, twin fuel tanks
Kenworth	N/A	1970	G	$4,250	3/07	NWIA	Cabover semi, Cat engine, 13 speed
Kenworth	N/A		G	$2,500	7/06	WCCA	Cummins diesel
Kenworth	N/A	1965	P	$600	6/06	WCMN	Semi, day cab, 6V71 Detroit, 15 speed
Kenworth	N/A	1967	F	$3,000	11/06	SCCA	Side dump, Detroit diesel, 9 speed
Kenworth	N/A	1973	F	$1,000	3/06	NEKS	Cabover Cummins, 855 engine, Eaton Fuller 13 speed, twin screw
Kenworth	N/A	1974	G	$7,000	10/06	NCND	Tandem, 350 Cummins, 5×4 speed, 18.5' box, roll tarp
Kenworth	N/A	1965	G	$9,750	12/05	WCWA	Conventional water truck, 8V92, 5×4 speed, 4,000 gal., PTO pump, front, rear and side spray
Kenworth	N/A	1972	G	$9,750	12/05	SWMS	Day cab, diesel
Kenworth	N/A	1974	F	$2,000	11/05	WCKS	Semi tractor, 13 speed
Kenworth	T900	1969	G	$7,750	3/05	ECSD	Conventional, Cummins Big Cam 4, 13 speed, double frame, air slide fifth wheel, PTO
Kenworth	W900	1978	F	$3,900	3/08	ECNE	Conventional truck tractor, twin screw, Model NTC350BC diesel engine, Fuller Roadranger RT09513 trans., 4:33 axle ratio, GVW 48,000 lb., 36" sleeper, single upright exhaust
Kenworth	W900	1976	F	$2,600	12/07	ECNE	Tandem axle, Cummins N14, Fuller 9 speed
Kenworth	W900	1978	G	$11,000	3/07	SEND	Tri-axle air up/down dual pusher, 13 speed, 258,748 miles
Kenworth	W900	1971	G	$3,250	11/06	SCCA	Cummins, 10 speed
Kenworth	W900	1975	G	$6,500	3/06	NECO	Day cab, fifth wheel, 10K miles on Cummins 400 BC 1 engine, 13 speed, new rubber
Kenworth	W900	1977	G	$5,000	9/05	ECMN	
Kenworth	W900	1977	G	$7,000	3/05	ECID	Conventional, tandem axle, twin screw, 350 Cummins, fresh OH, 9 speed, 11R24.5, 1992 Logan 20' self-unloading bed, electric/hyd., PTO, 24" belt
Linn	626E	1926	G	$35,000	9/07	NEIL	Half track, logging truck
Mack	DM685	1978	G	$9,000	7/06	WCPA	Mack diesel, 6-speed lowhole trans.
Mack	DM685S		F	$1,400	3/08	NEAR	Tandem-axle tractor truck, 300 Mack, 6 speed, 10.00R-20 tires on spoke wheels
Mack	DM685S	1974	G	$7,000	6/05	ECNE	Tandem-axle dump truck, 15' steel box, 5×2 speed, Mack 300 diesel engine, budd wheels, air, end gate
Mack	DM685SX	1972	G	$7,500	1/07	WCNY	Rollback, diesel
Mack	N/A	1978	G	$5,900	2/08	SCMN	Semi tractor, Mack 300-hp. diesel, 5 speed, DOT inspected
Mack	N/A	1950	E	$125,000	9/07	NEIL	The refuse bed can be traced back to original owner, the Village of Shorewood, WI, continuous service in and around the city of Chicago until 1972, Leach refuse unit is now mounted on a 1950 B model Mack truck, restored

Trucks

Make	Model	Year	Cond.	Price	Date	Area	Comments
Mack	N/A	1953	G	$47,500	9/07	NEIL	Model W-71ST, 275-hp. Cummins supercharged engine
Mack	N/A	1974	G	$3,500	3/07	NETX	Water truck, 366 Mack engine, two-stick 5 speed, 4,000-gal.tank with sprayer
Mack	N/A	1976	G	$7,750	2/07	NEND	Cruiseliner twin screw, third axle, air, up/spring down pusher, 555 Cummins, 15 speed, 21'×8.5'×68" strong box, combo end gate, beet equipped, roll tarp, twin fuel tanks
Mack	N/A	1971	G	$1,600	12/06	ECMN	2 speed, 182" wheelbase, tri-axle, winch
Mack	N/A	1977	G	$9,000	11/06	WCMN	Tri-axle, 5 speed, 22' box, new roll tarp, new clutch
Mack	R		G	$9,000	7/07	SCMO	Mack 350 diesel engine, 5 speed hi/lo trans., tandem axle, full screw, 16' dump box
Mack	R	1976	G	$4,500	9/07	NEIL	
Mack	R	1976	G	$19,000	10/07	NWMN	Twin screw, vacuum truck, 300 Mack, 6 speed, 44,000-lb. rears, 168K miles
Mack	R	1977	G	$5,700	2/07	NENE	Gold Dog, 300 Maxadine 6 cyl., 5 speed
Mack	R	1978	G	$3,600	3/06	SWMN	Day cab, 237 Mack, 5 speed, PTO
Mack	R60	1968	F	$1,750	3/06	ECIN	Tandem axle
Mack	R600	1974	G	$3,000	6/07	WCMO	Dump truck, Mack diesel engine, 300 hp., 10 speed, AM/FM radio, 16' steel dump box, double steerable air tag, 12R-22.5 tires
Mack	R600	1971	G	$600	12/06	NEMN	Dump truck
Mack	R600	1971	F	$2,500	7/06	NWMN	Twin screw, 18' box and hoist, roll tarp, 5 speed
Mack	R600	1972	F	$3,000	7/06	NWMN	Tri-axle, twin screw, 22×8.5' box and hoist, roll tarp, 5 speed
Mack	R600	1973	G	$4,300	7/06	NWMN	Twin screw, day cab, 237 Mack, 5 speed, spring ride, wet kit, twin fuel tanks
Mack	R600	1973	G	$4,500	7/06	NWMN	Twin screw, 237 Mack, 5 speed, 20' Buffalo box, hoist, roll tarp, plumbed for drill fill, twin fuel tanks, PS, beet equipped
Mack	R600	1973	G	$10,500	7/06	NWMN	Twin screw, day cab, 237 Mack, 5 speed, spring ride, wet kit, PS, twin fuel tank
Mack	R600	1974	G	$4,100	7/06	NWMN	Twin screw, day cab, 273 Mack, 5 speed, spring ride, wet kit, twin fuel tanks
Mack	R600	1977	G	$9,000	7/06	NWMN	Twin screw, 300 Mack, 5 speed, 21' Buffalo box, hoist, beet equipped
Mack	R600		P	$475	8/05	WCTX	Rough, no batteries, needs rubber
Mack	R600	1976	F	$2,400	12/05	NCIA	300, 5 speed
Mack	R600	1978	G	$2,000	12/05	WCNJ	Dump truck, 300-hp. engine, 9 speed, steel body
Mack	R685LS	1972	G	$3,200	6/05	ECPA	
Mack	R685ST	1978	G	$5,500	12/06	NEMN	
Mack	R686ST	1977	G	$1,500	12/06	NEMN	466,207 miles
Mack	RD688S	1973	F	$2,200	12/05	SWMS	Diesel, tandem axle, water truck
Mack	RL600	1973	F	$1,900	6/05	ECNE	Mack engine, 5×3 speed, PS, (2) 70-gal. fuel tanks, twin screw, camelback spring suspension
Mack	RS686LS	1978	G	$9,000	1/08	ECNE	Tandem-axle truck tractor, Mack diesel, 5 speed, spring suspension, air ride, cab, air dryer
Mack	W-71	1953	G	$47,500	9/07	NEIL	Only 217 built, 275-hp. Cummins super-charged diesel engine

Trucks

Make	Model	Year	Cond.	Price	Date	Area	Comments
N/A	N/A	1954	G	$4,500	5/08	ECNE	Fire truck, Model F50T forward 500-gal. pumper, 6-cyl. Waukesha, 5 speed, 3,073 original one-owner miles
N/A	N/A	1918	G	$11,000	9/07	WCIA	Rare 1918 Old Reliable, dump box, chain drive
N/A	N/A	1962	G	$2,550	8/06	ECMN	Asphalt tanker
Peterbilt	N/A	1970	F	$900	3/08	WCMN	Cabover, 318 Detroit, 18 speed
Peterbilt	N/A	1974	G	$14,000	2/08	ECMT	Conventional, Cummins 400, 13 speed, strong box 19'×102"×54", Shur-Lok roll tarp, twin screw, air lift third axle, 11R24.5 tires
Peterbilt	N/A	1974	G	$2,500	7/07	ECKS	Conventional, 318 Detroit diesel engine, 4×4 trans., 3,000-gallon mild steel water tank, spray bars, tandem axle, full screw
Peterbilt	N/A	1978	F	$3,600	12/07	ECNE	Cabover, Cummins, 9 speed, PS
Peterbilt	N/A	1977	F	$4,500	4/06	NCOK	Cat 3406, 13 speed, twin screw, cabover semi, 1982 Wilson 42' aluminum grain trailer with tarp (grain trailer good)
Peterbilt	N/A	1976	F	$1,750	9/05	NWOR	Cabover, tandem-axle tractor, 350 Cummins
Peterbilt	N/A	1978	G	$4,500	9/05	ECMN	Dump truck
Peterbilt	N/A	1978	G	$37,000	3/05	NEMT	Peterbilt vacuum truck, 100-barrel steel tank, new Challenger pump, 350 Cummins, 5×4 speed, tandem, twin screw, new 11R24.5 recaps
REO	N/A		F	$6,500	11/07	WCIL	Speedwagon farm truck, not running, loose, unrestored, original, solid old farm truck, wood sideboards, wood spoke wheels, restoration opportunity, Kansas truck
REO	N/A	1938	E	$22,500	9/07	NEIL	Speedwagon, professional restoration, past AACA winner, runs great, ready for show
Studebaker	N/A	1959	G	$2,300	10/05	SWSD	18' box, hoist
White	4000	1968	G	$3,500	11/07	ECND	Twin screw, 220 Cummins, 10 speed, Knapheide 19' box, twin post hoist, roll tarp, pintle hitch
White	4000	1971	P	$2,200	8/06	NCKS	Really rough, 235 aspirated Cummins diesel, 2,800-gal. water tank with transfer pump, 9-speed Roadranger, twin screw
White	9000	1970	P	$1,800	2/06	NCKS	Twin screw truck, rough, Cummins 270, 13-speed Roadranger, 20' bed, 54" side, hoist
White	N/A	1925	G	$46,000	9/07	NEIL	Antique, tanker truck, SN 127750
White	N/A	1946	P	$350	6/07	ECND	Pusher second axle, box, hoist
White	N/A	1967	F	$950	6/06	SEND	Single-axle cabover, 6 cyl., 5 speed, 16' box, hoist, roll tarp
White	N/A	1968	F	$600	4/05	SCMI	Straight truck, 18' box, twin post hoist
White	T8164	1969	G	$2,200	3/07	WCNE	Tandem-axle cabover grain truck, 20' box with 52" sides, Detroit 671 diesel engine, 10 speed, twin screw, for parts
Winther Marwin	N/A	1918	G	$30,000	9/07	NEIL	4WD, professional restoration, one of the earliest 4WD trucks

Reviving steering wheels

By Ben Davidson

Dull, worn, and cracked steering wheels don't affect a tractor's performance. But seen against a gleaming new paint job, a ratty wheel becomes an eyesore.

You can send the wheel to a professional refurbisher and pay $100 to $200. And you can purchase a replacement for $100 to $300.

But for $30 to $80, you can get a restoration kit to do the job yourself plus have enough product left over for future projects.

Such kits won't restore the gripping knobs or ridges some wheels possessed coming out of the factory. But they do fill cracks and provide new luster to the component.

You need very little experience to work with the epoxy putty, which is the key ingredient. Much like working with body putty, the substance is very pliable and easy to work.

I took a chance on a kit and practiced on a truly hopeless steering wheel figuring that if I screwed it up, nothing would be lost but the putty. The first thing I discovered is that the putty is water-based and a dream to work with.

The substance has a long set-up time, unlike some epoxy remedies I've worked with. So I had time to not only fill cracks completely (a key to success) but also overfill the area for later sanding. I found that feathering out the overfill way beyond cracks is crucial to producing great results.

Finally, I discovered that sanding down the cured putty with very light sandpaper down to 600 grit or lower is essential. Rougher paper left swirls and grooves in the filler and old material on the wheel, which were later brought out by the glossy finished paint.

There is far more to the restoration process than can be detailed in this article. I recommend a step-by-step display of the effort found at the Second Chance Garage Web site. Visit www.secondchancegarage.com.

Resoration Supply Sources

• Eastwood Company, 800/343-9353 or www.eastwoodco.com. Kits range in price from $29.99 to $40.99 (with puller) or $9.99 for an 8-ounce can of putty. Eastwood also sells the Steering Wheel Restoration Book for $12.
• POF-15 Company, 800/726-0459 or www.por15.com. A complete kit is $83 and includes an instruction book, primer paint, and surfacer with hardener. Or you can buy 1-pound packs of putty for $21. A variety of epoxy-base putty can readily be applied to tractor wheels to eliminate cracks before a glossy paint finish goes on.

Combines

I've been asked the question a number of times over the years: "Pete, do you think there will ever be a collector's market for old combines?" Good question. It's natural to think that since the antique tractor market has exploded like it has over the last 10 to 15 years that the same thing could happen to antique harvesters.

But I wouldn't hold my breath. The only trend I'm seeing with older combines has to do with those fun combine demolition derbies. That said, older combines in the 20-plus-year-old range have definitely been rising in value since late 2007. It's just that buyers aren't collectors, they're folks looking for nice older combines they can use. Newer models going for $100K, $150K, $200K, or more is driving this trend.

Combines

Avery

Model	Year	Hours	Cond.	Price	Date	Area	Comments
Harvest-All	1939		F	$125	11/05	NCOH	Old pt

Case

Model	Year	Hours	Cond.	Price	Date	Area	Comments
660			F	$450	1/07	SEKS	13' grain head
660			F	$800	5/07	WCSD	13' header and pickup

Gleaner

Model	Year	Hours	Cond.	Price	Date	Area	Comments
A	1959		P	$150	2/06	WCNE	Gas, 14' wheat header, engine turns
A			G	$200	4/05	NCOK	LP, runs good, 14' header, pickup reel and attachment
C2	1966		G	$450	9/06	NENE	Gas engine, rubber tires
E			P	$450	4/05	NCOH	For parts
F		2,284	G	$860	3/07	ECIA	Diesel, A438 4R CH and 13' floating grain platform
F	1969		F	$400	9/07	NEND	Pickup head, stored inside, not running
F			P	$300	11/06	NEKS	Salvage
F			F	$550	8/06	NCIA	Cab, chopper, platform
F			E	$750	11/06	NEKS	
F	1969		G	$900	2/06	WCNE	Gas, 18' wheat header, 3R-30 CH, cab
F	1970		G	$750	3/06	NEKS	Gas, 12' grain head
F	1978			$0	3/06	NEKS	Diesel
F			P	$150	4/05	NCOH	13' grain head, salvage
F			F	$500	4/05	NCOH	13' flex head, variable-speed reel, straw chopper
F			G	$800	4/05	NCOH	Variable-speed reel, straw chopper
F2	1978		F	$1,000	8/06	NWOH	Gas, chopper, 13' grain head
F2	1978		G	$2,100	3/06	NEKS	Diesel
F2	1978		F	$4,700	3/06	NWOH	15' grain head, tach says 1,952 hours
G			F	$150	9/07	NCND	Gas, cab, straw chopper
K			F	$400	3/08	NWKS	12' platform
K			F	$450	1/07	NWOH	Gas, 12' head
K			G	$1,100	6/07	ECMN	Spike tooth, gas, very clean, always shedded, sold #238 CH and 13' head, heads in great condition
L			G	$2,000	5/07	WCSD	20' straight head pickup with Melroe pickup
L			G	$2,150	5/07	NCOK	24' platform
L	1974	2200	G	$3,500	6/07	SWNE	2,200 hours on tach, hydro-traction drive, AC 3500 6-cyl. diesel engine, 23.1-34 tires, monitor
L	1977		P	$2,050	2/06	NCCO	Gleaner A630 CH, 22' platform
L	1976		G	$1,300	7/05	ECND	Diesel, corn/soybean special, variable speed
L2	1978		G	$7,500	4/06	NCOK	24' header, big engine, 400 hours on OH, hydrostat, chopper, diesel
M	1976	2,257	F	$2,100	9/07	NECO	A630 CH, diesel
M			F	$450	12/06	WCMN	Gas, Gleaner 4R-30 CH
M		2,545	F	$1,100	1/06	SCMI	Gas, V-8, 13' grain table

Combines

Gleaner

Model	Year	Hours	Cond.	Price	Date	Area	Comments
M			G	$1,450	12/05	SEMN	
M			G	$2,700	8/05	SEMN	438 CH
M	1974		G	$3,000	9/05	WCNE	20' #980 header, Renn pickup
M	1978		P	$1,600	3/05	WCIL	Diesel, 6R-30 CH
M2	1977		F	$2,750	11/07	WCIL	Diesel
M2	1978		G	$1,875	1/05	SCNE	Gear drive, been shedded
M2	1978		G	$2,250	7/05	ECND	Corn and soybean special, 15' head, Melroe pickup
S	1951		G	$650	11/05	NCOH	5' cut, pt

IHC

Model	Year	Hours	Cond.	Price	Date	Area	Comments
101			G	$1,000	11/07	WCIL	Running, restored, SP
1420	1978		G	$4,500	2/06	SEIA	Chopper, always shedded, field-ready
1440	1978		G	$7,000	2/08	SEAL	
1440	1978	3,790	F	$4,000	3/06	NWOH	Spreader
1440	1978	2,444	F	$0	2/05	WCOK	No sale at $3,000, wanted $3,500, axial flow, dealer owned, 20' header, front tires good, rear tires fair, hours correct
1440	1978		G	$4,000	9/05	SEIA	850 hours on new motor
1460	1978		G	$11,000	2/08	NCOH	Axial flow, CAH, chaff spreader
1460	1978	4,991	F	$1,500	12/07	SCMT	
1460	1978	4,335	G	$2,800	12/07	SEND	23.1-26 singles
1460	1978	4,490	G	$4,800	11/07	NCIA	Cab, air, hydro, electric over hyd. for platform, OH, new fan, lots of new parts, Crary chaff spreader, 28L-26 tires
1460	1978	5,040	G	$5,750	1/07	WCIL	No heads
1460	1977	3,986	F	$3,400	3/06	SEIA	
1460	1978		F	$4,000	1/06	SESD	
1460	1978	3,122	G	$14,500	3/06	NWOH	IHC 820 20' grain head, rock trap, straw chopper
1460	1976	3,300	P	$3,000	2/05	NCIL	Rough
1460	1978	4,240	G	$3,750	4/05	ECND	
1480	1978		F	$4,000	3/07	NWIA	
403	1965		P	$500	12/07	WCNE	Gas, 14' pickup
403	1967		P	$450	12/07	WCNE	Gas, 14' pickup
403	1970		P	$450	12/07	WCNE	Gas, hydro, 14' pickup
715			F	$900	11/07	ECKS	17.5' wheat header, shows 1,400 hours but meter hasn't worked for three years, problems with water pump and variable speed, sells as is
715			G	$5,100	12/07	ECIL	Diesel, hydrostat, glow plugs, clean combine
715			F	$900	8/06	WCMN	IHC 844 CH, IHC 810 pickup head
715			G	$1,000	10/06	SEMN	
715			P	$800	8/05	NWIL	Diesel
715			G	$1,700	10/05	NCND	Windrow special, cab and pickup, diesel
715		2,415	G	$1,750	4/05	NCOH	Corn soybean special, hydro, German diesel, 810 13' grain head
815			F	$1,650	3/06	NWIL	Diesel
815	1977	2,832	G	$1,000	11/06	NEKS	Hydro, corn/soybean special

Combines

IHC

Model	Year	Hours	Cond.	Price	Date	Area	Comments
82			G	$975	2/06	NEIA	Pt
914			G	$675	4/08	SCND	Pt combine, Sund pickup
914	1973		G	$425	4/08	NEND	Pt combine
914			F	$1,050	11/07	SESD	Combine, 6-belt pickup
914			E	$2,400	7/05	SCMN	Pt, like new
915			P	$1,400	4/07	NCNE	Salvage, hydrostatic
915	1976		P	$350	12/07	WCNE	Gas, left front drive out
915	1978		G	$3,800	7/07	SEND	Hydro, engine OH, IHC 810 pickup head, IHC pickup
915			G	$1,800	4/05	NEKS	No heads, hydro, cab
915	1976	4,200	G	$1,300	7/05	ECND	Diesel, hydro, 810 pickup bean head

John Deere

Model	Year	Hours	Cond.	Price	Date	Area	Comments
105	1963		G	$575	6/07	SWMN	Corn special, square back, gas
105			F	$350	5/06	SEND	SP combine, cab, chopper
105			G	$825	9/06	NWMN	Square back, cab, chopper, pickup head
105			G	$1,000	4/06	WCMT	Cab, gas engine, 12' pickup header, pickup attachment
105			G	$1,800	12/05	NCOH	Gas, John Deere 216 grain head
30			P	$50	6/05	ECWI	
3300	1974		G	$1,800	10/07	SWOH	John Deere 213 grain platform
3300			G	$1,700	8/06	NWIL	2R-44" CH, diesel
3300			F	$3,200	3/06	NCCO	Test plot combine
3300			G	$800	4/05	NCOH	
3300			F	$1,300	12/05	WCWI	
3300	1976	2,050	G	$2,400	11/05	NEIA	Second owner, diesel, sharp
4400		3,461	G	$6,800	3/08	SCMN	Diesel
4400	1974	1,815	G	$2,100	3/08	ECNE	4 speed, 18.4R-26 front tires, 11L-16 rear tires, chaff spreader
4400	1977	3,527	F	$2,500	1/08	NWOH	Rotary screen
4400			G	$500	3/07	NEMI	Gas, spike cylinder, field-ready, no heads
4400			F	$1,900	8/07	SESD	Chopper
4400		2,686	G	$1,950	12/07	NCOH	No head, hole in oil pan
4400	1972	2,690	G	$675	2/07	ECNE	Gas
4400	1975	2,473	G	$1,900	8/07	SESD	
4400	1976		G	$2,200	1/07	ECNE	Diesel, gone through by local John Deere dealer in '06
4400	1977	2,697	G	$1,700	2/07	ECNE	Diesel
4400	1977	3,872	G	$5,750	12/07	NEMO	Chopper, diesel, 23.1-26 & 11.00-16 tires, nice straight combine
4400			F	$1,150	2/06	SWIN	Gas, John Deere 444 CH (low tin), John Deere 213 bean head
4400			F	$3,250	2/06	SWIN	John Deere 444 CH (low tin), John Deere 213 bean head, diesel, cab, rice tires, radio
4400			G	$3,750	3/06	NWIL	444 CH
4400	1970		F	$600	5/06	SEWY	Cab, gas
4400	1970		F	$700	8/06	ECMI	Gas, John Deere 13' floating grain platform, straw chopper

Combines

John Deere

Model	Year	Hours	Cond.	Price	Date	Area	Comments
4400	1974	3,938	G	$2,000	9/06	ECNE	Gear drive, straw chopper, 18.4-35 front tires, 11L-16 rear
4400		1,654	G	$1,400	1/05	NENE	Never used after $6,000 in repairs, gas
4400		2,200	G	$1,900	7/05	WCMN	Diesel, cab, no heads
4400			G	$2,400	8/05	SEMN	
4400	1971	3,573	G	$350	3/05	SWIN	Gas, chopper
4400	1971	1,807	F	$550	3/05	SEWY	Cab, gas
4400	1971	1,570	G	$600	9/05	NECO	Cab
4400	1974	3,660	G	$1,900	11/05	ECNE	Diesel
4400	1974	3,358	F	$2,000	4/05	SCMI	John Deere 16' grain head
4400	1974	2,687	G	$2,400	11/05	ECNE	Diesel
4400	1977	3,600	G	$2,950	12/05	SEMN	Straw chopper, monitor, air, AHHC
4400	1977	2,911	G	$4,200	8/05	NCOH	John Deere 215 flex grain head, CAH, 23.1-26 front rubber, 12.4-24 rear, rebuilt sieves/shaker
45			G	$3,900	7/07	ECND	Original condition, field-ready
45			F	$275	3/06	NCWI	45E, 10' grain head, round back
55			G	$400	1/07	ECNE	
55			F	$650	9/07	NWIL	Gas, John Deere 235 head, 12' platform
6600	1970		G	$3,500	2/08	WCOK	Cab, air, 20' header, hydrostat
6600	1974	3,406	G	$4,500	1/08	WCKS	Hydrostat, 20' platform
6600	1975		G	$1,250	2/08	SEIA	Cab, gear drive, chopper
6600		3,714	F	$1,100	1/07	SCNE	Diesel
6600		1,835	F	$5,000	4/07	NCIA	Started and ran good, hydro, rotary fan
6600	1972		F	$725	12/07	WCNE	Diesel, 18' pickup head
6600	1973		E	$3,800	2/07	SCKS	Cab, air, late chopper, 20', air, new motor
6600	1975	4,499	F	$1,250	2/07	WCOH	
6600	1975	4,668	G	$1,250	10/07	NCMI	Diesel, CAH, standard trans.
6600	1975		G	$1,500	2/07	NENE	Hydro
6600	1976	4,343	G	$2,900	2/07	SWIL	Mauer bin ext., chaff spreader
6600	1976	1,400	G	$10,750	8/07	SWSK	Canadian sale, 212 pickup, variable-speed belt driven, 1,400 hours, 128 hp., diesel engine, chopper
6600	1977	3,364	G	$7,300	9/07	NECO	John Deere 643 CH, diesel
6600	1978		G	$2,300	8/07	SESD	Hydro
6600	1978	4,700	G	$2,500	3/07	WCIL	Duals, diesel, no heads
6600			F	$500	12/06	SCMN	
6600			F	$1,200	8/06	NECO	
6600			F	$1,500	8/06	SEMN	Hydro
6600		4,799	G	$2,050	9/06	SCMN	No heads, diesel, hydro
6600		3,052	G	$2,700	3/06	NENE	23.1-26 front tires, 11:00-16 rear
6600		2,325	G	$3,750	8/06	NWIL	Diesel, chopper, spreader
6600			G	$3,900	10/06	WCWI	John Deere 215 bean and grain head
6600	1973	3,212	P	$400	9/06	NECO	Engine locked, diesel
6600	1974	4,813	G	$1,500	9/06	NENE	Diesel, variable speed, 18.4-26 front tires, 11L-16 rear tires, rear weights, straw spreader
6600	1974		F	$2,000	1/06	SCMI	300 hours on OH, 13' grain table
6600	1975		G	$1,100	6/06	NWMN	Hydro, pickup head
6600	1977		F	$1,000	3/06	NEMI	Diesel, 16' grain head
6600	1978		G	$3,350	1/06	SCMN	Hydro, cab, air, diesel
6600			F	$700	1/05	ECNE	
6600			F	$700	3/05	SEIA	

Combines

John Deere

Model	Year	Hours	Cond.	Price	Date	Area	Comments
6600			F	$1,300	12/05	WCWI	
6600			F	$1,600	9/05	NWOR	
6600			G	$1,700	8/05	NWIA	John Deere 444, 4RW CH and John Deere 454, 4RW crop head
6600			P	$2,100	11/05	NEIA	Rough, no heads
6600			G	$2,500	8/05	NEIA	
6600		3,175	F	$3,650	9/05	SEIA	
6600			G	$3,700	7/05	SCMN	
6600			F	$3,900	4/05	SCMI	John Deere 216 head, John Deere 443 CH, diesel, flat screen
6600	1971	4,053	F	$900	4/05	SCMI	Diesel, flat screen, no heads
6600	1975	3,875	G	$4,400	2/05	SWIN	Gear drive, chopper, diesel
6600	1976	2,443	G	$2,200	3/05	SWIN	Diesel, chopper, rice tires
6600	1976	3,823	G	$4,800	9/05	WCMN	All new rubber, 23.1-26 fronts, add-on auto header (robot), three years on new cylinder, concaves, new chopper
6600	1977	3,182	F	$3,600	8/05	NWIL	Gear drive, diesel, chopper
6600	1977		F	$3,750	10/05	NCOH	215 grain head, hydrostatic
7700			P	$1,300	12/07	NCIA	Gear drive, runs but needs work
7700			F	$2,300	12/07	NCIA	Turbo, hydro, chopper, tank extension, 28L-26 tires, engine & hydro OK but machine needs work
7700			G	$2,500	3/07	NEWI	
7700	1975		F	$2,400	3/07	WCIL	Turbo, hydro, 4WD, chopper, floater tires, runs
7700	1975	3,129	G	$3,900	8/07	WCIL	Hydro, 4WD, 28L-26 tires, air, diesel
7700	1977		F	$3,050	1/07	NWOH	Chopper
7700	1978	5,176	G	$4,600	7/07	WCKS	24' platform, hydro
7700			G	$1,000	12/06	SCMT	
7700			G	$1,300	3/06	SCIN	
7700			P	$2,050	8/06	NWIL	Hydro
7700			G	$2,500	3/06	NEKS	No heads
7700		6,000	G	$3,250	6/06	NCIL	1978 model, turbo, hydro, chopper, air
7700			G	$3,450	3/06	NWIL	Diesel, hydro
7700		4,795	F	$6,500	3/06	NWPA	4WD, turbo, hydro, chopper
7700	1973		F	$1,450	3/06	SWOH	Diesel, John Deere 216 grain head, 6R CH
7700	1973		F	$2,300	2/06	WCNE	Turbo, cab, monitor
7700	1976	3,089	G	$2,000	11/06	NEKS	Turbo, hydro, rear-wheel drive, chopper
7700	1976	3,356	F	$2,700	9/06	NWIL	Hyd. drive, nice tires
7700	1976	5,580	G	$6,400	3/06	SWOH	John Deere 218 grain head, diesel, hyd. drive
7700	1977		G	$3,900	8/06	NCIA	Turbo, hydro, chopper
7700	1978	4,819	G	$20,000	3/06	ECND	Turbo, hydro, converted edible bean combine, belted grain tank, unloading system, slow speed cup elevator, chain cyl. slow-down drive, spike tooth, chopper and spreader, 20' universal screened bottom head and pickup
7700			P	$300	12/05	WCCA	Hydro
7700			F	$950	12/05	SCND	
7700			F	$1,100	9/05	NWOR	
7700			F	$1,100	9/05	NWOR	

Combines

John Deere

Model	Year	Hours	Cond.	Price	Date	Area	Comments
7700			F	$2,800	12/05	NWIL	Diesel, no heads
7700		4,404	G	$9,500	8/05	SWOH	John Deere 915 grain table
7700	1972	3,126	G	$2,300	9/05	WCNE	Diesel, new sieves, straw chopper
7700	1974		G	$3,150	2/05	ECMI	Turbo, diesel, John Deere 220 20' flex head
7700	1976	5,700	P	$1,300	4/05	NCOH	Rough, hydro, turbo, diesel, 28-36 tires, straw chopper
7700	1978	3,200	G	$5,500	1/05	NEIA	Turbo, rear wheel door, diesel, fully equipped cab, large grain bin extension
7721	1978		G	$3,200	7/05	ECND	John Deere 6-belt pickup, pt, chopper
95	1967		G	$300	9/05	WCNE	Gas, cab
N/A	1929		F	$150	1/08	WCKS	

Massey-Harris

Model	Year	Hours	Cond.	Price	Date	Area	Comments
92	1958		P	$150	2/06	WCNE	14' wheat header, engine locked

McCormick

Model	Year	Hours	Cond.	Price	Date	Area	Comments
61	1939		F	$200	11/05	NCOH	Old pt combine

Massey Ferguson

Model	Year	Hours	Cond.	Price	Date	Area	Comments
300			F	$200	12/06	ECMO	2R CH, 13' rigid head
300			F	$600	8/05	SEMI	Gas, 13' grain head
410			F	$550	8/06	SEMN	
410	1972		G	$325	9/06	ECNE	Chevy 292 gas engine, Massey-Ferguson 44 4R-36 CH, 18.4-25 front tires, rear weights
510			G	$1,150	2/08	SCMN	Duals, diesel, 20' head
510			P	$75	1/07	SWKS	For parts, rigid header
510			P	$75	1/07	SWKS	For parts, rigid header
510			P	$250	10/06	SWNE	Diesel, platform head
510			G	$1,050	9/06	SCMN	4R ch
510			P	$350	4/05	SEND	Perkins engine
510			G	$2,000	11/05	NCOH	Diesel, no heads
510	1976		F	$650	5/05	WCNE	Diesel, no reverse gear
510	1977		G	$1,350	8/05	NWIA	Hydro, factory air, 15' bean platform, quick-cut sickle, 4R-30 offset CH
540	1978	1,766	G	$2,250	3/08	NEMO	Hydro, diesel, 23.1-26 & 11L-16 tires, only 1,766 hours, 13' grain head
540	1978		G	$3,500	3/06	NWIL	Perkins 318 diesel engine, 4R CH and platform
550	1978		G	$2,750	7/07	WCSK	Canadian sale
550			G	$1,350	1/05	ECNE	
550			G	$1,600	11/05	ECNE	Hydro, diesel, 23.1-26 rubber
550			G	$2,100	8/05	SEMN	
550		2,860	G	$6,000	3/05	NCOH	MF 1143 4R CH and 13' grain head, Gilchrest 4WD, 356 engine, hydrostatic drive
750			F	$1,200	10/07	NEKS	Cab, no heads
750			F	$2,200	8/07	NWIL	Diesel, no heads
750		1,590	G	$3,400	12/07	SEND	Gear drive
750	1974		F	$1,000	4/07	NECO	20' wheat header, diesel

Combines

Massey Ferguson

Model	Year	Hours	Cond.	Price	Date	Area	Comments
750			F	$1,700	8/06	NWIL	6R CH
750	1973		P	$850	9/06	NWIL	6R CH, diesel
750	1976		F	$500	3/06	NWKS	Perkins 372 diesel, 23.1-26 tires, MF 1859 24' rigid head, Milo guards
750			P	$300	2/05	WCIL	For salvage, diesel
750			F	$500	6/05	WCMN	20' head
750		2,381	G	$1,100	9/05	NWIL	Hydro, gray cab, one owner, nice
750			F	$1,700	6/05	SWMN	1859 20' flex head, U2 reel
750			G	$3,400	2/05	WCWI	Gray cab, no heads
750		3,000	G	$3,800	11/05	SCMI	MF 44 CH, combine
750	1975		F	$1,100	1/05	NCIA	Cab, newer motor, diesel, hydro
750	1977	2,716	G	$1,800	8/05	NCIA	Air, hydro, chopper, tank extension, motor rebuilt
760	1977		F	$1,000	12/07	WCMN	Perkins diesel, specialty rotor, chopper, rock trap, gear drive, 24.5-32 tires, 30% rubber
760	1978	4,841	G	$2,250	7/07	NCND	V-8 Perkins diesel, hydro, straw chopper, 378 6-belt Melroe pickup, gray cab, red auger
760	1972		G	$1,150	9/06	NWOH	Gear drive, chopper, Massey 16' grain head, U2 reel
760	1978		P	$425	11/06	WCMN	24.5-32 tires
760	1978	2,790	G	$1,500	7/06	NWMN	Chopper, tank extension, auger extension
760	1972		G	$1,700	3/05	NECO	
760	1973	3,188	F	$2,700	6/05	SEND	V-8, hydro, chopper, diesel
760	1974		P	$850	5/05	NCKS	24' header, really rough
760	1977		G	$1,300	12/05	NCCO	MF 63C 6R CH
760	1978		F	$1,000	8/05	NWIA	V-8 engine, hydrostat, 8R CH, 20' bean platform

Minneapolis-Moline

Model	Year	Hours	Cond.	Price	Date	Area	Comments
H3			G	$275	8/07	NWOH	Wisconsin ZV4 engine, pt
H3			G	$275	8/07	NWOH	Wisconsin V-4 engine, original, pt

New Holland

Model	Year	Hours	Cond.	Price	Date	Area	Comments
1400	1976		F	$1,000	4/07	NECO	6R-30 CH, diesel
1400	1977		G	$1,400	7/07	WCSK	Canadian sale, diesel, new traction drive belts, new bearings in pickup
1500			G	$1,400	7/07	WCSK	Canadian sale, shedded, new belts, field ready, SN 3906203
TR 70		3,169	F	$1,700	5/07	NWSK	Canadian sale, 3208 Cat diesel, rake-up 12' pickup, twin S-cube rotor, hopper extension, factory spreader
TR 70			F	$1,000	9/06	NCIA	
TR 70		4,071	G	$1,500	8/06	NCOH	Chopper, 15' grain head
TR 70		3,057	F	$2,000	2/05	WCOH	15.5 grain table, floating cutter bar, Cat 3208 diesel
TR 70			F	$2,500	9/05	NENE	4R CH

White

Model	Year	Hours	Cond.	Price	Date	Area	Comments
8600	1977	1,200	G	$5,600	3/06	ECMI	Harvest Boss 2WD, hydro, White 15' floating cutter bar grain platform and White 704N 4R CH

Construction Equipment

This is actually one of the largest categories I track auction sale price data on. We lump most all types of heavy equipment in this category, things like crawlers, dozers, backhoes, trenchers, road graders, wheel loaders, cranes, and forklifts. So it's expansive.

Interesting time to track used equipment values on this stuff. For years I've noticed how older crawlers, dozers, and backhoes from the 1960s and earlier were still considered viable, valuable, productive assets for all different types of farming operations. Now we're beginning to see this trend with tractors. It's no longer odd to find serviceable tractors with 10,000 to 15,000 hours selling on the used market.

With the price of newer late-model crawlers, dozers, and backhoes not getting any cheaper, I think demand for stuff from the 1950s, '60s and '70s should still be in demand in the coming decade.

Construction Equipment

Allis-Chalmers

Model	Year	Hours	Cond.	Price	Date	Area	Comments
100			G	$8,400	1/06	NWIA	M100 road grader, 12' blade, double snow blade, nice
545			G	$7,000	11/06	SWCA	Rubber-tired loader
545			G	$7,000	11/06	SWCA	Rubber-tired loader, articulated frame, enclosed cab
545		2,488	G	$8,000	11/05	SWCA	Rubber-tired loader, canopy
645			G	$3,000	6/07	WCMO	Diesel engine, 3-speed trans., 6' bkt., articulating, 20.5-25 tires
645			G	$9,500	11/05	NECO	Wheel loader, articulating
840C		7,847	G	$3,250	5/06	SWCA	Rubber-tired loader, canopy
ACP-80	1975	3,704	G	$4,250	1/06	NECO	2 stage, 7' mast
D			F	$4,500	12/07	NWMO	Motor grader, diesel, orange in color
D			G	$3,200	9/06	NEIN	Grader, 4-cyl. gas, 12' moldboard
D			G	$3,400	3/06	ECIN	Grader, scarifier, hyd. side shift, tilting wheels
D			G	$7,000	6/06	NEIN	Cab, scarifier, ps side shift
G6			G	$6,250	8/06	WCMN	Crawler/loader, 13" tracks, 6.5' bkt., TS-6
HD 10			F	$2,000	3/08	ECNE	Cable dozer, 14' blade
HD 11			G	$3,100	9/06	NEIN	Crawler, canopy, straight blade, tilt
HD 11			G	$4,500	11/05	NCCA	Crawler
HD 16			G	$3,000	4/08	ECND	Dozer, manual trans., cable lift
HD 16			G	$1,750	6/07	SWIA	AC 16000H 250-hp. diesel engine, PS, open ROPS canopy, sweeps, Garwood cable blade, drawbar
HD 16			G	$6,000	5/06	SCCA	
HD 6	1958		F	$6,750	8/07	NWIL	Dozer, some new parts, 8' hyd. lift blade, cage, tracks rebuilt
HD 6	1956		G	$4,600	8/06	NEKS	HD6-HA crawler tractor, 4-cyl. diesel engine, 16" tracks, 10' dozer blade, recent OH
HD 6	1960		F	$2,950	4/06	SWIN	Bulldozer
HD 7			P	$600	5/07	ECND	Crawler, 3 cyl., for parts
HD 5			G	$3,700	6/05	SEND	Crawler, hyd. dozer, GM engine
K			F	$900	6/07	NCIN	Crawler with hyd. front bkt. loader, unrestored
K			G	$2,200	8/06	SCIL	4 cyl., gas, crawler
N			G	$50	5/06	SCCA	Ditch closer attachment
N			G	$6,500	3/05	ECMN	Grader
N/A		4,148	F	$750	6/07	NCIN	Forklift, FPL60-24, LP, forks, no brakes
N/A			F	$600	3/06	WCMI	3,000 gas forklift
N/A			F	$900	11/06	ECNE	Forklift, C1000
N/A			G	$2,800	4/06	NWSD	Road grader
N/A			F	$950	3/05	NCIL	Motor grader, diesel

Case

Model	Year	Hours	Cond.	Price	Date	Area	Comments
1000			G	$3,000	1/07	SEKS	Crawler dozer, 9' hyd. straight blade
1450			G	$11,000	6/06	NEIN	Crawler loader, gp bkt.
1450			G	$13,000	12/06	ECIN	Crawler, canopy, straight blade
1450	1976		G	$13,500	9/06	ECMN	24" pads, 119" blade, power tilt, OROPS
1450		2,200	G	$17,500	9/05	NWOR	Crawler dozer, brush rake, 4-way blade
310			G	$5,500	3/05	ECMN	Crawler dozer

Construction Equipment

Case

Model	Year	Hours	Cond.	Price	Date	Area	Comments
350			G	$8,000	12/06	ECIN	Diesel, canopy, sweeps, 6-way blade
430	1965		G	$6,200	2/07	ECIL	Construction King backhoe/loader, gas, 14.9-24 tires, 6' front bkt., 3' backhoe bkt., nice for age
450			G	$4,900	3/07	SEMI	Crawler loader, diesel, gp bkt.
450		7,283	F	$5,500	2/07	NEIN	Bulldozer, rollover bar, 8' adjustable blade
450			G	$11,000	9/07	ECMN	Crawler dozer, 6-way blade, OROPS
450			G	$5,200	11/06	SEIA	Dozer, 4-way blade
450			G	$5,500	9/06	NEIN	Crawler, canopy, 16" pads
450			G	$5,750	6/06	NEIN	Crawler loader
450		1,622	G	$6,500	12/06	NEMN	
450			G	$6,600	6/06	NEIN	Canopy, 6-way blade
450			G	$7,000	3/06	SCIN	6-way blade
450			G	$7,000	12/06	SWNY	Crawler, diesel, EROPS, gp bkt.
450			G	$8,250	3/06	ECIN	Crawler loader, John Deere 207 turbo diesel, new sprockets, rails and starter
450			F	$8,250	3/06	NWIL	6' bkt.
450			G	$5,250	12/05	SWMS	OROPS, 6-way blade
450			G	$9,000	3/05	ECMN	Crawler loader, 4-in-1 bkt.
580	1966		G	$1,500	6/07	SCNE	Construction King, diesel engine, front end loader, Henry breaker
580B	1976	7,500	G	$9,250	12/05	SEND	Cab, heat, 57 hp., shuttle shift trans.
580C			F	$7,000	1/08	NEMO	Loader
580C	1976		F	$3,000	5/08	ECMN	Industrial loader, shuttle shift, diesel
580C		2,240	G	$13,000	11/07	WCOH	Backhoe, cab
580C			G	$13,000	1/07	WCOH	4-in-1 bkt., tractor/loader/backhoe, diesel
580C			G	$6,000	5/06	SCCA	Loader, EROPS, 12"
580C		5,740	G	$7,500	11/06	SETN	24" bkt.
580C			G	$8,500	3/06	SCIN	Standard hoe
580C			G	$8,500	9/06	NEIN	Forklift, diesel, cab
580C			G	$9,000	7/06	WCCA	Bkt., canopy, 24" hoe bkt.
580C		5,822	F	$9,100	8/06	NWOH	Backhoe, cab
580C			G	$9,500	11/06	SWCA	Canopy
580C			G	$13,000	3/06	WCWA	Loader backhoe, 80" gp bkt., OROPS
580C			G	$4,500	12/05	WCNJ	Tractor/loader/backhoe, diesel, OROPS, digging bkt.
580C			G	$7,000	11/05	SETX	OROPS, diesel, digging bkt.
580C			G	$7,000	12/05	ECIN	4-in-1 bkt.
580C			G	$9,750	3/05	WCCA	GPS bkt., diesel, loader/backhoe, canopy
580C			G	$11,500	11/05	SWCA	Canopy
580C	1978	6,707	G	$7,000	8/05	NEIN	ROPS, Extend-A-Hoe, tractor/loader/backhoe
750			G	$4,000	12/06	WCIL	Crawler, bkt., street pads, uses some oil
850			G	$9,250	9/06	NWTN	Crawler dozer
850			G	$15,500	6/06	NEIN	6-way blade
850	1975		G	$9,500	2/05	WCIL	Dozer
850B			G	$5,000	12/06	NEMN	Excavator
850B			G	$16,500	9/06	NWTN	Crawler dozer, 6-way blade, OROPS, brush guard
850B	1978		G	$15,500	7/06	NWOH	Dozer, PS, 6-way blade, newer undercarriage and trans.
W11			G	$10,000	6/05	ECMN	Rubber-tired loader
W14			G	$5,000	2/08	ECMT	Wheel loader, 4WD, articulating, cab, 4-cyl. diesel, 1.5-yard bkt.

Construction Equipment

Case

Model	Year	Hours	Cond.	Price	Date	Area	Comments
W14			G	$10,000	11/05	SWCA	Rubber-tired loader, canopy
W36			F	$6,000	6/06	NECO	Payloader, cab, 3.5 yard bkt., diesel
W4			G	$2,750	4/07	NEMO	Loader, 4×4, diesel hydrostat, articulate steer, 5' bkt.
W4			G	$4,000	3/05	ECMN	Rubber-tired loader
W7			G	$6,000	8/06	WCMN	Loader, all-wheel drive, 1½- to 2-yard bkt., cab heat, locally owned, 15.5 rubber
W7			F	$4,800	6/05	WCMN	Payloader, 6' bkt., no cab, 4WD
W7	1970	2,895	G	$5,500	8/05	NCMI	1¼-yard bkt., 75 hp.
W9		3,009	G	$2,000	8/05	NCMI	Straight frame wheel loader, diesel, 17.5-25 tires
W9			F	$2,900	9/05	NECO	Roll cage, payloader

Caterpillar

Model	Year	Hours	Cond.	Price	Date	Area	Comments
112			F	$2,400	4/07	SCOK	Road grader
112			G	$4,000	8/06	WCTX	Motor grader, diesel, EROPS
112			G	$14,000	8/06	WCTX	Motor grader, EROPS
12			P	$1,250	1/08	WCKS	Motor grader, bad pony motor
12			F	$1,500	4/07	SCKS	Road grader, 12' blade, pony motor, needed work, diesel
12			F	$2,600	7/07	WCSD	Road grader, 14' moldboard
12			G	$3,250	11/07	SEID	Road grader, 12' moldboard, pony motor start
12			G	$7,500	8/07	ECMO	Cat diesel engine, 6-speed hi/lo trans., 14' hyd. blade, cab, heat, wheel tilt, scarifier, 13.00-24 TG tires
12			F	$1,500	2/06	WCSD	Road grader, yellow
12			P	$2,000	3/06	SWNE	Motor grader, needs work, 12' blade, 13:00-24 tires
12			G	$5,500	8/06	WCTX	Motor grader, diesel, cab, air, scarifier
12			G	$300	6/05	ECWI	Motor grader, diesel, EROPS, moldboard
12			F	$2,300	9/05	SEIA	Pony start, grader
12			F	$3,800	2/05	NENE	Motor grader, diesel, pony motor, 13:00-24 tires, 14' blade
12			G	$4,000	3/05	ECCO	Motor grader, pony start
12			G	$4,250	10/05	NWMT	Motor grader
12			G	$5,500	3/05	NEMT	12' moldboard, mechanical
120	1977		G	$13,000	1/06	NECO	14' boards
120			G	$8,000	12/05	SWMS	EROPS, 14' moldboard
14			G	$7,200	2/07	NECA	Motor grader, front rippers
16			G	$3,250	2/07	SEFL	Motor grader, cat diesel
20			F	$2,000	3/08	ECNE	Dozer, 4 cyl., gas, no blade, late 1920s to early 1930s
212	1953		G	$5,000	3/07	NWIL	Road grader, diesel, cab
212			F	$800	9/06	NEIN	Grader, diesel, pony motor, cab
22			F	$1,000	4/08	NENE	Crawler tractor, 4-cyl. gas engine, 10" tracks, drawbar
22			F	$2,500	9/05	NENE	Crawler tractor, 4-cyl. gas engine, 16" tracks, drawbar
22			F	$750	8/06	SCIL	
22			F	$2,100	6/05	ECWI	
225	1978	15,000	G	$12,500	3/08	ECNE	Excavator, SN 51U2766, Cat 3208 diesel engine, hydroshuttle trans., cab, standard stick, 42" bkt., teeth, good undercarriage and track

Construction Equipment

Caterpillar

Model	Year	Hours	Cond.	Price	Date	Area	Comments
40			P	$300	8/06	SCIL	Parts unit, dozer
44			F	$225	8/06	SCIL	Pt road grader
44			F	$325	8/06	SCIL	Pt road grader
621	1967		G	$23,000	11/07	SEID	Paddle scraper, 3406 Cat, 21-yard, new cutting edge, good tires
621			F	$6,500	3/06	SEIA	Scraper, 3306 6-cyl. engine conversions, PS, 23H Series, good rubber
621			G	$6,500	12/05	ECIN	New turbo injectors
70			G	$10,500	2/08	ECNE	Cable dirt scraper, converted to hyd., hyd. push off, 18.00-24 tires
910			F	$5,100	1/08	WCTN	Front end loader, 3204 Cat motor
910			G	$7,500	3/07	SCTX	Rubber-tired loader, SN 80U4740 powered by Cat 3204 diesel engine, 65 hp., equipped with 3F/1R ps trans., OROPS, general purpose bkt., excellent rubber
910	1978		G	$17,000	7/07	SENE	Wheel loader, cab, 15.5-25 tires, 1¼-cubic-yard bkt.
910		1,773	G	$8,000	5/06	SWCA	Rubber-tired loader, cab
910		683	G	$12,750	5/06	SWCA	Rubber-tired loader, grapple bkt., canopy
910		777	G	$14,000	11/05	SWCA	Rubber-tired loader, canopy
910	1977		G	$13,000	6/05	ECNE	Loader, gp bkt., cab
920	1973		G	$21,500	3/08	WCKS	Payloader, extended scoop with loading grain
920			G	$14,000	5/06	SWCA	Rubber-tired loader, cab
920			G	$12,500	9/05	NCTN	Wheel loader, 4-in-1 bkt., canopy
920			G	$13,500	11/05	NECO	Wheel loader, articulating
920			G	$14,000	9/05	NCTN	Wheel loader, forks, cab
922			F	$3,500	4/08	NEIA	Rubber-tired loader
922			G	$6,000	9/05	NECO	Cab, 3 hyd., payloader
922A			G	$6,200	10/05	ECMN	Loader
922B	1966		G	$8,500	12/07	SEND	Payloader
922B			G	$3,000	10/06	SENJ	Front end loader
922B	1967		G	$8,000	11/05	NWMN	Payloader, rear steer
930	1974		G	$9,500	1/07	WCIL	4WD, loader, diesel, shuttle trans., ps, hyd. controls, 8' bkt., 4,000-lb. capacity
930			G	$13,500	7/06	ECNY	Rubber-tired loader, diesel, EROPS, gp bkt.
930			G	$15,000	5/06	SWCA	Rubber-tired loader, canopy
930			G	$15,500	5/06	SWCA	Rubber-tired loader, canopy
930			G	$20,000	9/06	WCTX	Rubber-tired loader, diesel, EROPS, gp bkt.
930			G	$20,500	1/06	WCPA	Rubber-tired loader, diesel, EROPS, gp bkt.
930	1970		F	$16,200	2/06	NCCO	4WD loader, brakes out, OH 7 years ago
930		933	G	$26,000	9/05	ECMN	Third valve, CB radio, 2½-yard bkt.
930	1973		G	$16,500	6/05	ECNE	Loader, gp bkt., cab
950	1974		G	$19,500	3/08	ECMN	Wheel loader
950			G	$13,000	3/07	WCNE	Wheel loader, 4-cylinder diesel engine, 2-yard bkt., canopy, 20.5-25 tires
950			G	$18,000	1/07	WCOH	Rubber-tired loader, Cat 3304 diesel, gp bkt.
950			G	$18,000	1/07	WCNY	Rubber-tired loader
950			G	$20,000	3/07	NETX	2-yard bkt. cab
950			G	$17,750	12/06	ECIN	Rubber-tired loader, 4-in-1 bkt.
950			G	$24,500	3/06	ECIN	Rubber-tired loader
950			G	$12,500	3/05	ECMN	Rubber-tired loader
950			G	$12,500	3/05	ECMN	Rubber-tired loader
950			G	$18,000	12/05	ECIN	Rubber-tired loader
950	1970		G	$18,250	2/05	NWKS	Wheel loader, 3 yard bkt.
950	1971		G	$16,500	6/05	ECNE	Loader, gp bkt., cab

Construction Equipment

Caterpillar

Model	Year	Hours	Cond.	Price	Date	Area	Comments
955			G	$17,500	8/07	ECMO	Crawler loader, Cat diesel engine, ps trans., ROPS canopy, 7' bkt., fresh steering clutches and brakes, excellent undercarriage, 15" pads
955			G	$3,700	6/06	NEIN	Crawler loader
955H			G	$2,250	11/06	SCCA	Crawler
955H			F	$11,000	3/05	WCNJ	Track loader, motor and undercarriage rebuilt five years ago, used very little
955L			G	$17,500	1/08	NEMO	Track loader
955L	1978		G	$30,000	1/08	WCIL	High lift, ROPS, new undercarriage, pedal steer, PS
955L	1978	9,455	G	$16,000	7/07	SCMO	Diesel engine, PS, OROPS canopy
955L			G	$4,500	7/06	WCPA	Crawler loader, Cat diesel, OROPS, gp bkt., recent engine work
955L			G	$9,000	8/06	SETN	Crawler loader, OROPS
955L		6,253	G	$21,000	9/06	NEIN	Crawler loader, canopy, 4-in-1 bkt.
955L			G	$8,000	6/05	ECPA	Track Loader
D2			G	$600	8/06	SCIL	Loader, bkt., blade
D2			G	$2,500	11/05	NCCA	Crawler
D2			F	$3,100	7/05	NCMN	Diesel, hyd. bkt., hydraulics weak
D3			G	$11,500	3/07	SEAR	6-way, OROPS
D3			G	$14,500	7/06	WCPA	Crawler, diesel, OROPS, 6-way blade
D3		623	G	$50,000	8/06	SETN	Dozer, OROPS, 6-way blade
D3			G	$9,000	9/05	NCTN	6-way blade, canopy
D3	1975		G	$8,500	6/05	ECNE	Tractor, Cat 3204 diesel engine, canopy
D4			G	$5,000	9/07	NCND	Crawler, dozer, stored inside
D4	1951		G	$4,800	11/07	ECKS	Dozer, hyd. lift, runs good
D4			G	$650	8/06	SCIL	
D4			G	$1,250	8/06	SCIL	Crawler loader, tracks, 4 cable loader
D4			G	$2,250	9/06	NEIN	Crawler loader, pony motor, diesel
D4	1948		F	$2,500	5/06	SEWY	Crawler, 8' dozer
D4	1959		E	$5,000	9/06	ECMI	Pony motor, hyd.
D4			G	$3,900	11/05	NCOH	Dozer, straight blade
D46U			G	$4,100	11/07	SWIN	Repainted, hyd. dozer, PTO, rebuilt pony, 30% sprockets, some new rollers
D6			F	$6,500	4/07	SCOK	D6 (8U) dozer, ROPS
D6			F	$3,200	4/06	SWMB	Canadian sale, straight cable dozer, likely running
D6	1960		G	$9,400	9/06	NWIL	Dozer, new radiator, 9.5' hyd. blade, new pins and bushings, diesel
D6			P	$600	2/05	SCCA	Crawler
D6			P	$740	12/05	NWIL	Cable lift dozer, for parts
D6				$3,000	3/05	NEMT	Crawler; 4R3017, 11' hyd. dozer, 18" tracks, pony start
D6B			G	$5,000	10/06	NCWI	Crawler, Cat diesel, ROPS, hyd. straight blade
D6B			G	$11,250	12/06	ECIN	Crawler, Leco canopy, sweeps, straight blade, tilt
D6B			F	$2,000	11/05	ECWA	Williams CAH, direct start, 3 live remotes, 6-roller, 20" pads, hyd. track adjusters, extended fuel tank
D6B			F	$2,250	11/05	ECWA	Crawler, cab, air, direct start, 6-roller, 20" pads, 5 hyd.
D6B			F	$3,250	11/05	ECWA	Crawler, CAH, direct start, 3 live remotes, 6-roller, 20" pads, hydraulic track adjusters, extended fuel tank, second owner machine, 3,000 hours on new finals and rear end

Construction Equipment

Caterpillar

Model	Year	Hours	Cond.	Price	Date	Area	Comments
D6C	1965	8,431	G	$18,000	1/07	SEKS	Crawler dozer, PS, 10' straight blade, tree guard
D6C			G	$9,500	7/06	ECNY	Crawler, diesel, OROPS, angle blade, twin tilt
D6C			G	$12,000	3/06	ECMI	Lgp, wide track, 140 hp.
D6C			G	$15,750	3/06	ECIN	Crawler, turbo diesel, new rollers, canopy
D6C			G	$24,000	12/06	NEMN	Dozer
D6C	1965		G	$12,000	11/06	ECND	S dozer, tilt, excellent undercarriage, 24" hd. pads, 1,500 hours on new engine, Cat 25 CCU, third valve to back
D6C	1973		G	$41,000	5/06	ECNE	Dozer, straight blade with hyd. tilt, sweeps, OROPS canopy, 300 hours on new undercarriage, new starter
D6C	1974		G	$22,500	9/06	NEMO	D6C, dozer, ROPS, 12' 4-way blade, Cat 3306 engine
D6C	1978		G	$26,000	12/06	WCIL	Crawler, 1,400 hours on OH, three-way blade, winch, diesel
D6C			F	$8,000	2/05	SCCA	Trunion-mounted dozer blade, canopy
D6C			G	$8,250	12/05	ECWA	Crawler, Prentice knuckle boom, Onan generator and air compressor
D6C		4,642	G	$8,500	12/05	SWMS	OROPS, straight blade
D7	1968		G	$20,500	3/08	ECND	Pony start, hyd. dozer, double cable, rear winch, ROPS, PS
D7			G	$6,100	4/07	NEMO	Cable, stick, turbo, oil clutch, electric start, tight blade, good undercarriage
D7			G	$6,250	4/07	NCNE	Dozer blade
D7			G	$2,750	6/06	NEIN	Brush cab, cable 10' blade
D7			G	$3,200	1/06	WCOH	Crawler, diesel, canopy, 122" street blade
D7			G	$3,500	9/06	NEIN	Crawler, diesel, hyd. brakes, hyd. blade
D7			F	$3,500	4/06	WCMT	Dozer, electric start pony motor, 12' cable angle dozer with a 2-spool Cat cable control, main OH 3,622 hours ago, new grouser bars, pins turned once
D7	1950		F	$4,900	9/06	NWIL	Dozer, direct electric start, diesel
D7			G	$2,250	12/05	ECWA	Crawler
D7			G	$6,500	12/05	SEND	Cable dozer, canopy, shuttle shift, good starting engine
D7			G	$24,000	3/05	NEMT	11' tilt dozer, 22" tracks, Kelly RC7 ripper, ROPS cab
D7E	1965		G	$17,250	3/07	NWIL	Dozer, 200 hours on new rebuilt engine and turbo, pony motor start, Balderson 10' U-blade
D7E	1968		G	$25,000	7/07	SCMO	Diesel engine, PS, pedal steer, ROPS canopy, semi U-blade, all hyd., winch, sweeps
D7E			G	$13,000	12/06	ECIN	Crawler, OH by Cat 10/98, 500 hours since
D7E			G	$13,000	6/06	NEIN	ROPS, sweeps, straight blade, new trans.
D7E			F	$6,000	12/05	NENE	Crawler tractor, cable dozer, canopy, S-blade, Cat 25 winch
D7E			G	$16,000	7/05	ECND	Canopy, hyd. lift tilt
D7F			G	$17,500	1/08	NEMO	Dozer, winch
D7F			G	$37,500	7/07	SCMO	Diesel engine, PS, pedal steer, ROPS canopy, AM/FM/cassette, semi U-blade, all hyd., winch with 210' of cable, rear end OH, deep grousers, sweeps

Construction Equipment

Caterpillar

Model	Year	Hours	Cond.	Price	Date	Area	Comments
D7F			G	$15,000	11/06	SETN	Hyd. tilt blade
D7F			G	$21,000	3/06	SCIN	Straight blade, tilt, ROPS
D7F			G	$23,000	12/05	ECIN	Straight blade, tilt
D7G			G	$32,000	2/07	SEFL	Crawler Cat diesel, rebuilt engine
D7G			G	$40,500	3/07	SCTX	Crawler tractor powered by Cat 3306-T diesel engine, 200 hp., equipped with 3F/3R ps trans., OROPS, engine enclosures, sweeps, hyd. straight blade with tilt, 3-shank ripper
D7G		2,526	G	$45,000	6/07	SWIA	250-hp. diesel engine, pst trans., open ROPS canopy, fully enclosed engine covers, hyd. blade with tilt, single stick controls, Grace Track GT90-6 hyd. scraper system, drawbar
D7G	1977		G	$39,000	8/07	ECMO	All good glass, fresh paint, 11' semi-U 6-way blade, joystick controls, good undercarriage, new rollers, 26" pads, approximately 1,000 hours on complete OH, 200 hours on rebuilt trans. and clutches
D7G			G	$27,000	6/06	NEIN	Dozer, cab, straight blade, tilt, lpg
D7G			G	$28,000	3/06	SCIN	Dozer, ROPS
D7G			F	$10,500	2/05	SCCA	Canopy, 2 hyd.
D7G			G	$25,000	12/05	WCCA	Canopy, 6-way direct drive, 2 hyd.
D7G			G	$33,000	11/05	NEOH	Crawler, diesel, OROPS, rear screen, 7S blade, tilt
D8			G	$5,250	8/06	SCIL	
D8			G	$5,900	3/06	NENE	Dozer, 2-U straight stick, Cat 10' blade, tree guard, OROPS
D8			G	$21,000	2/06	WCME	Crawler dozer, ripper
D8H			G	$15,000	9/07	ECMN	Crawler dozer
D8H			G	$18,000	2/07	SEFL	Crawler, diesel, D89C rear-mount winch
D8H			G	$25,000	7/07	ECKS	Dozer, 270-hp. Cat diesel engine, PS, open ROPS canopy, hydraulic semi U-blade with tilt, drawbar, 70% undercarriage, new battery, new starter, new generator
D8H			G	$26,000	2/07	SEFL	Crawler, diesel, D89C rear-mount winch
D8H			G	$29,500	7/07	ECKS	Dozer, 270-hp. Cat diesel engine, PS, open ROPS canopy, hydraulic semi U-blade with tilt, drawbar, 70% undercarriage, new brakes, new clutches
D8H			G	$33,000	2/07	SEFL	Crawler, diesel, 2-barrel multishank, Cat ripper, 8A blade
D8H			G	$33,000	2/07	SEFL	Crawler, diesel, 2-barrel multishank, Cat ripper, 8A blade
D8H			G	$11,000	12/06	NEMN	Dozer
D8H			G	$9,500	9/05	ECMN	Crawler dozer
D8H			G	$9,500	11/05	NEOH	Crawler, diesel, OROPS, 8SU blade, tilt, rear drawbar
D8H			G	$10,000	6/05	ECMN	Crawler dozer, electric start
D8K			G	$26,000	2/07	SEFL	Crawler, diesel, OROPS, 8A blade
D8K	1978		G	$32,500	7/07	SCMO	D342 diesel engine, PS, pedal steer, ROPS canopy, semi U-blade, all hydraulic, sweeps
D8K			G	$16,500	9/05	NCTN	Crawler tractor, tilt, canopy

Construction Equipment

Caterpillar

Model	Year	Hours	Cond.	Price	Date	Area	Comments
D8K			G	$31,000	9/05	NCTN	Crawler tractor, tilt, canopy, screen recent out of frame
RD4			G	$425	8/06	SCIL	
RD4			G	$450	8/06	SCIL	Crawler
RD4			G	$750	8/06	SCIL	

Cletrac

Model	Year	Hours	Cond.	Price	Date	Area	Comments
BDH	1947		E	$3,900	10/07	NEKS	Dozer blade, electric start
HG42			F	$900	6/07	SEID	Crawler, tracks in fair condition, restorable
N/A			P	$500	4/05	NCOH	Dozer, parts only
N/A			F	$1,500	9/05	NENE	Oliver Cletrac crawler tractor

Cleveland

Model	Year	Hours	Cond.	Price	Date	Area	Comments
H	1920		G	$7,250	9/07	NEIL	Crawler, runs great

Dresser

Model	Year	Hours	Cond.	Price	Date	Area	Comments
TD8	1976		G	$7,000	8/05	NCMI	Crawler, 8.5 ton, 78 hp., 7'8" blade, 16" pads

Fiat-Allis

Model	Year	Hours	Cond.	Price	Date	Area	Comments
100			G	$4,500	12/05	SWMS	
10C			G	$14,500	9/05	NCTN	Crawler tractor, straight blade, canopy, sweeps, screen
14C			F	$9,200	12/05	NWIL	Crawler, 4-way blade, diesel, nice ROPS, hyd. tilt
21B			F	$5,000	11/06	ECND	Hyd. blade dozer, PS, ROPS canopy
345B			G	$10,000	6/05	ECMN	
345B			G	$13,000	1/05	NENE	Payloader, articulating, aux. hyd. 7.5' bkt., 15.5-25 tires
545			G	$8,000	4/07	NWOH	Rubber-tired loader
545			G	$8,000	3/05	ECMN	Rubber-tired loader
545			G	$9,250	9/05	NCTN	Wheel loader, gp bkt., forks, canopy
545B			G	$14,300	8/07	WCIL	Articulated wheel loader, diesel, EROPS
545B			G	$7,500	6/06	NEIN	Cab, 2-yard bkt.
545B		5,500	G	$15,000	1/06	NECO	1¾-yard bkt., grapple, articulating
545B	1972	7,682	G	$11,000	3/06	ECIL	Payloader, 2½-yard bkt., hi/lo and reverse
545B			G	$7,500	6/05	ECPA	
645B			G	$9,000	3/06	ECIN	Rubber-tired loader
645B			G	$13,000	11/06	SCCA	7 to 8 yard
645B	1976		G	$20,500	2/06	NCCO	4WD loader, 4 years on engine and trans. OH
7GB			G	$6,500	11/06	SETN	OROPS, crawler

Construction Equipment

Ford

Model	Year	Hours	Cond.	Price	Date	Area	Comments
3400				$0	8/05	ECMI	Industrial loader, low hours on OH, new bkt. and front tires
3500	1975		G	$7,400	11/05	WCIL	Backhoe
4500			F	$4,250	3/07	SCMI	Industrial backhoe, diesel, 6' material bkt., 2' hoe
4500		10,000	G	$4,500	10/07	NCMI	Backhoe, 2WD, diesel, 2' dig bkt., 10,000-plus hours
4500			G	$3,250	9/06	NEIN	Gas
4500			G	$3,900	3/06	SCIN	Standard hoe
A62			G	$10,000	1/07	WCIL	Loader, diesel, shuttle trans., hyd. brakes and controls
A62		8,665	G	$10,750	11/07	ECSD	Payloader, cab, heat
A62			G	$5,750	6/06	NEIN	Rubber-tired loader, cab, diesel
A62			G	$11,500	3/05	ECMN	Rubber-tired loader

Galion

Model	Year	Hours	Cond.	Price	Date	Area	Comments
118			G	$3,000	12/06	ECIN	Grader, Detroit diesel, hyd. side shift
118		1,522	G	$4,600	8/06	WCTX	Motor grader, diesel, EROPS, 12' moldboard
118B			G	$2,250	9/06	NEIN	Grader, diesel, scarifier
150T	1972		G	$16,000	3/05	ECND	Hyd. truck crane, 63' plus 15 jib, two winches, hook block

Hough

Model	Year	Hours	Cond.	Price	Date	Area	Comments
50	1970		G	$3,000	8/07	ECIA	Payloader, 2.5 yard bkt., 806 gas engine
60		7,144	G	$8,400	1/06	NECO	Articulating
60			G	$5,500	3/05	ECMN	Rubber-tired loader
65			G	$7,000	9/05	ECMN	Payloader
90			F	$2,500	8/06	SCMI	Payloader, 12' front blade
90			G	$4,500	11/06	SCCA	Wheel loader
H			F	$2,600	9/05	NWOR	Payloader
H50C			F	$500	1/06	WCNH	Rubber-tired loader, EROPS
H65C			G	$7,000	9/05	WCWA	Wheel loader, gp bkt.
H90C			G	$8,000	5/06	ECMI	Articulation front end loader, Detroit diesel
N/A	1973		G	$18,000	8/07	SESK	Canadian sale, wheel loader, V-6 Cummins, 3-speed shuttle shift, 2-yard bkt., SN 17408
N/A			G	$1,400	6/06	NEIN	Payloader, Hercules inline 6 engine, gas, 1½-yard bkt.
N/A			F	$500	8/05	NCMI	Rigid frame loader, 470 Detroit diesel
N/A	1942		F	$375	8/05	NEIN	Payloader, no brakes, engine knock

Huber

Model	Year	Hours	Cond.	Price	Date	Area	Comments
N/A			G	$5,750	11/06	SCCA	Motor grader, 14', Detroit diesel, 4 cyl.
N/A			G	$3,500	3/05	NEMT	12' moldboard plus 2' ext., scarifier, Detroit diesel, hydraulic controls, tandem

Hyster

Model	Year	Hours	Cond.	Price	Date	Area	Comments
N/A	1978		G	$1,300	3/08	ECND	Forklift

Construction Equipment

IHC

Model	Year	Hours	Cond.	Price	Date	Area	Comments
165	1977		P	$9,000	1/07	NCIL	Crawler loader, ROPS, not running
H65C	1973		P	$4,100	8/05	WCTX	Payloader, rough, bkt. sold separately, hyd. leak, rubber 40% to 50%
T20			F	$1,550	9/05	NEIN	Crawler, repainted
T5			G	$3,000	3/05	ECCO	Crawler loader, dual bkt., gas
T6			G	$6,750	8/06	SCIL	Crawler, new paint
T6	1940		G	$2,600	9/06	ECMI	Crawler, blade, runs excellent
T6	1955		F	$2,800	9/06	ECMI	Crawler
TD14			G	$8,100	4/06	SEND	Crawler, 9' hyd. dozer, winch
TD14	1952		F	$1,800	3/06	ECNE	Crawler dozer, diesel engine, gas
TD14A			P	$700	6/05	NEND	Hyd. dozer
TD 15			F	$4,000	3/08	SWNE	Crawler tractor, 12' dozer blade
TD 15B			G	$5,000	8/06	WCNC	Bulldozer
TD 15B			G	$13,000	6/06	NEIN	Crawler, canopy, straight blade, tilt
TD 15B			G	$5,500	3/05	ECMN	
TD 15B			G	$16,000	11/05	NECO	Crawler, angle blade, EROPS
TD 15C			G	$23,500	3/06	SEIA	Dozer, ROPS, sweeps
TD 15C			G	$13,250	3/05	NCAL	Dozer
TD 18			F	$2,400	7/07	SCSK	Canadian sale, not running
TD 18			F	$1,350	3/06	SEIA	Crawler tractor, no blade
TD 18			F	$3,500	9/05	SEIA	Crawler tractor
TD 20			G	$8,500	5/07	SCKS	Crawler dozer, Series B, 1968
TD 20E			G	$15,000	1/07	NENY	Crawler, diesel, hyd., straight blade
TD 24			F	$3,500	6/05	ECWI	
TD 25		1,004	G	$15,500	12/06	NEMN	Dozer
TD 25C			G	$8,000	12/06	ECIN	Crawler, single shank ripper, sweeps, straight blade
TD7			F	$4,750	3/06	ECMI	6-way blade, ROPS, sweeps, weak steering
TD7			G	$10,000	3/06	ECMI	6-way blade, PS
TD7E			G	$7,250	3/07	SEAR	6-way, OROPS
TD8			F	$6,000	2/08	WCMN	TD-8C, dozer, diesel, ROPS, PS, 6-way blade
TD8	1978		G	$9,100	8/06	SETN	Crawler, 6-way blade, ROPS
TD8E			G	$10,250	3/05	ECCO	Crawler dozer, cable plow, ROPS
TD9			G	$3,000	3/08	SCIL	Dozer, gas engine, 12 volt, 8' hyd. blade, rear winch
TD9	1968		G	$4,500	3/08	NWOH	Dozer, straight blade, brush rake
TD9			F	$3,250	6/06	NWMN	Crawler, dozer
TD9			G	$4,200	4/06	NWMN	Crawler, hyd. dozer
TD9	1954		P	$500	9/06	ECMI	Crawler
TD9			G	$7,300	6/05	SEND	Crawler, Bucyrus Erie hyd. dozer, 2 hyd.

John Deere

Model	Year	Hours	Cond.	Price	Date	Area	Comments
1010			F	$3,600	12/06	WCMN	Crawler, high-rise loader
1010			F	$3,550	8/05	SEMN	Crawler
300	1965		F	$2,500	10/07	NECO	Loader backhoe, diesel, no 3 pt., no PTO, no backhoe
300B			G	$4,700	12/06	ECMN	Loader, PTO, 3 pt., 8' back blade
300B	1975		G	$6,500	3/06	SWOH	300-BD backhoe, WF, diesel, 12' bkt., Model 250 rear unit
310A			G	$6,000	1/07	WCOH	Tractor/loader/backhoe, John Deere diesel, standard stick, ROPS, gp bkt.

Construction Equipment

John Deere

Model	Year	Hours	Cond.	Price	Date	Area	Comments
310A			G	$6,750	11/06	SWCA	Extend-A-Hoe
310A			G	$6,750	9/06	NEIN	Loader backhoe, canopy
310A			G	$7,500	6/05	ECMN	Tractor/loader/backhoe, 2WD, standard hoe, Wain Roy coupler, 24" hoe bkt.
310A	1977		G	$7,250	6/05	ECNE	Loader, backhoe
350	1966		G	$6,000	9/06	NEIN	Crawler loader, backhoe, forks, gas, 3 cyl.
350B			G	$8,400	9/05	SCON	Canadian sale, blade, roll cage, dozer
350B	1973	1,110	E	$14,500	12/05	NECO	Crawler, diesel, 6.5' 6-way hyd. blade
40			G	$1,200	3/07	NWIA	Forklift, gas, pneumatic tires
40	1955		G	$4,750	9/07	NEIL	Crawler, starts, runs, older restoration
400			G	$8,500	2/08	SCMN	Industrial loader backhoe
400			F	$1,400	12/06	WCMN	Industrial loader, 7' bkt.
410			G	$14,000	1/07	WCOH	Tractor/loader/backhoe, 4WD, John Deere diesel, standard hoe, gp bkt.
410		4,428	G	$5,000	9/06	SECA	
410		7,232	G	$6,750	12/06	NEMN	
410		7,232	G	$14,000	12/06	NEMN	
410			G	$7,000	8/05	SEMN	410 tractor/loader/backhoe
410			G	$8,500	3/05	WCCA	Loader/backhoe, diesel, GP bkt., canopy
410A		1,613	G	$10,000	3/07	NEIN	2WD backhoe, ROPS, standard 2' hoe bkt., 7' front bkt.
410A			G	$7,200	6/05	ECNE	Loader backhoe
410B		5,062	G	$8,500	9/07	SCMI	Backhoe, cab, 8' bkt., 2' hoe bkt.
410B			G	$9,500	1/07	WCOH	Tractor/loader/backhoe, John Deere diesel, 4-speed reversing trans., ROPS GP bkt.
410B			G	$4,500	6/06	SEID	Backhoe, OROPS, 24" digging bkt., 7' loader bkt.
410B			G	$7,500	11/06	SETN	24" bkt., OROPS
410B			G	$9,500	11/06	SCCA	EROPS
410C			G	$10,500	9/07	ECMN	4×4, Extend-A-Hoe
420			F	$6,500	3/06	SEPA	Dozer, rebuilt undercarriage
420	1956		G	$3,100	6/05	ECIL	6' blade, recent repairs
420C			P	$1,600	11/07	SWIN	Crawler loader, rough, undercarriage rough, Model 90 loader
440	1958		F	$3,500	12/07	ECNE	440LC, track loader, manual shift, shop-built boom, forks
440	1959		G	$5,000	3/07	NWIL	Backhoe, gas, Model 50 backhoe, Model 71 loader
440		4,705	G	$3,700	8/05	NWID	Loader/backhoe
450		4,492	F	$15,000	1/08	WCTN	Dozer, 6-way blade, new rails not installed
450	1968		G	$9,500	2/07	NENE	Crawler loader, OH engine with less than 200 hours
450			G	$13,000	8/06	WCIL	Crawler, bkt., forks
450			G	$12,000	3/05	NCAL	Crawler dozer, 6-way blade
450B			G	$9,000	12/06	SWNY	Crawler, John Deere diesel
450B	1974		G	$2,500	9/06	NEIN	Crawler loader, diesel, rubber pads, runs good, one steering clutch froze
450B			G	$5,500	10/05	NWMT	Crawler
450B			G	$10,250	12/05	SWMS	OROPS, 6-way blade
450B			F	$10,500	3/05	NEKS	Crawler dozer
450C			G	$8,600	3/07	SEAR	6-way, OROPS
450C			G	$7,000	12/05	ECIN	Crawler, canopy, sweeps, 6-way blade
450C			F	$7,850	3/06	SEPA	Dozer, ROPS, tilt blade

Construction Equipment

John Deere

Model	Year	Hours	Cond.	Price	Date	Area	Comments
450C			E	$16,500	12/06	SWWI	Dozer, undercarriage like new
450C	1974		G	$12,500	12/06	WCIL	Crawler, diesel, ROPS, 8.5' blade
450C			G	$6,000	6/05	ECMN	Crawler loader
450C			G	$8,500	12/05	ECIN	Dozer, cable plow, 6-way blade
450C			G	$8,500	12/05	ECIN	Cable plow, 6-way blade
450C			G	$10,000	12/05	SWMS	OROPS, 6-way
450C			G	$10,000	12/05	SWMS	OROPS, 6-way blade
450C			G	$10,500	8/05	ECMI	Dozer, 6-way blade, undercarriage fair
450C			G	$11,000	9/07	ECMN	Crawler dozer, 6-way bkt.
450C			F	$11,000	10/05	NEPA	Dozer
5010	1963	7,829	G	$7,500	4/07	NEND	Elevating motor scraper, 5010 industrial offset power unit with 401 scraper unit, cab, 23.5-25 drive tires, shows 7,829 hours
510	1978		F	$10,000	3/07	NEOH	Backhoe, loader
544			G	$8,500	11/06	SECA	ROPS
544			F	$15,000	8/06	NECO	Wheel loader
544	1970		G	$14,500	4/06	SEND	Wheel loader, cab, heater, 1½-yard bkt.
544A			G	$18,000	11/07	SEID	Wheel loader, 4WD articulating, diesel, shop-built quick attach, 17.5-25 tires, 2-yard quick-attach bkt., 5-prong bale fork
544A			G	$28,200	2/07	NCCO	4WD loader
544A			G	$2,500	10/06	WCFL	Rubber-tired loader, diesel, OROPS, pin on bkt.
544A			G	$7,000	12/06	SWNY	Rubber-tired loader, John Deere diesel, gp bkt., EROPS
544A			G	$9,500	9/05	ECMN	Rubber-tired loader, 2-yard bkt.
544A	1973	6,791	G	$12,500	6/05	ECNE	Wheel loader
544B			F	$8,600	12/07	WCNY	Wheel loader, cab
544B			F	$8,750	12/07	WCOK	Forks, bkt.
544B			G	$15,000	3/07	SEMI	4×4 articulating loader, EROPS, gp bkt.
544B	1976	5,463	G	$15,500	3/07	WCMI	Payloader, 8' material bkt.
544B			G	$11,000	9/07	ECMN	Closed cab, 8' bkt.
544B			G	$8,000	9/05	NCTN	Wheel loader, forks, cab
544B			F	$10,000	12/05	NWIL	Wheel loader, diesel
544B			F	$10,500	3/05	SCNY	Payloader, poor rubber
544B			G	$11,000	9/07	ECMN	Rubber-tired loader, third valve, 2-yard bkt.
600	1967		G	$8,400	4/06	SEND	Tractor loader backhoe, PS, roll bar
644			G	$8,500	4/07	NWOH	Rubber-tired loader
644			G	$5,500	7/06	ECNY	Rubber-tired loader, diesel, EROPS, gp bkt.
644			G	$9,000	3/05	ECMN	Rubber-tired loader
644A		7,784	G	$16,500	2/08	ECKS	Wheel loader, 20.5-25 12-ply tires, 4-in-1 bkt.
644A			G	$10,250	11/07	SEND	Payloader, 2½-yard bkt.
644A			G	$15,000	3/07	SEND	Payloader, cab, heat, 2-yard bkt., farm use only
644A	1973		G	$14,500	11/06	SCSD	Payloader, 3-yard bkt., newer trans.
644B			G	$22,000	12/07	SCMN	Wheel loader, 4-in-1 bkt.
644B	1977		G	$16,500	4/06	SEND	Wheel loader, cab, heater, 2½-yard bkt.
644B	1978	5,635	G	$34,000	11/06	ECND	Payloader, 4-in-1 bkt., one owner, always farm loader
750			F	$9,000	9/07	ECMN	Crawler dozer, straight blade
MC			G	$3,500	10/06	SCMN	Crawler, push blade, original condition, looks and runs good
MC	1950		F	$4,900	4/06	SEMI	Crawler, gas, AW55 5' blade

Construction Equipment

Komatsu

Model	Year	Hours	Cond.	Price	Date	Area	Comments
D21A-6			G	$12,500	9/05	NCTN	D21A, 6-way blade, good undercarriage
D21A-6			F	$14,175	12/07	WCOK	Dozer
D31			G	$15,250	3/06	NCMN	D31A, 6-way dozer
D31			G	$9,000	6/05	ECPA	
D31A			G	$10,000	4/07	NWNY	Crawler tractor SN 32777 powered by diesel engine, equipped with hydrostatic trans., ROPS, 6-way blade
D31A			G	$11,000	10/06	SWOH	Crawler, diesel, OROPS, 6-way blade
D31A	1977		G	$12,500	7/06	SEND	Dozer, fresh OH out of frame, nice tight machine
D57	1978		G	$7,250	11/06	ECPA	Crawler, loader, diesel, PS, gp bkt., teeth
D65E			G	$22,500	12/06	WCIL	6" dozer, 4-way 11' hyd. blade, ROPS, 24' pads
D65E			G	$7,500	8/05	NCMI	OROPS
D65E-6	1971		F	$17,000	9/05	SEIA	Dozer
D65E-6	1977		F	$21,000	9/05	SEIA	Dozer

Massey Ferguson

Model	Year	Hours	Cond.	Price	Date	Area	Comments
44	1972		F	$2,100	11/06	ECND	Wheel loader, 2-yard, 4WD, 4-wheel steer, diesel
450S				$0	6/05	NESD	Excavator, 33,510 lb.
450S	1970		F	$7,750	6/05	NESD	Excavator, 33,510 lb.
470	1965		G	$5,300	9/05	NECO	New paint, rebuilt
50			G	$1,200	11/06	ECPA	Rubber-tired loader, gas, 3 speed
50			G	$10,500	11/06	SWCA	4WD, 3 valve, rear scraper
50			G	$8,500	6/05	ECPA	4WD

Michigan

Model	Year	Hours	Cond.	Price	Date	Area	Comments
125			G	$3,750	6/05	ECMN	Rubber-tired loader
125B			G	$5,250	9/06	NEIN	Rubber-tired loader, diesel, cab, gp bkt.
125B			G	$4,000	3/05	ECMN	Rubber-tired loader, 4-yard bkt., cab, heat
125B			G	$6,500	3/05	ECMN	Rubber-tired loader
175			F	$4,000	11/07	SEND	Payloader, rebuilt 471 Detroit
175			G	$4,500	9/06	NEIN	Rubber-tired loader, Detroit diesel, cab, gp bkt.
175B			G	$4,250	1/07	WCOH	Rubber-tired loader, diesel, gp bkt.
175B			G	$4,800	3/06	SCIN	Articulated loader, 4-yard bkt.
175B			G	$5,500	3/06	ECIN	Rubber-tired loader, diesel, 5-yard bkt.
175B			G	$6,500	12/05	ECIN	Rubber-tired loader, diesel, 5-yard bkt.
175B			G	$6,500	3/05	ECMN	
210H			G	$2,600	10/06	SWOH	Motor scraper, Cummins diesel, Hancock self-loading rear scraper
275A			G	$8,000	12/05	SEND	Payloader, Cummins 350, high dump bulk material bkt.
35			G	$3,100	9/06	ECMN	Front end loader, Detroit diesel, all-wheel steer, 4WD
35AWS			G	$6,000	3/07	WCIL	Wheel loader, OH, cab, air, hyd. brakes
55			G	$5,200	2/08	WCMN	Payloader, cab, heat, 453 Detroit diesel, ROPS, 8' material bkt.

Construction Equipment

Michigan

Model	Year	Hours	Cond.	Price	Date	Area	Comments
55B			G	$9,000	3/06	ECIN	Rubber-tired loader
75			G	$4,500	12/06	ECMN	Front end loader, 8.5' bkt., closed cab
75A			F	$2,900	8/06	WCMN	Waukesha diesel engine, 1⅓-yard with round bale attachment, 14:00×24 rubber

N/A

Model	Year	Hours	Cond.	Price	Date	Area	Comments
N/A	1966		G	$8,500	2/05	SCMN	Trackmobile
N/A	1975		F	$4,250	2/05	SCMN	Trackmobile

Oliver

Model	Year	Hours	Cond.	Price	Date	Area	Comments
N/A	1951		E	$4,200	9/06	ECMI	HG-68 crawler, wide gauge, complete restoration

P&H

Model	Year	Hours	Cond.	Price	Date	Area	Comments
N/A	1969		G	$36,000	3/07	NETX	Crane, 40 ton, 200' reach with scales

Pettibone

Model	Year	Hours	Cond.	Price	Date	Area	Comments
N/A	1968		F	$6,000	1/07	SEAL	30-ton crane truck, aux. winch is nonfunctional

Swinger

Model	Year	Hours	Cond.	Price	Date	Area	Comments
200	1977		F	$4,250	11/06	ECND	Articulating loader, 4-cyl. gas engine, ROPS, dirt bkt. with pallet fork ext., new paint and tires, recent engine and pump work

Terex

Model	Year	Hours	Cond.	Price	Date	Area	Comments
TS-14B	1973		G	$8,500	7/07	ECKS	TS-14 scraper, twin Detroit diesel engines, 6-speed ps, 12 volt, hyd. bowl, low hours on rear OH

Toyota

Model	Year	Hours	Cond.	Price	Date	Area	Comments
FGC20	1970		F	$600	3/08	ECNE	Forklift, Model FGC20, 2 stage, 2,000 lbs., gas engine

Wabco

Model	Year	Hours	Cond.	Price	Date	Area	Comments
101	1974		G	$7,000	12/05	ECIN	Motor scraper

Yale

Model	Year	Hours	Cond.	Price	Date	Area	Comments
GLC050		5,587	G	$1,500	9/06	NEIN	Forklift, LP, 5,000 lbs.
GLP050	1978		G	$3,300	10/06	WCMA	5,000 pneumatic-tired forklift, 2 stage

OILS ARE BACKDATED

Internet discussions reveal a mountain of misinformation about oil and older engines.

"Everyone knows straight-weight oils are best," was a typical point made at one Web site. **"Never use detergent oils in any old tractor,"** came a reply.

The across-the-board response to these and similar opinions is simply that today's engine oils are not only backdated to service all older engines but are also far better at both lubricating and keeping the engine clean.

Many collectors get hung-up over the oil recommendations made in service manuals. These guidelines should not be seen as gospel but rather as a reflection of the best advice on oil quality at the time the tractors were built.

"Oil quality and its abilities have advanced far beyond that of oils that were sold 50 years ago, let alone just a decade ago," says Dan Arcy of Shell Oil.

Today's oils are superior when it comes to lubrication (engine oil's original requirement). And they're also better at removing operating contaminants such as carbon, ash, acids, and sludge. "Older oil formulations are best kept in squirt cans," Arcy adds.

Arcy acknowledges that today's oils are more expensive. But he quickly points out that not only do they last far longer in use, but they're also far more stable when an engine is not operating. "This is where modern lubricants excel in older engines in collectibles that are often stored for long periods of time," he adds. "Quality oils help prevent moisture from condensing on parts, which could lead to flash rust."

The quality of oil you burn in any engine (antique or modern, gas or diesel) has a huge impact on piston carbon deposits or oil pan sludge buildup.

Corn Pickers

How the escalating price of corn in late 2007 and into the first half of 2008 affected used farm equipment values, specifically harvesting equipment. Let me present exhibit A:

On a huge consignment auction in southeast Alabama on February 1, 2008, a one-row New Idea 323 picker sold for $2,000. That's the highest selling 323 picker I've seen since March 23, 1996, when one sold for $2,100 in northeast Ohio.

Exhibit B? That would be the New Idea 325 two-row picker with 12-roll husking bed that sold for $5,000 in northwest Ohio on a farm auction January 19, 2008. The highest sale price I've seen in the last 12 years on a 325 picker. No other even comes close. Next highest price was $4,250 back on March 7, 1997, on a sale in northwest Pennsylvania.

Corn Pickers

AC

Model	Year	Hours	Cond.	Price	Date	Area	Comments
N/A			F	$150	4/05	SCMI	Mounted corn picker for AC D17 tractor

IHC

Model	Year	Hours	Cond.	Price	Date	Area	Comments
N/A			G	$40	8/07	SWIL	Corn sheller

John Deere

Model	Year	Hours	Cond.	Price	Date	Area	Comments
101			F	$110	2/07	SCKS	1R corn picker
1A			G	$400	4/07	NEMO	Floor model corn sheller, restored
227			G	$200	1/05	NENE	Corn picker
237			F	$50	6/07	NENE	Corn picker, wagon elevator, John Deere 10/20 mounts
237			G	$450	2/05	NCIN	Mounted corn picker
300			G	$600	8/07	NENE	Husker corn picker
300			G	$1,250	12/07	WCMN	Picker, 2R-36 head, long elevator, 8-bolt hubs
300			G	$900	9/06	SEIA	Pt corn picker, 2R-38 head
300			G	$1,050	9/06	NEIA	Husker, 2R Wilman SS, 5-ton fertilizer spreader
300			G	$1,100	1/06	WCNE	3R-30, pt, 344 head
300			G	$1,150	10/06	SEMN	Corn picker
300			G	$1,150	9/06	NECO	Corn picker, John Deere 343 3RN CH, pt, PTO
300			F	$300	2/05	NEIN	Picker
300			G	$1,600	4/05	NCIA	Husker, John Deere 343 3RN CH
300			G	$1,900	8/05	SEMN	
43			G	$1,325	8/07	NEIA	Sheller
43			F	$350	4/05	NCOH	Corn sheller
71			G	$325	2/05	WCIN	Corn sheller, 40' of drags
N/A			F	$120	6/07	NENE	Corn sheller
N/A			F	$175	2/07	SEMI	Corn sheller
N/A			F	$35	11/06	NEKS	Corn sheller
N/A			G	$95	9/06	NCIA	Hand crank corn sheller
N/A			G	$625	10/06	SCMN	2A corn sheller on cart

Massey-Harris

Model	Year	Hours	Cond.	Price	Date	Area	Comments
N/A			G	$2,600	8/05	WCIL	2R corn picker

McCormick

Model	Year	Hours	Cond.	Price	Date	Area	Comments
N/A			G	$700	11/05	NCOH	Old, pt picker

Minneapolis-Moline

Model	Year	Hours	Cond.	Price	Date	Area	Comments
D			G	$400	9/06	NEIN	Corn sheller, stationary
D			G	$500	3/05	NCIL	Corn sheller
E			P	$125	12/07	WCNE	Corn sheller, PTO, pt, all there
N/A			F	$425	12/07	WCNE	Corn sheller, mounted on 1940s Dodge truck (not running)
N/A			F	$150	5/06	SEND	LD 2R picker/sheller, PTO

Corn Pickers

N/A

Model	Year	Hours	Cond.	Price	Date	Area	Comments
N/A			G	$55	9/06	ECNE	Hand crank corn sheller
N/A			G	$180	2/06	SEMI	Wood corn sheller
N/A			G	$975	6/05	SWOH	Cast corn sheller

New Idea

Model	Year	Hours	Cond.	Price	Date	Area	Comments
318			G	$800	8/06	ECIL	2R picker
319			F	$250	2/07	SEMI	2R mounted picker
322			G	$275	9/06	NENE	2R mounted corn pickup, John Deere 4020 mounts, always shedded
323			G	$2,000	2/08	SEAL	1R picker
323			E	$0	9/06	NEIN	
323			G	$950	2/06	SWIN	1R pt picker
323			G	$1,250	9/06	NEIN	1R picker
323			G	$300	4/05	NCOH	1R picker
323			G	$600	12/05	NCOH	1R picker
323			G	$675	3/05	SWIN	1R ear picker
323			G	$900	8/05	NEIN	1R picker
323			G	$1,300	8/05	NEIN	1R picker
323			E	$1,800	8/05	NEIN	1R r
324			P	$375	2/08	WCMN	2R
324			G	$525	8/07	NWIL	2R pull picker, 329 sheller
324			G	$700	8/07	NWIL	2R picker
324			E	$900	11/07	ECIL	8-roll husking bed
324			G	$1,200	8/07	ECIL	New Idea 327 bed, 12 roll
324			F	$185	9/06	NWIL	2R picker, pt
324			G	$200	12/06	NWIL	2R picker
324			F	$320	3/06	NEOH	2R pt
324			G	$1,000	10/06	SEMN	Picker, 12R bed
324			G	$1,950	9/06	NEIN	2R
324			G	$300	8/05	SEMN	
324			F	$400	7/05	SESD	
324			F	$400	7/05	SESD	Sheller
324			G	$725	7/05	SESD	2R
324			G	$1,000	8/05	NEIN	
324			E	$2,400	3/05	WCWI	Picker, 12R husking bed
324/327			F	$250	4/05	SCMI	2R, 12R husking bed

Corn Pickers

New Idea

Model	Year	Hours	Cond.	Price	Date	Area	Comments
325			E	$2,300	1/07	NEOH	2R picker
325			G	$200	10/06	SWOH	2R, single-axle transport, rear-discharge conveyor
325			G	$900	4/06	SCMN	2R-30 picker, 8-roll bed
325			G	$1,350	9/06	NEIN	2R picker
325			F	$200	4/05	SCMI	2R picker, 12R husking bed
325			G	$1,850	2/05	NEIN	2R picker
325			G	$2,200	4/05	SCMI	2R picker, 12R husking bed
325			E	$2,700	8/05	NEIN	2R picker
326			G	$325	9/07	SEMN	2R pull picker
326			G	$1,900	8/05	NCIA	2R-30 pull picker
327			G	$1,250	10/07	WCWI	2RN picker, 12-roll husking bed
328			G	$2,700	1/07	WCIL	3R-30 pt picker
328			G	$1,100	7/06	NCIL	Super picker, 12R husking bed, 536 acres since OH
N/A			G	$375	3/08	NWWI	1R picker
N/A			G	$2,700	2/08	SEAL	2R picker
N/A			F	$100	3/07	SCNE	1R picker
N/A			G	$2,750	8/07	NWIL	3R picker
N/A			G	$125	1/06	WCNE	2R-30, pt
N/A			F	$160	2/06	NECO	2R-30, pt
N/A			G	$350	2/06	NECO	3R-30, pt
N/A			G	$400	4/06	NENE	2R pt picker, shedded
N/A			G	$1,500	11/06	SWOH	1R picker
N/A			F	$150	9/05	NECO	2R-30, conveyor twisted
N/A			F	$200	9/05	NECO	2R-30, pt
N/A			G	$200	9/05	NECO	2R-30, pt
N/A			G	$1,250	10/05	NEPA	2R picker

Rosenthal

Model	Year	Hours	Cond.	Price	Date	Area	Comments
40			G	$200	10/06	SCMI	Corn husker

Drills

On April 1, 2008, a 2004 model John Deere 1890 42.5-foot air seeder drill sold on an auction in north-central North Dakota. Final sale price? $146,000. Nope, not an April Fool's joke – $146,000.

What do you suppose your father, your grandfather, or great-grandfather would have thought of $146,000 for a used drill? Hmm, I'd love to know what Great-Granddad would have thought. Maybe he was using a Van Brunt drill back in the day. I track sale prices on those too. Average auction sale price in 2008 on a Van Brunt drill? $287. Not quite all the gizmos you'd find on that 1890, I guess.

Drills

Case

Model	Year	Hours	Cond.	Price	Date	Area	Comments
10'			G	$500	7/07	NWIL	Grain drill, old, Model G or L
10'			G	$500	11/05	NEIA	Grain drill, grass seed, low rubber

Crustbuster

Model	Year	Hours	Cond.	Price	Date	Area	Comments
24'	1977		E	$120	4/05	NEKS	Hoe drill, 8" spacing

Great Plains

Model	Year	Hours	Cond.	Price	Date	Area	Comments
30'	1978		F	$600	2/05	NWKS	Grain drill, 12", steel press wheels, 3-section front fold, 2 rows of openers

IHC

Model	Year	Hours	Cond.	Price	Date	Area	Comments
10			G	$1,100	1/08	WCIL	Grain drill, rubber tires, grass and clover seeder
10			E	$250	11/06	NEKS	16-hole grain drill, White box, press wheels, fertilizer
10			F	$275	2/06	SCMI	16-run grain drill, seeder
10			F	$300	2/06	WCNE	14'×8", pt
10			P	$90	4/05	NEKS	Rough
10			F	$350	3/05	SCMI	16-run grain drill
10			F	$400	3/05	SCMI	16-run grain drill
10			F	$425	8/05	NWIL	
10			G	$825	4/05	WCWI	8' grain drill with grass
10'			G	$275	3/07	ECIL	
10'			F	$115	3/06	NEMI	16-hole grain drill
10'			F	$250	2/06	ECMN	Double disk, grass seeder, hyd. lift
10'			G	$700	1/06	SCMN	
10'			G	$400	4/05	WCWI	Grain drill, grass seed attachment
100			F	$100	5/06	SEND	16' press drill
100			G	$925	2/06	WCNE	12' double disk drill with 6" spacing
11			F	$60	4/05	SWMN	Double disk grain drill, low rubber, grass seeder, hyd. lift
11'			F	$500	2/05	NWIL	Grain drill, grass seed
12'			F	$325	2/08	SEMN	Drill on steel, grass seed
12'			F	$100	8/05	SWMN	Grain drill on low rubber
12'			F	$110	12/05	NWIL	Grass seed
14'			F	$0	5/07	WCSD	Hoe drill
150			G	$150	2/07	NENE	
150			F	$480	3/07	WCSD	(2) 14' deep furrow drills, tandem hitch, with dry fertilizer, 28' total, 12" spacing
150			G	$725	5/07	WCSD	Shovel drill, grass seeder
150			F	$100	2/06	WCNE	(2) 12'×14" drills, hitches
150			P	$300	10/06	NCKS	Hoe drill
150			F	$500	2/06	WCNE	(2) 12' grain drills, 14" spacing, pt, hitch
150			G	$500	2/06	WCNE	(2) 10' shoe drills, 12" spacings, pt
150			F	$550	1/06	NCCO	15×10 shoe drill, seeder
150			F	$200	2/05	NWKS	Grain drill, hyd. hitch, spoke press wheels, Case openers
150			F	$250	6/05	NENE	16-hole hoe drill, press wheels
150			F	$350	9/05	NECO	14'

Drills

IHC continued

Model	Year	Hours	Cond.	Price	Date	Area	Comments
510			G	$1,950	3/08	WCMN	12' grain drill, grass seeder
510			G	$2,500	1/08	SCKS	Grain drill
510			G	$250	10/07	NEKS	
510			F	$600	11/07	ECKS	20-hole drill
510			G	$600	4/07	NCNE	16'×8", alfalfa box
510			G	$850	2/07	WCOH	Soybean special
510			G	$300	6/06	SEID	12', double disks, grass seed boxes
510			G	$300	6/06	SEID	12', double disks, grass seed boxes
510			E	$700	9/06	NEIN	Grain drill
510			G	$850	2/07	WCOH	21-hole grain drill
510			G	$950	1/06	SCMI	18-run grain drill, seeder
510			G	$1,050	9/06	NECO	12', pt, seeder
510			G	$1,600	2/06	WCIL	18-7 grain drill with grass seeder, used very little
510			P	$450	4/05	NCOH	10", rubber press wheels, rough
510			F	$550	4/05	SCMI	21-run grain drill
510			F	$600	2/05	WCIL	
510			E	$725	3/05	SCIL	Grain drill, 21 flute
510			G	$800	8/05	NWIL	15', double disk
510			G	$1,050	2/05	NCIN	
510			P	$1,100	3/05	WCNY	Single disk, 13 hole, no grass seed
510			G	$1,500	11/05	NCIN	7", 21-hole drill, grass seed
510			G	$1,500	12/05	WCMN	
510			G	$2,700	3/05	WCWI	12' grain drill, double disk, grass and brome
510			G	$3,750	8/05	SEMN	12' grain drill with 24-6" openings, press wheel
620			G	$700	2/08	ECIL	14' grain drill, grass seed
620			G	$3,000	2/08	SWMN	(2) 14' press drill, transports and markers
620			F	$300	1/07	NECO	10'×10", pt
620			F	$400	4/07	NEND	Press grain drills, 24', two 12' drills, Erskine transport
620			F	$450	3/07	ECND	(2) 10' drills, 6" spacing, no dry fertilizer, rubber press
620			F	$1,200	1/07	NWOH	24×7 grain drill, soybean special
620			G	$1,400	3/07	WCNE	14' grain drill, 7" spacing, seeder
620			G	$2,000	3/07	ECND	Press drill, 14', dry fertilizer, grass seeder
620			G	$2,100	11/07	SWNE	20' grain drill, 7.5" spacing
620			G	$3,250	11/07	NEND	Press drills, 36', dry fertilizer, end hitch, folding markers
620			F	$200	3/06	SEMN	Two-section press wheel drill, 28', 7" spacing
620			F	$325	8/06	NCOH	24×7
620			F	$350	5/06	SEND	Press drills, 28', transport
620			G	$400	5/06	SEND	20' press drill, fertilizer and rubber press, track wacker
620			P	$450	7/06	NWMN	Press drills, two 14's, 6" space, dry fertilizer, rubber press, hyd. markers
620			P	$600	7/06	NWMN	(3) 12' press drills, 6" space, rubber press, dry fertilizer, markers, transport, tarps
620			F	$650	4/06	NWMN	(3) 10' press drills, grass seeder, dry fertilizer, rubber press, 6" space, markers
620			F	$200	4/05	SCMI	Grain drill and seeder, 24 run
620			G	$500	9/05	NECO	14', pt
620			G	$650	9/05	NECO	14', pt, seeder

Drills

Model	Year	Hours	Cond.	Price	Date	Area	Comments
620			G	$900	3/05	SENE	20×7
7'			G	$150	2/06	NECO	Seeder
N/A			F	$100	3/07	SCNE	8-16 low-end wheel drill, alfalfa seeder
N/A			F	$155	3/05	WCNE	Pt drill, transport
N/A			G	$525	3/05	NCCO	16'×7" double disk grain drill, seeder

John Deere

Model	Year	Hours	Cond.	Price	Date	Area	Comments
10'			F	$375	3/07	SEWY	10'×8", pt, seeder
10'			F	$200	2/06	WCNE	7", pt, seeder, high wheels
11'			G	$225	5/07	WCSD	Galvanized, 7" spacing, alfalfa seeder
12'			G	$1,500	9/07	NECO	7", pt, seeder, agitator
12'			F	$150	9/05	NECO	8"×12'., pt, scales
12'			F	$300	1/07	NECO	3 pt.
12'			G	$475	2/06	NWSD	
12'			G	$1,100	7/06	NCIL	Grain drill, grass seed
12'			F	$48	4/05	SEND	Drill
12'			G	$160	8/05	NCIL	Drill, double disk
14'			G	$675	2/08	ECIL	7.5" spacing, hyd. markers
14'			G	$500	3/07	SCNE	
14'			F	$300	1/07	NECO	Pt
14'			G	$300	3/06	WCSD	RR-F grass seeder disk drills, double disk, 6" spacing, dry fertilizer
14'			G	$475	11/06	SCNE	Grain drill
14'			F	$475	2/06	NECO	Pt
14'			P	$10	4/05	SEND	Drill, grass seeder
16'			F	$475	2/06	NECO	Seeder
24'			G	$1,000	11/07	NEND	LLA grain drill, pt, grass, 7.5" spacing, markers
616			F	$2,250	11/05	ECWA	(4) HZ 616 grain drills, 6" spacing × 16 drop, like-new packer rings, Telecky hitch
8'			G	$150	2/06	NECO	Pt, seeder
8000			G	$2,400	3/08	WCMN	13', markers, lift cyl.
8000			F	$600	8/07	NWIL	
8000			G	$1,250	2/07	NENE	Pt drill, single disk openers, closing wheels
8000			G	$2,000	3/07	SCNE	Grain drill, 20-8, kept inside
8000			G	$1,225	12/06	NCNE	
8000			F	$700	2/05	WCOK	20×8, single disk
8000			F	$850	2/05	WCOK	20×8, single disk
8100			G	$2,150	4/07	NEIA	14×7, grass seed
8100			G	$1,000	11/06	WCSD	Single disk drill, grass seeder, 7" spacing, 12'
8100			G	$1,500	5/06	SWWI	Grain drill
8250			G	$1,600	10/07	WCWI	10' grain drill
8250			F	$975	4/05	SEPA	Seeder
8250			G	$2,000	4/05	SEPA	18×7, single disk, grass seed
8250			G	$4,200	10/05	NWMT	
B			G	$800	2/08	WCIL	20-7, grass seeder
B			G	$1,375	3/08	NEMO	16×7 grain drill, grass, nice
B			G	$550	1/07	NECO	7"×12', pt, seeder
B			G	$300	9/06	NWOH	20×7 grain drill, seeder, single disk
B			G	$400	3/06	SENE	18-7

Drills

John Deere continued

Model	Year	Hours	Cond.	Price	Date	Area	Comments
B			F	$485	1/06	NECO	12', seeder, pt
B			G	$650	9/05	NECO	8"×12', pt, seeder
B			G	$825	2/06	ECIL	20/7 grain drill, grass seed
B			F	$75	2/05	SCCA	Grain drill
B			G	$275	1/05	SENE	16×7
B			G	$400	3/06	SENE	20 hole
B			G	$400	9/05	WCNE	12', pt, seeder, rubber tires
B			G	$400	12/05	NEIA	10' grain drill, grass seed
B			E	$625	1/05	SENE	167
B			G	$800	9/05	NECO	12', seeder
B			G	$1,450	12/05	NCIL	Grain drill, 12', grass seed, double disk
B246B			F	$600	3/08	WCMN	12' end wheel drill, 6" spacing, grass seed, tall wheels
DFB			G	$30	4/05	NEKS	
DR			F	$75	6/07	SWNE	16-hole grain drill, 10" spacing, single disk opener, press wheels, grass seeder, rubber tires
DR			F	$150	4/07	SCKS	DR5803, single disk, no press wheels
DRA			F	$35	4/06	NCOK	Single disk drill on rubber
DRB			F	$1,000	3/06	NCOK	DR208B, 20-8 single disk
DRB			F	$1,000	3/06	NCOK	DR208B, 20-8 single disk
FBB			F	$125	11/06	NEKS	15-hole grain drill, fertilizer
FBB			G	$700	3/06	NEMI	17-hoe grain drill, seeder
FBB			F	$25	11/05	SCMI	13-run grain drill
FBB			F	$400	3/05	SCMI	15 run, grain drill
L166			F	$500	6/05	WCMN	24' press drill (3) 8' drills and hitch
LL			F	$150	2/07	SENE	16' grain drill, 7" spacing, steel wheels
LL			G	$550	1/07	NECO	12'×8", pt, seeder
LL			G	$1,500	2/07	NCCO	LL 207A grain drill, double disk
LL			G	$700	3/06	ECND	(3) 8' grain drills, grass seeder, dry fert.
LL			F	$1,500	3/06	NCCO	20×7, double disk, seeder and rear-mount ditcher
LL			F	$1,500	3/06	NCCO	20×7, double disk, seeder, press wheels
LL-246A			F	$350	7/05	WCMN	Press drill, 12', transport
LL166			F	$400	3/06	NESD	24', (3) 8' drills
LLA			F	$400	1/08	ECNE	Dual front dolly wheels, grass seed attachment, pt, 20-shoe, rear steel press wheels (always shedded)
LLA			G	$750	2/08	SENE	24-hole high wheel grain drill, 7" spacing, fertilizer attachment
LLA			F	$225	9/07	NCND	12' grain drill
LLA			F	$350	9/07	NCND	(2) 12' press drills
LLA			F	$100	9/07	WCMN	12' press drill, grass seeder
LLA			G	$1,050	8/06	ECNE	12' press drill
LLA			G	$1,250	11/06	SENE	24-hole grain drill, 7" spacing, double disk, rubber press wheels, shop-built 3 pt. transport, shedded
LLA			F	$150	9/05	NECO	14' seeder, hyd. lift
LLA			G	$2,500	2/05	ECSD	12' press drill
LZ			G	$725	3/08	NWKS	24', 10×10, transport
LZ			F	$100	1/07	SWNE	14' grain drill, 10" spacing
LZ			F	$400	9/07	NECO	(3) 8' drills, 12", pt
LZ			F	$450	9/06	NECO	14', pt, seeder
LZ			F	$700	3/06	NWKS	LZ 812 24' hoe drill, SN 5437
LZ			F	$700	3/06	NWKS	LZ 812 24' hoe drill, SN 64111
LZ			G	$750	9/06	NECO	12', pt, seeder

Drills

Model	Year	Hours	Cond.	Price	Date	Area	Comments
LZ			G	$3,900	4/06	WCMT	14' hoe drill, 7" spacing, steel packers, fertilizer, grass seeder, hyd. lift, excellent paint
LZ1010			G	$1,700	3/07	WCNE	Drills, 3×10', 10" spacings, solid press wheels, hitch
LZ1010			F	$660	1/06	NECO	3 drill units
LZ1010			G	$50	4/05	NCOK	Older hoe drills, sowed wheat last fall, 20-hole opening
N/A			F	$140	3/08	SCKS	8×16
N/A			F	$100	5/07	WCSD	Set of (3) 7' drills with hitch
N/A			G	$110	2/07	WCNE	
N/A			G	$400	4/06	NENE	Rubber-tired grain drill
N/A			G	$500	2/07	ECMI	17-hole grain drill, towable
N/A			F	$650	1/07	NEOH	FB-13 grain drill, seeder
N/A			F	$85	3/06	ECNE	18-hole grain drill, seeder, press wheels, rubber tires
N/A			G	$100	9/06	NECO	End gate seeder, old, 2 rotors
N/A			F	$175	2/06	SWIN	Single disk 13-hole grain drill, seed attachment
N/A			F	$250	3/06	ECNE	18-hole drill on rubber tires
N/A			F	$375	3/06	NWOH	18×78, 18-run drill
N/A			F	$475	4/06	SWSD	2-12' deep furrow drills, 12" spacing
N/A			G	$650	3/06	ECNE	18-hole drill, seeder on rubber tire
N/A			G	$900	8/06	ECNE	18-hole drill, single disk openers, rubber tires, grass seeder, shedded
N/A			F	$175	9/05	NECO	8' hyd. lift
N/A			G	$425	6/05	ECWI	Pony drill, packer
N/A			G	$475	10/05	NCOH	Older 18×7 grain drill, seeder, galvanized boxes
Van Brunt			F	$100	1/07	SWNE	13' grain drill, 9" spacing
Van Brunt			F	$125	1/08	ECNE	20-hole grain drill, 7" spacing, rubber tires, press wheels
Van Brunt			F	$200	3/08	NCIL	10' grain drill
Van Brunt			G	$400	1/08	WCIL	7×16 drill, grass seeder
Van Brunt			F	$60	1/07	ECNE	18-7, clutch lift
Van Brunt			G	$200	6/07	SCIL	Wooden-wheeled Van Brunt drill, very early
Van Brunt			G	$275	4/07	NCNE	14' grain drill, grass seed attachment
Van Brunt			F	$375	10/07	NECO	8', 7", pt, both seeders
Van Brunt			G	$550	9/07	ECNE	20×7, seeder, high press wheels
Van Brunt			G	$675	9/07	NECO	10'×7", pt, seeder
Van Brunt			G	$675	3/07	NENE	18-hole grain drill, rubber tires, 7" spacing
Van Brunt			P	$25	3/06	SWNE	10-hole 7" grain drill
Van Brunt			F	$35	3/06	ECNE	20-hole drill on steel
Van Brunt			F	$125	3/06	NENE	10'
Van Brunt			F	$125	12/06	SEMN	
Van Brunt			F	$200	11/06	SWIA	Grain drill, hyd. lift
Van Brunt			G	$250	12/06	NWIL	
Van Brunt			F	$300	2/06	WCNE	14' grain drill, 10" spacings
Van Brunt			G	$420	8/06	SWWI	6' grain drill on steel, metal boxes
Van Brunt			G	$460	3/06	SEIA	Grain drill
Van Brunt			G	$475	9/06	NENE	18-hole grain drill, 7" spacing, rubber tires, seeder, drag chains
Van Brunt			G	$675	6/06	NCIA	10' grain drill, grass seed
Van Brunt			G	$700	10/06	SCMI	13-run grain drill, seeder
Van Brunt			G	$1,000	6/06	NWIL	12' grain drill, grass seeder
Van Brunt	1939		F	$25	7/06	NCIL	Grain drill, 10', grass seed
Van Brunt			F	$100	2/05	NENE	18-hoe grain drill on rubber, grass seeder
Van Brunt			G	$175	3/05	NCIL	12', double disk, grass seed

Drills

John Deere continued

Model	Year	Hours	Cond.	Price	Date	Area	Comments
Van Brunt			F	$225	1/05	NWIL	12' grass seed
Van Brunt			F	$250	2/05	WCIL	10' grain drill, grass seeder
Van Brunt			G	$325	11/05	ECNE	Model B
Van Brunt			G	$500	4/05	NCIA	12' drill, grass seed, hyd.
Van Brunt			G	$560	2/05	WCIA	12' grain drill on rubber
Van Brunt			G	$625	9/05	NCIA	12' drill, grass seed, hyd. lift
Van Brunt			G	$950	8/05	WCIA	

McCormick

Model	Year	Hours	Cond.	Price	Date	Area	Comments
10'			F	$125	3/08	NWIL	Grain drill, grass seed, hyd. lift
12'			G	$300	8/05	SEMN	Grain drill
M			G	$450	6/07	ECMN	10' grain drill, grass, low rubber, trip
N/A			F	$115	3/07	ECNE	Grain drill
N/A			G	$200	6/07	SCIL	McCormick-Deering wood wheel grain drill
N/A			G	$230	8/06	NCMI	16-hole grain drill, seeder

Massey Ferguson

Model	Year	Hours	Cond.	Price	Date	Area	Comments
10'			F	$375	2/06	NWIL	Grain drill, grass seed, 15" tires
33			F	$90	3/05	SCMI	15R
424			G	$1,200	3/07	SEWY	12'×7", pt, seeder
424			G	$1,200	1/06	NEIN	Grain drill, markers, seeder
427			G	$5,400	11/05	NCOH	No-till drill, 18×7
43			G	$425	12/07	NCOH	16×10, end wheel drill
43			F	$475	2/07	ECMI	Grain drill, 21 hole
N/A			G	$375	4/07	SWIN	13-hole double disk grain drill
N/A			G	$900	1/07	SEKS	20-hole grain drill, 8" spacing, flute feed, chain drags, one owner, always shedded

Minneapolis-Moline

Model	Year	Hours	Cond.	Price	Date	Area	Comments
10'			F	$400	8/07	SCMN	Grain drill
12'			G	$575	2/07	NEIL	Grain drill, grass seed, double disk
14'			E	$1,000	1/06	NECO	6" disk, front and rear wheels
N/A			F	$60	11/06	SCNE	Grain drill, steel wheels

Oliver

Model	Year	Hours	Cond.	Price	Date	Area	Comments
12'			G	$130	7/07	NWIL	On rubber
12'			F	$200	9/05	NECO	8", pt, seeder
12'			G	$175	9/05	WCIL	On steel
76			F	$135	9/07	SCMI	13-run grain drill
N/A			F	$60	11/06	SCNE	Grain drill, steel wheels
Superior			F	$100	9/07	NWIL	10', wood box

YOU KNOW YOU'RE IN TROUBLE WHEN THE KID BEHIND
THE PARTS COUNTER IS YOUNGER THAN THE TRACTOR
YOU'RE BUYING PARTS FOR.

Hay Balers

Old balers. Nobody looking for these relics, right? Wrong. Actually, I get tons of calls from folks specifically searching for older, smaller square balers that would perfectly handle the baling they need to do on their hobby farms or acreages.

Explains why I see things like the average auction sale price on John Deere 336 balers going from $1,749 back in 2000, all the way up to an average auction price of $3,042 in 2008. That's an impressive 73.9% increase in value.

Hay Balers

Make	Model	Cond.	Price	Date	Area	Comments
Allis-Chalmers	302	G	$700	10/06	SCMI	Square baler
Allis-Chalmers	302	F	$350	2/05	NCIN	
Allis-Chalmers	N/A	G	$100	9/07	NWIL	Small Roto-Baler
Allis-Chalmers	N/A	F	$350	2/07	WCKY	Small round baler
Allis-Chalmers	N/A	E	$100	11/06	NEKS	Roto-Baler
Allis-Chalmers	Roto-Baler	G	$100	3/08	NEKS	Old round baler
Case	NI	G	$1,100	11/05	NCOH	Old baler, wire type
Ford	530	F	$325	9/06	NECO	Tie, PTO
Ford	530	F	$150	4/05	SCMI	Square baler
Ford	530	G	$635	10/05	WCIL	
Ford	552	F	$500	8/05	ECMI	Round baler, missing knotter
IHC	2400	F	$850	4/08	SCMN	Round baler
IHC	2400	F	$300	9/06	NEIA	Round baler
IHC	2400	F	$625	8/05	SEMN	Round baler
IHC	241	P	$200	1/08	NEMO	Round baler
IHC	241	P	$150	9/05	SEIA	Round baler
IHC	241	F	$375	8/05	NEIN	Round baler
IHC	241	G	$500	12/05	SEMN	Round baler
IHC	241	G	$525	12/05	SEMN	
IHC	425	E	$1,600	10/05	NCOH	Square baler
IHC	430	G	$475	2/07	NCIL	Hay baler
IHC	430	G	$600	2/07	SEIA	Square baler
IHC	430	G	$850	2/07	NEIL	
IHC	430	E	$1,800	3/07	NCOH	Twine baler, looks great
IHC	435	F	$700	8/07	NEIA	
IHC	435	F	$800	8/07	NEIA	
IHC	435	G	$875	1/05	WCIL	Square baler
IHC	440	F	$600	4/07	NECO	PTO, pt, twine
IHC	440	G	$3,200	2/07	ECNE	Loft-twist, wire tie, small square baler, 540 PTO, 5' pickup
IHC	440	P	$250	3/06	ECMN	Square baler, #10 bale thrower on back (bad thrower clutch) rough shape all around, +10% buyers premium
IHC	440	F	$300	1/07	NECO	Wire tie, PTO
IHC	440	F	$650	4/05	SEPA	Kicker
IHC	445	G	$2,800	8/07	SWOK	Square baler, wire, shedded
IHC	445	G	$1,400	3/05	NEOH	Thrower
IHC	46	F	$250	11/06	NEKS	
IHC	46	G	$525	8/06	NCIA	PTO baler
IHC	46	G	$370	8/05	NCIA	PTO baler
IHC	47	G	$1,000	11/07	WCIL	String tie baler
IHC	47	G	$500	3/05	NESD	Small square baler
IHC	N/A	G	$650	8/05	NEIN	Small square baler
John Deere	14T	E	$425	6/07	ECMN	Pristine condition
John Deere	14T	F	$700	12/07	NCOH	
John Deere	14T	G	$1,000	11/07	ECIL	
John Deere	14T	F	$350	4/06	NWSD	

Hay Balers

Make	Model	Cond.	Price	Date	Area	Comments
John Deere	14T	F	$360	2/06	NEIA	
John Deere	14T	G	$600	12/06	SEMN	
John Deere	14T	G	$750	7/06	WCOH	Twine
John Deere	14T	G	$1,025	8/06	ECIL	
John Deere	14T	F	$375	11/05	SCMI	Square baler
John Deere	14T	G	$525	1/05	SENE	
John Deere	14T	G	$850	1/05	SENE	
John Deere	24T	G	$1,900	2/08	NCOH	Twine tie baler
John Deere	24T	G	$800	2/07	NENE	Small square twine tie baler
John Deere	24T	G	$1,150	11/07	ECNE	Small square baler, 540 PTO
John Deere	24T	G	$1,325	8/07	NEIA	Baler, chute
John Deere	24T	F	$500	6/05	WCMN	
John Deere	24T	F	$650	6/06	SWMN	
John Deere	24T	G	$800	12/06	SEMN	
John Deere	24T	G	$1,050	8/06	ECNE	Small square baler
John Deere	24T	G	$1,300	3/06	NCWI	Bale ejector
John Deere	24T	F	$250	9/05	NWIA	
John Deere	24T	F	$400	8/05	SEMN	
John Deere	24T	G	$500	2/05	WCOH	Square baler
John Deere	24T	G	$525	7/05	ECND	Square baler
John Deere	24T	F	$550	7/05	SESD	Square baler
John Deere	24T	G	$625	8/05	SWOH	Square baler
John Deere	24T	G	$750	11/05	SWMN	
John Deere	24T	F	$800	8/05	NWIL	
John Deere	24T	G	$825	4/05	NCOH	
John Deere	24T	F	$850	2/05	NWIL	
John Deere	24T	G	$900	8/05	NEIN	
John Deere	24T	G	$1,350	2/05	WCIL	Square baler
John Deere	24T	G	$1,450	8/05	NWIL	
John Deere	24T	G	$1,500	1/05	NENE	Twine tie square baler
John Deere	336	G	$900	2/08	WCMN	Small square baler
John Deere	336	G	$3,700	2/08	SEMN	
John Deere	336	G	$4,200	1/08	NEMO	
John Deere	336	G	$4,400	4/08	SEIA	No kicker, very little use, always shedded
John Deere	336	E	$4,500	3/08	NWIL	
John Deere	336	G	$1,500	2/07	WCOH	Square baler, twine
John Deere	336	F	$1,700	5/07	WCWI	Ejector
John Deere	336	G	$1,750	3/07	NEMI	Square baler
John Deere	336	F	$1,850	5/07	WCWI	Ejector
John Deere	336	G	$1,900	8/07	NWIL	
John Deere	336	G	$2,400	8/07	NWIL	
John Deere	336	F	$2,700	8/07	NEIA	
John Deere	336	G	$2,750	1/07	WCIL	Small square wire baler
John Deere	336	G	$2,900	1/07	NWOH	Twine baler
John Deere	336	G	$3,100	12/07	WCMN	Baler, thrower
John Deere	336	G	$3,500	6/07	SCMI	
John Deere	336	G	$4,000	2/07	NEMO	Twine tie

Hay Balers

Make	Model	Cond.	Price	Date	Area	Comments
John Deere	336	F	$800	4/06	WCWI	Baler, thrower
John Deere	336	F	$1,000	8/06	SEPA	Kicker
John Deere	336	F	$1,000	7/06	WCMN	Square baler
John Deere	336	G	$1,600	8/06	SEMN	Baler, thrower
John Deere	336	G	$1,650	10/06	WCWI	Baler, thrower
John Deere	336	G	$1,800	2/06	WCOH	Wire tie baler
John Deere	336	G	$1,800	2/06	WCOH	Wire tie baler
John Deere	336	G	$2,000	2/06	NECO	PTO, twine, small square
John Deere	336	G	$2,200	5/06	SWWI	Baler, #30 thrower
John Deere	336	G	$2,600	1/06	NEIN	
John Deere	336	G	$3,000	1/06	SCMI	Thrower, extra thrower
John Deere	336	G	$3,000	4/06	NEIA	One owner, chute, no thrower
John Deere	336	G	$3,750	12/06	WCIL	Wire tie square baler, always shedded
John Deere	336	G	$1,500	9/05	SCON	
John Deere	336	G	$1,900	7/05	WCMN	Thrower
John Deere	336	G	$2,000	7/05	NEIA	Baler, thrower
John Deere	336	G	$2,000	9/05	SCON	Canadian sale
John Deere	336	F	$2,200	11/05	NEIA	Thrower
John Deere	336	G	$2,200	4/05	SCMI	Square baler, thrower
John Deere	336	G	$2,500	3/05	WCMN	
John Deere	336	G	$2,700	4/05	NEIN	Kicker
John Deere	336	E	$3,500	3/05	WCWI	Square baler, John Deere 30 kicker
John Deere	336	G	$3,500	1/05	SWIL	Wire baler
John Deere	336	F	$3,800	4/05	SEPA	#30 ejector
John Deere	336	G	$4,000	2/05	NCIN	Hay baler, thrower
John Deere	336	G	$4,500	4/05	WCWI	Baler, thrower with auto. 30-gal. liquid applicator
Massey Ferguson	10	G	$475	9/07	NCND	Square baler, PTO, stored inside
Massey Ferguson	12	F	$160	1/06	SESD	Square
Massey-Ferguson	12	G	$600	8/05	SCMI	
Massey Ferguson	124	G	$750	6/07	SWNE	Small square baler, twine tie
Massey Ferguson	124	G	$1,025	8/07	NEIA	Thrower
Massey Ferguson	124	G	$1,225	8/07	NEIA	Thrower
Massey Ferguson	124	G	$725	3/06	ECIA	
Massey Ferguson	124	F	$850	2/05	NWIL	Small square baler
Massey Ferguson	124	P	$260	4/05	SCSD	Salvage
Massey Ferguson	3	G	$225	3/06	ECMN	Square baler, +10% buyers premium
Massey Ferguson	3	F	$350	2/05	NEIN	
Minneapolis-Moline	760	G	$1,900	8/07	NWOH	PTO drive, nice
Minneapolis-Moline	K3	G	$600	8/07	NWOH	Wisconsin V-4 engine, original
New Holland	268	F	$375	3/06	NWOH	Baler

Hay Balers

Make	Model	Cond.	Price	Date	Area	Comments
New Holland	268	G	$600	10/07	WCWI	Bale slide
New Holland	268	G	$2,100	8/05	SEMN	Hayliner baler, shedded, good shape
New Holland	268	P	$175	8/06	NEIA	Thrower, rough
New Holland	268	F	$300	9/06	NEIA	Thrower
New Holland	268	F	$625	2/06	NECO	PTO, pt, twine
New Holland	268	G	$1,250	8/06	SETN	
New Holland	268	G	$1,800	9/06	NEIN	Small square baler, PTO
New Holland	268	F	$200	4/05	SCMI	Square baler
New Holland	268	F	$500	6/05	SEND	Square baler, PTO
New Holland	268	G	$800	9/05	NECO	Small square baler
New Holland	269	F	$550	4/07	ECND	
New Holland	269	G	$2,000	3/07	SCNE	Small square baler
New Holland	269	F	$275	3/06	ECNE	Hayliner square baler, twine tie
New Holland	269	F	$600	6/06	SWMN	Hayliner baler
New Holland	269	G	$1,075	3/06	SENE	
New Holland	269	G	$1,000	1/05	ECNE	Square baler
New Holland	269	E	$1,100	9/05	NENE	Small square twine tie baler, sharp, shedded
New Holland	269	G	$2,600	4/05	SCMI	
New Holland	271	F	$160	3/06	WCMN	Small square baler
New Holland	271	G	$500	9/06	NEIN	Small square baler
New Holland	276	G	$1,700	3/06	NCCO	
New Holland	66	P	$230	3/08	SCMN	Rough shape
New Holland	66	G	$350	2/05	NEIN	Small square baler
New Holland	67	F	$1,300	10/07	NWKY	Square baler
New Holland	67	G	$1,050	1/06	WCIL	String tire square baler
New Holland	67	F	$290	3/05	SCMI	
New Holland	68	G	$500	3/07	SCNE	Wire tie baler
New Holland	68	E	$750	1/07	SENE	
New Holland	68	F	$100	1/07	SWNE	Twine tie
New Holland	68	P	$225	2/06	NWIL	Hayliner baler
New Holland	68	F	$350	7/06	NCIL	Hay baler
New Holland	68	G	$625	9/06	NEIN	Small square baler
New Holland	68	G	$1,575	4/06	SWIN	Hayliner
New Holland	68	P	$100	8/05	SEMN	
New Holland	68	F	$400	11/05	NCOH	
New Holland	68	F	$500	3/05	SCMI	Square baler
New Holland	78	F	$150	5/06	SEND	Super hay liner model PTO baler
New Holland	850	P	$400	12/07	SEND	Round baler
New Holland	Super 66	G	$350	11/06	NEOH	Old square baler
New Holland	Super 66	F	$1,100	3/06	NWOH	
New Holland	Super 66	F	$550	11/05	NEIA	
New Holland	Super 67	F	$400	11/05	NCOH	
Oliver	62T	F	$95	3/05	SCMI	Square baler

Successfully Selling Iron Online

With billions of dollars of transactions every year, eBay has become a cultural phenomenon. And the latest group of partipants are antique tractor collectors who have discovered the Internet sale site as a great place to find collectibles.

But now auctioneers are getting into the Internet act by providing online auctions. There is the intimacy of an auction, but you have buyers from across the country using live online Webcasting technology to bid in real time. The sellers have peace of mind knowing they are getting the highest value possible.

If you want to list an item on eBay, be sure to realize a few rules before you do. For example, when you post an item, you are committed to sell it unless you can place a reserve price on it.

But any reserve price should be reasonable, otherwise, most buyers will turn away immediately. Also know that eBay charges higher listing fees with reserve auctions since, in many cases, the item doesn't sell because the reserve was set too high.

Whether you are at a traditional auction or using eBay, the term "reserve" will always scare away certain bidders because they feel you are not truly serious about selling. If you've ever been at an auction where they announce the reserve has been met, you know that many times it brings out a new bidder or two to compete for the item, and then the sale really takes off.

Even if you list a tractor on eBay, that doesn't mean you must completely depend on the Web site to promote the sale for you. It can be worthwhile to place classified ads both online and offline that point to your eBay auction.

And if you want to make it even more visible, go to www. domainsite.com, buy a $9 domain name, and point to your eBay auction or online classified ad. Pick a name like www.iowa4020. com to promote your 4020, for example.

An Internet site like eBay also allows a large allowance for text as well as photographs. So be sure to be as descriptive and detailed as possible. Think like a buyer. List those things you would want to know about a tractor if you were serious about buying.

And, for good measure, add as many photos as necessary.

By Dwayne Leslie

Frustrated by being unable to easily find auction listings, Portage la Prairie, Manitoba, farmer Dwayne Leslie created the www.FarmAuctionGuide.com Web site. He later joined forces with seven other Web sites to create the now massive www.GlobalAuctionGuide.com.

Serial Numbers

Advance-Rumely

Advance-Rumely Advance Thresher steam engines

Year	Beginning number
1885	101
1886	130
1887	231
1888	401
1889	652
1890	969
1891	1410
1892	1724
1893	2151
1894	2625
1895	2915
1896	3304
1897	3707
1898	4016
1899	4423
1900	4957
1901	5512
1902	6237
1903	6951
1904	7732
1905	8463
1906	9192
1907	9926
1908	10578
1909	11134
1910	11690
1911	12359
1912	13005
1913	13466
1914	14190
1915	14438
1916	14453
1917	14638

Meinrad Rumely steam traction engines

Year	Beginning number
1895	2600 – 2725
1896	2726 – 2878
1897	2879 – 3049
1898	3050 – 3218
1899	3219 – 3297

Year	
1900	3438 – 3640
1901	3641 – 3855
1902	3856 – 4136
1903	4137 – 4374
1904	4375 – 4435
1904	4437 and 4452
1904	4439 – 4444
1904	4458 – 4462
1904	4466 – 4468
1904	4473 – 4477
1904	4482 – 4483
1904	4506 – 4529
1904	4532 – 4539
1905	4436
1905	4445 – 4450
1905	4452 and 4478
1905	4455 – 4457
1905	4469 – 4472
1905	4480 – 4481
1905	4484 – 4505
1905	4530 – 4531
1905	4540 – 4628
1906	4438
1906	4453 – 4454
1906	4463 – 4465
1906	4629 – 4764
1907	4479
1907	4765 – 4952
1908	4953 – 5184
1908	5186 – 5195
1909	5185
1909	5196 – 5587
1910	5588 – 5920
1911	5921 – 6287
1912	6288 – 6588
1913	6589 – 6714
1914	6715 – 7036
1915	7037 – 7038

Rumely and Advance-Rumely OilPull tractors

B 25-45 OilPull

Year	Number range
1910	1 - 100
1911	2101 - 2269
1912	2270 - 2936

E 30-60 OilPull

Year	Number range
1910	101 – 236
1911	237 – 746

Serial Numbers

1912............................ 747 – 1678
1913............................ 1679 – 1787
1914................................None built
1915............................ 1819 – 2018
1916............................ 2019 – 2100
1917............................ 2997 – 8724
1918............................ 8725 – 8902
1919.......................... 11500 – 11596
1920............................ 2252 – 2351
1921............................ 2352 – 2402
1922............................ 2404 – 2453
1923............................ 2454 – 2503

F 15-30 & later (1918) 18-35 OilPull

Year	Number range
1911	5001 – 5680
1912	5681 – 6738
1913	6739 – 7487
1914	7500 – 7856
1915	None built
1916	7857 – 8084
1917	8085 – 8591
1918	8903 – 9177

G 20-40 OilPull

Year	Number range
1918	10425 – 10750
1919	10751 – 15221
1919	G741 – G948
1920	G949 – G1727
1921	G1728 – G2241
1922	G2242 – G2689
1923	G2690 – G3558
1924	G3559 – G3894

H 16-30 & earlier (pre-1919) 14-28 OilPull

Year	Number range
1917	8627 – 8699
1918	9178 – 10710
1919	10711 – 16284
1919	H3751 – H4392
1920	H4393 – H7239
1921	H7240 – H7395
1922	H7396 – H8645
1923	H8646 – H9045
1924	H9046 – H9645

K 12-20 OilPull

Year	Number range
1918	12000 — 12100
1919	12101 – 13656
1920	13657 – 15100
1920	16836 – 17639
1921	17640 – 18648
1922	18649 – 19268
1923	19269 – 20510
1924	20511 – 21018

L 15-25 OilPull

Year	Number range
1924	1 – 10
1925	11 – 1606
1926	16074213
1927	4214 – 4855

M 20-35 OilPull

Year	Number range
1924	1
1925	2 – 1013
1926	1014 – 3084
1927	3085 – 3671

R 25-45 OilPull

Year	Number range
1924	1
1925	2 – 138
1926	139 – 647
1927	648 – 761

S 30-60 OilPull

Year	Number range
1924	1 – 4
1925	5 – 34
1926	35 – 234
1927	235 – 434
1928	435 – 514

W 20-30 OilPull

Year	Number range
1928	1 – 2128
1929	2129 – 3733
1930	3734 – 3952

X 25-40 OilPull

Year	Number range
1928	X-1 – X-1545
1929	1546 – 2259
1930	2260 – 2400

Serial Numbers

Y 30-50 OilPull

Year	Number range
1929	1 – 245

Z 40-60 OilPull

Year	Number range
1929	1 – 215

Do-All Tractor

Year	Number range
1928	501 – 700
1929	701 – 2115
1930	2116 – 3513
1931	3514 – 3693

Rumely 6-A Tractor

Year	Number range
1930	501 – 502
1931	503 – 1302

The following Model X tractors were converted from Model M 20-35 OilPull tractors built earlier: 1-125, 757, 773, 1616, 1637, 1800, 1955-57, 1970 2020–2022, 2040, 2046, 2080–2082, 2100, 2150, 2200–2212, 2220 & 2225.

Tractors Y-1 to Y-100 and Y-221 to Y-245 were converted from Model R to Model Y tractors.

Tractors Z-1 to Z-62 were converted from Model S tractors to Model Z tractors and later converted back again to Model S tractors.

Allis-Chalmers

Location of serial numbers

Models 20-35, E20-35, and E25-40:

Tractor – top of transmission ahead of shift lever quadrant
Engine – left side of block

Models U, United All-Crop, UC, and IU:

Tractor – rear axle housing near PTO
Engine – Continental right rear of block
(Model UM - right side of block)

Model A:

Tractor – rear axle housing near PTO
Engine – left side of block

Models WC (prior to serial number 74330) and WF (prior to serial number 1904) and RC:

Tractor – rear side of differential near oil filler plug
Engine – rear of engine block

Models WC (serial number 74330 and up) and WF (serial number 1904 and up):

Tractor – rear side of differential near oil filler plug
Engine – left side of block to rear of carburetor

Models B, IB, and C:

Tractor – top of transmission case in front of gear shift lever
Engine – rear side of block or flange to clutch housing

Models WD and WD45:

Tractor – rear side of differential housing

Serial Numbers

Engine – gas left side of block; diesel left side of engine

Model CA:
Tractor – top of transmission next to shift pattern
Engine – rear left side of engine

Model G:
Tractor – top of transmission
Engine – near starter

Models D14, D15, D17, D10, D12, D15, and D21:
Tractor – left front end of torque housing
Engine – left side of block

Model D19:
Tractor – top of rear main housing
Engine – top left side of block

Model ED40:
Tractor – under seat on left side of transmission
Engine – top right side of block

Models 170G, 175G, 180, 190, 190XT, 210, and 220:
Tractor – left front end of torque housing
Engine – left side

Model 170D:
Tractor – left front of torque housing
Engine – right side of block

Model 175D:
Tractor – left front end of torque housing
Engine – some left side of block some right side of block

Models 185 and 200:
Tractor – right front of torque housing
Engine – left side of block

Model 440:
Tractor – left side of instrument panel
Engine – left front and left rear of engine block

Model 160:
Tractor – plate ahead of fuel tank
Engine – left side of block

Model 6040:
Tractor – plate ahead of fuel tank
Engine – right side of block under intake manifold

Models 5040, 5045, and 5050:
Tractor – rear of console under steering wheel
Engine – left side of block

Models 5015, 5020, and 5030:
Tractor – nameplate on rear side of steering gearbox; 5015 also has product identification number (PIN)
Engine – left side of block
Transmission – right side of engine block

Models 6060, 6070, and 6080:
Tractor – right side of adapter housing
Engine – left side of block; after 1982 PIN plate is on rear of left side frame

Model 6140:
Tractor – product identification number on plate at left side of clutch housing
Engine – left side of block

Models 7010, 7020, 7030, 7040, 7045, 7050, 7060, and 7080:
Tractor – on rear main housing near PTO
Engine – upper left side of block

Models 7580 and 8550:
Tractor – rear of left front frame
Engine – upper left side of block

Serial Numbers
Allis-Chalmers serial number location continued

Models 8010, 8030, 8050, and 8070:
Tractor – product identification number on top rear of left side frame
Engine – upper left side of block

Models 4W220 and 4W305:
Tractor – product identification number on top rear of left side frame
Engine – upper left side of block

Monarch 50:
Tractor – on front of main frame and on instruction plate on dash

Models K and KO:
Tractor – on right near top of transmission case and on instrument panel

Models L and LO:
Tractor – on shelf that runs out to rear of transmission and on instrument panel

Model M:
Tractor – on transmission case on right side behind clutch inspection cover
Engine – right side of block

Models S and SO:
Tractor – top right rear face of transmission case
Engine – lower center of left side of block

Model HD7:
Tractor – top right rear face of transmission case and on master clutch inspection cover
Engine – top front of right side of engine

Model HD10:
Tractor – top right rear face of transmission case and on steering lever stop angle, which is on top of transmission

Model HD14:
Tractor – right side of shelf that runs out to rear of transmission case and on right side of front floor plate

Models WC Speed Maintainer and W Speed Patrol:
Tractor – on right differential housing
Engine – left side of block
Maintainer – plate on left drawbar member
Patrol – plate on right front frame extension

Models 42, 54, and KO 54 Speed Patrols:
Grader – plate on front of left frame
Tractor – rear axle housing near PTO
Engine – left side of block

Models I40 and I400:
Same as Models D10, I60, I600, and 600 FL

Model 500 FL:
Location same as D15

Models 190 Beachmaster and 918:
Location same as 190

Models H3 and HD3:
Location same as D15, H4, and HD4

10-18
Year	Number range
1914-1923	No information

6-12
Year	Number range
1918-1926	No information

15-30/18-30
Year	Number range
1918	5000 – 5005
1919	506 – 5160
1920	5161 – 6014

Serial Numbers

1921........................... 6015 – 6160
1922..None

20-35

Year	Number range
1923	6161 – 6396
1924	6397 – 6754
1925	6755 – 7368
1926	7369 – 8069
1927	8070 – 9869
1928	9870 – 10000

E20-35

Year	Number range
1928	1201 – 17661
1929	16762 – 20250
1929	22001 – 23251
1930	23252 – 24185

E25-40

Year	Number range
1930	24186 – 24842
1931	24843 – 24971
1932	24972 – 25023
1933	25024 – 25061
1934	25062 – 25308
1935	25309 – 25581
1936	25582 -25611

L12-20/15-25

Year	Number range
1921	20001 - 20334
1922	20335 – 20497
1923	20498 – 20905
1924	20906 – 20995
1925	20996 – 21370
1926	21371 – 21681
1927	21682 – 21705

A

Year	Number range
1936	25701 – 25725
1937	25726 – 26304
1938	26305 – 26613
1939	26614 – 26781
1940	26782 – 26895
1941	26896 – 26914
1942	26915 – 26925

B

Year	Number range
1937	1 – 96
1938	97 – 11799
1939	11800 – 33501
1940	33502 – 49720
1941	49721 – 56781
1942	56782 – 61400

B (125-cubic-inch engine)

Year	Number range
1943	64501 – 65501
1944	65502 – 70209
1945	70210 – 72264
1946	72265 – 73369
1947	73370 – 74079
1947	75080 – 80555
1948	80556 – 85883
1948	87834 – 92294
1949	92295 – 102392
1950	102393 – 103578
1950	106579 – 114526
1951	114527 – 118673
1952	118674 -122309
1953	122310 – 124200
1954	124201 – 124710
1955	124711 – 126496
1956	126497 – 127185
1957	126186 – 127461

Model IB

Year	Number range
1946	1001 – 102
1947	1003 – 1009
1948	1010 – 1281
1949	1282 – 1555
1950	1556 – 1878
1951	1879 – 2118
1952	2219 – 2567
1953	2570 – 2847
1954-58	INA

Model C

Year	Number range
1940	1 – 111
1941	112 – 12388
1942	12389 – 18781
1943	18782 – 23907
1944	23908 – 30694
1945	30695 – 36377

Serial Numbers
Allis-Chalmers continued

1946	36378 – 39167
1947	39168 – 51514
1948	51515 – 68280
1949	68281 – 80517
1950	80518 – 84030

CA

Year	Number range
1950	14- 321
1951	322 – 10538
1952	10539 – 22180
1953	22181 – 31423
1954	31424 – 32906
1955	32907 – 37202
1956	37203 – 38617
1957	38618 – 38976
1958	38977 – 39513

ED40

Year	Number range
1963-64	No information

G

Year	Number range
1948	6 – 10960
1949	10961 – 23179
1950	23180 – 24005
1951	24006 - 25268
1952	25269 – 26496
1953	26497 – 28035
1954	28036 – 29035
1955	29036 – 29976

Model RC

Year	Number range
1939	4 – 4391
1940	4392 - 5416
1941	5417 – 5504

United U (Continental engine)

Year	Number range
1929	U1 – U1974
1930	U1975 – U6553
1931	U6554 – U7261
1932	U7262 – U7404

Model U (UM engine)

Year	Number range
1932	U7405 – U7418
933	U7419 – U7684
1934	U7685 – U8062

1935	U8063 – U9470
1936	U9471 – U9988

U (4.5-inch bore engine)

Year	Number range
1936	U12001 – U12821
1937	U12822 – U14854
1938	U14855 – U15586
1939	U15587 – U16077
1940	U16078 – U16721
1941	U16722 – U17136
1942	U17137 – U17469
1943	U17470 – U17801
1944	U17802 – U17819
1945-46	No information
1947	18309 – 18612
1947	20174 – 20178
1948	18613 – 19421
1948	20179 – 20300
1948	21000 – 21035
1949	20301 – 20421
1949	21036 – 21986
1949	22018 – 22023
1950	20422 – 20512
1950	22024 – 22547
1951-52	No information

All-Crop/UC
(Continental engine)

Year	Number range
1930	UC1 – UC38
1931	UC39 – 1099
1932	UC1100 – UC1231
1933	UC1232 – UC1268

UC (UM engine)

Year	Number range
1933	UC1269 – UC1293
1934	UC1294 – UC1551
1935	UC1552 – UC2000
1936	UC2001 – UC2281

UC (4.5-inch bore engine)

Year	Number range
1936	UC2282 – UC2770
1937	UC2771 – UC3756
1938	UC3757 – UC4546
1939	UC4547 – UC4769
1940	UC4770 – UC4971
1941	UC4972 – UC5037

Serial Numbers

Model UC Cane

Year	Number range
1944	5038 – 5067
1945-46	None
1947	5068 – 5267
1948	5268 – 5525
1949	5526 – 5643
1950	5644 – 5805
1951	5806 – 5938
1952	5939 – 6142
1953	6143 – 6217

WC (square radiator)

Year	Number range
1933	WC1 – WC28
1934	WC29 – WC3126
1935	WC3127 – WC13869
1936	WC13870 – WC31783
1937	WC31784 – WC60789
1938	WC60790 – WC74329

WC (styled)

Year	Number range
1938	74330 – 75215
1939	75216 – 91533
1940	91534 – 103516
1941	103517 – 114533
1942	114534 – 123170
1943	None
1944	304 – 3194
1945	3195 – 3509
1946	3510 – 3747
1947	3748 – 4110
1948	4111 – 5499
1949	5500 – 7317
1950	7318 – 8315
1951	8316 – 8353

WC Speed Maintainer

Year	Number range
1938	IC1 – IC150
1939	IC151 – 1C400
1940	IC401 – IC407
1941	IC408 – IC411

W Speed Patrol

Year	Number range
1940	IE1 – IE466
1941	IE467 – IE634
1942	IE635 – IE678
1943-44	None
1945	IE681 – IE819
1946	IF820 – IF1437
1947	IE1438 – IE2278
1948	IE2279 – IE3515
1949	IE3516 – IE3746
1950	IE3547 – IE3753

WD

Year	Number range
1948	7-9249
1949	9250 – 35444
1950	35445 – 72327
1951	72328 – 105181
1952	105182 – 126931
1952	127007 – 131242
1953	131243 – 146606

WD45

Year	Number range
1953	146607 – 160385
1954	160386 – 190992
1955	190993 – 217991
1956	217992 – 230294
1957	230295 – 236958

D10

Year	Number range
1959	1001 – 1933*

D10 (Persian orange 2)

Year	Number range
1960	1950 – 2702
1961	2801 – 3262

D10 (149-cubic-inch engine)

Year	Number range
1961	3501 – INA
1962	INA – 4511*

D10 Series II

Year	Number range
1963	6801 – 7674
1964	7675 – 7850

D10 Series III

Year	Number range
1964	9001 – 9203
1965	9204 – 9485
1966	9486 – 9794
1967	9795 – 9978
1968	9979 – 10100

Serial Numbers
Allis-Chalmers continued

D12
Year	Number range
1959	1001 – 1734*

D12 (Persian 2)
Year	Number range
1960	1950 – 2428*
1961	2801 – 2919

D12 (149-cubic-inch engine)
Year	Number range
1961	3001 – INA
1962	INA – 3638

D12 Series II
Year	Number range
1963	5501 – 6011
1964	6012 – 6144

D14
Year	Number range
1957	1001 – 9399
1959	9400 – 14899
1960	14900 – 18230

D14 (black bars)
Year	Number range
1959	19001 – 21799
1960	21800 – 24050

D15
Year	Number range
1960	1001 – 1899
1961	190 – 6469
1962	6470 – 8169

D15 Series II
Year	Number range
1963	13001 – 16927
1964	16928 – 19680
1965	19681 – 21374
1966	21375 – 23733
1967	23734 – 25126
1968	25127 – 25419

D17
Year	Number range
1957	1001 – 4299
1958	4300 – 16499
1959	16500 – 23363

D17 (black bars)
Year	Number range
1959	24001 – 28199
1960	28200 – 31625

D17 (Persian orange 2)
Year	Number range
1960	32001 – 32099
1961	33100 – 38069
1962	38070 – 41540

D17 Series III
Year	Number range
1962	42001 – 43358
1963	65001 – 70610
1964	70611 – 72768

D17 Series IV
Year	Number range
1964	75001 – 77089
1965	77090 – 80532
1966	80533 – 86060
1967	86061 – 89213

D19
Year	Number range
1961	1001 – 1249
1962	1250 – 7331
1963	12001 – 14944
1964	14945 – 16266

D21
Year	Number range
1963	1001 – 1416
1964	1417 – 2078
1965	2079 – 2129

D21 Series II
Year	Number range
1965	2201 – 2407
1966	2408 – 2862
1967	2863 – 3776
1968	3777 – 4497
1969	4498 – 4609

One Sixty
Year	Number range
1969-71	No info.

Serial Numbers

160

Year	Number range
1972-73	No info.

One-Seventy

Year	Number range
1967	1005 – 2720
1968	2721 – 5373
1969	5374 – 6368
1970	6369 – 6987
1971	6988 – 7384

170 (black hood decal)

Year	Number range
1971	7500 – 7796
1972	7797 – 8820
1973	8821 – 10300

175

Year	Number range
1970	1001 – 1476
1971	1477 – 1623
1972	1624 – 1739
1973	1740 – 2152
1974	2153 -3254
1975	3255 – 3739
1976	3740 – 4811
1977	4812 – 5662
1978	5663 – 6320
1979	6321 – 6998
1980	6999 – 7502

One-Eighty

Year	Number range
1967	1007 – 2681
1968	2682 - 6093
1969	6094 – 9234
1970	9235 – 10560
1971	10561 – 11728
1972	11729 – 12446
1973	12447 – 12985

185

Year	Number range
1970	1001 – 1951
1971	1952 – 2934
1972	2935 – 3762
1973	3763 – 4960
1974	4961 – 6541
1975	6542 – 8366

1976	8367- 10024
1977	10025 – 11625
1978	11626 – 13159
1979	13160 – 14671
1980	14672 – 15647
1981	15648 – 15961

One-Ninety and XT (bar grille)

Year	Number range
1964	1001 – 2484
1965	2485 – 8218
1966	8219 – 8626
1966	904 – 13272
1967	13273 – 17784

One-Ninety and XT Series II (no bars)

Year	Number range
1967	1901 – 19261
1968	19262 – 22153

One-Ninety and XT Series III

Year	Number range
1968	23001 – 23233
1969	23234 – 2590
1970	25901 – 29135
1971	29136 – 31056
1972	31057 – 31117
1973	31118 – 31140

200

Year	Number range
1972	1004 – 3343
1973	3344 – 3558
1973	4001 – 6293
1974	6294 – 9249
1975	9250 – 11521

Two-Ten

Year	Number range
1970	104 – 1106
1971	1107 – 2081
1972	2082 – 2469

Two-Twenty

Year	Number range
1969	1004 – 1937
1970	1938 – 2450
1971	2451 – 2625
1972	2626 – 2866

Serial Numbers
Allis-Chalmers continued

440

Year	Number range
1972	1001 – 1139
1973	1140 – 1339
1974	1440 – 1650
1975	1651 – INA
1976	INA – 2010

5015

Year	Number range
1982	1001 – 1726
1983	1727 – 3276
1984	3277 – 4231
1985	4232 – INA

5020

Year	Number range
1977	1001 – 2219
1978	2220 – 3090
1979	3091 – 4114
1980	4115 – 5789
1981	5790 – 7033
1982	7034 – 8387
1983	8388 – 8733
1984	8734 – 9216
1985	9217 – INA

5030

Year	Number range
1978	1001 – 2004
1979	2005 – 2254
1980	2255 – 2975
1981	2976-3519
1982	3520 – 4065
1983	4066 – 4213
1984	4214 – 4368
1985	4369 – INA

5040

Year	Beginning number
1976	408445
1977	462148
1978	410384
1979	47300
1980	474000

5045

Year	Beginning number
1981	988501

6060

Year	Number range
1980	1001 – 1296
1981	1297 – 2462
1982	2463 – 3893
1983	3894 – 4571
1984	4572 – 5027

5050

Year	Beginning number
1977	573461
1978	579832
1979	584000
1980	591000
1981	596014
1982	597730
1983	599290

6040

Year	Beginning number
1974	No info.

6070

Year	Number range
1984	1001 – 1608
1985	1609 – 1972

6080

Year	Number range
1980	1001 – 1151
1981	1152 – 3001
1982	3002 – 4566
1983	4567 – 5779
1984	5780 – 6852
1985	6853 – 7698

6140

Year	Number range
1982	1001 – 1446
1983	1447 – 1850
1984	1851 – 2710
1985	2711 – 2714
1985	3545 – 3732

7000 (maroon)

Year	Number range
1975	1001 – 1647
1976	1648 – 5935
1977	5036 – 6372
1978	6373 – 6760

Serial Numbers

7000 (black)

Year	Number range
1978	8000 – 8962
1979	8963 – 9503

7010

Year	Number range
1979	1001 – 1924
1980	1925 – 2805
1981	2806 – 3433

7020

Year	Number range
1977	1001 – 1316
1978	1317 – 2731
1979	2732 – 3841
1980	3842 - 4709
1981	4710 – 5209

7030

Year	Number range
1973	1001 – 2594
1974	2596 – 4398

7040

Year	Number range
1974	1001 – 1302
1975	1303 – 4128
1976	4129 – 6839
1977	6840 – 8250

7045

Year	Number range
1977	1001 – 1233
1978	1234 – 2151
1979	2152 – 3398
1980	3399 – 4224
1981	4225 – 4888

7050

Year	Number range
1973	1001 – 1687
1974	1688 – 3300

7060 (maroon)

Year	Number range
1974	1001 – 1298
1975	1299 – 2748
1976	2749 – 4581
1977	4582 – 5644

7060 (black)

Year	Number range
1978	6001 – 6788
1979	6789 – 7692
1980	7693 – 8441
1981	8442 – 9142

7080 (maroon)

Year	Number range
1974	1001 – 1006
1975	1007 – 1571
1976	1572 – 2500
1977	2501 – 2930

7080 (black)

Year	Number range
1978	3001 – 3267
1979	3268 – 3647
1980	3648 – 3953
1981	3954 – 4225

7580 (maroon)

Year	Number range
1976	1001 – 1287
1977	1288 – 1604
1978	1605 only

8550

Year	Number range
1977	1001 – 1082
1978	1083 – 1341
1979	1342 – 1552
1980	1553 – 1722
1981	1723 – 2021

8010

Year	Number range
1981	1001 – 1019
1982	1020 – 1711
1983	1712 – 2265
1984	2266 – 2608
1985	2609 – 2832

8030

Year	Number range
1981	1001 – 1008
1982	1009 – 2092
1983	2093 – 2700
1984	2701 – 3145
1984	3146 – 3328

Serial Numbers
Allis-Chalmers continued

8050

Year	Number range
1981	1001 – 1015
1982	1016 – 1923
1983	1924 – 2495
1984	2496 – 3186
1985	3187 – 3336

8070

Year	Number range
1981	1001 – 1003
1982	1004 – 1429
1983	1430 – 2089
1984	2090 – 2902
1985	2903 – 3354

4W220

Year	Number range
1981	101 – 1002
1982	1003 – 1080
1983	1081 – 1144
1984	1145 – 1175

4W305

Year	Number range
1981	1001 – 1003
1982	1004 – 1111
1983	1112 – 1175
1984	1176 – 1337
1985	1338 – 1412

West Allis Industrial Tractors

140

Year	Number range
1964	1055 – INA
1965	INA-INA
1966	INA-1608

160

Year	Number range
1965	1005 – INA
1966	INA – 2340

1600

Year	Number range
1966	1002 – 1746
1967	1747 – 2708
1968	2709 – 3056

600 Forklift

Year	Number range
1968	1001 – 1114
1969	1115 – 1714

918 LBH

Year	Number range
1968	1001 – 1031

H3/HD3

Year	Number range
1960	1001 – 1195
1961	1250 – INA
1962	3199 – 4294
1963	6001 – 6944
1964	6945 – 7889
1965	7890 – 8855
1966	8856 – 9482
1967	9483 - 9699
1968	9700 – 9949

H4/HD4

Year	Number range
1965	1001 – 1243
1966	1244 – 2689
1967	2690 – 2937
1968	3001 – 3905
1969	3906 – 4332

Speed Patrol H

Year	Number range
1932	2001 – 2049
1933	2050 only

42S*

Year	Number range
1933	2501 – 2643
1934	2644 – 2775
1935	2766 – 2941
1936	3031 – 3160
1937	3276 – 3325
1938	3455 – 3516
1939	3517 – 3567
1940	3568 – 3575

*42S and 42T SNs are mixed.

42T*

Year	Number range
1936	3161T – 3199T
1937	3200T – 3361T

Serial Numbers

1938	3362T – 3488T
1939	3489T – 3625T
1940	3636T – 3675T

*42S and 42T SNs are mixed.

54S**

Year	Number range
1934	6501 – 6502
1935	6503 – 6597
1936	6628 – 6763
1937	6821 – 7002

54T**

Year	Number range
1936	6639 – 6777
1937	6825T – 7108T
1938	7021T – 7186T
1939	7187T – 7227T
1940	7261T – 7277T

KO54S**

Year	Number range
1937	681700 – 71270

K054T**

Year	Number range
1937	6778OT–6977OT
1938	7053OT–7120OT
1939	7728OT–7260OT

TW Speed Ace (Springfield)

Year	Number range
1935	1 – 10
1936	13 – 37
1937	38 – 87

Monarch crawlers

Monarch F 10 Ton

Year	Number range
1926	10001 – 10081
1927	10082 – 10240
1928	10241 – 10250

Monarch F-75

Year	Number range
1928	70001 – 70207
1929	70208 – 70654
1930	70655 – 70894
1931	70895 – 71066

Monarch G 5 Ton

Year	Number range
1926	7001 – 7090
1927	7091 – 7117

Monarch H 6-Ton

Year	Number range
1927	60001 – 60206
1928	60207 – 60297

Monarch H-50

Year	Number range
1928	60298 – 60610
1929	60611 – 61558
1930	61559 – 62146
1931	62147 – 62297

Monarch K/KO

Year	Number range
1929	1 – 48
1930	49 – 1372
1931	1373 – 2333
1932	2334 – 2653
1933	2654 – 3045
1934	3046 – 3593
1934	6001 – 6018
1935	3594 – 4792
1936	4793 – INA
1936	6019 – 6335
1937	6336 – 7707
1938	7708 – 8017
1939	8018 – 8522
1940	8523 – 8956
1941	8957 – 9268
1942	9269 – 9393
1943	9394 – 9468

Monarch L/LO

Year	Number range
1931	1 – 34
1932	35 - 498
1933	499 – 660
1934	661 – 887
1934	401 – 4021
1935	888 – 1438
1936	1439 – 2136
1937	2137 – 2705
1938	2706 – 2943
1939	2944 – 3232
1939	LD 8-17

Serial Numbers
Allis-Chalmers continued

1940	3233 – 3251
1941	3252 – 3272
1942	3273 – 3357

Monarch M

Year	Number range
1932	1 – 41
1933	42 – 401
1934	402 – 841
1935	842 – 1941
1936	1942 – 3841
1937	3842 – 7066
1938	7067 – 8126
1939	8127 – 9539
1940	9540 – 11379
1941	11380 – 12946
1942	12947 – 14524

Monarch S/SO

Year	Number range
1937	3 – 412
1938	413 – 584
1939	585 – 1084
1940	1085 – 1127
1941	1128 – 1211
1942	1212 – 1227

GM Two-Cycle Diesels HD5

Year	Number range
1946	1 – 6
1947	6 – 1357
1948	1358 – 4315
1949	4316 – 7498
1950	7499 – 11070
1951	11071 – 14289
1952	14290 – 17557
1953	17558 – 21836
1954	21837 – 25563
1955	25564 – 29255

HD7 and HD7W Military

Year	Number range
1940	3 – 502
1941	503 – 1136
1942	1137 – 2980*
1943	2977 – 5952
1944	5285 – 8948*
1945	7998 – 12090*
1946	12091 – 13077
1947	13078 – 15121*
1948	15117 – 16751

1949	16752 – 18085
1950	18086 – 18505

HD9

Year	Number range
1950	1
1951	2 – 737
1952	738 – 1882
1953	1883 – 3590
1954	3591 – 5208
1955	5209 – 5850

HD10 and HD10W Military

Year	Number range
1940	2 – 641
1941	642 – 1445
1942	1446 – 2081
1942	2452 – 2855
1943	2082 – 2338
1943	2856 – 3651
1944	2339 – 2451
1944	3652 – 4094
1944	4152 – 4336
1944	4452 – 4896
1945	4095 – 4151
1945	4337 – 4451
1945	4897 – 5963
1946	5964 – 6462
1947	6463 – 7601
1948	7602 – 8675
1949	8676 – 9630
1950	9631 – 10198

HD14 and HD14C

Year	Number range
1939	18 – 25
1940	26 – 548
1941	549 – 1165
1942	1166 – 2112
1943	2113 – 3137
1944	3138 – 4258
1945	4259 – 5454
1946	5455 – 5814
1947	5815 – 6422

HD15

Year	Number range
1950	1
1951	2 – 810
1952	811 – 1857
1953	1858 – 2855

Serial Numbers

Year	Number range
1954	2856 – 3684
1955	3685 – 3909

HD19
Year	Number range
1947	4 – 120
1948	121 – 1195
1949	1196 – 2001
1950	2002 – 2654

HD20
Year	Number range
1951	3001 – 3827
1952	3828 – 4922
1953	4923 – 5736
1954	5737 – 6100

HD3 (Springfield)
Year	Number range
1942	3 – 32

(No 20 or 30)

HD6, HD6G, and HD6GB
Year	Number range
1955	101 – 1146
1956	1147 – 6465
1957	6466 – 7947
1958	7948 – 10053
1959	10054 – 12505
1960	12506 – 13776
1961	13777 – 14897
1962	14898 – 16041
1963	16042 – 17104
1964	17105 – 18188
1965	18189 – 19400
1966	19401 – 20200
1967	20201 – 20808
1968	20809 – 21431
1969	21432 – 22271
1970	22272 – 23354
1971	23355 – 24011
1972	24012 – 24671
1973	24672 – 25011
1974	25012 – 25271

HD11
Year	Number range
1955	101 – 1057
1956	1058 – 3254
1957	3255 – 4114
1958	4115 – 4767
1959	4768 – 5801
1960	5802 – 6447
1961	6448 – 6994
1962	6995 – 7869
1963	7870 – 8605
1964	8606 – 10532
1965	10533 – 11450
1966	11451 – 12250
1967	12251 – 13130
1968	13131 – 13656
1969	13657 – 14680
1970	Begins 14681

HD11 Series B
Year	Number range
1971	16001 – 16615
1972	16615 – 17342
1973	17343 – 18015
1974	18016 – 18798

HD16 and HD16G
Year	Number range
1955	101 – 1008
1956	1009 – 2462
1957	2463 – 2756
1958	2757 – 443
1959	4144 – 4725
1960	4726 – 5097
1961	5098 – 5447
1962	5448 – 5731
1963	5732 – 6269
1964	6270 – 6964
1965	6965 – 7467
1966	7468 – 8075
1967	8076 – 8424
1968	8425 – 8793
1969	8794 – 9564
1970	9565 – 9619

HD16 Series B
Year	Number range
1970	10301 – 10461
1971	10462 – 11703
1972	11704 – 12698
1973	12699 – 13142
1974	13143 – 14000
1975	14001 – 14300
1976	14301 – 14911
1977	14912 – 15021
1978	15022 – 15149
1979	15150 – 15365

Serial Numbers

1980	15366 – 15439
1981	15440 – 15521

HD21 HD21G

Year	Number range
1954	7001 – 7012
1955	7013 – 7915
1956	7916 – 9090
1957	9091 – 9301
1958	9302 – 11028
1959	11029 – 11681
1960	11682 – 11938
1961	11939 – 12259
1962	12260 – 12456
1963	12457 – 13249
1964	12250 – 13736
1965	13737 – 14158
1966	14159 – 14604
1967	14605 – 14883
1968	14884 – 15110
1969	15511 – 15207

HD21 Series B

Year	Number range
1969	16001 – 16560
1970	16561 17656
1971	17657 – 18504
1972	18505 – 19113
1973	19114 – 19443
1974	19444 – 19946
1975	19947 – 21000

Aultman-Taylor

Steam engines

Year	Serial number
1897	5148-5285
1898	5286-5492
1899	5493-5758
1900	5759-5939
1901	5940-6126
1902	6127-6347
1903	6348-6547
1904	6548-6805
1905	6806-7056
1906	7057-7373
1907	7374-7615
1908	7616-7702
1909	7703-7875
1910	7876-8123
1911	8124-8290
191	8291-8438
1913	8439-8617
1914	8618-8896
1915	8897-9089
1916	9090-9185
1917	9186

Model A

Year	Beginning number
1945	4A786
1946	7A305
1947	9A867
1948	13A427
1949	17A456
1950	19A366

Model BF

Year	Beginning number
1950	R500
1951	R1839
1952	R4460

Model BFW

Year	Beginning number
1953	R6538

Model BFD

Year	Beginning number
1953	57700001

Model BFS

Year	Beginning number
1953	57600001

Model BFH

Year	Beginning number
1953	58000001

Model BG

Year	Beginning number
1953	57900001
1954	57900601
1955	57900769

Model V

Year	Beginning number
1946	1V5
1947	1V144
1948	2V577

Serial Numbers

1949	4V490
1950	5V501
1951	6V207
1952	6V422

Case

For all J.I. Case except for the Models 70, 90, 94, and 96 Series serial numbers can be found stamped on the nameplate fastened to the instrument panel. Serial numbers can also be found on the toolbox.

B, C, CC, CI, CO, R, RC, D, DC, DI, DO, LA, LAI, S, SC, SI, SO, V, VC, VI, VO, VAC, VAH, VAI, VAO, and W

Year	Beginning number
1929	300201
1930	300301
1931	300401
1932	300501
1933	300601
1934	300701
1935	300801
1936	300901
1937	301001
1938	4200001
1939	4300001
1940	4400001
1941	4500001
1942	4600001
1943	4700001
1944	4800001
1945	4900001
1946	5000001
1947	5100001
1948	5200001
1949	5300001
1950	5400001
1951	5500001
1952	5600001
1953	5700001
1954	5800001
1955	5900001
1956	6000001

9-18

Year	Beginning number
1912	100
1913	891
1914	2496
1915	2842
1916	3691
1917	7492
1918	13285

10-18

Year	Beginning number
1918	13285
1919	22223
1920	32841
1921	42256
1922	43943

10-20

Year	Beginning number
1915	2842
1916	3691
1917	7492
1918	13285

12-20

Year	Beginning number
1921	42256
1922	43943
1923	45281
1924	48227
1925	48402
1926	55919
1927	62409

12-25

Year	Beginning number
1914	2496
1915	2842
1916	3691
1917	7492
1918	13285

15-27

Year	Beginning number
1919	22223
1920	42435
1921	42835
1922	42852
1923	43435
1924	48413

Serial Numbers
Case continued

18-32

Year	Beginning number
1925	51678
1926	55919
1927	62409

20-40

Year	Beginning number
1912	100
1913	691
1914	2496
1915	2842
1916	3691
1917	7492
1918	13285
1919	22223

22-40

Year	Beginning number
1919	22223
1920	32841
1921	42256
1922	43943
1923	45281
1924	48413
1925	51678

30-60

Year	Beginning number
1912	100
1913	691
1914	2496
1915	2842
1916	3691

40-72

Year	Beginning number
1921	40256
1922	43943
1923	45281

40-80

Year	Beginning number
1915	2842

200B and 210B

Year	Beginning number
1958	6095001
1959	6120001

211B, 311B, 411B, 511B, and 611B

Year	Beginning number
1958	5095001
1959	6120001

300 and 320

Year	Beginning number
1956	605000301
1957	6075001

300B, 310B, and 320B

Year	Beginning number
1958	6095001
1959	6120001

300 and 400

Year	Beginning number
1957	3000001

320

Year	Beginning number
1956	6050301
1957	6075001
1958	6095001
1959	6120001

400

Year	Beginning number
1955	8060001
1956	8080001
1957	8100001

400B

Year	Beginning number
1958	6095001
1959	6120001

420B

Year	Beginning number
1958	6095001
1959	6120001

Models 430 and 530 Utility

Year	Beginning number
1960	3012275

Models 430 and 530

Year	Beginning number
1961	Utility

Serial Numbers

Models 430, 530, 530B, 630, 730, 830, 930, and 1030

Year	Beginning number
1960	8160001
1961	8168801
1962	8190001
1963	8208001
1964	8229001
1965	8253501
1966	8279001
1967	8306501
1968	8332101
1969	8356251

Models 470, 570, 590B, 770, 870, 970, 1070, 1090, and 1170

Year	Beginning number
1970	8650001
1971	8674001
1972	8693001
1973	8712001
1974	8736601
1975	8770001

For Case 70 Series tractors the serial number is stamped just above the platform floor and next to the steering wheel on the left-hand side of the tractor.

500

Year	Beginning number
1953	5700001
1954	8035001
1955	8060001
1956	8080001
1957	8100001

500B and 600B

Year	Beginning number
1958	6095001
1959	6120001

600

Year	Beginning number
1957	8100001

680C

Year	Beginning number
1969	9103226
1970	9104651

1971	9106001

700 and 800

Year	Beginning number
1958	8120001
1959	8140001

900

Year	Beginning number
1957	8100001
1958	8120001
1959	8140001

1031 and 1032

Year	Beginning number
1966	8279001
1967	8306501
1968	8332101
1969	8356251

1175, 1270, and 1370

Year	Beginning number
1972	8693001
1973	8712001
1974	8736601
1975	8770001

1200TK

Year	Beginning numberr
1966	9802101
1967	9806101
1968	9808000
1969	9808276

1470TK

Year	Beginning number
1969	9810000
1970	9811301
1971	8674001
1972	8691901

2470 and 2670

Year	Beginning number
1971	8674001
1972	8692381
1973	8712001
1974	8762001
1975	8767001

Serial Numbers

A, AE, and AI
Year	Beginning number
1928	69004
1929	69803

B
Year	Beginning number
1928	69004

D and LA
Year	Beginning number
1953	5700001
1954	5800001

K
Year	Beginning number
1928	69004
1929	69803

L and LI
Year	Beginning number
1929	303201
1930	303301
1931	303401
1932	303501
1933	303601
1934	303701
1935	303801
1936	303901
1937	304001
1938	4200001
1939	4300001
1940	4400001

S
Year	Beginning number
1953	5700001
1954	8035001

Caterpillar

Holt 2 Ton
Years	Beginning number
1925-1928	25003, 70001

Holt 5 Ton
Years	Beginning number
1919-1926	19001, 42001, 43001

Holt 10 Ton
Years	Beginning number
1925	15001, 34001

Ten
Years	Beginning number
1928-1933	PT1

Fifteen
Years	Beginning number
1929-1933	PV1
1932-1933	7G1

Fifteen-HC
Years	Beginning number
1932-1939	1D1

Twenty
Years	Beginning number
1932-1934	8C1
1927-1933	L1, PL1

Twenty-Two
Years	Beginning number
1934-1939	2F1, 1J2

Twenty-Five
Years	Beginning number
1931-1933	3C1

Twenty-Eight
Years	Beginning number
1933-1935	4F1

Thirty
Years	Beginning number
1925-1932	S1001, PS1

Thirty-R4
Years	Beginning number
1935-1944	6G1

Thirty-Five
Years	Beginning number
1932-1934	5C1

Thirty-Five Diesel
Years	Beginning number
1933-1934	6E1

Serial Numbers

Forty
Years	Beginning number
1934-1936	5G1

Forty Diesel
Years	Beginning number
1934-1936	3G1

Fifty
Years	Beginning number
1931-1937	5A1

Fifty Diesel
Years	Beginning number
1933-1936	1E1

Sixty
Years	Beginning number
1925-1931	101A, PA1

Sixty-Five
Years	Beginning number
1932-1933	2D1

Sixty-Five Diesel
Years	Beginning number
1931-1932	1C1

Seventy
Years	Beginning number
1933-1937	8D1

Seventy Diesel
Year	Beginning number
1933	3E1

Seventy-Five Diesel
Years	Beginning number
1933-1935	2E1

D2
Years	Beginning number
1938-1947	3J1, 5J1
1947 On	4U, 5U

D4 and RD4
Years	Beginning number
1936-1947	4G1, 2T1, 7J1, 5T1

D4
Years	Beginning number
1947 and on	6U, 7U

D5NG
Year	Beginning number
1939	M

D6 and RD6
Years	Beginning number
1935-1941	5E801, 2H1

D6
Years	Beginning number
1941-1947	4R, 5R
1947	8U, 9U

D7 and RD7
Years	Beginning number
1935-1940	5E7501, 9G1

D7
Years	Beginning number
1940 and on	7M, 3T

D8 and RD8
Years	Beginning number
1935 and on	5E, 1H, 8R, 2U

R-2
Years	Beginning number
1934-1937	5E3501
1938-1942	6J1, 4J1

R-3
Years	Beginning number
1934-1935	5E2501

R-5
Years	Beginning number
1934-1940	4H501, 3R1, 5E3001

Cletrac

AG
Serial number is located in the upper right-hand corner of instrument panel.

Serial Numbers
Cletrac continued

Year	Beginning number
1939	2X0886
1940	2X1702
1941	2X2474
1942	2X3102

AG-6

Year	Beginning number
1944	3X0000
1945	3X1512
1946	3X3260
1947	3X4586
1948	3X6062
1949	3X7644
1950	3X8542
1951	3X8456
1952	3X8626
1953	3500000
1954	4500000
1955	4X020
1956	4X100
1957	4X216

AD
Serial number is located on the right-hand side of panel behind the engine.

Year	Beginning number
1939	5Z36
1940	9Z32
1941	1Z106
1942	1Z690
1943	1Z836
1944	1Z838
1945	2Z172
1946	2Z672
1947	3Z700
1948	4Z912
1949	6Z214
1950	7Z390
1951	7Z422
1952	8Z296
1953	3500000
1954	4500000
1955	8Z992
1956	9iZ128
1957	9Z178
1958	9Z212
1959	9Z225

AD2

Year	Beginning number
1939	4N60
1940	5N76

BD
Serial number is located on the right-hand side of panel behind the engine.

4-Cylinder

Year	Beginning number
1939	3D708
1940	4D206

6-Cylinder

Year	Beginning number
1940	5D000
1941	5D422
1942	6D032
1943	6D286
1944	6D948
1945	7D714
1946	8D474
1947	9D680
1948	11D224
1949	13D820
1950	15D658
1951	16D514
1952	18D032
1953	3500802
1954	4500937
1955	12C934

BG
Serial number is located on the upper right-hand side of panel behind the engine.

BG 4-Cylinder

Year	Beginning number
1939	2C578
1940	2C798

BG 6-Cylinder

Year	Beginning number
1940	3C00
1941	3C338
1942	3C654
1943	3C738
1944	3C97

Serial Numbers

BGS

Year	Beginning number
1944	6C000
1945	6C630
1946	8C370
1947	9C062
1948	10C238
1949	11C462
1950	11C950
1951	12C230
1952	12C504
1953	3500802
1954	4500937
1955	12C934

CG

Year	Beginning number
1939	5M260
1940	5M312
1941	5M544
1942	5M604

DD

DD serial numbers are located on the upper left-hand side of panel behind the engine.

DD 4-Cylinder

Year	Beginning number
1939	1L4220

DD 6-Cylinder

Year	Beginning number
1939	1L5000
1940	1L5222
1941	1L5644
1942	1L6332
1943	1L6560
1944	1L6820
1945	1L7422
1946	1L8012
1947	1L8792
1948	2L0194
1949	2L2142
1950	2L3528
1951	2L4130
1952	2L5350
1953	3500000
1954	4500000
1955	3L226
1956	3L404
1957	3L498
1958	3L596

DG

Serial number is located on the upper right-hand side of panel behind the engine.

DG 4-Cylinder

Year	Beginning number
1939	4E66

DG 6-Cylinder

Year	Beginning number
1939	7E00
1940	7E68
1941	1E200
1942	1E364
1943	1E392
1944	1E464
1945	1E748
1946	2E356
1947	2E620
1948	2E852
1949	3E186
1950	3E298
1951	3E362
1952	3E402
1953	3500000
1954	4500000
1955	3E502
1956	3E516
1957	3L558***
1958	3L600***

E31

Year	Beginning number
1939	1B816

E42

Year	Beginning number
1939	5H130
1940	5H428
1941	5H574

ED-38 and ED-42

Year	Beginning number
1939	1AA78
1940	7AA24
1941	8AA74

Serial Numbers

Cletrac continued

ED2-38/42

Year	Beginning number
1939	5S92
1941	6S74

ED2/62/68/76

Year	Beginning number
1939	2V28
1940	3V58

EHG-62/68/76

Year	Beginning number
1939	5R20
1940	7R78
1941	8R18

FD

Serial number is located on the right side of panel beside the fuel tank.

Year	Beginning number
1939	8Y094
1940	8Y366
1941	8Y476
1942	8Y684
1943	8Y772
1944	8Y910
1945	9Y116

FDLC

Year	Beginning number
1941	1HA00
1942	1HA076
1943	1HA096
1944	1HA110
1945	1HA196

FDE

Year	Beginning number
1945	10Y000
1946	10Y200
1947	10^480
1948	10^794
1949	11Y122
1950	11Y582
1951	11Y790
1952	12Y124

FG6

Year	Beginning number

Year	Beginning number
1939	1CA532
1940	1CA608
1941	1CA796
1942	1CA846
1943	1CA880

GG

Year	Beginning number
1939	1FA000
1940	5FA388
1941	1FA0086
1941	1FA1000
1942	1FA6532

HG

Serial number is located on right-hand side of engine main frame rear of footrest (early) or front of footrest (late).

Year	Beginning number
1939	1GA000
1940	2GA884
1941	5GA022
1942	7GA484
1943	8GA702
1944	9GA632
1945	13GA422
1946	19GA994
1947	25GA240
1948	33GA146
1949	41GA896
1950	48GA428
1951	55GA858

HGF

Year	Beginning number
1947	26GA108
1948	35GA340
1949	46GA508

HGR

Year	Beginning number
1945	1NA000
1947	1NA050
1948	1NA706

MG1

Year	Beginning number
1941	1JA000
1942	1JA236

Serial Numbers

1943	7JA542
1944	12A710

MG1

Year	Beginning number
1942	1DA000
1943	1DA062

MG2

Year	Beginning number
1942	1KA000

MG3

Year	Beginning number
1942	1LA000

Oliver Cletrac crawlers

R

Year	Number range
1916-1917	1 – 1000

H

Year	Number range
1917-1919	1001 – 13755

W (W-12)

Years built	Number range
1919-1932	13756 – 30971

F

Years built	Number range
1920-1922	1 – 3000

20K

Years built	Number range
1925-1932	101 – 10207

30A (A30)

Years built	Number range
1926-1928	6 – 1421

30B (B30)

Years built	Number range
1929-1930	1601 – 3057

40

Years built	Number range
1928-1931	101 – 1833

55-40

Years built	Number range
1931-1932	1835 – 1889

55

Years built	Number range
1932-1938	1890 – 3852

100

Years built	Number range
1927-1930	50 – 158

15

Years built	Number range
1931-1933	76 – 11999

20C

Years built	Number range
1933-1936	12000 – 14547

AG

Years built	Number range
1936-1937	14548 – 20201

AG

Years built	Number range
1937-1942	2X0202 – 2X3398

AG-6

Years built	Number range
1944-Up	3X0000 – Up

AD

Years built	Number range
1937-Up	1Z00 – Up

BD Four Speed

Years built	Number range
1936-1939	1D00 – 4D236

BD Six Speed

Years built	Number range
1939-Up	5D000 – Up

BD2

Years built	Number range
1937-1938	1P00 – 1P16

Serial Numbers
Cletrac continued

25

Years built	Number range
1932-1935	76 – 1372

30G

Years built	Number range
1935-1936	1C00 – 2C79

BG Four Speed

Years built	Number range
1937-1939	2C80 – 2C798

BG Six Speed

Years built	Number range
1939-Up	3C000 – Up

BGS

Years built	Number range
1944-Up	6C000 – Up

40-30

Years built	Number range
1930-1931	76 – 399

35

Years built	Number range
1932-1936	400 – 3835

CG

Years built	Number range
1936-1942	28365M000– 3246 5M608

35D

Years built	Number range
1934-1935	10000 – 10217

40D

Years built	Number range
1935-1936	10218 – 10831

DD Four Speed

Years built	Number range
1936-1939	10832 1L3000 – 11581 1L4460

DD Six Speed

Years built	Number range
1939 – Up	1L5000 Up

DG Four Speed

Years built	Number range
1936-1939	1E00 – 5E86

DG Six Speed

Years built	Number range
1939 and up	7E00 and up

EN (E31)

Years built	Number range
1934-1939	1B00 – 1B946

E38 (pre-streamlined)

Years built	Number range
1934-1936	7B30 – 1B318

E38 (streamlined)

Years built	Number range
1936-1938	2H000 – 3H168

E42

Years built	Number range
1938-1942	5H000 – 5H604

E-62-68-76

Years built	Number range
1934-1941	1A00 – 5A330

EHG-62-68-76

Years built	Number range
1937-1941	1R00 – 8R52

ED-38-42

Years built	Number range
1938-1941	1AA00 – 9AA00

ED2-38

Years built	Number range
1937-1941	S00 – 1S52

ED2-42

Years built	Number range
1937-1941	5S00 – 6S80

ED2-62-68-76

Years built	Number range
1938-1941	1V00 – 4V13

80D

Years built	Number range
1933-1936	6000 – 6321

Serial Numbers

FD Four Speed
Years built	Number range
1936-1938	6322 – 6699

FD Six Speed
Years built	Number range
1938-Up	8Y000 – Up

FDLC
Years built	Number range
1941-Up	1HA000 – Up

30-60
Years built	Number range
1930-1932	113 – 409

80
Year	Beginning number
1932-1932	420 – 499

80G
Years built	Number range
1932-1936	500 – 846

FG Four Speed
Years built	Number range
1936-1938	1CA046 – 1CA054

FG Six Speed
Years built	Number range
1938-Up	1CA500 – Up

GG (sold as Twin Row General and Co-Op)
Years built	Number range
1939-1942	1FA0001FA1000–1FAO1641FA6886

HG-31-42-68
Years built	Number range
1939-Up	1GA000 – Up

60 Standard
Year	Number range
1942	410001-410500
1943	410501-410510
1944	410511-410616
1945	410617-410910
1946	410911-411310
1947	411311-411960
1948	411961-413605

Cockshutt

Serial numbers for Cockshutt Models 20, 30, 40, and 50 are found on the top of the left frame side next to the engine. Models 35, 35L, Golden Eagle 40D4, and 40PD serial numbers are on the top of the right frame side. Serial numbers for Models 550, 560, 570, and 570 Super are on the top right of the frame side. It's on the top left frame side of the 540.

20
Year	Beginning number
1952	101
1953	1657
1954	2568
1955	10001
1956	20001
1957	30001
1958	40001

30 and Gamble's Farmcrest)
Year	Beginning number
1946	1
1947	442
1948	6705
1949	17370
1950	26417
1951	28524
1952	32389
1953	32389
1954	35974
1955	40001
1956	50001

35 Deluxe and 35L Deluxe
Year	Beginning number
1956	1001
1957	10001
1958	20001

Golden Arrow
Year	Beginning number
1957	16001

Serial Numbers
Cockshutt continued

40

Year	Beginning number
1949	101
1950	194
1951	4101
1952	6901
1953	10501
1954	11401
1955	20001
1956	30001
1957	40001
1958	50001

40PD and Golden Eagle 40D4

Year	Beginning number
1955	27001
1956	30001
1957	40028
1958	50001

50

Year	Beginning number
1953	101
1954	1801
1955	10001
1956	20001
1957	30001
1958	40001

540

Year	Beginning number
1958	AM1001
1959	AN5001
1960	None built
1961	AP1001
1962	AR1001

550

Year	Beginning number
1958	BM1001
1959	BN5001
1960	BO1001
1961	BP1001
1962	BR1001

560

Year	Beginning number
1958	CM1001
1959	CN5001
1960	CO7001
1961	CP1001

570

Year	Beginning number
1958	DM1001
1959	DN5001
1960	DO7001

570 Super

Year	Beginning number
1961	DP1001
1962	DR1001

18-28

Year	Serial number
1930	800 001
1931	800 460
1932	800 964
1933	800 985
1934	800 051
1935	801 241
1936	801 990
1937	802

28-44

Year	Serial number
1930	500 001
1931	503 600
1932	506 185
1933	506 212
1934	506 255
1935	506 401
1936	507 176
1937	508 016

70 RC

Year	Serial number
1935	200 001
1936	200 686
1937	208 729
1938	219 645
1939	223 255
1940	231 116
1941	236 356
1942	241 391
1943	243 640
1944	244 711
1945	250 180
1946	252 780
1947	258 140
1948	262 840

Serial Numbers
Cockshutt continued

1350

Year	Beginning number
1966	28302844
1967	28303141
1968	28304546

2

Year	Beginning number
1954	2601
1955	10001
1956	20001

3

Year	Beginning number
1953	35601
1955	40001

18-28

Year	Number range
1930	800001-800459
1931	800460-800963
1932	800964-800984
1933	800985-801050
1934	801051-801240
1935	801241-801989
1936	801990-802937
1937	802938-803928

28-44

Year	Number range
1930	500001-503599
1931	503600-506184
1932	506185-506211
1933	506212-506254
1934	506255-506400
1935	506401-507175

60 Row Crop

Year	Number range
1940	600001-600070
1941	600071-606303
1942	606304-607394
1943	607395-608525
1944	608526-612046
1945	612047-615627
1946	615628-616706
1947	616707-620256
1948	620257-625131

60 Standard

Year	Number range
1942	410001-410500
1943	410501-410510
1944	410511-410616
1945	410617-410910
1946	410911-411310
1947	411311-411960
1948	411961-413605

70 Row Crop

Year	Number range
1935	200001-200685
1936	200686-208728
1937	208729-216925
1937	220426-220694
1938	219645-220425
1938	220695-223254
1939	223255-231115
1940	231116-236355
1941	236356-241390
1942	241391-243639
1943	243640-244710
1944	244711-250179
1945	250180 252779
1946	252780-258139
1947	258140-262839
1948	262840-267866

70 Standard

Year	Number range
1936	300001-300633
1937	300634-301802
1937	301803-302083
1938	302084-303464
1939	303465-305361
1940	305362-306593
1941	306594-307579
1942	307580-308187
1943	308188-308483
1944	308484-310217
1945	310218-311115
1946	311116-312689
1947	312690-314220
1948	314221-315420

80 Row Crop

Year	Number range
1937	109152-109166
1938	109167-109782
1939	109783-110220

Serial Numbers

Year	Number range
1940	110221-110614
1941	110615-110944
1942	110945-111218
1943	111219-111390
1944	111391-111928
1945	111929-112878
1946	112879-114143
1947	114144-114943
1948	114944-115373

80 Standard

Year	Number range
1937	803929-803990
1938	803991-805376
1939	805377-806879
1940	806880-808124
1941	808125-809050
1942	809051-809990
1943	809991-810469
1944	810470-811990
1945	811991-813066
1946	813067-814563
1947	814564-815215
1948	815216-816241

90 and 99

Year	Number range
1937	508918-508934
1938	508935-509611
1939	509612-510067
1940	510068-510563
1941	510564-510976
1942	510977-511295
1943	511296-511473
1944	511474-512043
1945	512044-512043
1946	512821-513105
1947	513106-513855
1948	513856-514855
1949	514856-516275
1950	516276-516887
1951	516888-517873
1952	517874-518212

Co-op

Serial numbers for Co-op Models No. 1, 2, and 3 are located on the top edge of the left frame rail near the distributor. Throughout the years of production they are present in at least three separate forms:
A. X XXX
B. X*XXX
C. XXXX-X

One might expect this variation to denote manufacturing location. The first digit in A and B above would be model number, e.g., 1 for Model No. 1, 2 for a No. 2, and 3 for No. 3. The third example appears on the Farmers Union production models from St. Paul. Numbering on the West Virginia version is unknown because no specific tractors can be traced to the West Virginia plant.

Post-war Co-op No. 3S serial numbers are found on a plate affixed to the dash and start with 3001.

The B-2, B-3, and B-2 Jr. serial numbers are located on the right frame rail near the brake pedal.

Model C serial numbers are located on the top of the right side frame rail near the belt pulley. Two forms of numbering are known to exist:
A. C – XXX
(square-axle housing)
B. C XX-XXX
(round-axle housing)

The first 53 produced had square axles; therefore the example in B directly above would start C54-XXX.

E-2 (Cockshutt built)

Year	Beginning number
1952	101
1953	1657
1954	2568
1955	10001

Serial Numbers

1956	20001
1957	30001
1958	40001

E-3 (Cockshutt built)

Year	Beginning number
1946	0
1947	442
1948	6705
1949	17370
1950	26417
1951	28524
1952	32389
1953	32389
1954	35974
1955	40001
1956	50001

E-4 (Cockshutt built)

Year	Beginning number
1949	101
1950	194
1951	4101
1952	6901
1953	10501
1954	11401
1955	20001
1956	30001
1957	40001
1958	50001

E-5 (Cockshutt built)

Year	Beginning number
1953	101
1954	1801
1955	10001
1956	20001
1957	30001
1958	40001

David Brown

VAK 1

Years	Serial numbers
1939-1944	1001-6000

VAK 1A

Years	Serial numbers
1945-1947	6351-9852

Cropmaster

Years	Serial numbers
1947-1953	P10000-44000

30C First Series

Years	Serial numbers
1953-1954	P44500-46058

30C Second Series

Years	Serial numbers
1954-1957	P/30 10001-12766

Cropmaster Diesel

Years	Serial numbers
1949-1953	PD10001-18499

30D Diesel First Series

Years	Serial numbers
1953-1954	PD18500-20797

30D Second Series

Years	Serial numbers
1954-1957	PD/30 10001-19452

Super Cropmaster

Years	Serial numbers
1950-1952	SP10001-10600

Prairie Cropmaster

Years	Serial numbers
1951-1954	SPP10001-14881

Prairie Cropmaster Diesel

Years	Serial numbers
1952-1954	PDP10001-10245

Cropmaster Vineyard

Years	Serial numbers
1947-1953	N10026-10205

Cropmaster Diesel Vineyard

Years	Serial numbers
1951-1952	ND10001-10084

50D

Years	Serial numbers
1953-1958	VAD 10001-11260

25C

Years	Serial numbers
1953-1958	P25 10001-21318

Serial Numbers
David Brown continued

25D

Years	Serial numbers
1953-1958	PD25 10001-23424

2D

Years	Serial numbers
1957-1961	VAD12V 10001-11644

2DV

Years	Serial numbers
1957-1961	VAD12V 10001-10364

900 Petrol

Years	Serial numbers
1956-1958	900G 10001-10305

900 Kerosene

Years	Serial numbers
1956-1958	900K 10001-10265

900 Diesel

Years	Serial numbers
1956-1958	900D 10001-17437

900 J & L Series

Years	Serial numbers
1957-1958	J900GL900G J900K L900K J900D L900D 50001-55445

950

Years	Serial numbers
1958-1959	T950GU950G T950K U950K T950D U950D 57000-62934

950 Implematic First Series

Years	Serial numbers
1959-1961	V950GW950G V950K W950K V950D W950D 63000-81126

950 Implematic Second Series

Years	Serial numbers
1961-1962	950/3A950/3B 4400001-401489

850 Implematic First Series

Years	Serial numbers
1960-1961	A850DB850D A850G B850G 300001-306334

850 Implematic Second Series

Years	Serial numbers
1961-1965	850C850D 850DM 310001-317439

850 Implematic Narrow

Years	Serial numbers
1961-1965	850CV850DV N390001-390469

880 Implematic First Series

Years	Serial numbers
1961-1964	880C880D 350001-362382

880 Implematic Narrow

Years	Serial numbers
1961-1965	880CV880DV N395001-395303

900 Implematic

Years	Serial numbers
1961-1965	990A990B 440001-480600

880 Implematic Second Series

Years	Serial numbers
1964-1965	880E880F 521001-527521

Oliver 500

Years	Serial numbers
1960-1964	100001-102000

Oliver 600

Years	Serial numbers
1961-1964	449800-453700

770 Selectamatic

Years	Serial numbers
1965-1970	580001-592375

Serial Numbers

780 Selectamatic
Years   Serial numbers
1967-1971 600001-611551

780 Selectamatic Narrow
Years   Serial numbers
1969-1971 645001-645647

880 Selectamatic
Years   Serial numbers
1965-1971 530001-563379

990 Selectamatic
Years   Serial numbers
1965-1968 482001-505286

990 Selectamatic
Years   Serial numbers
1968-1971 800001-831351

1200 Selectamatic
Years   Serial numbers
1967-1971 700001-718990

3800 Selectamatic
Years   Serial numbers
1968-1971 650001-650522

4600 Selectamatic
Years   Serial numbers
1968-1971 900001-900582

775 Synchromesh
Years   Serial numbers
1972-1976 651001-651870

885 Gasoline
Years   Serial numbers
1972-1976 594001-594235

885 Synchromesh
Years   Serial numbers
1971-1976 620001-640365

885 Synchromesh Narrow
Years   Serial numbers
1971-1976 64601-647360

990 Synchromesh
Years   Serial numbers
1971-1976 850001-868541

995 Synchromesh
Years   Serial numbers
1971-1976 920001-936004

996 Synchromesh
Years   Serial numbers
1971-1976 980001-989042

1210 Synchromesh
Years   Serial numbers
1971-1976 720001-731855

1212 Hydra Shift
Years   Serial numbers
1971-1976 1000001-1005283

1410 Synchromesh and 1412 Synchromesh
Years   Serial numbers
1974-1976 1050001-1051058

8 Series
Years   Serial numbers
1976-1983 11000001-11021851

9 Series
Years   Serial numbers
1976-1980 11070001-11106274

12 Series
Years   Serial numbers
1976-1980 11150001-11167079

14 Series
Years   Serial numbers
1976-1980 11200001-11206445

3800
Year	Beginning number
1968	650001
1969	650277
1970	650358

4600
Year	Beginning number
1968	900001
1969	900312
1970	900422

Serial Numbers
David Brown continued

VAK on tracks
Years	Number produced
1940	550

DB4
Years	Number produced
1942-1949	110

Trackmaster
Years	Serial numbers
1950-1953	10001-10700

Trackmaster Diesel
Years	Serial numbers
1950-1953	10001-10500

Trackmaster 30T
Years	Serial numbers
1953-1958	10750-11153

Trackmaster 30TD
Years	Serial numbers
1953-1959	10550-11203

Trackmaster 50D
Years	Serial numbers
1952-1953	10001-10150

Trackmaster 50TD
Years	Serial numbers
1953-1956	10175-10686

30ITD
Years	Serial numbers
1953-1959	10001-10343

50ITD
Years	Serial numbers
1953-1956	10001-10298

50TD MkII
Years	Serial numbers
1957-1963	20001-20471

40TD
Years	Serial numbers
1960-1963	30001-30143

Eagle

Eagle 6A
Year	Starting number
1930	1987
1931	2226
1932	2256
1933	2283
1934	2314
1935	2354

6A, 6B, and 6C
Year	Starting number
1936	2355
1937	2392
1938	2457

Empire

Model 88
Years	Serial numbers
10/1946 to 8/1947	#0001 - #3000 (estimate)

Model 90
Years	Serial numbers
9/1947 to 1/1948	#3001 - #6800 (estimate)

Euclid
Serial numbers for all crawler tractor models

Year	Beginning number
1934	01
1935	69
1936	256
1937	481
1938	644
1939	752
1940	958
1941	1253
1942	1945
1943	2667
1944	3322
1945	3901
1946	4594
1947	5698
1948	7626
1949	9140
1950	10085
1951	11394

Serial Numbers

Year	Number
1952	13224
1953	15154
1954	16923
1955	18442
1956	20685
1957	23849
1958	26238
1959	27571
1960	29700
1961	30888
1962	32487
1963	34434
1964	36809
1965	39451
1966	42631
1967	48296
1968	50801
1969	60398
1970	61329
1971	61900
1972	62572
1973	63154
1974	64211
1975	65437
1976	66721
1977	67667
1978	68298
1979	69164
1980	69967
1981	70573
1982	71386
1983	71898
1984	72141
1985	72579
1986	73035
1987	73641
1988	74153
1989	74591
1990	75047
1991	75290
1992	75530
1993	75765
1994	75995

Ferguson

Ferguson Model A

Year	Number range
1936 - 1938	1 - 1350

Ford-Ferguson

Year	Beginning number
1939	1
1940	14644
1941	47843
1942	92363
1943	107755
1944	131783
1945	174638
1946	204129
1947	267289
end	306221

Ferguson TE-20

Year	Beginning number
1946	1
1947	315
1948	20895
1949	77773
1950	116551
1951	167923

Ferguson TE 20-85

Year	Beginning number
1951	172588
1952	241585
1953	311009

Ferguson TE-30

The serial number is located behind the steering wheel and above the throttle.

Year	Beginning number
1950	116462
1951	167837
1952	241336
1953	310780
1954	367999

Ferguson TO-35 Gas Deluxe

Year	Beginning number
1958	178216
1959	188851

Ferguson TO-35 Gas Special

Year	Beginning number
1958	183348
1959	185504

Serial Numbers
Ferguson continued

Ferguson TO-35 Diesel

Year	Beginning number
1958 Diesel	I180742
1959 Diesel	I187719

Ferguson TO20 and TO30

The serial number is located behind the steering wheel and above the throttle.

Year	Beginning number
1948	1
1949	1808
1950	14660
1951	39163

Ferguson TO-30

The serial number is located behind the steering wheel and above the throttle.

Year	Beginning number
1951	60001
1952	72680
1953	108645
1954	125959

Ferguson TO-35

The serial number is located on the dash a 2×3-inch plate attached with rivets.

Year	Beginning number
1954	140001
1955	140006
1956	167157
1957	171741

Ferguson 35

(also the Massey-Ferguson 35)

Year	Beginning number
1956	1001
1957	9226
1958	79553
1959	125068
1960	171471

Massey-Ferguson 65

Year	Beginning number
1958	500001
1959	510451
1960	520569
1961	533180

Ferguson 40

Year	Beginning number
1956	400001
1957	405671

Ford
Serial number location

9N, 2N, and 8N tractors produced from 1939 to 1952

The serial number is located on the mid-left side of the engine block on an approximately ½×3-inch vertical flat. The serial number is hand-stamped onto the vertical flat not cast into the engine block. Format is model designation before sequential production number. Serial numbers for the 9N and 2N Ford begin with the designation "9N." Serial numbers for the 8N Ford begin with the designation "8N."

Golden Jubilee and NAA tractors produced from 1953 to 1954

On early production models before serial number NAA-22239, the serial number is located on the left-front corner of the engine block just below the manifold.

On models numbered NAA-22239 and beyond, the serial number is located on the left side of the transmission housing above and to the rear of the starter just below a 3×6-inch horizontal flat on a ⅜×3-inch semivertical surface.

The serial number is hand-stamped onto the semivertical flat not cast into the transmission housing. Format is model designation (e.g., "NAA" or "NAB") before sequential production number.

Serial Numbers

Hundred Series tractors produced from 1955 to 1962
The serial number is located on the left side of the transmission housing above and to the rear of the starter on a 3×6-inch horizontal flat. The serial number is hand-stamped onto the horizontal flat not cast into the transmission housing. Format is model designation above or below sequential production number.

Hundred Series tractors produced from 1955 to 1958
Workmaster and Powermaster Ford tractors were produced from 1958 to 1962. The numerical designations 501, 600, 601, 700, 701, 800, 801, 900, and 901 represent the series designation. The serial number reflects the model designation and sequential production number, e.g., a 641 Ford tractor is Series 601 Model 641.

Thousand Series tractors produced from 1962 to 1964
The serial number is located on the left side of the transmission housing above and to the rear of the starter or directly below the starter on a 3×6-inch horizontal flat. The serial number is hand-stamped onto the horizontal flat not cast into the transmission housing. Format is model designation above or below sequential production number. The numerical designations 2000, 4000, 6000, etc., represent the series designation. The serial number reflects the model designation and sequential production number.

Additional designations for Ford tractors produced from 1939 to 1962
A "star" before and/or after the serial number indicates hardened steel-cylinder sleeves. Hardened steel-cylinder sleeves were used in production up to tractor serial number 8N-433578.

A "diamond" before and/or after the serial number indicates cast-iron cylinder sleeves that are .098-inch larger in bore. Suffix-AN, e.g., "8NAN" distillate model 1939-1952.

Suffix-B, e.g., "NAB" distillate model 1953-1954.

Suffix-L, e.g., "640L" LP-gas model 1955-1962.

Suffix-D, e.g., "851D" diesel model 1958-1962.

Ford tractors have no separate engine serial number. Engine type may be determined by the casting code found on the right side of the engine. The casting code is cast into the engine block rather than hand-stamped like the serial number. Casting code EAE signifies the 134-cubic-inch Red Tiger OHV engine found on the Golden Jubilee/NAA 600, 700, 501, 601, 701, and early four-cylinder 2000 Series Ford tractors.

Casting code EAF signifies the 172-cubic-inch Red Tiger OHV engine found on the 800, 900, 801, 901, and early four-cylinder 4000 Series Ford tractors.

Fordson

Year	Beginning number
1917	1
1918	255
1919	34422
1920	91712
1921	162667
1922	199448
1923	268433
1924	370331
1925	453341
1926	557509
1927	645611
1928	739583

Serial Numbers
Fordson continued

Year	Beginning number
1929	747584
1930	757265

Fordson Dexta

Year	Beginning number
1958	16066
1959	22427
1960	46216
1961	72003

Fordson Super Dexta

Year	Beginning number
1961	09A-312001-M
1962	09B-070000-A
1963	09C-731454-A
1964	09D-900000-A

Fordson Major (diesel)

Year	Beginning number
1951	1217101
1952	1217104
1953	1247381
1954	1276857
1955	1322525
1956	1371418
1957	1412409
1958	1458281

Fordson Power Major

Year	Beginning number
1958	1481091
1959	1494448
1960	1538065
1961	1583906

Fordson Super Major

Year	Beginning number
1961	08A-300001M
1962	08B-740000-A
1963	08C-781370-A
1964	08D-900000-A

9N

Year	Beginning number
1939	1
1940	10234
1941	45976
1942	88888

2N

Year	Beginning number
1942	99003
1943	105375
1944	126538
1945	169982
1946	198731
1947	258504

8N

Year	Beginning number
1947	1
1948	37908
1949	141370
1950	245637
1951	343593
1952	442035

600, 700, 800, and 900

Year	Beginning number
1954	1
1955	10615
1956	77271
1957	116368

501, 601, 701, 801, and 901

Year	Beginning number
1957	1001
1958	11997
1959	58312
1960	105943
1961	131427
1962	15553

2000, 4000, and 6000

Year	Beginning number
1962	1001
1963	11948
1964	38931

2000 (three-cylinder) 9000

Year	Starting number
1965	C100000
1966	C124200
1967	C161300
1968	C190200
1969	C226000
1970	C257600
1971	C292100
1972	C327200
1973	C367300

Serial Numbers

1974................................C405200
1975................................C450700

Hundred Series and Thousand Series

The model number consists of three digits followed in some cases by a suffix consisting of a number and/or letter. The first digit designates the engine size and tractor type. The second digit designates the transmission type, and the third digit designates the production year range of the series.

The numerical designations 501, 600, 601, 700, 701, 800, 801, 900, and 901 represent the series designation. Hundred Series Ford tractors (Series 600, 800, 900, etc.) were produced from 1955 to 1958. Workmaster and Powermaster Series Ford tractors (Series 601, 801, 901, etc.) were produced from 1958 to 1962. The serial number reflects the series model designation and sequential production number (e.g., the 641 Ford tractor is Series 601 Model 641.)

5** Single-row 8-inch offset tractor with 134-cubic-inch gasoline/LP-gas or 144-cubic-inch diesel engine

6** Four-wheel adjustable axle with 134-cubic-inch gasoline/LP-gas or 144-cubic-inch diesel engine

7** High-clearance row crop with 134-cubic-inch gasoline LP-gas or diesel engine

8** Four-wheel adjustable axle with 172-cubic-inch engine

9** High-clearance row-crop type with 172-cubic-inch engine

1 Select-O-Speed transmission without PTO

2 Four-speed transmission without PTO or hydraulic lift

3 Four-speed transmission without PTO

4 Four-speed transmission

5 Five-speed transmission with transmission PTO

6 Five-speed transmission with live PTO

7 Select-O-Speed transmission with single-speed PTO

8 Select-O-Speed transmission with two-speed and ground drive

PTO

**0 Series designation 1955 to 1958

**1 Series designation 1958 to 1962

***-1 Tricycle type with single front wheel

***-4 High-clearance four-wheel adjustable axle type

***-D Diesel engine

***-L LP-gas engine

***-37 Equipped with Sherman Reversing Transmission

***-21 Equipped with Sherman Combination Transmission

Thousand Series Model Numbers

2***134-cubic-inch gasoline or 144-cubic-inch diesel engine

4***172-cubic-inch gasoline LPG or diesel engine

*0**Industrial models produced prior to 1963

*1**Industrial and agricultural models produced after 1963

**3*Utility-type tractor with nonadjustable front axle

**4*Heavy-duty industrial-type with sub frame

***1 Low center of gravity model

***1 Four-speed without PTO or hydraulic system

***2 Four-speed with hydraulic system without PTO

***3 Four-speed with hydraulic system and PTO

Serial Numbers

Gaar-Scott

Steam traction engine serial numbers

Year	Beginning number
1885	4158
1886	4331
1887	4666
1888	4817
1889	5118
1890	5519
1891	5861
1892	6362
1893	No data available
1894	7233
1895	7519
1896	7842
1897	8216
1898	8568
1899	9061
1900	9603
1901	10109
1902	10775
1903	11475
1904	11967
1905	12548
1906	13071
1907	13597
1908	14052
1909	14363
1910	14738
1911	15288
1912	15608
1913	16182
1914	16544

Hart Parr
Little Red Devil

Year	Beginning number
1914	6219
1915	6245
1916	6244

10-20 B

Year	Beginning number
1921	35001
1922	35217

10-20 C

Year	Beginning number
1922	35501
1923	35528
1924	35761

12-24 E

Year	Beginning number
1924	36001
1925	36075
1926	36601
1927	37195
1928	38119

12-24 H

Year	Beginning number
1928	39602
1929	39687
1930	42278

12-27

Year	Beginning number
1914	5816
1915	5993

15-30

Year	Beginning number
1910	2332
1911	2353
1912	2383

15-30 A

Year	Beginning number
1918	8401
1919	9384
1921	18470
1922	18851

15-30 A

Year	Beginning number
1920	13026-17915

15-30 C

Year	Beginning number
1922	21001
1923	21393
1924	21899

16-30 E

Year	Beginning number
1924	22501

Serial Numbers

1925	22602

16-30 F

Year	Beginning number
1926	24001

17-30

Year	Beginning number
1901	1205
1902	1206
1903	1208-1219
1904	None built
1905	1346-1347
1906	1435-1454

18-30

Year	Beginning number
1903	1207 only

18-36 G

Year	Beginning number
1926	26001
1927	26360

18-36 H

Year	Beginning number
1927	28851
1928	29636
1929	24567
1929	85001
1930	89159

20-40

Year	Beginning number
1912	4112-4211
1913	4714
1914	4764

22-40

Year	Beginning number
1923	70001
1924	70021
1925	70114
1926	70251
1927	70494

22-40

Year	Beginning number
1923	70001
1924	70021
1925	70114

1926	70251
1927	70494

28-50

Year	Beginning number
1923	70001
1927	70501
1928	70719
1929	70968
1930	71401

40-80

Year	Beginning number
1923	70001
1908	2015
1909	2019
1910	2025
1911	2101
1912	2201
1913	2276
1914	2301

30-60

Year	Beginning number
1907	1605
1908	1811-2014
1909	2325-2331
1910	2432
1911	3311-3999
1912	4212-4711
1913	4814
1914	5262
1915	5441
1916	5522
1917	5552
1918	5641

60-100

Year	Beginning number
1911	4000
1912	4101

Huber

Huber tractor models shared a unified serial number system where the year any model was made is determined by its serial number regardless of make. Huber models included the Model B,

Serial Numbers
Gaar-Scott continued

Farmers Tractor, HK 32-45, HS 27-42, K, Light Four 12-25, Light Four 20-36, Mast Four 25-50, Modern Farmer, Modern Farmer L, Modern Farmer LC, Modern Farmer SC, OB Orchard, Super Four, 15-30 Super Four, 18-36 Super Four, 20-40 Super Four, 25-50 Super Four, 32-45 Super Four 40-62, Farmer Tractor 13-22, 20-40, 30-60, and the 35-70.

Year	Number range
1911	100-181
1212	182-217
1913	218-320
1914	321-402
1915	403-511
1916	512-639
1917	640-1612
1918	1613-2179
No record of nos.	2180-4501
1919	4502-5426
1920	5427-6499
1921	6500-7029
1922	7030-7256
1923	7257-7427
1924	7428-7702
1925	7703-8023
1926	8024-8399
1927	8400-8957
1928	8958-9399
1929	9400-10187
1930	10188-10707
1931	10708-10964
1932	10965-11044
1933	11045-11077
1934	11078-11143
1935	11144-11487
1936	11488-11958
1937	11959-12845
1938	12846-13460
1939	13461-13796
1940	13797-13930
1941	13931-14018
1942	14019-14258
1943	14259-14303

IHC

The numbers listed here include those for tractors built under the names of International Harvester Company, International, Farmall, McCormick, and McCormick-Deering.

Except where indicated, number nameplate is generally found on the left or right side of transmission housing or clutch housing.

International and Farmall

8-16 Mogul

Year	Number range
1915	SB501-SB3750
1916 1917	SB3751-SB15000

10-20 Mogul

Year	Number range
1917-1919	BC501-?

12-25 Mogul

Year	Number range
1913-1918	F501-F750 (flat head)
1913-1918	F751-F2100 (round)

10-20 Titan with barrel

Year	Number range
1916	TV116-TV2356
1917	TV2357-TV11397
1918	TV11398-TV29072
1919	TV29073-TV46306
1920	TV46307-TV50235
1920	TY50236-TY67810
1921	TY67811-TY75539
1922	TY75540-TY78464

I-130 and F-130

Year	Beginning number
1956	501
1957	1120
1958	8363

Serial Numbers

I-966 and F-966

Year	Beginning number
1971	7101
1972	11815
1973	17794
1974	22526
1975	28119

I-140 and F-140

Year	Beginning number
1958	501
1959	2011
1960	8082
1961	11168
1962	16637
1963	21181
1964	25387
1965	28408
1966	31285
1967	34818
1968	37352
1969	39906
1970	42300
1971	44424
1972	46605
1973	48507
1974	50720
1975	54273

I-1066 and F-1066

Year	Beginning number
1971	7101
1972	12677
1973	24205
1974	34849
1975	46855

I-1466 and F-1466

Year	Beginning number
1971	7101
1972	10408
1973	15533
1974	19746
1975	25404

I-1468 and F-1468

Year	Beginning number
1971	7201
1972	7239
1973	9109
1974	9670

I-1566 and F-1566

Year	Beginning number
1974	7101
1975	7837

I-1568 and F-1569

Year	Beginning number
1974	7201
1975	7821

10-20 Gear Drive – KC

Year	Beginning number
1923	501
1924	7641
1925	18869
1926	37728
1927	62824
1928	89470
1929	119823
1930	159111
1931	191486
1932	201213
1934	204239
1935	206179
1936	207275
1937	210235
1938	212425
1939	214886

10-20 Gear Drive – NC and NT

Year	Beginning number
1926	501
1927	649
1928	832
1929	1155
1930	1543
1931	1750
1932	1833
1933	1912
1934	1952

Cub

Number nameplate is located on the right side of the steering gear housing.

Year	Beginning number
1947	501
1948	11348

Serial Numbers
IHC continued

Year	Beginning number
1949	57831
1950	99536
1951	121454
1952	144455
1953	162284
1954	179412
1955	186441
1956	193658
1957	198231
1958	204389
1959	211441
1960	214974
1961	217382
1962	220038
1963	221383
1964	223453
1965	225110
1966	227209
1967	229225
1968	231005
1969	232981
1970	234868
1971	238506
1973	240581
1974	242746
1975	245651

Cub Cadet

Year	Beginning number
1961	590
1962	23675
1963	49846
1964	73875
1965	104307
1966	137051
1968	225056
1969	280840
1970	322858
1971	373301
1972	417643
1973	457305
1974	495749
1975	536073

Cub Lo-Boy
Number nameplate is located on the right side of the steering gear housing.

Year	Beginning number
1955	501

Year	Beginning number
1956	2555
1957	3929
1958	5582
1959	10567
1960	12371
1961	13904
1962	15506
1963	16440
1964	17928
1965	19406
1966	21176
1967	23115
1968	24481

Cub 154 Lo-Boy and 185 Lo-Boy
Number nameplate is located on the top left front side of the main frame.

Cub 154 Lo-Boy

Year	Beginning number
1968	3505
1969	3773
1970	15502
1971	20332
1972	23343
1973	27538
1974	31766

Cub 185 Lo-Boy

Year	Beginning number
1974	37001
1975	37316

International B-275

Year	Beginning number
1959	501
1960	1720
1961	3415

Hydro 70

Year	Beginning number
1973	7501
1974	7570
1975	8681

Hydro 100

Year	Beginning number
1973	7501
1974	7727

Serial Numbers

1975...................................... 10915

I-6, ID-6, I-9, and ID-9
Number nameplate is located on left side of clutch compartment cover.

I-6 and ID-6

Year	Beginning number
1940	501
1941	1225
1942	3718
1943	5057
1944	6371
1945	9518
1946	14198
1947	17317
1948	24021
1949	28868
1950	35472
1951	38518
1952	44318
1953	45274

I-9 and ID-9

Year	Beginning number
1940	501
1941	578
1942	3003
1943	3651
1944	5501
1945	11459
1946	17294
1947	22715
1948	29207
1949	36350
1950	45689
1951	51739
1952	59407
1953	64014

I-100

Year	Beginning number
1954	501
1955	504
1956	575

I-240

Year	Beginning number
1958	501

Year	Beginning number
1959	4835
1960	8628
1961	10079
1962	10727

I-300U

Year	Beginning number
1955	501
1956	20219

I-330

Year	Beginning number
1957	501
1958	1488

I-340

Year	Beginning number
1958	501
1959	2467
1960	5741
1961	8736
1962	11141
1963	12032

I-350

Year	Beginning number
1956	501
1957	1963
1958	15049

IW-400

Year	Beginning number
1955	501
1956	2187

I-424

Year	Beginning number
1964	501
1965	1402
1966	7841
1967	13627

I-444

Year	Beginning number
1967	501
1968	1190
1969	5720
1970	9010
1971	12357

Serial Numbers
IHC continued

I-404 and 2404

Year	Beginning number
1961	501
1962	1045
1963	4205
1964	6452
1965	8292
1966	9548
1967	10534
1968	11032

IW-450

Year	Beginning number
1956	501
1957	568
1958	1661

I-454

Year	Beginning number
1970	501
1971	508
1972	4908
1973	8064

I-460

Year	Beginning number
1958	501
1959	2711
1960	6883
1961	9420
1962	11619
1963	11898

I-464

Year	Beginning number
1973	100003
1974	102196
1975	104775

I-544 and 2544

Year	Beginning number
1968	10250
1969	12699
1970	14589
1971	16018
1972	16838
1973	17341

I-504 and 2504

Year	Beginning number
1961	501
1962	512
1963	3376
1964	6797
1965	10996
1966	14695
1967	17992
1968	20392

I-560

Year	Beginning number
1958	501
1959	1210
1960	3103
1961	4032
1962	4944
1963	5598

I-574

Year	Beginning number
1970	504
1971	650
1972	3329
1973	7074
1974	102961
1975	107880

I-600

Year	Beginning number
1956	501

I-606 and 2606

Year	Beginning number
1961	501
1962	503
1963	1702
1964	3214
1965	5041
1966	6960
1967	7922

I-650

Year	Beginning number
1956	501
1957	688
1958	11659

I-656 and 2656

Year	Beginning number
1966	7501

Serial Numbers

Year	Beginning number
1967	7842
1968	9929
1969	11802
1970	13353
1971	14194
1972	14952
1973	15746

I-660

Year	Beginning number
1959	501
1960	3398
1961	4259
1962	5883
1963	6995

I-664

Year	Beginning number
1972	2501
1973	3212

I-666

Year	Beginning number
1972	7500
1973	8200
1974	11585
1975	13131

I-674

Year	Beginning number
1973	100001
1974	101862
1975	103172

I-706 and 2706

Year	Beginning number
1963	501
1964	1251
1965	3478
1966	4789
1967	5316

I-756 and 2756

Year	Beginning number
1967	7501
1968	7672
1969	8164
1970	8424
1971	8427

I-766

Year	Beginning number
1971	7101
1972	7416
1973	9611
1974	12378
1975	14360

I-806 and 2806

Year	Beginning number
1963	501
1964	1403
1965	3758
1966	5917
1967	7409

I-826 and 2826

Year	Beginning number
1969	7501
1970	7518
1971	7719

I-856 and 2856

Year	Beginning number
1967	7501
1968	7904
1969	9016
1970	9544
1971	9653

I-1026 and 21026

Year	Beginning number
1970	7501
1971	7550

I-1206 and 21206

Year	Beginning number
1965	7501
1966	7772
1967	8492

I-1256 and 21256

Year	Beginning number
1967	7501
1968	7703
1969	8444

I-1456 and 21456

Year	Beginning number
1969	10001

Serial Numbers
IHC continued

1970	10025
1971	10249

I-4100
Year	Beginning number
1966	8001
1967	8723
1968	8986

I-4156
Year	Beginning number
1969	9219
1970	9365

I-4166
Year	Beginning number
1972	10001
1973	10769
1974	11255
1975	11684

I-4366 and I-4568
Number nameplate is located on left side on top step.

I-4366
Year	Beginning number
1973	7501
1974	7780
1975	8616

I-4568
Year	Beginning number
1975	8001

Farmall 100
Year	Beginning number
1954	501
1955	1720
1956	12895

200
Year	Beginning number
1954	501
1955	1032
1956	10904

230
Year	Beginning number
1956	501

1957	815
1958	6827

240
Year	Beginning number
1958	501
1959	1777
1960	3415
1961	3989

300
Year	Beginning number
1954	501
1955	1779
1956	23224

340
Year	Beginning number
1958	501
1959	2723
1960	5411
1961	6642
1962	7276
1963	7699

350
Year	Beginning number
1956	501
1957	1004
1958	14175

400
Year	Beginning number
1954	501
1955	2588
1956	29065

404
Year	Beginning number
1961	501
1962	826
1963	1936
1964	2259
1965	2568
1966	2790
1967	2980
1968	3170

450
Year	Beginning number

Serial Numbers

1956	501
1957	1734
1958	21871

460

Year	Beginning number
1958	501
1959	4765
1960	16902
1961	22622
1962	28029
1963	31552

504

Year	Beginning number
1961	501
1962	810
1963	5000
1964	7732
1965	10696
1966	13596
1967	15113
1968	16115

544

Year	Beginning number
1968	10253
1969	12541
1970	13585
1971	14507
1972	15262
1973	15738

560

Year	Beginning number
1958	501
1959	7341
1960	26914
1961	36125
1962	47798
1963	60278

656

Year	Beginning number
1965	8501
1966	15505
1967	24372
1968	32007
1969	38861
1970	42518

1971	45497
1972	47951

666

Year	Beginning number
1972	7500
1973	8200
1974	11585
1975	13131

706

Year	Beginning number
1963	501
1964	7073
1965	21162
1966	30288
1967	38521

756

Year	Beginning number
1967	7501
1968	9940
1969	14125
1970	17832
1971	18374

806

Year	Beginning number
1963	501
1964	4709
1965	15946
1966	24038
1967	34943

826

Year	Beginning number
1969	7501
1970	8153
1971	16352

856

Year	Beginning number
1967	7501
1968	9854
1969	19554
1970	32099
1971	32420

1026

Year	Beginning number

Serial Numbers
IHC continued

1970.. 7501
1971.. 9707

1206

Year	Beginning number
1965	7501
1966	8626
1967	12731

1256

Year	Beginning number
1967	7501
1968	8849
1969	13140

1456

Year	Beginning number
1969	10001
1970	10405
1971	14149

A, AV, B, and BN
Number nameplate is located on left side of seat support.

Year	Beginning number
1939	501
1940	6744
1941	41500
1942	8739
1944	96390
1945	113218
1946	146700
1947	182964

Super A
Number nameplate is located on right side of toolbox and seat support.

Year	Beginning number
1947	250001
1948	250082
1949	268196
1950	281269
1951	300126
1952	324470
1953	336880
1954	353348

C

Year	Beginning number
1948	501
1949	22624
1950	47010
1951	71880

Super C

Year	Beginning number
1951	100001
1952	131157
1953	159130
1954	187788

F-12

Year	Number range
1932	501-525
1933	526-4880
1934	4881-17410
1935	17411-48659
1936	48660-81836
1937	81837-117517
1938	117518-123942

F-14

Year	Number range
1938	124000-139606
1939	139607-155902

F-20

Year	Beginning number
1932	501
1933	1251
1934	3001
1935	6382

F-30

Year	Beginning number
1931	501
1932	1184
1933	4305
1934	5526
1935	7032
1936	10407
1937	18684
1938	27186
1939	29007

H and HV

Serial Numbers

Year	Beginning number
1939	501
1940	10653
1941	52387
1942	93237
1943	122091
1944	150251
1945	186123
1946	214820
1947	241143
1948	268991
1949	300876
1950	327975
1951	351923
1952	375861
1953	390500

Super H and HV

Year	Beginning number
1953	501
1954	22202

M, MD, MDV, and M

Year	Beginning number
1939	501
1940	7240
1941	25371
1942	50988
1943	60011
1944	67424
1945	88085
1946	105564
1947	122823
1948	151708
1949	180514
1950	213579
1951	147518
1952	290923

Super M, MD, MDV, MTA, and MV

Year	Beginning number
1952	F501
1952	L500001
1953	F12516
1953	L501906
1954	F51977
1954	L600001

Regular and Fairway

Year	Beginning number

Year	Beginning number
1924	501
1925	701
1926	1539
1927	5969
1928	15471
1929	40370
1930	75691
1931	117784
1932	131872

McCormick

O-4, OS-4, W-4, and Super W-4
Number nameplate is located on left side of transmission housing.

O-4 ,OS-4, and W-4

Year	Beginning number
1940	501
1941	943
1942	4056
1943	5693
1944	7593
1945	11171
1946	13934
1947	16022
1948	18880
1949	21912
1950	24470
1951	28167
1952	31214
1953	33067

Super W-4

Year	Beginning number
1953	501
1954	2668

O-6, OS-6, ODS-6, W-6, and WD-6
Number nameplate is located on left side of clutch compartment cover.

Year	Beginning number
1940	501
1941	1225
1942	3718
1943	5057
1944	6313

Serial Numbers
IHC continued

1945	9396
1946	14153
1947	17792
1948	22961
1949	28704
1950	33698
1951	28518
1952	44318
1953	45274

1933	522
1934	548
1935	3182
1936	9723
1937	15095
1938	23834
1939	29922
1940	32482

Super W-6, W-6TA, and WD-6
Number nameplate is located on left side of transmission housing.

Year	Beginning number
1952	501
1953	2908
1954	8997

W-9, WD-9, WDR-9, and WR-9
Number nameplate is located on the right side of the fuel tank support located above the clutch housing.

Year	Beginning number
1940	501
1941	578
1942	2993
1943	3651
1944	5394
1945	11459
1946	17289
1947	22714
1948	29207
1949	36159
1950	45551
1951	51739
1952	59407
1953	64014

WR-9-S

Year	Beginning number
1953	501
1954	550
1955	722
1956	755

W-30

Year	Beginning number
1932	50

McCormick-Deering

O-12, O-14, and Fairway 12 & 14

Year	Beginning number
1934	512
1935	1092
1936	1626
1937	2277
1938	3261
1939	3882

W-12

Year	Beginning number
1934	503
1935	1356
1936	2031
1937	2768
1938	3799

W-14

Year	Beginning number
1938	4134
1939	4610

W-40, WD-40, and WK-40

Year	Beginning number
1935	501
1936	1441
1937	5120
1938	7665
1939	9756
1940	10323

International Crawlers

100 Series C

Year	Beginning number
1970	501
1971	599
1972	946

Serial Numbers

1973	1152
1974	1473

100 Series E

Year	Beginning number
1974	4001
1975	4287

125 Series C

Year	Beginning number
1970	501
1971	647
1972	1077
1973	1427
1974	1854

125 Series E

Year	Beginning number
1974	6001
1975	6384

500

Year	Beginning number
1965	501
1966	891
1967	2501
1968	3544
1969	4065

500C

Year	Beginning number
1970	1377
1971	2577
1972	3189
1973	4018
1974	4961

500E

Year	Beginning number
1974	5501
1975	6103

T-6 and TD-6

Year	Beginning number
1940	?
1941	17373
1949	21146
1950	24862
1951	27692
1952	31056

1953	33157
1954	35746
1955	37570
1956	38706
1956 TD-61	38951
1957	39817
1958	40822
1959	41448
1959 TD-62	501
1960	1063
1961	1629
1962	2056
1963	2404
1964	2758
1965	3177
1966	3631
1967	3991
1968	4221
1969	4424

T-20

Year	Beginning number
1931	501
1932	551
1933	2051
1934	2528
1935	3276
1936	5387
1937	8501
1938	12519
1939	14550

TD-7 Series C

Year	Beginning number
1979	501
1971	694
1972	1040
1973	1471
1974	2007

TD-7 Series E

Year	Beginning number
1974	3001
1975	3379

TD-8 Series C

Year	Beginning number
1970	501
1971	704
1972	1023

Serial Numbers
IHC continued

1973	459
1974	1912

TD-8 Series E

Year	Beginning number
1974	5001
1975	5521

Unless otherwise indicated tractor number is stamped on the instrument panel nameplate.

T-9

Year	Beginning number
1940	Begin?
1948	29519
1949	32978
1950	36927
1951	40465
1952	44710
1953	47855
1954	52600
1955	56276
1956	58819

TD-9

Year	Beginning number
1940	Begin?
1948	29512
1949	32973
1950	36963
1951	40400
1952	44710
1953	48848
1954	52599
1955	56208
1956	56436
1956 TD-91	60301
1957	63946
1958	65802
1959	57290
1959 TD-92	201
1960	3353
1961	5895
1962	6090
1962 TD-9B	7502
1963	6455
1964	8898
1965	10446
1966	11685

1970	51530
1971	15965
1972	16243
1973	16499
1974	16732

TD-14 and TD-14A

Year	Beginning number
1939	501
1940	575
1941	2045
1942	3856
1943	5380
1944	7451
1945	10728
1946	12704
1947	14437
1948	21027
1949	26001
1950	27082
1951	29361
1952	32427
1953	34525
1954	37344
1955	39061
1956 TD-141	39301
1956	40862
1956 TD-142	41551
1957	43968

TD-15

Year	Beginning number
1958	501
1959	577
1960 TD-150	2456
1961	3501
1961 TD-151	4001
1962	4680
1963 TD-15B	6001
1964	7100
1965	8935
1966	20377

TD-15C

Year	Beginning number
1972	501
1973	1130
1974	1849
1975	2750

Serial Numbers

TD-18

Year	Beginning number
1939	520
1940	1240
1941	2292
1942	3607
1943	5804
1944	6899
1945	9011
1946	11092
1947	12644
1948	14626
1949	16708

TD-18A

Year	Beginning number
1939	520
1940	1240
1941	2292
1942	3607
1943	5804
1944	6899
1945	9011
1946	11092
1947	12644
1948	14626
1949	16708

TD-24

Year	Beginning number
1947	506
1948	515
1949	1501
1950	1967
1951	2915
1952	4120
1953	5075
1954	6049
1955	6556

TD-241

Year	Beginning number
1955	8001
1956	8793
1957	9655
1958	10330
1959	10631

TD-24 – 241 Series

Year	Beginning number
1955	601
1956	826
1957	1249
1958	1420
1959	1598

TD-40

Year	Beginning number
1934	2501
1935	2785
1936	4393
1938	7719
1939	8599

TD-340

Year	Beginning number
1960	1979
1961	5370
1962	6169
1963	6397
1964	7338
1965	8525

Stationary engines

IHC Type M
1½-hp. 650-rpm kerosene

Year	Beginning number
1917	A101
1918	A1970
1919	A19492
1920	A42765
1921	A70745
1922	A82430
1923	A91826

IHC Type M
1½-hp. 500-rpm gasoline

Year	Beginning number
1923	Ab101
1924	AB153
1925	AB1061
1926	AB1786
1927	AB1880
1928	AB1923
1929	AB2006
1930	AB2031
1931	AB2039
1932	AB2046

Serial Numbers
IHC continued

IHC Type M
1½-hp. 500-rpm kerosene

Year	Beginning number
1924	AX101
1925	AX530–
1926	AX3407
1927	AX5344
1928	AX7452
1929	AX10274
1930	AX13178
1931	AX16092
1932	AX16333
1933	AX16474

IHC Type M
3-hp. kerosene

Year	Beginning number
1918	B101
1919	B3038
1920	B15813
1921	B35786
1922	B42783
1922	B45951
1923	B49813
1924	B59660
1925	B64819
1926	B65032
1927	B65294
1928	B65322
1929	B65362
1930	B65381
1931	B65386
1932	B65392

IHC Type M
6-hp. kerosene

Year	Beginning number
1925	CW101
1926	CW2297
1927	CW7201
1928	CW11202
1929	CW14401
1930	CW17972
1931	CW19989
1932	CW20623
1933	CW21203
1934	CW21222
1935	CW21471
1936	CW21783
1937	CW21855

IHC Type M
6-hp. kerosene

Year	Beginning number
1918	C101
1919	C752
1920	C6873
1921	C16143
1922	C20458
1922	C22211
1923	C25768
1924	C31469
1924	C34575
1925	C34730
1926	C34824
1927	C34971
1928	C34994
1929	C35026
1930	C35044
1931	C35058
1932	C35068
1934	C35073

IHC Type M
10-hp. kerosene

Year	Beginning number
1920	D101
1921	D1390
1922	D2749
1922	D3076
1923	D3350
1924	D3930
1925	D3955
1926	D4228
1927	D4680
1928	D4783
1929	D4862
1930	D4981
1931	D5052
1932	D5069
1933	D5082

IHC Type M
10-hp. kerosene

Year	Beginning number
1927	DW157
1928	DW672
1929	DW848
1930	DW1184
1931	DW1331
1932	DW1402

Serial Numbers

IHC Type L
1½ hp.
Year	Beginning number
1929	EW101

McCormick Deering Type LA
1½ to 2½ hp.
Year	Beginning number
1934	LA101
1935	LA133
1936	LA14655
1936	LAA17853
1937	LAA29505
1938	LAA41947

McCormick Deering Type LA
3 to 5 hp.
Year	Beginning number
1935	LA501
1936	LA981
1936	LAB1726
1937	LAB4471
1938	LAB9030

McCormick Deering Type LB
1½ - 2½ hp.
Year	Beginning number
1941	62614
1942	66394
1943	72944
1944	76061
1945	86648
1946	100999
1947	114788
1948	130394

McCormick Deering Type LB
3 to 5 hp.
Year	Beginning number
1941	22287
1942	25274
1943	30923
1944	33509
1945	40101
1946	48767
1947	56536
1948	63626

John Deere

Years are product years (usually August 1 of the preceding calendar year to the following July 31) unless noted as follows: *Calendar Year; **Fiscal Year (November 1 to October 31). Tractors are listed alphabetically and numerically rather than chronologically.

A
Year	Beginning number
1934	410 000
1935	412 869
1936	424 025
1937	442 151
1938	466 787
1939	477 000
1940	488 000
1941	499 000
1942	514 127
1943	523 133
1944	528 778
1945	548 352
1946	555 334
1947	578 516
1948	594 433
1949	620 843
1950	648 000
1951	667 390
1952	689 880

AO and AR
Year	Beginning number
1936	250 000
1937	253 521
1938	255 416
1939	257 004
1940	258 045
1941	260 000
1942	261 558
1943	262 243
1944	263 223
1945	264 738
1946	265 870
1947	267 082
1948	268 877

Serial Numbers
John Deere continued

Year	Beginning number
1949	270 646
1950	272 985
1951	276 078
1952	279 770
1953	282 551

AO (Styled)

Year	Beginning number
1937	AO-1000
1938	AO-1539
1939	AO-1725
1940	AO-1801

B

Year	Beginning number
1935	1 000
1936	12 012
1937	27 389
1938	46 175
1939	60 000
1940	81 600
1941	96 000
1942	126 345
1943	143 420
1944	152 862
1945	173 179
1946	183 673
1947	199 744
1948	209 295
1949	237 346
1950	258 205
1951	276 557
1952	299 175

BO and BR

Year	Beginning number
1936	325 000
1937	326 655
1938	328 111
1939	329 000
1940	330 633
1941	332 039
1942	332 427
1943	332 780
1944	333 156
1945	334 219
1946	335 641
1947	336 746

BO (Lindeman) Crawler

Year	Beginning number
1943	332 901
1944	333 110
1945	333 666
1946	335 361
1947	336 441

D

Year	Beginning number
1924	30 401
1925	31 280
1926	35 309
1927	43 410
1928	54 554
1929	71 561
1930	95 367
1931	109 944
1932	115 477
1933	115 665
1934	116 273
1935	119 100
1936	125 430
1937	130 700
1938	138 413
1939	143 800
1940	146 566
1941	149 500
1942	152 840
1943	155 005
1944	155 426
1945	159 888
1946	162 598
1947	167 250
1948	174 879
1949	183 516
1950	188 420
1951	189 701
1952	191 180
1953	191 439

G/GM/G

Year	Beginning number
1938	1 000
1939	7 734
1940	9 321
1941	10 489
1942	12 069
1943	12 941
1944	13 748
1945	13 905

Serial Numbers

Year	Beginning number
1946	16 94
1947	20 527
1948	28 127
1949	34 587
1950	40 761
1951	47 194
1952	56 510
1953	63 489

GP Standard

Year	Beginning number
1928	200 111
1929	202 566
1930	216 139
1931	224 321
1932	228 666
1933	229 051
1934	229 216
1935	230 515

GP – Wide Tread

Year	Beginning number
1929	400 000
1930	400 936
1931	402 741
1932	404 770
1933	405 110

GP-O

Year	Beginning number
1931	15 000
1932	15 226
1933	15 387
1934	15 412
1935	15 589

H

Year	Beginning number
1939	1 000
1940	10 780
1941	23 654
1942	40 995
1943	44 755
1944	47 796
1945	48 392
1946	55 956
1947	60 107

L

Year	Beginning number
1937	621 000
1938	621 079
1939	626 265
1940	630 160
1941	634 191
1942	640 000
1943	640 738
1944	641 038
1945	641 538
1946	641 958

LA

Year	Beginning number
1941	1 001
1942	5 361
1943	6 029
1944	6 159
1945	9 732
1946	11 529

M

Year	Beginning number
1947	10 001
1948	13 734
1949	25 604
1950	35 659
1951	43 525
1952	50 580

MC Crawler

Year	Beginning number
1949	10 001
1950	11 630
1951	13 630
1952	16 309

MT

Year	Beginning number
1949	10 001
1950	18 544
1951	26 203
1952	35 845

R

Year	Beginning number
1949	1 000
1950	3 541
1951	6 368
1952	9 293
1953	15 720

Serial Numbers
John Deere continued

1954.................................... 19 485

Waterloo Boy L and LA
Year	Beginning number
1914	1 000

Waterloo Boy N
Year	Beginning number
1917	10 020
1918	10 221
1919	13 461
1920	18 924
1921	27 026
1922	27 812
1923	28 119
1924	29 520

Waterloo Boy R
Year	Beginning number
1915	1 026
1916	1 401
1917	3 556
1918	6 982
1919	9 056

40 Hi-Crop
Year	Beginning number
1954	60 001
1955	60 060

40 Special
Year	Beginning number
1955	60 001

40 Standard
Year	Beginning number
1953	60 001
1954	67 359
1955	69 474

40 Tricycle
Year	Beginning number
1953	60 0
1954	72 167
1955	75 531

40 2-Row Utility
Year	Beginning number
1955	60 001

40 Utility
Year	Beginning number
1953	60 001
1954	60 202
1955	63 140

40C Crawler
Year	Beginning number
1953	60 001
1954	63 358
1955	66 894

50
Year	Beginning number
1952	5 000 001
1953	5 01 254
1954	5 016 041
1955	5 021 977
1956	5 030 600

60
Year	Beginning number
1952	6 000 001
1953	6 007 694
1954	6 027 995
1955	6 042 500
1956	6 057 650

70
Year	Beginning number
1953	7 000 001
1954	7 005 692
1955	7 017 501
1956	7 034 950

80
Year	Beginning number
1955	8 000 001
1956	8 000 755

320
Year	Beginning number
1956	320 001
1957	321 220
1958*	325 127

Serial Numbers

330

Year	Beginning number
1958	330 001
1959	330 171
1960	330 935

420

Year	Beginning number
1956	30 001
1957	107 813
1958	127 782

420C Crawler

Year	Beginning number
1956	80 001
1957	107 813
1958	127 782

430

Year	Beginning number
1958	140 001
1959	142 671
1960	158 632

430C Crawler

Year	Beginning number
1958	140 001
1959	142 671
1960	158 632

435

Year	Beginning number
1959	435 001
1960	437 655

520

Year	Beginning number
1956	5 200 000
1957	5 202 982
1958	5 209 029

530

Year	Beginning number
1958	5 300 000
1959	5 301 671
1960	5 307 749

620

Year	Beginning number
1956	6 200 000
1957	6 203 778
1958	6 215 048

630

Year	Beginning number
1958	6 300 000
1959	6 302 749
1960	6 314 381

650

Year	Beginning number
1981	1 000
1982	3 539
1983	6 250
1984	10 543
1985	15 001
1986	19 001
1987	22 501

655

Year	Beginning number
1986	360 001
1987	420 001

720

Year	Beginning number
1956	7 200 000
1957	7 203 420
1958	7 217 368

730

Year	Beginning number
1958	7 300 000
1959	7 303 761
1960	7 322 075
1961	7 328 801

750

Year	Beginning number
1981	1 000
1982	3 448
1983	5 613
1984	8 597
1985	13 001
1986	18 501
1987	22 601

755

Year	Beginning number
1986	360 001

Serial Numbers
John Deere continued

1987...................................... 420 001

820 (two-cylinder)
Year	Beginning number
1956	8 000 000
1957	8 200 565
1958	8 203 850

820 (three-cylinder)
Year	Beginning number
1968	10 000
1969	23 100
1970	36 000
1971	54 000
1972	71 850
1973	90 200

830 (two-cylinder)
Year	Beginning number
1958	8300 000
1959	8 300 727
1960	8 305 301
1961	8 306 892

830 (three-cylinder)
Year	Beginning number
1974	108 507
1975	155 914

850
Year	Beginning number
1978	1 024
1979	3 859
1980	7 389
1981	11 338
1982	12 481
1983	14 183
1984	16 006
1985	18 001
1986	22 001
1987	25 501

855
Year	Beginning number
1986	360 001
1987	420 001

900 HC
Year	Beginning number
1986	1 001

1987...................................... 1 701

950
Year	Beginning number
1978	1 024
1979	5 229
1980	10 453
1981	14 893
1982	16 204
1983	18 204
1984	20 007
1985	23 001
1986	26 001
1987	28 501

1010
Year	Beginning number
1961	10 001
1962	23 630
1963	32 188
1964	43 900
1965	53 722

1010C Crawler
Year	Beginning number
1960	10 001
1961	13 692
1962	23 630
1963	32 188
1964	43 900
1965	52 722

1020
Year	Beginning number
1965	14 501
1966	14 682
1967	42 715
1968	65 184
1969	82 409
1970	102 039
1971	117 500
1972	134 700
1973	157 109

1050
Year	Beginning number
1980	1 000
1981	5 280
1982	6 572
1983	9 001

Serial Numbers

1984	11 006
1985	14 001
1986	17 001
1987	19 501

1250

Year	Beginning number
1982	1 001
1983	1 256
1984	3 001
1985	4 001
1986	5 001
1987	5 501

1450

Year	Beginning number
1984	1 020
1985	2 201
1986	3 001
1987	3 501

1520

Year	Beginning number
1968	76 112
1969	82 405
1970	102 061
1971	117 500
1972	134 700
1973	157 109

1530

1974	108 811L
	176 601T
1975	145 500L

1650

Year	Beginning number
1984	1 021
1985	2 401
1986	3 001
1987	3 501

2010

Year	Beginning number
1961	10 001
1962	21 087
1963	31 250
1964	44 036
1965	58 186

2010C Crawler

Year	Beginning number
1960	10 001
1961	10 999
1962	21 087
1963	31 250
1964	44 036
1965	58 186

2020**

Year	Beginning number
1965	14 502
1966	14 680
1967	42 721
1968	65 176
1969	82 404
1970	102 032
1971	117 500

2030

Year	Beginning number
1972	134 700T
1973	157 109T
1974	140 000L
	187 301T
1975	145 500L
	213 350T

2040

Year	Beginning number
1976	179 963
1977	221 555
1978	266 057
1979	304 165
1980	336 935
1981	392 026
1982	419 145

2150

Year	Beginning number
1983	433 467
1984	505 001
1985	532 000
1986	562 001

2155

Year	Beginning number
1987	600 000

Serial Numbers
John Deere continued

2240

Year	Beginning number
1976	179 298
1977	221 716
1978	277 267
1979	305 307
1980	337 767
1981	392 292
1982	418 608

2255

Year	Beginning number
1983	468 228
1984	505 001
1985	532 000
1986	562 001

2350

Year	Beginning number
1983	433 474
1984	505 001
1985	532 000
1986	562 001

2355

Year	Beginning number
1987	600 000

2355N

Year	Beginning number
1987	600 000

2440

Year	Beginning number
1976	235 210
1977	258 106
1978	280 789
1979	305 501
1980	335 625
1981	362 173
1982	376 746

2510

Year	Beginning number
1966	1 000
1967	8 958
1968	14 291

2520

Year	Beginning number
1969	17 000

1970	19 416
1971	22 000
1972	22 911
1973	23 865

2550

Year	Beginning number
1983	433 480
1984	505 001
1985	532 000
1986	562 001

2555

Year	Beginning number
1987	600 000

2630

Year	Beginning number
1974	188 601
1975	213 360

2640

Year	Beginning number
1976	235 313
1977	258 106
1978	280 789
1979	305 505
1980	335 628
1981	362 175
1982	376 744
1983	388 347

2750

Year	Beginning number
1983	433 494
1984	505 001
1985	523 000
1986	562 001

2755

Year	Beginning number
1987	600 000

2840

Year	Beginning number
977	214 909
1978	264 711
1979	304 654

Serial Numbers

2855N

Year	Beginning number
1987	600 000

2940

Year	Beginning number
1980	350 586
1981	390 496
1982	418 953

2950

Year	Beginning number
1983	433 508
1984	505 001
1985	532 000
1986	562 001

2955

Year	Beginning number
1987	600 000

3010

Year	Beginning number
1961	1 000
1962	10 801
1963	32 400

3020

Year	Beginning number
1964	50 000
1965	68 000
1966	84 000
1967	97 286
1968	112 933
1969	123 000
1970	129 897
1971	150 000
1972	154 197

3150

Year	Beginning number
1985	532 000
1986	562 001
1987	587 950

4000

Year	Beginning number
1969	211 422
1970	222 143
1971	250 000
1972	260 791

4010

Year	Beginning number
1961	1 000
1962	20 201
1963	38 200

4020

Year	Beginning number
1964	65 000
1965	91 000
1966	119 000
1967	145 660
1968	173 982
1969	201 000
1970	222 143
1971	250 000
1972	260 791

4030

Year	Beginning number
1973	1 000
1974	6 700
1975	10 153
1976	13 022
1977	15 417

4040

Year	Beginning number
1978	1 000
1979	3 199
1980	6 033
1981	8 707
1982	11 727

4050

Year	Beginning number
1983	1 000
1984	3 501
1985	5 001
1986	6 501
1987	7 001

4230

Year	Beginning number
1973	1 000
1974	13 000
1975	22 074
1976	28 957

Serial Numbers
John Deere continued

1977.................................. 35 588

4240
Year	Beginning number
1978	1 000
1979	7 434
1980	14 394
1981	20 186
1982	25 670

4250
Year	Beginning number
1983	1 000
1984	6 001
1985	9 001
1986	11 001
1987	12 501

4320
Year	Beginning number
1971	6 000
1972	17 031

4430
Year	Beginning number
1973	1 000
1974	17 500
1975	33 050
1976	47 222
1977	62 960

4440
Year	Beginning number
1978	1 000
1979	14 820
1980	29 539
1981	42 665
1982	56 346

4450
Year	Beginning number
1983	1 000
1984	11 001
1985	18 001
1986	22 001
1987	24 001

4520
Year	Beginning number

1969.................................. 1 000
1970.................................. 7 005

4620
Year	Beginning number
1971	10 000
1972	13 692

4630
Year	Beginning number
1973	1 000
1974	7 022
1975	11 717
1976	18 392
1977	25 794

4640
Year	Beginning number
1978	1 000
1979	7 422
1980	13 860
1981	19 459

4650
Year	Beginning number
1983	1 000
1984	7 001
1985	10 001
1986	12 501
1987	14 001

4840
Year	Beginning number
1978	1 000
1979	4 233
1980	7 539
1981	11 042
1982	4 933

4850
Year	Beginning number
1983	1 000
1984	5 001
1985	8 001
1986	10 001
1987	11 001

5010
Year	Beginning number
1963	1 000

Serial Numbers

1964	4 500
1965	8 000

5020

Year	Beginning number
1966	12 000
1967	15 650
1968	20 399
1969	24 038
1970	26 624
1971	30 000
1972	30 608

6030

Year	Beginning number
1972	33 000
1973	33 550
1974	34 586
1975	35 400
1976	36 014
1977	36 577

7020

Year	Beginning number
1971	1 000
1972	2 006
1973	2 700
1974	3 156
1975	3 579

7520

Year	Beginning number
1972	1 000
1973	1 600
1974	3 054
1975	4 945

8010

Year	Beginning number
1960	1 000

8020

Year	Beginning number
1964	1 000

8430

Year	Beginning number
1975	1 000
1976	1 690
1977	3 962

1978	5 323

8450

Year	Beginning number
1982	1 000
1983	2 000
1984	3 501
1985	5 001
1986	5 501
1987	6 001

8440

Year	Beginning number
1979	1 001
1980	2 266
1981	3 758
1982	5 235

8630

Year	Beginning number
1975	1 000
1976	2 382
1977	5 222
1978	7 626

8640

Year	Beginning number
1979	1 500
1980	3 198
1981	5 704
1982	7 960

8650

Year	Beginning number
1982	1 500
1983	3 000
1984	5 001
1985	7 001
1986	8 001
1987	8 501

8850

Year	Beginning number
1982	2 000
1983	4 000
1984	5 101
1985	6 001
1986	6 501
1987	7 001

Serial Numbers

Massey-Harris-Ferguson

Serial number locations

Massey-Harris 20
Numbers are located on plate on left side of main frame and stamped on frame on top of transmission cover.

Colt Mustang 22, 30, 33, 44-4, 44-6, and 55
Numbers are located rear left-hand side of tractor frame just forward of the transmission case.

333, 444, and 555
Numbers are located rear left-hand side of tractor frame just forward of the transmission case.

35 through 204 Industrial
Unless otherwise indicated numbers are stamped on instrument panel nameplate.

MH-50
Numbers are stamped on instrument panel nameplate.

230, 235, 245, 255, 265, 275, and 285
Numbers are located on left-hand side of tractor just forward of transmission.

1100 and 1130
Numbers are located on left-hand side of tractor just forward of transmission.

1085, 1105, 1135, and 1155
Numbers are on left-hand side just ahead of instrument panel.

20 R.C.

Year	Beginning number
1946	1001
1947	1580
1948	3584

20 Standard

Year	Beginning number
1946	1001
1947	1002
1948	2230

20K R.C.

Year	Beginning number
1947	1001
1948	1354

20K Standard

Year	Beginning number
1947	1001
1948	1819

21-Colt

Year	Beginning number
1952	1001
1953	1417

22 R.C.

Year	Beginning number
1948	1001
1949	2096
1950	4580
1951	7624
1952	10145
1952	20046
1953	20585

22 Standard

Year	Beginning number
1948	1001
1949	1542
1950	3208
1951	4532
1952	5717
1952	20046
1953	20585

22K R.C.

Year	Beginning number
1948	1001
1949	1154

Serial Numbers

1950	1336
1951	1558
1952	1776
1952	20046
1953	20585

22K Standard

Year	Beginning number
1948	1001
1949	1317
1950	1488
1951	1570
1952	1748
1952	20046
1953	20585

23 – Mustang

Year	Beginning number
1952	1001
1953	1666
1954	4346
1955	4553
1956	4773

25

Year	Beginning number
1932-35	69001
1936	69970
1937	71045
1938 A Gears	73112
1938 B Gears	85001

25 Industrial

Year	Beginning number
1937	90001

30K R.C.

Year	Beginning number
1947	1001
1948	1225
1949	2010
1950	2393
1951	2731
1952	30001
1953	30596

30 R.C.

Year	Beginning number
1946	1001
1947	1002

1948	3386
1949	6825
1950	9345
1951	13816
1952	17934
1952	30001
1953	30596

30 Standard

Year	Beginning number
1946	1001
1947	1002
1948	2120
1949	3194
1950	5567
1951	7491
1952	8696
1952	30001
1953	30596

30K Standard

Year	Beginning number
1947	1001
1948	1894
1949	3251
1950	3531
1951	3861
1952	30001
1953	30596

33

Year	Beginning number
1952	1001
1953	2055
1954	6617
1955	9782

44 R.C.

Year	Beginning number
1946	1001
1947	1002
1948	2048
1949	5318
1950	13822
1951	21815
1952	31275
1952	40001
1953	43700
1954	51364
1955	58067

Serial Numbers
Massey-Harris-Ferguson continued

44 Standard
Year	Beginning number
1946	1001
1947	1141
1948	1871
1949	4528
1950	9581
1951	13726
1952	17059
1952	40001
1953	43700
1954	51364
1955	58067

44D Orchard
Year	Beginning number
1950	1001
1951	1002

44D R.C.
Year	Beginning number
1949	1001
1950	1004
1951	2483
1952	4704
1952	40001
1953	43700
1954	51364
1955	58067

44D Standard
Year	Beginning number
1948	1001
1949	1023
1950	2180
1951	3989
1952	5639
1952	40001
1953	43700
1954	51364
1955	58067

44 GRA (high-altitude R.C.)
Year	Beginning number
1951	1001
1952	1164

44 GSA (high-altitude Standard)
Year	Beginning number

| 1951 | 1001 |
| 1952 | 1055 |

44 Orchard
Year	Beginning number
1950	1001
1951	1101
1952	40001
1953	43700

44 Special G & D
Year	Beginning number
1953	50001
1954	51364
1955	58067

44 Vineyard
Year	Beginning number
1950	101
1951	1031
1952	40001
1953	43700

44K R.C.
Year	Beginning number
1947	1001
1948	1079
1949	1856
1950	2599
1951	3329
1952	40001
1953	43700
1954	51364
1955	58067

44 – LP R.C.
Year	Beginning number
1952	1001

44 – LP Standard
Year	Beginning number
1952	1001

44K Standard
Year	Beginning number
1946	1001
1947	1011
1948	1441
1949	3598
1950	4827

Serial Numbers

Year	Beginning number
1951	6019
1952	6787
1952	40001
1953	43700
1954	51364
1955	58067

44-6 R.C.

Year	Beginning number
1946	1001
1947	1002
1948	2983
1949	4755
1950	5255
1951	5509

44-6 Standard

Year	Beginning number
1947	1001
1948	2001
1950	2601

55

Year	Beginning number
1946	1001
1947	1116
1948	2132
1949	3581
1950	5468
1951	6399
1952	10001
1953	13017
1954	15299
1955	17059

55D Standard

Year	Beginning number
1949	1001
1950	1022
1951	2058
1952	2822
1953	10001
1954	13017
1954	15299
1955	17059

55D Riceland & Hillside

Year	Beginning number
1950	1001
1951	1152

Year	Beginning number
1952	1452
1952	10001
1953	13017
1954	15299
1955	17059

55 DISH – DSW (diesel Western)

Year	Beginning number
1951	1001
1952	1190
1952	10001
1953	13017
1954	15299

55 – GISH - GSW (gas Western)

Year	Beginning number
1951	1002
1952	1083
1952	10001
1953	13017
1954	15299

55 – GSA, GSWA, and GSHA (high altitude)

Year	Beginning number
1951	1001
1952 GSWA	1025
1952	10001
1953	13017
1954	15299

55 – GSH & GIWH Riceland and Hillside

Year	Beginning number
1949	1001
1950	1035
1951	1216
1952	10001
1953	13017
1954	15299
1955	17059

55K Standard

Year	Beginning number
1946	1001
1947	1013
1948	1554
1949	3033
1950	4078
1951	4808

Serial Numbers
Massey-Harris-Ferguson continued

Year	Beginning number
1952	5503
1952	10001
1953	13017
1954	15299
1955	17059

55K Standard - Riceland & Hillside

Year	Beginning number
1949	1001
1950	1013
1951	1110
1952	10001
1953	13017
1954	15299
1955	17059
1956	20001
1957	22649

81 R.C.

Year	Beginning number
1941	400001
1942	403168
1944	403354
1945	403364
1946	404664

81 Standard

Year	Beginning number
1941	425001
1942	425678
1944	425757
1945	425780
1946	426803

82 R.C.

Year	Beginning number
1941	420001
1942	420055
1945	420274
1946	420307

82 Standard

Year	Beginning number
1941	435001
1942	435279
1943	435452
1945	435458
1946	435738

101 Jr. R.C.

Year	Beginning number
1939	375001
1940	37618
1941	395570
1942	397637
1943	398596
1944	500003
1945	502434
1946	503779

101 Jr. Standard

Year	Beginning number
1939	377001
1940	377928
1941	379550
1942	379815
1943	379855
1944	380641
1945	382569
1946	384298

101 Senior R.C.

Year	Beginning number
1938	255001
1939	256085
1940	257281
1941	258769
1942	259762
1943	260430
1944	260796
1945	263020
1946	270145

101 Senior Standard

Year	Beginning number
1938	355001
1939	355603
1940	356792
1941	358188
1942	358869
1943	358975
1944	359457
1945	360927
1946	362520

102 Junior R.C.

Year	Beginning number
1939	387001
1940	37031
1941	387127

Serial Numbers

Year	Beginning number
1942	387419
1943	387601
1944	387844
1945	388240
1946	388995

102 Junior Standard

Year	Beginning number
1939	385001
1940	385204
1941	385450
1942	36099
1943	386662
1944	390008
1945	390994
1946	391913

102 Senior R.C.

Year	Beginning number
1942	265001
1943	265044
1944	265078

102 Senior Standard

Year	Beginning number
1941	365001
1942	365202
1943	366062
1944	366183
1945	367353

201

Year	Beginning number
1940	91201
1941	91541
1942	91691
1943	98674
1944	98807
1946	99689
1947	100120

201 – Diesel

Year	Beginning number
1940	99501

202

Year	Beginning number
1940	95001
1941	95002
1942	95182

203

Year	Beginning number
1944	95223
1945	
1946	95295
1947	95338

203 – Diesel

Year	Beginning number
1940	98001
1941	98028
1942	98364
1943	98674
1944	98807
1945	n/a
1946	99689
1947	100120

303 – Industrial

Year	Beginning number
1956	1001
1957	1076
1958	1194

333

Year	Beginning number
1956	20001
1957	22649

404 – Industrial

Year	Beginning number
1956	1001
1957	1051

444

Year	Beginning number
1956	70001
1957	73989
1958	77000

555

Year	Beginning number
1955	20001
1956	20133
1957	21133
1958	22950

744 – Great Britain

Year	Beginning number
1948	201
1949	401

Serial Numbers
Massey-Harris-Ferguson continued

1950.. 1401
(After 1950, year is indicated by a
letter, e.g., F=1951 G=1952 etc.)

Challenger

Year	Beginning number
1936-37	130001

Challenger Distillate

Year	Beginning number
1936	140001

Challenger Twin Power Gas

Year	Beginning number
1938	133367

GP 4WD

Year	Beginning number
1930-35	300001
1936-38	303001

I-162

Year	Beginning number
1953	1001

FSI-244 (USAF w/ PTO)

Year	Beginning number
1955	2001
1956	2219

I-244 (no PTO)

Year	Beginning number
1955 Navy	1001
1956 Navy	1020
1956 USAF	3001
1957 USAF	3181

I-330 (Navy)

Year	Beginning number
1954	1001

MH-50

Year	Beginning number
1955	500001
1956	500473
1957	510764

Pacemaker

Year	Beginning number
1936-37	107001

Pacemaker Distillate

Year	Beginning number
1936	120001

Pacemaker Distillate OPA

Year	Beginning number
1936	204001

Pacemaker Distillate VPA

Year	Beginning number
1936	201501

Pacemaker Twin Power Gas

Year	Beginning number
1938	109838

Pacemaker Twin Power OPA Gas

Year	Beginning number
1938	200403

Pacemaker Twin Power VPA Gas

Year	Beginning number
1938	201042

Pacer – 16

Year	Beginning number
1954	50001
1955	51613
1956	52771

Pony – 11 & 14

Year	Beginning number
1947	PGS 1001
1948	PGS 1321
1949	PGA1571A
1950	PGA10817A
1951	PGA 13726
1952	PGA 18225
1953	PGA 22007
1954	PGA 22669

Model 35

Year	Beginning number
1960	204181
1961	211071
1962	222207
1963	235123
1964	247605

Serial Numbers

50

Year	Beginning number
1956	500001
1957	510764
1958	515708
1959	522693
1960	528163
1961	528419
1962	529821
1963	533422
1964	536063

65

Year	Beginning number
1957	650001
1958	650024
1959	661164
1960	671379
1961	680210
1962	685370
1963	693040
1964	701057
1965	710788

85

Year	Beginning number
1958	800001
1959	800048
1960	804355
1961	807750
1962	808564

88

Year	Beginning number
1959	880001
1960	881453
1961	882229
1962	882496

90 and Super

Year	Beginning number
1962	810000
1963	813170
1964	816113
1965	819342

90WR and Super

Year	Beginning number
1961	885000
1962	885010

1963	885870
1964	886829
1965	888238

97D

Year	Beginning number
19i62	25200001
1963	25200506
1964	25202005
1965	25203504

97 LPG

Year	Beginning number
1962	25300001
1963	25300096
1964	25300397
1965	25300399

135

Year	Beginning number
1964	641000001
1965	641001909
1966	641014871
1967	9A10001
1968	9A39386
1969	9A63158
1970	9A87325
1971	9A107519
1972	9A128141
1973	9A152025
1974	9A182761
1975	9A207681

150

Year	Beginning number
1964	642000001
1965	642000015
1966	642000505
1967	9A10001
1968	9A39836
1969	9A63158
1970	9A87325
1971	9A107519
1972	9A128141
1973	9A152025
1974	9A182761
1975	9A207681

Serial Numbers
Massey-Harris-Ferguson continued

165

Year	Beginning number
1964	643000001
1965	643000003
1966	643000149
1967	9A10001
1968	9A39836
1969	9A63158
1970	9A87325
1971	9A107519
1972	9A128141
1973	9A152025
1974	9A182761
1975	9A207681

175

Year	Beginning number
1965	644000001
1966	644000004
1967	644000214
1967	9A10001
1968	9A39836
1969	9A6158
1970	9187325
1971	9A107519
1972	9A128141
1973	9A152025
1974	9A182761
1975	9A207681

230, 235, 245, 255, 265, 275, and 285

Year	Beginning number
1975	9A182761
1975	9A207681

180

Year	Beginning number
1964	645000001
1965	645000002
1966	645000047
1967	9A10001
1968	9A39836
1969	9A63158
1970	9A87325
1971	9A107519
1972	9A128141
1973	9A152025
1974	9A182761
1975	9A207681

202 Industrial

Year	Beginning number
1958	301172
1959	303158
1960	305108
1961	30619
1962	306779
1963	307923
1964	309222
1965	310067
1966	311084

203 Industrial

Year	Beginning number
1961	659000001
1962	659000681
1963	659001379
1964	659001767
1965	659002054
1966	659002667

204 Industrial

Year	Beginning number
1959	340001
1960	341381
1961	341986
1962	342648
1963	343182
1964	343836
1965	344309
1966	345103

1080

Year	Beginning number
1969	9B18673
1970	9B23486
1971	9B28227
1972	9B31958

1100

Year	Beginning number
1964	650000001
1965	650000003
1966	650000831
1967	9B10001
1968	9B14693
1969	9B18673
1970	9B23486
1971	9B28227
1972	9B31958

Serial Numbers

1130

Year	Beginning number
1964	651500001
1965	651500004
1966	651500049
1967	9B10001
1968	9B14693
1969	9B13673
1970	9B23486
1971	9B28227
1972	9B31958

1085, 1105, 1135, and 1155

Year	Beginning number
1972	9B36563
1973	9B36841
1974	9B43432
1975	9B50494
1976	9B58735

1500 and 1800

Year	Beginning number
1971	9C1007
1972	9C1912
1973	9C2462
1974	9C3184

1505 and 1805

Year	Beginning number
1974	9C003184
1975	9C004227
1976	9C006086
1945	388240
1946	388995

102 Junior Standard

Year	Beginning number
1939	385001
1940	385204
1941	385450
1942	36099
1943	386662
1944	390008
1945	390994
1946	391913

102 Senior R.C.

Year	Beginning number
1942	265001
1943	265044
1944	265078

102 Senior Standard

Year	Beginning number
1941	365001
1942	365202
1943	366062
1944	366183
1945	367353

201

Year	Beginning number
1940	91201
1941	91541
1942	91691
1943	98674
1944	98807
1946	99689
1947	100120

201 – Diesel

Year	Beginning number
1940	99501

202

Year	Beginning number
1940	95001
1941	95002
1942	95182

203

Year	Beginning number
1944	95223
1946	95295
1947	95338

203 – Diesel

Year	Beginning number
1940	98001
1941	98028
1942	98364
1943	98674
1944	98807
1945	n/a
1946	99689
1947	100120

303 – Industrial

Year	Beginning number
1956	1001
1957	1076
1958	1194

Serial Numbers
Massey-Harris-Ferguson continued

333
Year	Beginning number
1956	20001
1957	22649

404 – Industrial
Year	Beginning number
1956	1001
1957	1051

444
Year	Beginning number
1956	70001
1957	73989
1958	77000

555
Year	Beginning number
1955	20001
1956	20133
1957	21133
1958	22950

744 – Great Britain
Year	Beginning number
1948	201
1949	401

Minneapolis-Moline

Twin City tractors

Twin City 16-30
Year	Number range
1936	5501 - 6203

Twin City 17-28 TY
Year	Number range
1930	30104 - 30281
1931	30282 - 30298
1932	30299 - 30309
1933	30310 - 30333
1934	30334 - 30762
1935	30763 - 30808

TC 17-28 Industrial
Year	Beginning number
1934	43001

Twin City 27-44 AT
Year	Number range
1926 - 1928	May fall within 250001 – 250672
1929	250001 - 250730
1930	250731 - 250796
1931	250797- 250799
1932	None produced
1933	None produced
1934	250800 - 250805
1935	250806 - 250839

Twin City 21-32
Year	Number range
1926 - 1928	150001 - 150302

Twin City KT Orchard
Year	Number range
1931	301863 - 301890

Twin City 21-32 FT
Year	Number range
Pre-1930	150303 - 151796
1930	151797 - 154073
1931	154074 - 154123
1932	154124 - 154129
1934	154130 - 154275

Twin City 21-32 FTA
Year	Number range
1935	154300 - 155381
1936	154382 - 156247
1937	156124 - 154129
1938	156909 - 154275

Twin City FT Industrial
Year	Number range
1932	46001 - 46004
1934	46005 - 46029

Twin City FTA Industrial
Year	Number range
1935	46030
1936	46031 - 46046
1937	46047 - 46074

Twin City KT
Year	Number range
1929	300001 - 300079
1930	300080 - 301583

Serial Numbers

1931	301584 - 301862
	301865 - 301866
	301882
1932	301 957 - 301 981
1933	301 982 - 201 987
1934	301 988 - 302 078

Twin City KT Industrial

Year	Number range
1932	40001 only
1933	40002 - 40004
1934	40005 - 40008
1935	40009 only

Twin City LT

Year	Number range
1934	500001 - 500010

Twin City KTA

Year	Number range
1934	302 200 - 302 371
1935	302 372 - 303 825
1936	303 826 - 304 701
1937	304 702 - 306 281
1938	306 282 - 306 751

Twin City MT

Year	Number range
1930	525 001 - 525 020
1931	525 021 - 526 118
1932	526 096 - 525 334
1933	525 335 - 525 345
1934	525 346 - 525 420

Twin City Universal MTA

Year	Number range
1934	525 421 - 525 490
1935	525 491 - 526 118
1936	526 119 - 526 960
1937	526 961 - 528 049
1938	528 050 - 528 645

Twin City Universal JT

Year	Number range
1934	550 001 - 550 025
1935	550 026 - 551 762
1936	551 763 - 554 554
1937	554 555 - 556 244

Twin City JT Standard

Year	Number range
1936	600 001 - 600 322
1937	600 323 - 600 469

Twin City JT Orchard

Year	Number range
1936	625 001 - 625 103
1937	625 104 - 625 156

Twin City LT

Year	Number range
1930	500 001 - 500 010

Minneapolis-Moline tractors
GT (GE or 403 engine)

Year	Number range
1938	160 001 - 160 076
1936	160 077 - 160 545
1940	160 546 - 160 878
1941	160 879 - 161 253

GTA (LE engine)

Year	Number range
1942	162 001 - 162 300
1943	162 301 - 162 302
1944	162 303 - 162 659
1945	162 660 - 162 869
1946	162 870 - 163 219
1947	163 220 - 163 610

GTB (403 engine)

Year	Number range
1947	164 001 - 164 178
1948	164 179 - 164 214
	Early
1948	016 480 0001 - 016480 0600
	Late
1949	016 490 0001 - 016 490 1205
1950	016 500 0001 - 016 501 863
1951	01601864 - 01603396
1952	01603397 - 01604889
1953	01604890 - 01605972
1954	01605973 - 01606289

GTB-D   D425-6

Year	Number range
1953	06800001 only
1954	06800002 - 06800850

Serial Numbers
Minneapolis-Moline continued

GTC 340-4

Year	Number range
1951	04700001 - 04700018
1952	04700019 - 04700676
1953	04700677 - 0470110

GB 403 C - 4

Year	Number range
1955	08900001 - 0890150
1956	08901501 - 08902601
1957	08902602 - 08903401
1958	08903402 - 08904251
1959	08904252 - 08904442

GB-D

Year	Number range
1955	09000001 - 09000850
1956	09000851 - 09001525
1957	09001526 - 09002145
1958	09002146 - 0900265
1959	09002656 - 09002790

RT

Year	Number range
1939	400 001 - 402 200
1940	402 201 - 405 575
1941	405 576 - 407 950
1942	407 951 - 408 825
1943	408 826 - 409 357
1944	409 358 - 410 747
1945	410 748 - 413 754
1946	413 755 - 416 544
1947	416 545 - 422 057

RTE

Year	Number range
1948	0044800001 - 0044800501
1949	0044900001 - 0044900315
1950	0045000001 - 0045000204
1951	00400202 - 00400281
1952	00400282 only
1953	00400283 - 00400287

RTN

Year	Number range
1948	0034800001 - 0034800100
1949	0034900001 - 0034900200
1950	0035000001 - 0035000093
1951	00300094 - 00300173

RTI

Year	Number range
1948	0054800001 - 0054800700
1949	0054900001 - 0054900450
1950	0055000001 - 0055000115
1951	00500116 - 00500598
1952	00500599 - 00501000
1953	00501001 - 00501311
1954	00501312 - 00501511
1955	00501512 - 00501579

RTS

Year	Number range
1949	0024900001 - 0024900375
1950	0025000001`- 0025000300
1951	00200301 - 00200401
1952	00200402 - 00200551
1953	00300552 - 00200701

RTU

Year	Number range
1948	0014800001 - 0014802402
1949	0014900001 - 0014903039
1950	0015000001 - 0015002155
1951	00102156 - 00103972
1952	00103973 - 00104823
1953	
1954	00104824 - 00104831

RTI-M

Year	Number range
1953	05500001 - 0550249

U

Year	Number range
1938	310 026 - 310 645
1939	310 646 - 312 450
1940	312 451 - 314 892
1941	314 893 - 316 500
1942	316 501 - 317 701
1943	317 702 - 318 162
1944	318 163 - 321 101
1945	321 102 - 325 231
1946	325 231 - 329 751
1947	329 752 - 337 412
1948	337 418 - 339 682

UDLX

Year	Number range
1938	310001-310025(early)
1938	310501-310625(late)

Serial Numbers

UTC (6-volt ignition)

Year	Number range
1948	0154800001 - 0154800300
1949	0154900001 - 0154900100
1951	01500101 - 01500180
1952	01500181 - 01500265
1954	01500266 - 01500271

UTC (12-volt ignition)

Year	Number range
1954	08800001 - 08800060
1955	08800061 - 08800110

UTE

Year	Number range
1951	04300001 - 04300111
1952	04300112 - 04300261
1953	04300262 - 04300264
1954	04300265 - only

UTN

Year	Number range
1950	0385000001 - 0385000101
1951	03800102 - 03800204
1952	03800205 - 03800354

UTS

Year	Number range
1948	0124800001 - 0124802276
1949	0124900001 - 0124903901
1950	0125000001 - 01203850
1951	01203851 - 01207138
1952	01207139 - 01210570
1953	01210571 - 01213219
1954	01213220 - 01213325
1955	01213326 - 01214125
1956	01214126 - 01215100
1957	01215101 - 01215150

UTU

Year	Number range
1948	0114800001 - 0114802053
1949	0114900001 - 0114905000
1950	0115000001 - 01105383
1951	01105384 - 01110117
1952	01110118 - 01113449
1954	01113450 - 01113453
1955	01113454 - 01113456

UDU

Year	Number range
1952	04900001 only
1953	04900002 - 04900030

UTS-D

Year	Number range
1952	05000001 - 05000018
1954	05000019 - 05000755
1955	05000955 - 05001154
1956	05002105 - 05002404

UDS-M

Year	Number range
1954	05000756 - 05000954
1955	05001155 - 05002104

UBU

Year	Number range
1953	05800001 - 05802912
1954	05802913 - 05804002
1955	05804003 - 05805077

UBE

Year	Number range
1953	05900001 - 05900896
1954	05900897 - 05901068
1955	05901069 - 05901421

UBN

Year	Number range
1953	06000001 - 06000202
1954	06000203 - 06000207
1955	0600208 - 06000241

UBU Diesel

Year	Number range
1954	07800001 - 07800746
1955	07800747 - 07801041

UBE Diesel

Year	Number range
1954	07000001 - 07000231
1955	07000232 - 07000362

UBN Diesel

Year	Number range
1954	06900001 – 06900048

UB Special

Year	Number range
1955	09700001 - 09701475

Serial Numbers
Minneapolis-Moline continued

UB Special - Diesel
Year	Number range
1955	09800001 - 09800300
1956	09800301 - 09800464
1957	09800465 - 09800520

ZTI
Year	Number range
1936	599 001 - 599 003
1937	599 004 - 599 016
1936	599 017 - 599 018
1936	559 019 - 559 022

ZTU - ZTN
Year	Number range
1936	560 001 - 560 037
1937	560 038 - 562 974
1938	562 975 - 565 406
1939	565 407 - 567 154
1940	567 155 - 568 754
1941	568 755 - 570 821
1942	570 822 - 571 421
1943	571 422 - 572 967
1944	572 968 - 575 712
1945	575 713 - 576 813
1445	572 713 - 576 813
1946	576 814 - 578 013
1947	578 041 - 581 814
1948	581 815 - 585 817

ZTS
Year	Number range
1937	610 001 - 610 035
1938	610 036 - 610 388
1939	610 389 - 610 684
1940	610 685 - 611 087
1941	611 088 - 611 342
1942	611 343 - 611 446
1943	611 447 - 611 965
1944	611 966 - 612 485
1945	612 486 - 612 885
1946	612 886 - 613 085
1947	613 086 - 613 490

ZAU
Year	Number range
1949	0064900001 - 0064903013
1950	0065000001 - 00605435
1951	00605436 - 00609939
1952	00609940 - 00614658

ZAS
Year	Number range
1949	0074900001 - 0074900150
1950	0075000001 - 00700480
1951	00700481 - 00701285
1952	00701286 - 00701910
1953	00701911 - 00702610

ZAN
Year	Number range
1949	0084900001 - 0084900150
1950	0085000001 - 00800238
1951	00800239 - 00800442
1952	00800443 - 00800618
1953	00800619 - 00800620

ZAE
Year	Number range
1949	094900001 - 0094900301
1950	0095000001 - 00900373
1951	00900374 - 0000576
1952	00900577 - 0900997
1953	00900998 - 00901122

ZBE
Year	Number range
1953	06300001 - 06300075
1954	06300076 - 06300306
1955	06300307 - 06300501

ZBN
Year	Number range
1953	06300001 - 0630075
1954	06300076 - 06300306
1955	06300307 - 06300501

ZBN
Year	Number range
1954	06400001 - 06400072
1955	06400073 - 06400106

ZBA
Year	Number range
1953	06200001 - 06200957
1954	06200958 - 06202479
1955	06202480 - 06203059

335 Utility
Year	Number range
1956	10400001 - 10400101

Serial Numbers

1957...........10400102 - 10402087
1958...........10402088 - 10402336
1959...........10402337 - 10402439
1960...........10402440 - 10402489
1961...........10402490 - 10402539

335 Universal

Year	Number range
1957	11600001 - 11600301
1958	11600302 - 11600305
1959	11600306 - 11600334

335 Industrial

Year	Number range
1957	11300001 - 11300440
1958	11300441 - 11300521
1959	11300522 - 11300596
1960	11300597 - 11300746

445 Universal

Year	Number range
1956	10100001 - 10102854
1957	10102855 - 10104125
1958	10104126 - 10104804
1959	10104805 - 10104847

445 Utility

Year	Number range
1956	10200001 - 10201445
1957	10201446 - 10202101
1958	10202102 - 10202242
1959	10202243 - 10202249

445 Utility (diesel)

Year	Number range
1959	15400001 - 15400018

445 Industrial

Year	Number range
1956	11100001 - 11100075
1957	11100076 - 11100388
1958	11100389 - 11100645

445 Industrial (diesel)

Year	Number range
1958	15200001 - 15200025

445 (military)

Year	Number range
1958	15700001 - 15700074

445 Universal (diesel)

Year	Number range
1958	15200001 - 15200190

Big Mo 400 (gas)

Year	Number range
1961	16700001 - 16700100
1962	16700101 - 16700210
1963	16700211 - 16700360
1963	16700361 - 16700410

Big Mo 400 (military)

Year	Number range
1959	17000001 - 17000356
1960	17000357 - 17000632
1961	17000633 - 17000648
1962	17000649 - 17000652
1963	17000653 - 17000757

Big Mo 500 (gas)

Year	Number range
1960	16800001 - 16800160
1961	16800161 - 16800391
1962	16800392 - 16800606
1963	16800607 - 16800681
1964	16800682 - 16800746
1965	16800747 - 16800866
1966	16800867 - 16800881

Big Mo 500 (diesel)

Year	Number range
1960	17800001 - 17800065
1963	17800066 - 1780090
1964	16800091 - 17800115
1965	17800116 - 17800145

Big Mo 600

Year	Number range
1960	18400001 - 18400060

4 Star Series (gas)

Year	Number range
1959	16600001 - 16600890
1960	11600891 - 16601685
1961	16601686 - 16601860
1962	16601861 - 16602407
1963	16602408 - 16602537

4 Star Series (diesel)

Year	Number range
1960	18200001 - 18200050

Serial Numbers
Minneapolis-Moline continued

1961............18200051 - 18200072
1962............18200073 - 18200097

5 Star Universal (gas)
Year	Number range
1957............	11000001 - 11001057
1958............	11001058 - 11002067
1959............	11002068 - 11002914

5 Star Universal (diesel)
Year	Number range
1957............	14400001 - 14400203
1958............	14400204 - 14400785
1959............	11400786 - 14401295

5 Star Standard (gas)
Year	Number range
1958............	11200001 - 11200380

5 Star Industrial (diesel)
Year	Number range
1958............	14500001 - 14500165
1959............	14500166 - 14500188

5 Star Industrial (gas)
Year	Number range
1957............	11700002 - 11700006
1958	11700007 - 11700025
1959............	11700026 - 11700084

5 Star Industrial (diesel)
Year	Number range
1958............	14600001 - 14600010
1959............	14600011 - 14600028
1960............	14600029 - 14600060

2 Star Crawler
Year	Number range
1958............	12000001 - 12000051

Motrac Crawler (gas)
Year	Number range
1960............	18500001 - 18500030
1961............	18500031 - 18500038

Motrac Crawler (diesel)
Year	Number range
1960............	18600001 - 18600160
1961......................	18600161 only

Jet Star 2 (gas)
Year	Number range
1963............	25800001 - 25801100

Jet Star (gas)
Year	Number range
1959............	16500001 - 16500284
1960............	16500285 - 16500834
1961............	16500835 - 16501701
1962............	16501702 - 16502439

Jet Star (diesel)
Year	Number range
1960............	17500001 - 17500060
1961............	17500061 - 17500135
1962............	17500136 - 17500196

Jet Star 2 (diesel)
Year	Number range
1963............	25700001 - 25700113

Jet Star 3 (gas)
Year	Number range
1963............	28300001 - 28301000
1964............	28301001 - 28301984

Jet Star 3 Super (gas)
Year	Number range
1965............	28301985 - 28302055
1966............	28302056 - 28302843
1967............	28302844 - 28303565
1968............	28303566 - 28304800
1969............	28304801 - 28305085
1970............	28305086 - 28305335

Jet Star 3 (diesel)
Year	Number range
1964............	28400001 - 28400050
1965............	28400051 - 28400200
1966............	28400201 - 28400385
1967............	28400386 - 28400466
1968............	28400467 - 28400526
1969............	28400527 - 28400601
1970............	28400602 - 28400711

Jet Star Orchard (gas)
Year	Number range
1965............	30700001 - 30700050
1966..........................	30700051--
1967............	30700052 - 30700070

Serial Numbers

Jet Star Orchard (diesel)

Year	Number range
1967	34400001 - 34400020

Jet Star 3 (LP gas)

Year	Number range
1970	36000001 - 36000010

Jet Star 3 Industrial (diesel)

Year	Number range
1966	30900001 - 30900050

U302 (gas)

Year	Number range
1964	27600001 - 27601000
1965	27601001 - 27601300

U302 Super (gas)

Year	Number range
1966	27601301 - 27602300
1967	27602301 - 27602425
1968	27602426 - 27602759
1969	27602760 - 27602859
1970	27602860 - 27602969

U302 Super (diesel)

Year	Number range
1967	27700001 - 27700100
1968	27700101 - 27700150
1969	27700151 - 27700164
1970	27700165 - 27700190

U302 Super (LP gas)

Year	Number range
1969	36100001 - 36100025
1970	36100026 - 36100050

M5 (gas)

Year	Number range
1960	17100001 - 17101535
1961	17101536 - 17103495
1962	17103496 - 17104707
1963	17104708 - 17105157

M5 (diesel)

Year	Number range
1960	17200001 - 17201040
1961	17201041 - 17201000
1962	17202000 - 17202506
1963	17202507 - 17202656

M504 4-Wheel Drive (gas)

Year	Number range
1962	24300001 - 24300010

M504 4-Wheel Drive (diesel)

Year	Number range
1962	24200001 - 24300010

M602 (gas)

Year	Number range
1963	26600001 - 26601275
1964	26601276 - 26602957

M602 (diesel)

Year	Number range
1963	26700001 - 26700742
1964	26700743 - 26701772

M604 4-Wheel Drive (gas)

Year	Number range
1963	26800001 - 26800050
1964	26800051 - 26800053

M604 4-Wheel Drive (diesel)

Year	Number range
1963	26900001 - 26900050
1964	26900051 - 26900099

M670 (gas)

Year	Number range
1964	29900001 - 29900006
1965	29900007 - 29901891

M670 (diesel)

Year	Number range
1964	30000001 - 30000004
1965	30000005 - 30000819

M670 Super (gas)

Year	Number range
1966	29901892 - 29903579
1967	29903580 - 29904454
1968	29904455 - 29904594
1969	29904595 - 29905004
1970	29905005 - 29905104

M670 Super (gas)

Year	Number range
1966	29901892 - 29903579

Serial Numbers
Minneapolis-Moline continued

M670 Super (diesel)
Year	Number range
1966	30000820 - 30001634
1967	30001635 - 30002309
1968	30002310 - 30002569
1969	30002570 - 30002860
1970	30002861 - 30003085M670

Super (LP gas)
Year	Number range
1970	36200001 - 36200075

G VI (gas)
Year	Number range
1959	16000001 - 16000876
1960	16000877 - 16001675
1961	16001676 - 16002032
1962	16002033 - 16002352

G VI (diesel)
Year	Number range
1956	16200001 - 16200805
1960	16200806 - 16201890
1961	16201891 - 16202960
1962	16202961 - 16203235

G704 (LP gas)
Year	Number range
1962	23400001 - 23400081

G704 (diesel) L
1962	23500001 - 23500123

G705 (LP gas)
Year	Number range
1962	23800001 - 23800078
1963	23800079 - 23800590
1964	23800591 - 23801092
1965	23801093 - 23801223

G705 (diesel)
Year	Number range
1962	23900001 - 23900050
1963	23900051 - 23900898
1964	23900899 - 23901868
1965	23901869 - 23902094

G706 (LP gas)
Year	Number range
1962	24000001 - 24000072
1963	24000073 - 24000305
1964	24000306 - 24000350
1965	24000351 - 24000370

G706 (diesel)
Year	Number range
1962	24100001 - 24100106
1963	24100107 - 24100549
1964	24100550 - 24100795
1965	24100796 - 24100821

G707 (LP gas)
Year	Number range
1965	31200001 - 31200283

G707 (diesel)
Year	Number range
1965	31300001 - 31300415

G708 (LP gas)
Year	Number range
1965	31400001 - 31400031

G708 (diesel)
Year	Number range
1965	31500001 - 31500075

G900 (LP gas)
Year	Number range
1967	33000001 - 33000110
1968	33000111 - 33000550
1969	3300551 - 33000670

G900 (LP gas)
Year	Number range
1969	36300001 - 36300160

G900 (diesel)
Year	Number range
1967	33100001 - 33100316
1968	33100317 - 33101376
1969	33101377 - 33101946

G950 (LP gas)
Year	Number range
1969	43500001 - 43500060
1970	43500061 - 43500085
1971	43500086 - 43500186

Serial Numbers

G950 (diesel)

Year	Number range
1969	43600001 - 43600210
1970	43600211 - 43600415
1971	43600416 - 43600829
1972	43600830 - 43600834

G955

Year	Number range
1973	239825 - 243262
1974	244559 - 251357

G1000 Row Crop (gas LP gas)

Year	Number range
1965	30500001 - 30500450
1966	30500451 - 30500926
1967	30500927 - 30501041
1968	30501042 - 30501051

G1000 Row Crop (diesel)

Year	Number range
1965	30600001 - 30600500
1966	30600501 - 3060112
1967	30601126 - 30601285
1968	30601289 - 30601300

G1000 Wheatland (LP gas)

Year	Number range
1966	32600001 - 32600515
1967	32600516 - 32600650
1968	32600651 - 32600652
1969	32600653 - 32600822

G1000 Wheatland (diesel)

Year	Number range
1966	32700001 - 32700796
1967	32700797 - 32701450
1968	32701451 - 32701774
1969	32701775 - 32702050

G1000 Vista (LP gas)

Year	Number range
1967	34500001 - 34500290
1968	34500291 - 34300390
1969	34500391 - 34500564

G1000 Vista (diesel)

Year	Number range
1967	34600001 - 34600735
1968	34600736 - 34601185

1969	34601186 - 34601610

G1050 (LP gas)

Year	Number range
1969	43000001 - 43000040
1970	43000041 - 43000060
1971	43000061 - 43000105
1972	43000106 - 43000111

G1050 (diesel)

Year	Number range
1969	43100001 - 43100285
1970	43100286 - 43100415
1971	43100544 - 43100544

G1350 Row Crop (LP gas)

Year	Number range
1969	43200001 - 43200022
1970	43255523 - 43200044
1971	43200045 - 43200097
1972	43200098 - 43200108

G1350 Row Crop (diesel)

Year	Number range
1970	43300001 - 43300042
1971	43300043 - 43300253
1972	43300254 - 43300322

G1350 Wheatland (LP gas)

Year	Number range
1969	45300001 - 45300005

G1355

Year	Number range
1973	236440 - 244184
1974	245258 - 252710

A4T-1400 (diesel)

Year	Number range
1969	43600001 - 43900102
1970	43900103 - 43900247

A4T-1600 (diesel)

Year	Number range
1970	45600001 - 45600187
1971	45600188 - 45600700
1972	45600701 - 45601190

Serial Numbers
Minneapolis-Moline continued

A4T-1600 (LP gas)
Year	Number range
1970	45700001 - 45700126
1971	45700127 - 45700197
1972	45700198 - 45700257

Uni-Tractor
Year	Number range
1951	75700001 - 75700254
1952	75700255 - 75701070
1953	75701071 - 75703118
1954	75703119 - 75704118
1955	08704119 - 08705418
1956	08705419 - 08706418
1957	08706419 - 08707687
1958	08707688 - 08708062
1959	08708063 - 08708488
1960	42200001 - 42200637
1961	42200638 - 42201134
1962	42201135 - 42201637

Minneapolis Steel & Threshing

Joy-McVicker 50-140
Year	Serial Number
1911	NA

Twin City 40-65A
Year	Number range
1910 - 1915	NA
1916-1924	1001 - 1820

Twin City 40-65B
Year	Number range
NA	1821 - 1825

Twin City 24-45
Year	Number range
Model A	2501 - 2646
Model B	2647 - 2673
Model C	2701 - 2797
Model D	2801 - 2815
Model E	2816 - 3126

Twin City 15-30
Year	Number range
1913 - 1917	50001 - 5478

Twentieth Century
Year	Number range
1914 - ?	50001 - 5478

Twin City 16-30
Year	Number range
1917	5501 - 6203
1918	6201 - 6503
1919-1920	NA

Twin City 12-20
Year	Number range
1919-1926	10201 - 19903

Twin City 20-35
Year	Number range
1920-1926	2101 - 4097

Oliver

18-27 Row Crop (single front wheel)
Year	Number range
1930	100001-102468
1931	102649-103300

18-27 (dual front wheels)
Year	Number range
1931	103301-103318
1932	103319-103617
1933	103618-104038
1934	104039-104850
1935	104851-107311
1936	107312-108573
1937	108574-109151

18-28
Year	Number range
1930	800001-800459
1931	800460-800963
1932	800964-800984
1933	800985-801050
1934	801051-801240
1935	801241-801989
1936	801990-802937
1937	802938-803928

Serial Numbers

28-44

Year	Number range
1930	500001-503599
1931	503600-506184
1932	506185-506211
1933	506212-506254
1934	506255-506400
1935	506401-507175

28-44 & High-Compression Special

Year	Number range
1936	507176-508015
1937	508016-508917

18 Industrial

Year	Number range
1931	900001-900005

28 Industrial

Year	Number range
1932	900006-900018
1933	900019-900021
1934	900022-900036
1935	900037-900072
1936	900073-900078
1937	900079-900086
1938	900087-900102
1939	900103-900112

HP 70 Row Crop

Tag located on left side of engine.

Year	Number range
1935	200001-200685
1936	200686-208728
1937	208729-216925

HP 70 Standard

Tag located on left side of engine.

Year	Number range
1936	300001-300633
1937	300634-301802

70 Row Crop

Tag located on left side of engine.

Year	Number range
1937	216926-219644
1937	220426-220694
1938	219645-220425
1938	220695-223254
1939	223255-231115
1940	231116-236355
1941	236356-241390
1942	241391-243639
1943	243640-244710
1944	244711-250179
1945	250180-252779
1946	252780-258139
1947	258140-262839
1948	262840-267866

70 Standard

Tag located on left side of engine.

Year	Number range
1937	301803-302083
1938	302084-303464
1939	303465-305361
1940	305362-306593
1941	306594-307579
1942	307580-308187
1943	308188-308483
1944	308484-310217
1945	310218-311115
1946	311116-312689
1947	312690-314220
1948	314221-315420

Hart-Parr Oliver 25 Industrial

Tag located on left side of engine.

Year	Number range
1937	400001-400002

25 Industrial

Tag located on left side of engine.

Year	Number range
1938	400003-400008
1939	400009-400016
1940	400017-400021
1941	400022-400047
1942	400048-400067
1943	400068-400096
1944	400097-400181

35 Industrial

Tag located on left side of engine.

Year	Number range
1939	900113-900127
1940	900128-900229
1941	900230-900315
1942	900316-900327
1943	900328-900339
1944	900340-900395
1945	900396-900440

Serial Numbers
Oliver continued

Industrial 80
Tag located on left side of engine.

Year	Number range
1945	900441-900633
1946	900634-900816
1947	900817-901124

44 Industrial
Tag located on left side of engine.

Year	Number range
1932	700001-700004
1933	700005-700033
1934	700034-700141
1935	700142-700239
1936	700240-700295
1937	700296-700326
1938	700327-700359
1939	700360-700367

50 Industrial

Year	Number range
1939	700368-700421
1940	700422-700604
1941	700605-700777
1942	700778-701001
1943	701002-701147
1944	701148-701163

80 Row Crop
Tag located on left side of engine.

Year	Number range
1937	109152-109166
1938	109167-109782
1939	109783-110220
1940	110221-110614
1941	110615-110944
1942	110945-111218
1943	111219-111390
1944	111391-111928
1945	111929-112878
1946	112879-114143
1947	114144-114943
1948	114944-115373

80 Standard
Tag located on left side of engine.

Year	Number range
1937	803929-803990
1938	803991-805376
1939	805377-806879
1940	806880-808124

Year	Number range
1941	808125-809050
1942	809051-809990
1943	809991-810469
1944	810470-811990
1945	811991-813066
1946	813067-814563
1947	814564-815215
1948	815216-816241

90-99
Tag located on left side of engine.

Year	Number range
1937	508918-508934
1938	508935-509611
1939	509612-510067
1940	510068-510563
1941	510564-510976
1942	510977-511295
1943	511296-511473
1944	511474-512043
1945	512044-512043
1946	512821-513105
1947	513106-513855
1948	513856-514855
1949	514856-516275
1950	516276-516887
1951	516888-517873
1952	517874-518212

99 Industrial
Tag located on left side of engine.

Year	Number range
1945	701164-701225
1946	701226-701265
1947	701266-701287

900 Industrial

Year	Number range
1946	710001-710077
1947	710078-710134
1948	710135-710227
1949	710228-710256
1950	710257-710281

60 Row Crop
Tag located on left side of engine.

Year	Number range
1940	600001-600070
1941	600071-606303
1942	606304-607394
1943	607395-608525

Serial Numbers

Year	Number range
1944	608526-612046
1945	612047-615627
1946	615628-616706
1947	616707-620256
1948	620257-625131

60 Standard
Tag located on left side of engine

Year	Number range
1942	410001-410500
1943	410501-410510
1944	410511-410616
1945	410617-410910
1946	410911-411310
1947	411311-411960
1948	411961-413605

66 Row Crop
Tag located on the right side of the transmission housing. In 1951 the tag moved to the lower left side of dash panel.

Year	Number range
1949	420001-423100
1950	423101-426010
1951	426011-429770
1952	429771-431472
1953	3503990-3510962
1954	4500309-4503563

66 Standard
Tag located on the right side of the transmission housing. In 1951 the tag moved to the lower left side of dash panel.

Year	Number range
1949	470001-471050
1950	471051-472390
1951	472391-474232
1952	474233-476408
1953	3504001-3511337
1954	4501624-4504476

77 Row Crop
Tag is located on the right side of the transmission. In 1951 the tag was moved to the lower left side of the dash panel.

Year	Number range
1948	320001-320240

Year	Number range
1949	320241-327900
1950	327901-337242
1951	337243-347903
1952	347904-354447
1953	3500001-3510830
1954	4501301-4504470

77 Standard
Tag is located on the right side of the transmission housing. In 1951 the tag moved to the lower left side of dash panel.

Year	Number range
1948	269001-269940
1949	269941-271266
1950	271267-272465
1951	272466-273375
1952	273376-274051

88 Row Crop (old style)

Year	Number range
1947	120001-120352
1948	120353-121300

88 Standard (old style)

Year	Number range
1947	820001-820136
1948	820136-820485

88 Row Crop
Tag is located on the right side of the transmission housing. In 1951 the tag moved to the lower left side of dash panel.

Year	Number range
1948	121301-123300
1949	123301-128652
1950	128653-132862
1951	132863-138183
1952	138184-143232
1953	3500977-3511566
1954	4500076-4505123

88 Standard
Tag is located on the right side of the transmission housing. In 1951 the tag moved to the lower left side of dash panel.

Year	Number range
1948	820486-821085

Serial Numbers
Oliver continued

1949	821086-824240
1950	824241-825810
1951	825811-826916
1952	826917-827966
1953	3501813-3511484
1954	4500080-4505081

99 (6-cylinder Fleetline)

Year	Number range
1953	518300-519244
1954	519245-519299

Super 44
Tag located on clutch shaft cover behind battery.

Year	Number range
1957	1001-1550
1958	1551-1775

Super 55
Tag located on the left side of the center frame.

Year	Number range
1954	6001-8290
1955	11837-31370
1956	35001-43647
1957	43916-56036
1958	56501-59033

Super 66
Tag is on lower left side of the dash panel.

Year	Number range
1954	7085-7284
1955	14099-27842
1956	39371-42430
1957	45846-55800
1958	57858-72824

Super 77
Tag is on lower left side of the dash panel.

Year	Number range
1954	8303-8988
1955	10001-29842
1956	38500-43637
1957	44167-55955
1958	56917-59008

Super 88

Tag is on lower left side of the dash panel.

Year	Number range
1954	6503-8302
1955	10075-29347
1956	36774-43715
1957	43901-55607
1958	56580-59001

Super 99
Tag is on left side of clutch cover.

Year	Number range
1954	519300-519675
1955	519676-520455
1956	520456-520943
1957	520944-521612
1958	521613-521635

OC-3

Year	Beginning number
1951	1WH000
1952	3WH712
1953	350000
1954	450000
1955	11WH760
1956	15WH306
1957	19WH090

OC-4

Year	Beginning number
1956	1TG002
1957	1TG004
1958	4TG077

OC-4-3-D

Year	Beginning number
1957	1WD002
1958	1WD120
1959	1WD950
1960	2WD824
1961	3WD594
1962	800270
1963	800431
1964	801280
1965	801795

OC-4-3-G

Year	Beginning number
1958	1WR002

Serial Numbers

Year	Beginning number
1959	1WR542
1960	4WR958
1961	6WR746
1962	800001000
1963	800431436
1964	801280
1965	801795

OC-6-D
Year	Beginning number
1953	3500000
1954	4500000
1955	1RC468
1956	1RC632
1957	1RC876
1958	2RC262
1959	2RC366
1960	2RC458

OC-6-G
Year	Beginning number
1953	3500000
1954	4500000
1955	1RM182
1956	1RM314
1957	1RM504
1958	1RM808
1959	2RM004
1960	2RM126

OC-9
Year	Beginning number
1959	1MA001
1960	1MA182

OC-96
Year	Beginning number
1959	1MB001
1960	1MB168
1961	2MB020
1962	800-270
1963	800431
1964	801-277
1965	801-856

OC-12-D
Year	Beginning number
1954	1JX001
1955	1JX042
1956	2JX350
1957	3JX636
1958	4JX652
1959	5JX140
1960	5JX506
1961	5JX828

OC-12-G
Year	Beginning number
1954	1JR001
1955	1JR002
1956	1JR062
1957	1JR178
1958	1JR202
1959	1JR21
1960	1JR228

Sears Handiman

Year	Serial or type numbers
1931	SR8355
1932	SR32538 - SR32251
1933	Type #60209
	Two models: Heavy-Duty and Model A
1934	1424
1935	Type #60500
	Two models: C-35 and B-35
1936	Type #60754
	Three models: Z-36, C-36, and B-36
1937	Type #60754
	Three models: Z-37, C-37, and B-37
1938	Model #917.5047 Type #20407
	Three models: Z-38, C-38, and B-38
1938-1939	Model #917.5032 Type #20430
	Three models: Z-38, C-38, and B-38
1939	Model #917.5044 Type #25862
	Three models: Z-38, C-38, and B-38
1939	Model #917.5161

Serial Numbers

194 ?................ Model #917.60754
Series 291
1940 ....No data
Three models: Z-40, C-40,
and B-40
1941 .... No data
Three models: Z-41, C-41,
and B-41
1942 ....No data
One model available

Handiman Junior

Year	Serial or type numbers
1938	917.5032
1940	917.50321
1941	917.503281 and 503282

Handiman R/T

Year	Serial or type numbers
1939	917.5151 and 917.5154
1940	917.5155
Late '50s	917.60120 and 21
Early '60s	917.60124 and 25

Silver King

Serial numbers for all models

Year	Starting number
1934	0
1935	326
1936	1001
1937	1986
1938	3025
1939	3876
1940	4245
1941	4906
1942	5256
1943	5594
1944	5710
1945	6161
1946	6449
1947	6947
1948	7475
1949	8245
1950	8395
1951	8545
1952	8627

1953	8708
1954	8717

Wallis

Bear

Year	Beginning number
1912	201-INA
1913	INA-210

Cub Models C and D

Year	Number range
1913 to 1917	1001-1660

Cub Model J

Year	Number range
1915	10001
1916	13505

Cub Model K

Year	Number range
1916	14001-INA
1916	OMA
1917	INA
1918	INA
1919	INA
1920	INA
1921	INA
1922	23156

Cub Model OKO

Year	Number range
1922	23200 to 23156
1926	40001 to 50000

Cub Certified Standard

Year	Number range
1926	25645 to 40000

White

Abbreviation guide:
2WD = two-wheel drive
FWD = front-wheel drive

2-50

Year	Number range
1976 (2WD)	516 625 – 518 781

Serial Numbers

1976 (FWD)	516 898 – 518 993
1977 (2WD)	518 782 – 520 783
1977 (FWD)	518 994 – 521 634
1978 (2WD)	520 784 – 525 267
1978 (FWD)	521 635 – 525 238
1979 (2WD)	525 268 – 525 725
1979 (FWD)	525 290 – 527 580
1980 (2WD)	525 726 – 527 626
1980 (FWD)	527 581 – 527 687

2-60

Year	Number range
1976 (2WD)	780 725 – 790 272
1976 (FWD)	782 037 – 790 272
1977 (2WD)	790 273 – 944 701
1977 (FWD)	790 273 – 946 284
1978 (2WD)	9i44 702 – 959 279
1978 (FWD)	946 285 – 959 302
1979 (2WD)	959 280 – 959 999
	480 187 – 491 334
1979 (FWD)	959 303 – 959 999
	480 307 - 486 531

2-70

Year	Number range
1976	266 173 – 273 088
1977	274 543 – 281 876
1978	283 917 – 284 276
1979	287 528 – 292 563
1980	293 819 – 294 062
1981	296 246 – 298 946
1982	299 887 – 300 091

2-85

Year	Number range
1975	263 341 – 265 402
1976	268 142 – 273 315
1977	274 287 – 281 504
1978	282 339 – 287 196
1979	287 469 - 293 408
1980	294 063 – 295 791
1981	297 751 – 299 123
1982	300 092 – 300 158

2-105

Year	Number range
1974	255 216 – 255 537
1975	255 538 – 265 927
1976	265 928 – 273 619
1977	273 760 – 280 588

1978	282 102 – 287 189
1979	287 197 – 293 357
1980	294 109 – 295 781
1981	296 878 – 299 731
1982	300 779 – 300 782

2-135

Year	Number range
1976	272 663 – 273 628
1977	273 629 – 282 078
1978	282 825 – 286 928
1979	288 201 – 293 818
1980	294 330 – 296 128
1981	296 611 – 299 632
1982	300 380 – 300 693

2-150

Year	Number range
1975	257 899 – 265 201
1976	266 783 – 271 312

2-155

Year	Number range
1976	272 595 – 272 812
1977	276 055 – 281 209
1978	282 280 – 286 929
1979	287 812 – 293 708
1980	296 160 – 296 244
1981	297 134 – 299 365
1982	300 259 – 300 429

2-180

Year	Number range
1977	281 993 – 282 087
1978	282 088 – 286 004
1979	289 447 – 292 891
1980	294 655 – 294 821
1981	296 571 – 299 002
1982	300 159 – 300 258

4-150

Year	Number range
1974	246 001 – 246 849
1975	246 871 – 262 243
1976	262 244 - 267 958
1977	275 051 – 275 405
1978	275 406 – 275 571

4-175

Year	Number range

Serial Numbers
White continued

Year	Number range
1979	292 187 – 292 334
1980	295 808 – 295 900
1981	297 293 – 299 848
1982	299 849 – 299 886

4-180

Year	Number range
1975	256 587 – 262 099
1976	262 524 – 268 111
1977	268 112 – 275 396
1978	275 450 – 275 502

4-210

Year	Number range
1978	275 572 – 275 943
1979	275 944 – 292 368
1980	295 391 – 296 205
1981	296 471 – 299 826
1982	300 694 - 300 778

2-30

Year	Number range
1979-84	001 418 – 101 461

2-32

Year	Number range
1985-86	000 007 – 000 315

2-35

Year	Number range
1979-84	05 822 – 004 001

2-45

Year	Number range
1979-81	000 001 – 000 548

2-55

Year	Number range
1982-87	000 097 – 000 807

2-62

Year	Number range
1979-81	000 001 – 001 143

2-65

Year	Number range
1982-87	000 099 – 001 202

2-75

Year	Number range
1982-87	000 177 – 000 955

2-88

Year	Number range
1982	301 457 – 301 717
1983	None built
1984	302 464 – 302 599
1985	None built
1986	400 001 – 400 599
1987	400 734 – 400 762

2-110

Year	Number range
1982	300 783 – 301 965
1983	301 998 – 302 158
1984	302 334 – 303 551
1985	303 552 – 303 614
1986	400 231 – 400 690
1987	400 764 – 401 005

2-135 Series 3

Year	Number range
1982	301 116 – 301 811
1983	302 159 – 302 233
1984	302 715 – 303 289
1985	None built
1986	400 167 – 400 733
1987	400 831 – 400 880

2-155 Series 3

Year	Number range
1982	300 928 – 301 921
1983	None built
1984	302 791 – 303 344
1985	None built
1986	400 107 – 400 718

2-180 Series 3

Year	Number range
1982	301 922 – 310 963
1983	301 966 – 301 997
1984	302 951 – 302 990
1985	None built
1986	400 082 – 400 230

4-225

Year	Number range
1983	302 234 – 302 273
1984	302 620 – 303 468
1985	None built
1986	400 344
1987	400 901 – 400 921

Serial Numbers

4-270

Year	Number range
1983	302 274 – 302 333
1984	302 655 – 303 423
1985	None built
1986	400 639 – 400 658
1987	400 922 – 400 941
1988	401 411 – 401 435

FB 16

Year	Number range
1986-89 (2WD)	02 314 – 02 811
1986-89 (FWD)	14 422 – 16 865

FB 21

Year	Number range
1986-89 (2WD)	00 595 – 01 181
1986-89 (FWD))	02 879 – 04 844

FB 31

Year	Number range
1986-89 (2WD)	00 126 – 01 981
1986-89 (FWD)	00 028 – 01 061

FB 37

Year	Number range
1986-89 (2WD)	00 083 – 00 315
1986-89 (FWD)	00 679 – 01 339

FB 43

Year	Number range
1986-89 (2WD)	00 060 – 00 156
1986-89 (FWD)	00 322 – 00 499

American 60

Year	Number range
1989	402 965 – 403 164
1990	404 299 – 404 454
1991	405 028 – 405 047

American 80

Year	Number range
1989	402 590 – 403 464
1990	404 266 – 404 541
1991	405 048 – 405 052

100

Year	Number range
1987	401 236 – 401 260
1988	401 361 – 401 970
1989	402 661 – 403 764

120

Year	Number range
1987	401 121 – 401 235
1988	401 296 – 402 520
1989	402 521 – 403 839

125

Year	Number range
1990	404 066 – 404 165
1991	404 601 – 404 969

140

Year	Number range
1987	401 151 – 401 200
1988	401 326 – 402 440
1989	402 736 – 404 064

145

Year	Number range
1991	404 671 – 404 923

160

Year	Number range
1987	401 096 – 401 120
1988	401 261 – 402 220
1989	403 640 – 404 024

170

Year	Number range
1990	404 228 – 404 265
1991	404 766 – 405 207

185

Year	Number range
1986 (FB 185)	400 659 – 400708
1987	400 881 – 401 095
1988	401 579 – 402 050
1989	402 761 – 404 014

195

Year	Number range
1990	404 166 – 404 227
1991	404 826 – 404 995

Tractor Test Results

The Nebraska Tractor Test's original self-contained mobile laboratory is seen below in a photograph taken in 1937. This vehicle rode on its own wheels and was loaded with what seemed at that time a wealth of testing equipment.

Over the years, this sleek-styled vehicle evaluated more than 1,500 tractors at the Nebraska Tractor Test's ⅓-mile-long track located on the University of Nebraska's east campus in Lincoln, Nebraska. Numerous upgrades were made to the mobile laboratory over the years; its latest design update was in 1963. The vehicle was retired in 2004 but can still be seen at the Nebraska Tractor Test museum.

For more information, contact the museum at 402/472-8389 or go to www.tractormuseum.unl.edu.

Tractor Test Results

Advance Rumely

Year tested	Name	Model	Company	Rated belt hp.	Drawbar hp.	Rated PTO hp.
1920	Rumely OilPull	E, 30-60	Advance-Rumely	75	49	
1920	Rumely OilPull	H, 16-30	Advance-Rumely	33	22	
1920	Rumely OilPull	K, 12-20	Advance-Rumely	25	15	
1920	Rumely OilPull	G, 30-40	Advance-Rumely	46	30	
1924	Rumely OilPull	S, 30-60	Advance Rumely Thresher Co.	70	40	
1925	Rumely OilPull	M, 20-35	Advance Rumely Thresher Co.	43	27	
1925	Rumely OilPull	L, 15-25	Advance Rumely Thresher Co.	30	19	
1925	Rumely OilPull	R, 25-45	Advance Rumely Thresher Co.	50	35	
1927	Rumely OilPull	W, 20-30	Advance Rumely	35	26	
1927	Rumely OilPull	X, 25-40	Advance Rumely	50	38	
1927	Rumely OilPull	Y, 30-50	Advance Rumely	63	47	
1928	Rumely	DO-All	Advance Rumely Thresher Co.	21	16	
1931	Rumely	6A	Advance Rumely Thresher Co.	48	33	

Allis-Chalmers

Year tested	Name	Model	Company	Rated belt hp.	Drawbar hp.	Rated PTO hp.
1920	Allis Chalmers	6-12	Allis-Chalmers	12	6	
1920	Allis Chalmers	18-30	Allis-Chalmers	33	20	
1921	Allis Chalmers	15-27	Allis-Chalmers	33	21	
1921	Allis Chalmers	12-20	Allis-Chalmers	33	21	
1921	Allis Chalmers	18-30	Allis-Chalmers	43	25	
1921	Allis Chalmers	22-38	Allis-Chalmers	43	25	
1928	Allis Chalmers	20-35 (A)	Allis-Chalmers	44	33	
1928	Allis Chalmers	20-35 (E)	Allis-Chalmers	44	33	
1929	Allis-Chalmers	U United	Allis-Chalmers	35	25	
1929	Monarch	35-30	Allis-Chalmers		40	
1930	Monarch	50	Allis-Chalmers	62	53	
1931	Allis-Chalmers	UC All Crop	Allis-Chalmers	36	24	
1931	Allis-Chalmers	EK	Allis-Chalmers	47	33	
1932	Allis-Chalmers	L	Allis-Chalmers	91	76	
1933	Allis-Chalmers	Special K	Allis-Chalmers	55	47	
1933	Allis-Chalmers	M	Allis-Chalmers	35	29	
1934	Allis-Chalmers	WC	Allis-Chalmers	21	Steel - 14 Rubber - 19	
1935	Allis-Chalmers	U	Allis-Chalmers	34	Steel - 24 Rubber - 28	
1935	Allis-Chalmers	UC	Allis-Chalmers	34	Steel - 23 Rubber -30	
1935	Allis-Chalmers	M	Allis-Chalmers	35	28	
1937	Allis-Chalmers	WK-0	Allis-Chalmers	59	50	
1937	Allis-Chalmers	S-0	Allis-Chalmers	74	62	
1937	Allis-Chalmers	L-0	Allis-Chalmers	91	76	
1938	Allis-Chalmers	B	Allis-Chalmers	15	13	
1938	Allis-Chalmers	WC	Allis-Chalmers	25	Steel - 24 Rubber - 28	
1938	Allis-Chalmers	WC	Allis-Chalmers	29	Steel - 22 Rubber - 24	
1939	Allis-Chalmers	RC	Allis-Chalmers	18	15	
1939	Allis-Chalmers	WK	Allis-Chalmers	62	53	

Tractor Test Results
Allis-Chalmers continued

Year tested	Name	Model	Company	Rated belt hp.	Drawbar hp.	Rated PTO hp.
1939	Allis-Chalmers	WS	Allis-Chalmers	84	68	
1939	Allis-Chalmers	L	Allis-Chalmers	108	91	
1940	Allis-Chalmers	HD-10W	Allis-Chalmers	98	82	
1940	Allis-Chalmers	HD-14	Allis-Chalmers	145	126	
1940	Allis-Chalmers	C	Allis-Chalmers	19	16	
1940	Allis-Chalmers	C	Allis-Chalmers	23	18	
1948	Allis-Chalmers	HD-5B	Allis-Chalmers	47	38	
1948	Allis-Chalmers	HD-19	Allis-Chalmers	129	110	
1948	Allis-Chalmers	G	Allis-Chalmers	10	9	
1948	Allis-Chalmers	WD	Allis-Chalmers	26	24	
1950	Allis-Chalmers	B	Allis-Chalmers	22	19	
1950	Allis-Chalmers	WD	Allis-Chalmers	34	30	
1950	Allis-Chalmers	CA	Allis-Chalmers	25	22	
1951	Allis-Chalmers	HD-9	Allis-Chalmers	79	67	
1951	Allis-Chalmers	HD-15	Allis-Chalmers	117	105	
1951	Allis-Chalmers	HD-20	Allis-Chalmers		116	
1953	Allis-Chalmers	WD-45	Allis-Chalmers	43	37	
1953	Allis-Chalmers	WD-45	Allis-Chalmers	33	29	
1953	Allis-Chalmers	WD-45	Allis-Chalmers	44	38	
1955	Allis-Chalmers	HD-21 AC	Allis-Chalmers		135	
1955	Allis-Chalmers	HD-16 AC	Allis-Chalmers		104	
1955	Allis-Chalmers	HD-16A	Allis-Chalmers	133	118	
1955	Allis-Chalmers	WD-45	Allis-Chalmers	43	39	
1956	Allis-Chalmers	HD-6B	Allis-Chalmers	60	49	
1956	Allis-Chalmers	HD-11B	Allis-Chalmers	89	73	
1957	Allis-Chalmers	D-14	Allis-Chalmers	34	30	
1957	Allis-Chalmers	D-17	Allis-Chalmers	52	48	
1957	Allis-Chalmers	D-17	Allis-Chalmers	51	46	
1958	Allis-Chalmers	D-17	Allis-Chalmers	50	46	
1958	Allis-Chalmers	D-14	Allis-Chalmers	31	28	
1958	Allis-Chalmers	HD-21A	Allis-Chalmers		147	
1959	Allis-Chalmers	D-12	Allis-Chalmers		24	28
1959	Allis-Chalmers	D-10	Allis-Chalmers		25	28
1961	Allis-Chalmers	H-3	Allis-Chalmers		27	32
1961	Allis-Chalmers	HD-3	Allis-Chalmers		27	32
1961	Allis-Chalmers	D-15	Allis-Chalmers		35	40
1961	Allis-Chalmers	D-15	Allis-Chalmers		33	36
1961	Allis-Chalmers	D-15	Allis-Chalmers		34	37
1962	Allis-Chalmers	D-19	Allis-Chalmers		63	71
1962	Allis-Chalmers	D-19	Allis-Chalmers		62	66
1962	Allis-Chalmers	D-10	Allis-Chalmers		29	33
1962	Allis-Chalmers	I-40	Allis-Chalmers		29	33
1962	Allis-Chalmers	D-12	Allis-Chalmers		29	33
1962	Allis-Chalmers	D-19	Allis-Chalmers		59	66
1963	Allis-Chalmers	D-15 Series II	Allis-Chalmers		39	46
1963	Allis-Chalmers	D-15 Series II	Allis-Chalmers		37	43
1963	Allis-Chalmers	D-21	Allis-Chalmers		95	103
1965	Allis-Chalmers	190	Allis-Chalmers		67	77
1965	Allis-Chalmers	200	Allis-Chalmers		84	93
1965	Allis-Chalmers	190XT	Allis-Chalmers		84	93
1965	Allis-Chalmers	190XT	Allis-Chalmers		77	85

Tractor Test Results

Year tested	Name	Model	Company	Rated belt hp.	Drawbar hp.	Rated PTO hp.
1965	Allis-Chalmers	190	Allis-Chalmers		65	75
1965	Allis-Chalmers	190XT	Allis-Chalmers		81	89
1967	Allis-Chalmers	180	Allis-Chalmers		56	64
1967	Allis-Chalmers	170	Allis-Chalmers		48	54
1967	Allis-Chalmers	170	Allis-Chalmers		48	54
1969	Allis-Chalmers	180	Allis-Chalmers		57	65
1969	Allis-Chalmers	220	Allis-Chalmers		121	135
1969	Allis-Chalmers	6040	Allis-Chalmers		38	40
1969	Allis-Chalmers	160	Allis-Chalmers		38	40

Aultman-Taylor

Year tested	Name	Model	Company	Rated belt hp.	Drawbar hp.	Rated PTO hp.
1920	Aultman-Taylor	30-60	Aultman-Taylor	75	55	
1920	Aultman-Taylor	30-60	Aultman-Taylor	80	58	
1920	Aultman-Taylor	15-30	Aultman-Taylor	34	21	
1920	Aultman-Taylor	22-45	Aultman-Taylor	46	28	

B.F. Avery

Year tested	Name	Model	Company	Rated belt hp.	Drawbar hp.	Rated PTO hp.
1920	Avery	7-14, C	Avery	14	8	
1920	Avery	8-15	Avery	15	8	
1920	Avery	12-20	Avery	24	17	
1920	Avery	14-28	Avery	31	21	
1920	Avery	25-50	Avery	56	32	
1920	Avery	40-80	Avery	69	49	
1920	Avery	5-10	Avery	11	6	
1920	Avery	18-36	Avery	44	27	
1921	Avery	12-25	Avery	25	13	
1921	Avery	8-16	Avery	16	9	
1923	Avery	15-25 Track Runner	Avery	29	20	
1923	Avery	20-35	Avery	37	22	

C.L. Best

Year tested	Name	Model	Company	Rated belt hp.	Drawbar hp.	Rated PTO hp.
1921	Best	60, 35-55	CL Best	56	50	
1921	Best	30, 18-30	CL Best	30	24	
1923	Best	60, 40-60	CL Best	65	56	
1923	Best	30, 20-30	Best Tractor Co.	32	25	
1924	Best	S30, 25-30	CL Best	37	33	
1924	Best	A60, 50-60	CL Best	72	61	

J.I. Case

Year tested	Name	Model	Company	Rated belt hp.	Drawbar hp.	Rated PTO hp.
1920	Case	10-18	J.I. Case	18	11	
1920	Case	15-27	J.I. Case	31	21	

Tractor Test Results
J.I. Case continued

Year tested	Name	Model	Company	Rated belt hp.	Drawbar hp.	Rated PTO hp.
1920	Case	22-40	J.I. Case	49	31	
1920	Case	10-20	J.I. Case	22	15	
1920	Case	20-40	J.I. Case	42	24	
1920	Wallis	15-25	J.I. Case	29		
1920	Wallis	15-25	J.I. Case	27	18	
1922	Case	12-20	J.I. Case	22	13	
1923	Case	40-72	J.I. Case	91	55	
1923	Case	12-20	J.I. Case	25	17	
1923	Wallis	OK,15-27	J.I. Case	28	18	
1924	Case	K, 18-32	J.I. Case	36	24	
1924	Case	25-45	J.I. Case	52	34	
1929	Case	L, 26-40	J.I. Case	44	30	
1929	Case	C, 17-27	J.I. Case	29	21	
1929	Case	CC	J.I. Case	28	22	
1936	Case	RC	J.I. Case	19	14	
1938	Case	R	J.I. Case	20	18	
1938	Case	L	J.I. Case	47	Steel - 31 Rubber - 40	
1940	Case	DC	J.I. Case	37	33	
1940	Case	VC	J.I. Case	24	18	
1940	Case	D	J.I. Case	35	30	
1941	Case	SC	J.I. Case	22	19	
1949	Case	VAC	J.I. Case	17	15	
1949	Case	VAC	J.I. Case	21	19	
1952	Case	LA	J.I. Case	58	51	
1952	Case	LA	J.I. Case	48	44	
1952	Case	LA	J.I. Case	59	51	
1953	Case	SC	J.I. Case	31	27	
1953	Case	SC	J.I. Case	24	23	
1953	Case	500	J.I. Case	63	56	
1955	Case	401	J.I. Case	49	43	
1955	Case	411	J.I. Case	53	44	
1957	Case	311	J.I. Case	33	29	
1957	Case	301	J.I. Case	30	28	
1957	Case	300B	J.I. Case	30	28	
1958	Case	511B	J.I. Case	45	39	
1958	Case	711B	J.I. Case	52	46	
1958	Case	811B	J.I. Case	53	49	
1958	Case	811B (TC)	J.I. Case	53	43	
1958	Case	801B	J.I. Case	54	51	
1958	Case	801B (TC)	J.I. Case	54	46	
1959	Case	611B	J.I. Case		37	44
1959	Case	611B (TC)	J.I. Case		32	44
1959	Case	211B	J.I. Case		26	30
1959	Case	411B	J.I. Case		31	37
1959	Case	411B (TC)	J.I. Case		27	37
1959	Case	701B	J.I. Case		47	51
1959	Case	711B	J.I. Case		48	53
1959	Case	811B (TC)	J.I. Case		46	55
1959	Case	811B	J.I. Case		50	55
1959	Case	940	J.I. Case		62	71

Tractor Test Results

Year tested	Name	Model	Company	Rated belt hp.	Drawbar hp.	Rated PTO hp.
1959	Case	910B	J.I. Case		62	71
1959	Case	310C	J.I. Case		27	33
1959	Case	1010	J.I. Case		55	
1959	Case	810	J.I. Case		40	
1959	Case	610	J.I. Case		33	38
1960	Case	831C (TC)	J.I. Case		52	63
1960	Case	831C	J.I. Case		58	63
1960	Case	831	J.I. Case		55	63
1960	Case	841	J.I. Case		53	64
1960	Case	940	J.I. Case		71	79
1960	Case	841	J.I. Case		54	63
1960	Case	930	J.I. Case		71	80
1960	Case	731C (TC)	J.I. Case		44	56
1960	Case	731C	J.I. Case		50	56
1960	Case	731	J.I. Case		48	56
1960	Case	541C (TC)	J.I. Case		33	41
1960	Case	541C	J.I. Case		37	41
1960	Case	640C (TC)	J.I. Case		39	49
1960	Case	640C	J.I. Case		43	49
1960	Case	541	J.I. Case		33	39
1960	Case	570	J.I. Case		33	39
1960	Case	531	J.I. Case		37	41
1960	Case	570	J.I. Case		37	41
1960	Case	640	J.I. Case		42	50
1960	Case	470	J.I. Case		29	33
1960	Case	441	J.I. Case		29	33
1960	Case	841C (TC)	J.I. Case		54	65
1960	Case	841C	J.I. Case		59	65
1960	Case	741C (TC)	J.I. Case		45	57
1960	Case	741C	J.I. Case		50	57
1960	Case	741	J.I. Case		51	57
1960	Case	841C	J.I. Case		57	65
1960	Case	841C (TC)	J.I. Case		53	65
1960	Case	741C	J.I. Case		52	57
1960	Case	741C (TC)	J.I. Case		46	57
1960	Case	741	J.I. Case		52	57
1961	Case	470	J.I. Case		31	34
1961	Case	431	J.I. Case		31	34
1961	Case	531C	J.I. Case		36	40
1961	Case	531C (TC)	J.I. Case		32	40
1961	Case	630C (TC)	J.I. Case		40	48
1961	Case	630C	J.I. Case		45	48
1961	Case	630	J.I. Case		40	48
1961	Case	310E	J.I. Case		31	36
1964	Case	1200	J.I. Case		106	119
1965	Case	831 CK	J.I. Case		58	64
1965	Case	931 GP	J.I. Case		76	85
1965	Case	841 CK	J.I. Case		59	65
1965	Case	941 GP	J.I. Case		72	86
1965	Case	841 CK	J.I. Case		59	64
1965	Case	941 GP	J.I. Case		75	85
1966	Case	1030	J.I. Case		92	101

Tractor Test Results
J.I. Case continued

Year tested	Name	Model	Company	Rated belt hp.	Drawbar hp.	Rated PTO hp.
1969	Case	1470	J.I. Case		132	144
1969	Case	870	J.I. Case		64	70
1969	Case	870	J.I. Case		63	70
1969	Case	770	J.I. Case		48	53
1969	Case	770	J.I. Case		50	56

Caterpillar

Year tested	Name	Model	Company	Rated belt hp.	Drawbar hp.	Rated PTO hp.
1928	Caterpillar	20, 20-25	Caterpillar	29	26	
1929	Caterpillar	15, 15-20	Caterpillar	24	21	
1929	Caterpillar	10, 10-15	Caterpillar	18	14	
1932	Caterpillar	25	Caterpillar	32	27	
1932	Caterpillar	50	Caterpillar	56	49	
1932	Caterpillar	20	Caterpillar	27	22	
1932	Caterpillar	35	Caterpillar	43	36	
1932	Caterpillar	15	Caterpillar	20	16	
1932	Caterpillar	60	Caterpillar	77	65	
1932	Caterpillar	65	Caterpillar	78	67	
1933	Caterpillar	70	Caterpillar	82	72	
1933	Caterpillar	50	Caterpillar	61	52	
1933	Caterpillar	35	Caterpillar	44	39	
1933	Caterpillar	75	Caterpillar	92	80	
1934	Caterpillar	R-5	Caterpillar	58	49	
1934	Caterpillar	R-2	Caterpillar	32	27	
1934	Caterpillar	22	Caterpillar	29	23	
1934	Caterpillar	R-3	Caterpillar	41	34	
1934	Caterpillar	22	Caterpillar	30	25	
1935	Caterpillar	50	Caterpillar	61	52	
1935	Caterpillar	50	Caterpillar	71	64	
1935	Caterpillar	40	Caterpillar	56	50	
1935	Caterpillar	RD-6	Caterpillar	48	42	
1935	Caterpillar	40	Caterpillar	48	42	
1935	Caterpillar	40	Caterpillar	48	41	
1935	Caterpillar	40	Caterpillar	56	48	
1936	Caterpillar	RD-7	Caterpillar	68	60	
1936	Caterpillar	RD-7	Caterpillar	77	65	
1936	Caterpillar	RD-7	Caterpillar	95	78	
1936	Caterpillar	RD-8	Caterpillar	103	91	
1936	Caterpillar	D-8	Caterpillar	103	91	
1936	Caterpillar	RD-8	Caterpillar	118	103	
1936	Caterpillar	30	Caterpillar	36	30	
1936	Caterpillar	R-4	Caterpillar	36	30	
1936	Caterpillar	R-4	Caterpillar	39	35	
1936	Caterpillar	30	Caterpillar	39	35	
1936	Caterpillar	D-4	Caterpillar	40	35	
1936	Caterpillar	RD-4	Caterpillar	40	35	
1938	Caterpillar	D-8	Caterpillar	109	96	
1939	Caterpillar	R-2	Caterpillar	29	23	
1939	Caterpillar	R-2	Caterpillar	28	22	
1939	Caterpillar	D-2	Caterpillar	29	25	

Tractor Test Results

Year tested	Name	Model	Company	Rated belt hp.	Drawbar hp.	Rated PTO hp.
1940	Caterpillar	D-8	Caterpillar	127	110	
1940	Caterpillar	D-7	Caterpillar	89	78	
1941	Caterpillar	D-6	Caterpillar	78	63	
1949	Caterpillar	D-8	Caterpillar		123	
1949	Caterpillar	D-6	Caterpillar	76	61	
1949	Caterpillar	D-4	Caterpillar	51	41	
1949	Caterpillar	D-2	Caterpillar	36	30	
1955	Caterpillar	D-2	Caterpillar	41	36	
1955	Caterpillar	D-4	Caterpillar	58	48	
1955	Caterpillar	D-6	Caterpillar	92	73	
1956	Caterpillar	D-7	Caterpillar	121	103	
1956	Caterpillar	D-8	Caterpillar		157	
1956	Caterpillar	D-9	Caterpillar		252	
1959	Caterpillar	D-7	Caterpillar		115	
1959	Caterpillar	D-8	Caterpillar		177	
1960	Caterpillar	D-4	Caterpillar	56	50	
1960	Caterpillar	D-6	Caterpillar		75	

Cletrac

Year tested	Name	Model	Company	Rated belt hp.	Drawbar hp.	Rated PTO hp.
1920	Cletrac	W, 12-20	Cleveland Tractor	24	15	
1922	Cletrac	F, 9-16	Cleveland Tractor	19	13	
1926	Cletrac	K, 15-25	Cleveland Tractor	20	23	
1926	Cletrac	K, 15-25	Cleveland Tractor	30	24	
1926	Cletrac	20-27	Cleveland Tractor	30	24	
1926	Cletrac	A, 30-45	Cleveland Tractor	48	38	
1928	Cletrac	40, 40-55	Cleveland Tractor	63	55	
1928	Cletrac	55	Cleveland Tractor	63	55	
1930	Cletrac	80-60	Cleveland Tractor	90	83	
1931	Cletrac	35	Cleveland Tractor	45	40	
1931	Cletrac	40-30	Cleveland Tractor	45	40	
1931	Cletrac	15	Cleveland Tractor	25	18	
1932	Cletrac	25	Cleveland Tractor	33	26	
1932	Cletrac	15	Cleveland Tractor	26	22	
1935	Cletrac	DD	Cleveland Tractor	63	57	
1935	Cletrac	40	Cleveland Tractor	63	57	
1936	Cletrac	CG	Cleveland Tractor	51	40	
1936	Cletrac	BG	Cleveland Tractor	39	28	
1936	Cletrac	EG	Cleveland Tractor	28	20	
1936	Cletrac	FG	Cleveland Tractor	104	87	
1936	Cletrac	FD	Cleveland Tractor	100	86	
1937	Cletrac	BD	Cleveland Tractor	41	34	
1937	Cletrac	CG	Cleveland Tractor	52	45	
1939	Cletrac	G General	Cleveland Tractor	19	14	
1939	Cletrac	HG	Cleveland Tractor	19	14	
1939	Cletrac	BD	Cleveland Tractor	45	36	
1939	Cletrac	FD	Cleveland Tractor	107	91	

Tractor Test Results

Cockshutt

Year tested	Name	Model	Company	Rated belt hp.	Drawbar hp.	Rated PTO hp.
1947	Co-op	E3	Cockshutt	31	27	
1947	Cockshutt	30	Cockshutt	31	27	
1947	Farmcrest	30	Cockshutt	31	27	
1950	Co-op	E4	Cockshutt	43	37	
1950	Cockshutt	40	Cockshutt	43	37	
1952	Co-op	E2	Cockshutt	28	25	
1952	Cockshutt	20	Cockshutt	28	25	
1952	Co-op	E5	Cockshutt	51	46	
1952	Cockshutt	50	Cockshutt	51	46	
1952	Co-op	E5	Cockshutt	55	51	
1952	Cockshutt	50	Cockshutt	55	51	
1958	Cockshutt	550	Cockshutt	38	34	
1958	Cockshutt	560	Cockshutt	48	45	
1958	Cockshutt	570	Cockshutt	60	52	

David Brown

Year tested	Name	Model	Company	Rated belt hp.	Drawbar hp.	Rated PTO hp.
1954	David Brown	25	David Brown		34	26
1959	David Brown	950	David Brown		36	39
1960	David Brown	850	David Brown		31	33
1960	David Brown	850	David Brown		28	32
1965	David Brown	880	David Brown		35	40
1965	David Brown	990	David Brown		45	51
1966	David Brown	770	David Brown		28	32
1966	David Brown	880	David Brown		36	42
1968	David Brown	1200	David Brown		56	65
1968	David Brown	990	David Brown		44	52
1969	David Brown	3800	David Brown		34	39
1969	David Brown	4600	David Brown		39	46

Ford

Year tested	Name	Model	Company	Rated belt hp.	Drawbar hp.	Rated PTO hp.
1920	Fordson	Fordson	Ford Motor Co.	19	9	
1926	Fordson	Fordson	Ford Motor Co.	22	12	
1930	Fordson	F	Ford Motor Co.	23	13	
1930	Fordson	F	Ford Motor Co.	29	18	
1937	Fordson	All-Around	Ford Motor Co.	22	Steel - 14 Rubber - 18	
1938	Fordson	All-Around	Ford Motor Co.	28	Steel - 19 Rubber - 21	
1940	Ford-Ferguson	9N	Ferguson-Sherman	23	17	
1940	Ford-Ferguson	2N	Ferguson-Sherman	23	17	
1947	Ford	8N	Ford Motor Co.	21	17	
1948	Ford	8N	Ford Motor Co.	25	21	
1950	Ford	8N	Ford Motor Co.	26	21	
1950	Ford	8NAN	Ford Motor Co.	21	18	

Tractor Test Results

Year tested	Name	Model	Company	Rated belt hp.	Drawbar hp.	Rated PTO hp.
1953	Ford	Jubilee	Ford Motor Co.	31	25	
1953	Ford	NAA	Ford Motor Co.	31	25	
1953	New Fordson	Major	Ford Motor Co.	38	35	
1953	New Fordson	Major	Ford Motor Co.	33	30	
1955	Ford	640	Ford Motor Co.	31	28	
1955	Ford	660	Ford Motor Co.	34	28	
1955	Ford	860	Ford Motor Co.	45	39	
1955	Ford	960	Ford Motor Co.	46	38	
1955	Ford	740	Ford Motor Co.	31	28	
1957	Ford	850	Ford Motor Co.	39	35	
1957	Ford	541	Ford Motor Co.	28	27	
1957	Ford	640	Ford Motor Co.	28	27	
1958	Ford	841	Ford Motor Co.	41	40	
1958	Ford	851	Ford Motor Co.	48	41	
1958	Ford	841	Ford Motor Co.	44	41	
1958	Ford	641	Ford Motor Co.	33	29	
1958	Ford	541	Ford Motor Co.	33	29	
1958	Ford	661	Ford Motor Co.	35	29	
1958	Ford	651	Ford Motor Co.	35	29	
1958	Ford	861	Ford Motor Co.	48	41	
1958	Ford	841	Ford Motor Co.	39	36	
1958	Ford	861	Ford Motor Co.	41	36	
1958	Ford	851	Ford Motor Co.	41	36	
1959	Fordson	Dexta	Ford Motor Co.		27	31
1959	Fordson	Power Major	Ford Motor Co.	47	43	
1959	Ford	741	Ford Motor Co.	31	29	
1959	Ford	641	Ford Motor Co.	31	29	
1959	Ford	631	Ford Motor Co.	31	29	
1959	Ford	621	Ford Motor Co	31	29	
1959	Ford	541	Ford Motor Co	31	29	
1959	Ford	881	Ford Motor Co.		37	46
1959	Ford	811	Ford Motor Co.		37	46
1959	Ford	681	Ford Motor Co.		27	34
1959	Ford	671	Ford Motor Co		27	34
1959	Ford	771	Ford Motor Co.		27	34
1959	Ford	611	Ford Motor Co.		27	34
1959	Ford	881	Ford Motor Co.		34	43
1959	Ford	871	Ford Motor Co.		34	43
1959	Ford	971	Ford Motor Co.		34	43
1959	Ford	811	Ford Motor Co.		34	43
1959	Ford	981	Ford Motor Co		34	43
1959	Ford	671	Ford Motor Co		26	32
1959	Ford	611	Ford Motor Co.		26	32
1959	Ford	771	Ford Motor Co.		26	32
1959	Ford	681	Ford Motor Co.		26	32
1959	Ford	881	Ford Motor Co.		33	41
1959	Ford	981	Ford Motor Co		33	41
1959	Ford	971	Ford Motor Co.		33	41
1959	Ford	811	Ford Motor Co.		33	41
1959	Ford	871	Ford Motor Co.		33	41
1959	Ford	771	Ford Motor Co		25	31

Tractor Test Results
Ford continued

Year tested	Name	Model	Company	Rated belt hp.	Drawbar hp.	Rated PTO hp.
1959	Ford	671	Ford Motor Co.		25	31
1959	Ford	681	Ford Motor Co.		25	31
1959	Ford	611	Ford Motor Co.		25	31
1961	Ford	6000	Ford Motor Co.		61	66
1961	Ford	6000	Ford Motor Co.		61	66
1963	Ford	6000	Ford Motor Co.		53	62
1963	Ford	6000	Ford Motor Co.		54	62
1963	Ford	2000 Super Dexta	Ford Motor Co.		33	38
1963	Ford	5000	Ford Motor Co.		41	47
1965	Ford	6000	Ford Motor Co.		58	66
1965	Ford	6000	Ford Motor Co.		63	66
1965	Ford	5000	Ford Motor Co.		49	55
1965	Ford	5000	Ford Motor Co.		45	54
1965	Ford	3000	Ford Motor Co.		36	39
1965	Ford	3000	Ford Motor Co.		33	38
1965	Ford	3000	Ford Motor Co.		36	39
1965	Ford	2000	Ford Motor Co.		27	30
1965	Ford	3000	Ford Motor Co.		30	36
1965	Ford	3000	Ford Motor Co.		33	37
1965	Ford	3000	Ford Motor Co.		34	39
1965	Ford	4000	Ford Motor Co.		39	45
1965	Ford	4000	Ford Motor Co.		41	46
1965	Ford	4000	Ford Motor Co.		39	45
1965	Ford	4000	Ford Motor Co.		42	46
1965	Ford	2000	Ford Motor Co.		27	30
1966	Ford	5000	Ford Motor Co.		53	60
1966	Ford	5000	Ford Motor Co.		50	58
1967	Ford	2000	Ford Motor Co.		28	32
1967	Ford	2000	Ford Motor Co.		28	31
1968	Ford	8000	Ford Motor Co.		91	105
1968	Ford	4000	Ford Motor Co.		46	51
1968	Ford	5000	Ford Motor Co.		60	67
1968	Ford	5000	Ford Motor Co.		54	65
1968	Ford	6600	Ford Motor Co.		57	67
1968	Ford	5000	Ford Motor Co.		57	67
1968	Ford	4000	Ford Motor Co.		42	50
1968	Ford	5000	Ford Motor Co.		58	66
1968	Ford	4600 (8×2)	Ford Motor Co.		44	52
1968	Ford	4610 (8×2)	Ford Motor Co.		44	52
1968	Ford	4000 (8×2)	Ford Motor Co.		44	52
1969	Ford	8000	Ford Motor Co.		94	105
1969	Ford	9000	Ford Motor Co.		117	131

Hart-Parr

Year tested	Name	Model	Company	Rated belt hp.	Drawbar hp.	Rated PTO hp.
1920	Hart-Parr	30, 15-30	Hart-Parr	31	19	
1921	Hart-Parr	20, 11-20	Hart-Parr	23	14	
1923	Hart-Parr	40, 20-40	Hart-Parr	46	28	
1924	Hart-Parr	E, 16-30	Hart-Parr	37	24	
1924	Hart-Parr	E, 12-24	Hart-Parr	26	17	

Tractor Test Results

Year tested	Name	Model	Company	Rated belt hp.	Drawbar hp.	Rated PTO hp.
1926	Hart-Parr	18-36	Hart-Parr	42	32	
1926	Hart-Parr	H,12-24	Hart-Parr	31	21	
1927	Hart-Parr	28-50	Hart-Parr	64	46	

Huber

Year tested	Name	Model	Company	Rated belt hp.	Drawbar hp.	Rated PTO hp.
1920	Huber	12-25	Huber	25	16	
1921	Huber	Super 4	Huber	39	26	
1926	Huber	18-36	Huber	43	* 30	
1926	Huber	20-40	Huber	50	40	
1927	Huber	40-62	Huber	69	50	
1929	Huber	20-36	Huber	42	29	
1937	Huber	LC	Huber	43	Steel - 30	
					Rubber - 31	
1937	Huber	B	Huber	27	Steel - 20	
					Rubber - 22	
1949	Global	B	Huber	42	37	
1949	Huber	B	Huber	42	37	

IHC

Year tested	Name	Model	Company	Rated belt hp.	Drawbar hp.	Rated PTO hp.
1920	International	10-20 Titan	International	28	15	
1920	International	15-30	International	36	25	
1920	International	8-16	International	18	11	
1922	McCormick-Deering	15-30	International	32	20	
1923	McCormick-Deering	10-20	International	21	15	
1925	McCormick-Deering	Farmall	International	20	13	
1926	McCormick-Deering	15-30	International	34	26	
1927	McCormick-Deering	10-20	International	24	19	
1929	McCormick-Deering	22-36	International	40	30	
1929	McCormick-Deering	15-30	International	40	30	
1931	McCormick-Deering	20 Industrial	International	29	23	
1931	McCormick-Deering	F-30 Farmall	International	32	24	
1931	McCormick-Deering	T-20 Tractractor	International	26	23	
1932	McCormick-Deering	W-30	International	33	24	
1932	McCormick-Deering	T-40 Tractractor	International	46	42	
1933	McCormick-Deering	F-12 Farmall	International	16	12	
1933	McCormick-Deering	F-12 Farmall	International	14	11	
1934	McCormick-Deering	F-20 Farmall	International	23	15	
1934	McCormick-Deering	W-12	International	15	12	
1934	McCormick-Deering	T-40	International	48	43	
1934	McCormick-Deering	W-12	International	17	13	
1935	McCormick-Deering	WD-40	International	48	37	

Tractor Test Results
IHC continued

Year tested	Name	Model	Company	Rated belt hp.	Drawbar hp.	Rated PTO hp.
1936	McCormick-Deering	F-20 Farmall	International	27	20	
1936	McCormick-Deering	WK-40	International	45	31	
1936	McCormick-Deering	WK-40	International	49	35	
1936	McCormick-Deering	F-20 Farmall	International	27	20	
1937	McCormick-Deering	TD-35 Tractractor	International	42	37	
1937	McCormick-Deering	T-35 Tractractor	International	42	35	
1937	McCormick-Deering	T-35 Tractractor	International	44	36	
1937	McCormick-Deering	T-40 Tractractor	International	49	42	
1937	McCormick-Deering	T-40 Tractractor	International	51	44	
1938	McCormick-Deering	F-14 Farmall	International	17	14	
1938	McCormick-Deering	TD-40 Tractractor	International	53	48	
1939	International	TD-18 Tractractor	International	80	72	
1939	McCormick-Deering	M Farmall	International	34	30	
1939	McCormick-Deering	M Farmall	International	36	33	
1939	McCormick-Deering	A Farmall	International	18	16	
1939	McCormick-Deering	A Super Farmall	International	18	16	
1939	McCormick-Deering	A Farmall	International	16	15	
1939	McCormick-Deering	A Super Farmall	International	16	15	
1939	McCormick-Deering	B Farmall	International	18	16	
1939	McCormick-Deering	B Super Farmall	International	18	16	
1939	McCormick-Deering	B Farmall	International	16	14	
1939	McCormick-Deering	B Super Farmall	International	16	14	
1939	McCormick-Deering	H Farmall	International	26	24	
1939	McCormick-Deering	H Farmall	International	23	21	
1940	McCormick-Deering	W-4	International	23	21	
1940	International	TD-14 Tractractor	International	61	51	
1940	International	TD-Tractractor	International	43	37	
1940	International	TD-6 Tractractor	International	34	28	
1940	International	T-6 Tractractor	International	36	30	
1940	International	T-6 Tractractor	International	34	29	
1940	McCormick-Deering	W-4	International	26	23	
1940	McCormick-Deering	W-6	International	34	31	
1940	McCormick-Deering	W-6	International	36	32	
1940	McCormick-Deering	WD-6	International	34	31	
1941	McCormick-Deering	MD Farmall	International	35	31	
1941	McCormick-Deering	W-9	International	49	44	
1941	McCormick-Deering	WD-9	International	46	42	
1941	McCormick-Deering	W-9	International	46	42	
1941	International	T-9 Tractractor	International	46	40	
1947	McCormick-Deering	Ccub Farmall	International	9	8	
1948	McCormick-Deering	C Farmall	International	21	18	

Tractor Test Results

Year tested	Name	Model	Company	Rated belt hp.	Drawbar hp.	Rated PTO hp.
1950	McCormick-Deering	WD-9	International	51	46	
1950	International	TD-14A	International	71	62	
1950	International	TD-18A	International	97	85	
1950	International	TD-24	International		142	
1951	McCormick-Farmall	C Super	International	23	20	
1951	McCormick-Deering	WD-6	International	37	33	
1951	McCormick-Farmall	MD	International	38	34	
1951	International	TD-9	International	46	39	
1951	International	TD-6	International	38	31	
1952	McCormick-Farmall	M Super	International	46	41	
1952	McCormick-Deering	W-6 Super	International	46	41	
1952	McCormick-Farmall	MD Super	International	46	42	
1952	McCormick-Deering	WD-6 Super	International	46	41	
1952	McCormick-Farmall	M Super	International	47	44	
1952	McCormick-Deering	W-6 Super	International	47	42	
1953	McCormick-Deering	W-4 Super	International	33	29	
1953	McCormick-Farmall	H Super	International	33	30	
1954	McCormick-Farmall	600	International	65	57	
1954	McCormick-Farmall	650	International	65	57	
1954	McCormick-Deering	WD-9 Super	International	65	57	
1954	International	TD-24	International		154	
1954	McCormick-Farmall	400	International	50	45	
1954	International	W-450	International	52	45	
1954	International	W-400	International	51	45	
1955	McCormick-Farmall	400	International	46	42	
1955	International	W-400	International	46	43	
1955	International	W-450	International	46	43	
1955	McCormick-Farmall	200	International	24	20	
1955	McCormick-Farmall	100	International	20	17	
1955	McCormick-Farmall	300	International	38	33	
1955	IHC	300 Utility	International	41	37	
1956	McCormick-Farmall	Cub	International	10	9	
1956	McCormick-Farmall	400	International	52	48	
1956	International	W-400	International	50	47	
1956	International	W-450	International	50	47	
1956	McCormick-Farmall	300	International	38	35	
1956	IHC	300 Utility	International	42	38	
1956	International	TD-14	International	91	76	
1956	International	TD-9	International	62	52	
1956	International	TD-6	International	48	39	
1956	International	TD-18	International	121	100	

Tractor Test Results

IHC continued

Year tested	Name	Model	Company	Rated belt hp.	Drawbar hp.	Rated PTO hp.
1956	International	T-6	International	48	39	
1957	McCormick-Farmall	450	International	48	45	
1957	McCormick-Farmall	350	International	38	36	
1957	IHC	350 Utility	International	42	40	
1957	McCormick-Farmall	350	International	40	37	
1957	McCormick-Farmall	450	International	55	51	
1957	IHC	350 Utility	International	43	39	
1957	McCormick-Farmall	230	International	28	25	
1957	McCormick-Farmall	130	International	22	19	
1957	IHC	650	International	62	56	
1957	IHC	350 Utility	International	45	41	
1957	McCormick-Farmall	450	International	54	49	
1957	IHC	650	International	63	58	
1957	McCormick-Farmall	350	International	41	38	
1957	International	TD-18	International		106	
1957	International	TD-24	International		168	
1957	IHC	330 Utility	International	34	31	
1957	IHC	340 Utility	International	34	31	
1958	IHC	340 Farmall	International	34	31	
1958	McCormick-Farmall	340	International	34	31	
1958	IHC	140	International	23	31	
1958	McCormick-Farmall	140	International	23	21	
1958	McCormick-Farmall	240	International	30	27	
1958	IHC	240 Utility	International	30	28	
1958	International	240	International	30	28	
1958	IHC	560	International	59	54	
1958	McCormick-Farmall	560	International	59	54	
1958	IHC	460	International	49	45	
1958	IHC	560	International	63	58	
1958	McCormick-Farmall	560	International	63	58	
1958	IHC	460	International	50	46	
1958	IHC	460 Utility	International	50	45	
1958	IHC	460 Utility	International	49	45	
1958	IHC	560	International	60	55	
1958	McCormick-Farmall	560	International	60	55	
1958	IHC	460	International	49	46	
1958	IHC	460 Utility	International	48	45	
1959	IHC	660	International		71	78
1959	IHC	660	International		71	81
1959	IHC	660	International		71	80
1959	International	T-340	International		31	36
1960	IHC	354	International		30	32

Tractor Test Results

Year tested	Name	Model	Company	Rated belt hp.	Drawbar hp.	Rated PTO hp.
1960	International	TD-15	International		84	
1960	International	TD-9	International		57	69
1960	International	TD-25	International		188	
1960	International	T-5	International		29	36
1960	International	T-4	International		27	32
1960	International	TD-5	International		30	35
1960	IHC	340 Farmall	International		36	38
1960	McCormick-Farmall	340	International		36	38
1960	International	TD-340	International		32	39
1960	IHC	B275	International UK		30	32
1961	International	TD-20	International		115	
1962	IHC	B414	International UK		32	35
1962	IHC	4300	Frank G. Hough Co.		214	
1962	IHC	B414	International UK		31	36
1962	IHC	504	International		40	45
1962	IHC	404 I	International		33	36
1962	IHC	504	International		41	46
1962	IHC	504	International		41	44
1962	IHC	I2606	International		46	53
1962	IHC	606	International		46	53
1962	IHC	606	International		48	54
1962	IHC	I2606	International		48	54
1963	IHC	706 Farmall	International		67	72
1963	International	706	International		67	72
1963	IHC	806 Farmall	International		86	94
1963	International	806	International		86	94
1963	IHC	706 Farmall	International		66	73
1963	International	706	International		66	73
1963	IHC	856	International		81	93
1963	IHC	806 Farmall	International		81	93
1963	International	806	International		81	93
1963	IHC	706 Farmall	International		67	73
1963	International	706	International		67	73
1963	IHC	856	International		84	93
1963	IHC	806 Farmall	International		84	93
1963	International	806	International		84	93
1965	IHC	424	International		32	36
1965	IHC	656 Farmall	International		54	63
1965	IHC	1206	International		99	112
1965	IHC	444	International		32	36
1965	IHC	424	International		32	36
1965	IHC	656 Farmall	International		54	61

Tractor Test Results
IHC continued

Year tested	Name	Model	Company	Rated belt hp.	Drawbar hp.	Rated PTO hp.
1965	IHC	4100	International	116		
1966	IHC	706 Farmall	International	69	76	
1966	IHC	756	International	69	76	
1966	IHC	706	International	68	76	
1966	IHC	756	International	68	76	
1966	IHC	756	International	68	76	
1966	IHC	706	International	68	76	
1966	International	500	International Canada	31	37	
1966	International	500	International Canada	31	36	
1967	IHC	656 Farmall	International	52	66	
1967	IHC	656 Farmall	International	51	65	
1968	IHC	856	International	90	100	
1968	IHC	1256 Farmall	International	105	116	
1968	IHC	544	International	45	52	
1968	IHC	544	International	45	52	
1968	IHC	444	International	33	38	
1969	IHC	2544	International	42	53	
1969	IHC	544	International	42	53	
1969	IHC	544	International	42	55	
1969	IHC	2544	International	42	55	

John Deere

Year tested	Name	Model	Company	Rated belt hp.	Drawbar hp.	Rated PTO hp.
1924	John Deere	D, 15-27	Waterloo Gas Engine Co.	30	22	
1927	John Deere	D, 15-27	John Deere	36	28	
1928	John Deere	10-20, GP	John Deere	24	17	
1931	John Deere	GP	John Deere	25	18	
1934	John Deere	A, GP	John Deere	24	18	
1935	John Deere	B, GP	John Deere	16	11	
1935	John Deere	D	John Deere	41	30	
1937	John Deere	G	John Deere	35	27	
1938	John Deere	B	John Deere	18	Steel - 14 Rubber - 16	
1938	John Deere	H	John Deere	14	12	
1938	John Deere	L	John Deere	10	9	
1939	John Deere	A	John Deere	29	26	
1940	John Deere	D	John Deere	42	38	
1941	John Deere	LA	John Deere	14	13	
1941	John Deere	2WD	John Deere	30	26	
1947	John Deere	B	John Deere	27	24	
1947	John Deere	B	John Deere	23	21	
1947	John Deere	G	John Deere	38	34	
1947	John Deere	A	John Deere	38	34	
1947	John Deere	330	John Deere	20	18	

Tractor Test Results

Year tested	Name	Model	Company	Rated belt hp.	Drawbar hp.	Rated PTO hp.
1947	John Deere	320	John Deere	20	18	
1947	John Deere	M	John Deere	20	18	
1949	John Deere	R	John Deere	48	43	
1949	John Deere	MT	John Deere	20	18	
1949	John Deere	AR	John Deere	37	34	
1950	John Deere	MC	John Deere	21	17	
1952	John Deere	60	John Deere	40	35	
1952	John Deere	50	John Deere	30	27	
1953	John Deere	60	John Deere	32	29	
1953	John Deere	70	John Deere	48	42	
1953	John Deere	40	John Deere	24	21	
1953	John Deere	40S	John Deere	23	21	
1953	John Deere	40C	John Deere	24	19	
1953	John Deere	70	John Deere	43	39	
1953	John Deere	50	John Deere	24	22	
1953	John Deere	60	John Deere	41	37	
1953	John Deere	70	John Deere	50	45	
1954	John Deere	70	John Deere	50	45	
1955	John Deere	50	John Deere	31	28	
1955	John Deere	40S	John Deere	20	18	
1955	John Deere	80	John Deere	65	60	
1956	John Deere	520	John Deere	37	33	
1956	John Deere	530	John Deere	37	33	
1956	John Deere	620	John Deere	49	44	
1956	John Deere	630	John Deere	49	44	
1956	John Deere	520	John Deere	25	23	
1956	John Deere	530	John Deere	25	23	
1956	John Deere	730	John Deere	57	52	
1956	John Deere	720	John Deere	57	52	
1956	John Deere	730	John Deere	56	51	
1956	John Deere	720	John Deere	56	51	
1956	John Deere	530	John Deere	37	33	
1956	John Deere	520	John Deere	37	33	
1956	John Deere	620	John Deere	46	42	
1956	John Deere	630	John Deere	46	42	
1956	John Deere	430W	John Deere	28	26	
1956	John Deere	420W	John Deere	28	26	
1956	John Deere	430S	John Deere	22	21	
1956	John Deere	420S	John Deere	22	21	
1956	John Deere	420C	John Deere	28	23	
1956	John Deere	430C	John Deere	28	23	
1956	John Deere	620	John Deere	34	32	
1956	John Deere	720	John Deere	57	53	
1956	John Deere	730	John Deere	57	53	
1956	John Deere	720	John Deere	44	40	
1956	John Deere	730	John Deere	44	40	
1957	John Deere	820	John Deere	72	67	
1957	John Deere	830	John Deere	72	67	
1959	John Deere	435	John Deere		28	32
1959	John Deere	440ID	John Deere		27	32
1959	John Deere	440I	John Deere		26	31

Tractor Test Results
John Deere continued

Year tested	Name	Model	Company	Rated belt hp.	Drawbar hp.	Rated PTO hp.
1959	John Deere	440ICD	John Deere	26	32	
1959	John Deere	440IC	John Deere	24	31	
1960	John Deere	4010	John Deere	72	80	
1960	John Deere	4010	John Deere	72	80	
1960	John Deere	4010	John Deere	73	84	
1960	John Deere	3010	John Deere	54	59	
1960	John Deere	3010	John Deere	52	55	
1960	John Deere	3010	John Deere	50	55	
1961	John Deere	1010C	John Deere	29	36	
1961	John Deere	2010RU	John Deere	41	46	
1961	John Deere	2010RU	John Deere	40	46	
1961	John Deere	1010C	John Deere	28	35	
1961	John Deere	1010RU	John Deere	30	36	
1961	John Deere	1010RU	John Deere	30	35	
1962	John Deere	5010	John Deere	108	121	
1962	John Deere	2010C	John Deere	39	47	
1962	John Deere	2010C	John Deere	39	47	
1963	John Deere	500	John Deere	57	65	
1963	John Deere	3020	John Deere	57	65	
1963	John Deere	4020	John Deere	78	91	
1963	John Deere	600	John Deere	78	91	
1963	John Deere	4020	John Deere	75	88	
1963	John Deere	600	John Deere	75	88	
1963	John Deere	500	John Deere	55	64	
1963	John Deere	3020	John Deere	55	64	
1963	John Deere	3020	John Deere	56	64	
1963	John Deere	4020	John Deere	79	90	
1965	John Deere	2510	John Deere	42	49	
1965	John Deere	2510	John Deere	46	53	
1965	John Deere	2510	John Deere	45	50	
1965	John Deere	2510	John Deere	48	54	
1965	John Deere	4020	John Deere	85	94	
1966	John Deere	4020	John Deere	86	94	
1966	John Deere	1020	John Deere	32	38	
1966	John Deere	2020	John Deere	44	53	
1966	John Deere	1020	John Deere	32	38	
1966	John Deere	2020	John Deere	47	54	
1966	John Deere	4020	John Deere	84	95	
1966	John Deere	3020	John Deere	63	70	
1966	John Deere	3020	John Deere	64	70	
1966	John Deere	5020	John Deere	116	133	
1968	John Deere	1520	John Deere	38	46	
1968	John Deere	2520	John Deere	54	61	
1968	John Deere	2520	John Deere	47	56	
1969	John Deere	2520	John Deere	48	56	
1969	John Deere	2520	John Deere	55	60	
1969	John Deere	1520	John Deere	38	47	
1969	John Deere	3020	John Deere	58	67	
1969	John Deere	3020	John Deere	63	71	
1969	John Deere	4020	John Deere	84	95	
1969	John Deere	4020	John Deere	87	96	

Tractor Test Results

Year tested	Name	Model	Company	Rated belt hp.	Drawbar hp.	Rated PTO hp.
1969	John Deere	4520	John Deere		111	122
1969	John Deere	4520	John Deere		112	123
1969	John Deere	4000	John Deere		85	96
1969	John Deere	4020	John Deere		84	95
1969	John Deere	5020	John Deere		126	141

Massey-Harris-Ferguson

Year tested	Name	Model	Company	Rated belt hp.	Drawbar hp.	Rated PTO hp.
1929	Massey-Harris	12	Massey-Harris	24	18	
1929	Wallis	12-20	Massey-Harris	24	18	
1930	Massey-Harris	Gen. Purp.	Massey-Harris	24	19	
1931	Massey-Harris	Gen. Purp.	Massey-Harris	22	16	
1933	Massey-Harris	3-4 Plow	Massey-Harris	44	33	
1936	Massey-Harris	Challenger	Massey-Harris	28	20	
1936	Massey-Harris	Pacemaker	Massey-Harris	29	19	
1937	Massey-Harris	Challenger Twin Power	Massey-Harris	34	Steel - 25 Rubber - 29	
1937	Massey-Harris	Pacemaker Twin Power	Massey-Harris	36	Steel - 26 Rubber - 30	
1938	Massey-Harris	101S	Massey-Harris	35	Steel - 23 Rubber - 30	
1938	Massey-Harris	101R	Massey-Harris	35	Steel - 25 Rubber - 31	
1939	Massey-Harris	101R Junior	Massey-Harris	26	20	
1940	Massey-Harris	20	Massey-Harris	30	24	
1940	Massey-Harris	101R Junior	Massey-Harris	30	24	
1941	Massey-Harris	81R	Massey-Harris	27	20	
1941	Massey-Harris	20	Massey-Harris	27	20	
1941	Massey-Harris	Colt	Massey-Harris	27	20	
1941	Massey-Harris	101R	Massey-Harris	46	34	
1947	Massey-Harris	44RT	Massey-Harris	45	39	
1948	Massey-Harris	55	Massey-Harris	58	52	
1948	Massey-Harris	16	Massey-Harris	11	10	
1948	Massey-Harris	Pony	Massey-Harris	11	10	
1948	Massey-Harris	22RT	Massey-Harris	31	22	
1949	Massey-Harris	30RT	Massey-Harris	34	26	
1949	Massey-Harris	44	Massey-Harris	41	37	
1949	Massey-Harris	44K	Massey-Harris	38	35	
1949	Massey-Harris	55K	Massey-Harris	52	47	
1950	Massey-Harris	555	Massey-Harris	59	52	
1950	Massey-Harris	55	Massey-Harris	59	52	
1951	Massey-Harris	55	Massey-Harris	66	57	
1951	Massey-Harris	555	Massey-Harris	66	57	
1953	Massey-Harris	33RT	Massey-Harris	39	35	
1953	Massey-Harris	44 Special	Massey-Harris	48	43	
1954	Massey-Harris	Pacer, 16	Massey-Harris-Ferguson	18	17	
1955	Ferguson	TO-35	Massey-Harris-Ferguson	33	30	
1955	Massey Ferguson	35	Massey-Harris-Ferguson	33	30	
1956	Massey-Harris	444	Massey-Harris-Ferguson	48	44	
1956	Massey-Harris	333	Massey-Harris-Ferguson	37	33	
1956	Massey-Harris	50	Massey-Harris-Ferguson	32	30	

Tractor Test Results
Massey-Harris-Ferguson continued

Year tested	Name	Model	Company	Rated belt hp.	Drawbar hp.	Rated PTO hp.
1956	Ferguson	40	Massey-Harris-Ferguson	32	30	
1956	Massey Ferguson	50	Massey-Harris-Ferguson	32	30	
1956	Massey-Harris	444	Massey-Harris-Ferguson	49	45	
1956	Massey-Harris	333	Massey-Harris-Ferguson	41	37	
1958	Massey Ferguson	65	Massey-Ferguson	42	38	
1958	Massey Ferguson	50	Massey-Ferguson	32	29	
1958	Massey Ferguson	65	Massey-Ferguson	46	41	
1959	Massey Ferguson	TO-35	Massey-Ferguson		30	32
1959	Massey Ferguson	85	Massey-Ferguson		52	61
1959	Massey Ferguson	90	Massey-Ferguson		52	61
1959	Massey Ferguson	85	Massey-Ferguson		56	62
1959	Massey Ferguson	90	Massey-Ferguson		56	62
1960	Massey Ferguson	35	Massey-Ferguson		33	37
1960	Massey Ferguson	65	Massey-Ferguson		42	48
1960	Massey Ferguson	88	Massey-Ferguson		55	63
1960	Massey Ferguson	85	Massey-Ferguson		55	63
1961	Massey Ferguson	90 Super	Motec Industries		68	78
1961	Massey Ferguson	50	Massey-Ferguson		34	38
1961	Massey Ferguson	65	Massey-Ferguson		45	50
1962	Massey Ferguson	90 Super	Massey-Ferguson		61	68
1963	Massey Ferguson	25	Massey-Ferguson		20	24
1965	Massey Ferguson	135	Massey-Ferguson		33	37
1965	Massey Ferguson	165	Massey-Ferguson		46	52
1965	Massey Ferguson	175	Massey-Ferguson		55	63
1965	Massey Ferguson	165	Massey-Ferguson		41	46
1965	Massey Ferguson	135	Massey-Ferguson		30	35
1965	Massey Ferguson	180	Massey-Ferguson		54	63
1965	Massey Ferguson	150	Massey-Ferguson		33	37
1966	Massey Ferguson	1130	Massey-Ferguson		109	120
1966	Massey Ferguson	130	Massey-Ferguson		23	26
1967	Massey Ferguson	1100	Massey-Ferguson		76	90
1969	Massey Ferguson	165	Massey-Ferguson		45	51
1969	Massey Ferguson	30 IND	Massey-Ferguson		45	51
1969	Massey Ferguson	150	Massey-Ferguson		32	37
1969	Massey Ferguson	135	Massey-Ferguson		32	37
1969	Massey Ferguson	175	Massey-Ferguson		51	61
1969	Massey Ferguson	180	Massey-Ferguson		54	62
1969	Massey Ferguson	1080	Massey-Ferguson		73	81

Minneapolis-Moline

Year tested	Name	Model	Company	Rated belt hp.	Drawbar hp.	Rated PTO hp.
1920	Minneapolis-Moline	D 9-18 Universal	Moline Plow	27	17	
1920	M-M Twin City	40-65	Minneapolis Steel & Machinery Co.	65	49	
1920	M-M Twin City	20-35	Minneapolis Steel & Machinery Co.	46	34	
1920	M-M Twin City	12-20	Minneapolis Steel & Machinery Co.	27	18	
1926	M-M Twin City	TY, 17-28	Minneapolis Steel & Machinery Co.	30	22	

Tractor Test Results

Year tested	Name	Model	Company	Rated belt hp.	Drawbar hp.	Rated PTO hp.
1926	M-M Twin City	AT, 27-44	Minneapolis Steel & Machinery Co.	49	34	
1926	M-M Twin City	FT, 21-32	Minneapolis Steel & Machinery Co.	35	31	
1928	M-M Twin City	FT, 21-32	Minneapolis Steel & Machinery Co.	39	30	
1930	M-M Twin City	KT, 11-20	Minneapolis-Moline	25	18	
1931	Minneapolis-Moline	MT Universal	Minneapolis-Moline	26	18	
1935	M-M Twin City	JT	Minneapolis-Moline	24	17	
1935	M-M Twin City	KTA	Minneapolis-Moline	33	24	
1935	M-M Twin City	MTA	Minneapolis-Moline	33	24	
1936	M-M Twin City	KTA	Minneapolis-Moline	41	Steel - 30 Rubber - 25	
1936	M-M Twin City	FTA	Minneapolis-Moline	44	35	
1937	M-M Twin City	ZT	Minneapolis-Moline	26	20	
1938	M-M Twin City	UTS	Minneapolis-Moline	42	39	
1938	M-M Twin City	UTS	Minneapolis-Moline	36	33	
1939	M-M Twin City	GTA	Minneapolis-Moline	55	47	
1939	M-M Twin City	GT	Minneapolis-Moline	55	47	
1939	M-M Twin City	GTB	Minneapolis-Moline	55	47	
1939	M-M Twin City	UTU	Minneapolis-Moline	42	36	
1940	M-M Twin City	RTU	Minneapolis-Moline	23	20	
1940	M-M Twin City	ZTU	Minneapolis-Moline	31	26	
1949	Minneapolis-Moline	U Standard	Minneapolis-Moline	46	41	
1950	Minneapolis-Moline	G	Minneapolis-Moline	58	49	
1950	Minneapolis-Moline	Z	Minneapolis-Moline	36	32	
1951	Minneapolis-Moline	R	Minneapolis-Moline	27	23	
1951	Avery	BF	Minneapolis-Moline	27	24	
1951	Minneapolis-Moline	BF	Minneapolis-Moline	27	24	
1954	Minneapolis-Moline	U	Minneapolis-Moline	37	33	
1954	Minneapolis-Moline	UB	Minneapolis-Moline	51	44	
1955	Minneapolis-Moline	GB	Minneapolis-Moline	70	62	
1955	Minneapolis-Moline	GB	Minneapolis-Moline	65	59	
1955	Minneapolis-Moline	GB	Minneapolis-Moline	62	55	
1956	Minneapolis-Moline	445 Universal	Minneapolis-Moline	41	38	
1956	Minneapolis-Moline	445 Utility	Minneapolis-Moline	41	38	
1957	Minneapolis-Moline	335	Minneapolis-Moline	33	29	
1958	Minneapolis-Moline	5 Star	Minneapolis-Moline	54	49	
1958	Minneapolis-Moline	5 Star	Minneapolis-Moline	54	49	
1960	Minneapolis-Moline	M602	Minneapolis-Moline		54	61
1960	Minneapolis-Moline	M5	Minneapolis-Moline		54	61
1960	Minneapolis-Moline	M5	Minneapolis-Moline		54	61
1960	Minneapolis-Moline	M602	Minneapolis-Moline		54	61
1960	Minneapolis-Moline	M5	Minneapolis-Moline		51	58
1960	Minneapolis-Moline	M602	Minneapolis-Moline		51	58
1961	Minneapolis-Moline	4 Star	Motec Industries		39	44
1961	Minneapolis-Moline	Jet Star 3	Motec Industries		39	44
1961	Minneapolis-Moline	Jet Star 3	Motec Industries		41	45

Tractor Test Results
Minneapolis-Moline continued

Year tested	Name	Model	Company	Rated belt hp.	Drawbar hp.	Rated PTO hp.
1961	Minneapolis-Moline	4 Star	Motec Industries		41	45
1961	Massey Ferguson 90 Super		Motec Industries		71	78
1961	Minneapolis-Moline	GVI	Motec Industries		71	78
1961	Minneapolis-Moline	GVI	Motec Industries		68	78
1963	Massey Ferguson	97	Minneapolis Moline		89	101
1963	Minneapolis-Moline	G706	Minneapolis-Moline		89	101
1963	Massey Ferguson	97	Minneapolis-Moline		90	101
1963	Minneapolis-Moline	G706	Minneapolis-Moline		90	101
1963	Massey Ferguson	97	Minneapolis-Moline		93	101
1963	Minneapolis-Moline	G705	Minneapolis-Moline		93	101
1963	Massey Ferguson	97	Minneapolis-Moline		92	101
1963	Minneapolis-Moline	G705	Minneapolis-Moline		92	101
1964	Minneapolis-Moline	U302	Minneapolis-Moline		49	55
1964	Minneapolis-Moline	U302	Minneapolis-Moline		49	55
1965	Minneapolis-Moline	M670 Super	Minneapolis-Moline		64	74
1965	Minneapolis-Moline	M670	Minneapolis-Moline		64	74
1965	Minneapolis-Moline	M670 Super	Minneapolis-Moline		64	73
1965	Minneapolis-Moline	M670	Minneapoli`s-Moline		64	73
1965	Minneapolis-Moline	M670	Minneapolis-Moline		64	71
1965	Minneapolis-Moline	M670 Super	Minneapolis-Moline		64	71
1966	Minneapolis-Moline	G1050	Minneapolis-Moline		110	102
1966	Minneapolis-Moline	G1000	Minneapolis-Moline		110	102
1966	Oliver	2055	Minneapolis-Moline		102	110
1966	Minneapolis-Moline	G1050	Minneapolis-Moline		99	110
1966	Minneapolis-Moline	G1000	Minneapolis-Moline		99	110
1966	Oliver	2055	Minneapolis-Moline		99	110
1968	Minneapolis-Moline	G1000 Vista	Minneapolis-Moline		100	111
1968	Minneapolis-Moline	G1000 Vista	Minneapolis-Moline		99	110
1968	Minneapolis-Moline	G950	Minneapolis-Moline		88	97
1968	Oliver	1865	Minneapolis-Moline		88	97
1968	Minneapolis-Moline	G950	Minneapolis-Moline		86	97
1968	Minneapolis-Moline	G900	Minneapolis-Moline		86	97
1968	Oliver	1865	Minneapolis-Moline		86	97
1968	Minneapolis-Moline	G900	Minneapolis-Moline		89	97
1968	Minneapolis-Moline	G950	Minneapolis-Moline		89	97

Oliver

Year tested	Name	Model	Company	Rated belt hp.	Drawbar hp.	Rated PTO hp.
1930	Hart-Parr	18-28	Oliver	30	23	
1930	Hart-Parr	28-44	Oliver	49	34	
1930	Oliver	90	Oliver	49	34	
1936	Oliver-Hart-Parr	70 HC Row Crop	Oliver	28	21	
1936	Oliver-Hart-Parr	70 KD Row Crop	Oliver	27	20	

Tractor Test Results

Year tested	Name	Model	Company	Rated belt hp.	Drawbar hp.	Rated PTO hp.
1937	Oliver-Hart-Parr	70 HC Standard	Oliver	27	19	
1937	Oliver-Hart-Parr	70 KD Standard	Oliver	26	19	
1938	Oliver	80 KD Row Crop	Oliver	38	29	
1938	Oliver	80 KD Standard	Oliver	39	28	
1940	Oliver	70 HC Row Crop	Oliver	31	28	
1941	Oliver	80 HC Standard	Oliver	41	35	
1941	Oliver	60 HC Row Crop	Oliver	18	16	
1947	Oliver	88 HC Row Crop	Oliver	41	36	
1947	Oliver	88 HC Standard	Oliver	43	37	
1948	Oliver	77 HC Row Crop	Oliver	33	28	
1948	Oliver	77 HC Standard	Oliver	33	28	
1949	Oliver	66 HC Row Crop	Oliver	24	21	
1949	Oliver	66 HC Standard	Oliver	24	21	
1949	Oliver	77 HC Row Crop	Oliver	37	32	
1949	Oliver	HG	Oliver	25	21	
1949	Oliver	DG	Oliver	69	59	
1949	Oliver	DD	Oliver	73	58	
1949	Oliver	OC-3	Oliver	25	21	
1950	Oliver	88 Row Crop	Oliver	43	38	
1950	Oliver	99	Oliver	62	52	
1951	Oliver	66 Row Crop	Oliver	25	22	
1952	Oliver	77 Row Crop	Oliver	36	32	
1952	Oliver	OC-18	Oliver		128	
1953	Oliver	OC-6	Oliver		31	
1953	Oliver	OC-6	Oliver		33	
1954	Oliver	55 HC Super	Oliver	34	29	
1954	Oliver	88 HC Super	Oliver	55	47	
1954	Oliver	55 Super	Oliver	33	27	
1954	Oliver	88 Super	Oliver	54	49	
1955	Oliver	66 HC Super	Oliver	33	27	
1955	Oliver	77 HC Super	Oliver	43	37	

Tractor Test Results
Oliver continued

Year tested	Name	Model	Company	Rated belt hp.	Drawbar hp.	Rated PTO hp.
1955	Oliver	77 Super	Oliver	44	38	
1955	Oliver	66 Super	Oliver	33	27	
1955	Oliver	OC-12	Oliver	57	50	
1955	Oliver	OC-12	Oliver	56	50	
1955	Oliver	99 GM Super	Oliver	78	73	
1955	Oliver	99 Super	Oliver	62	58	
1957	Oliver	OC-15	Oliver	101	91	
1958	Oliver	880	Oliver	61	54	
1958	Oliver	770	Oliver	50	42	
1958	Oliver	770	Oliver	48	44	
1958	Oliver	880	Oliver	59	52	
1958	Oliver	OC-4	Oliver	26	24	
1958	Oliver	OC-4	Oliver	25	23	
1958	Oliver	950	Oliver	67	61	
1958	Massey Ferguson	98	Oliver	84	77	
1958	Oliver	990 GM	Oliver	84	77	
1959	Oliver	550	Oliver		35	41
1959	Oliver	550	Oliver		35	39
1960	Oliver	1800	Oliver		63	73
1960	Oliver	1800	Oliver		62	70
1960	Oliver	1900	Oliver		82	89
1962	Cockshutt	1900	Oliver		86	98
1962	Oliver	1900	Oliver		86	98
1962	Cockshutt	1800	Oliver		67	77
1962	Oliver	1800	Oliver		67	77
1962	Oliver	1800	Oliver		62	76
1962	Oliver	1800	Oliver		64	76
1963	Oliver	1800	Oliver		70	80
1963	Oliver	1600	Oliver		48	57
1963	Oliver	1600	Oliver		48	56
1963	Oliver	1800	Oliver		69	80
1963	Oliver	1900	Oliver		90	100
1964	Oliver	1850	Oliver		84	92
1964	Oliver	1850	Oliver		81	92
1964	Oliver	1950	Oliver		99	105
1964	Oliver	1950	Oliver		97	105
1964	Oliver	1650	Oliver		57	66
1964	Oliver	1650	Oliver		59	66
1964	Oliver	1855	Oliver		77	92

Tractor Test Results

Year tested	Name	Model	Company	Rated belt hp.	Drawbar hp.	Rated PTO hp.
1964	Oliver	1850	Oliver		77	92
1964	Oliver	1850	Oliver		86	92
1966	Minneapolis-Moline	G550	Oliver		45	53
1966	Oliver	1550	Oliver		45	53
1966	Oliver	1555	Oliver		45	53
1966	Minneapolis-Moline	G550	Oliver		46	53
1966	Oliver	1555	Oliver		46	53
1966	Oliver	1550	Oliver		46	53
1967	Oliver	1750	Oliver		68	80
1967	Oliver	1750	Oliver		68	80
1967	Oliver	1950T	Oliver		93	105
1968	Oliver	1950T	Oliver		91	105
1968	Oliver	2150	Oliver		114	131
1968	Oliver	2050	Oliver		104	118

Sears-David Bradley

Year tested	Name	Model	Company	Rated belt hp.	Drawbar hp.	Rated PTO hp.
1951	David Bradley	Garden Tractor	David Bradley	1	1	
1953	David Bradley	Super Power	David Bradley	2	1	
1959	David Bradley	Suburban	David Bradley	3	2	
1959	David Bradley	575 Super	David Bradley	3	3	
1959	David Bradley	300 Super	David Bradley	1	1	
1959	David Bradley	Handiman	David Bradley	2	0.6	

Lesser-Knowns

A

Year tested	Name	Model	Company	Rated belt hp.	Drawbar hp.	Rated PTO hp.
1920	Allwork	14-28	Electric Wheel Co.	28	19	

B

Year tested	Name	Model	Company	Rated belt hp.	Drawbar hp.	Rated PTO hp.
1931	Bradley	Gen. Purp.	Bradley	24	20	
1920	Bates	Steel Mule "D"	Bates	24	20	
1920	Bates	Steel Mule F, 15-22	Bates	29	23	
1931	Bates	35 Steel Mule	Foote Bros.	52	43	
1931	Bates	45 Steel Mule	Foote Bros.	66	54	
1923	Bear	B, 25-35	Bear	49	35	
1924	Bear	B, 25-35	Bear	55	44	

Tractor Test Results
Lesser-Knowns continued

Year tested	Name	Model	Company	Rated belt hp.	Drawbar hp.	Rated PTO hp.
1920	Beeman	G	Beeman	2	1	
1952	Bolens	12BB	Bolens	1	1	
1929	Baker	43-67	A.D. Baker Co.	75	55	

C

Year tested	Name	Model	Company	Rated belt hp.	Drawbar hp.	Rated PTO hp.
1948	Centaur	KV-48	LeRoi	24	20	
1950	Choremaster	B	The Lodge & Shipley Co.	1	.7	
1920	Coleman	B, 16-30	Coleman	30	15	
1927	Continental Cultor	32	Continental Cultor		8	
1936	Co-op	#3	Duplex Machinery	42	37	
1936	Co-op	#2	Duplex Machinery	33	29	
1949	Corbitt	G-50	Corbitt	34	30	

D

Year tested	Name	Model	Company	Rated belt hp.	Drawbar hp.	Rated PTO hp.
1950	Dodge	T137 Power Wagon	Chrysler	42	40	

E

Year tested	Name	Model	Company	Rated belt hp.	Drawbar hp.	Rated PTO hp.
1921	Eagle	H, 16-30	Eagle	31	19	
1921	Eagle	F, 12-22 HP	Eagle	23	14	
1930	Eagle	6A	Eagle	40	29	
1952	Economy	Special	Engineering Products	6	5	
1957	Eimco	105	Eimco		72	
1946	Ellinwood	3000-1 Bear Cat	Ellinwood	2	Steel - 1 Rubber - 1	
1947	Ellinwood	Tiger Cat	Ellinwood	4	2	
1920	Emerson-Brantingham	12-20	Emerson-Brantingham	25	17	

F

Year tested	Name	Model	Company	Rated belt hp.	Drawbar hp.	Rated PTO hp.
1949	Farmaster	FD-33	Farmaster	23	21	
1949	Farmaster	FG-33	Farmaster	28	25	
1953	Federal	DF	Intercontinental	33	31	
1948	Ferguson	TE-20	Harry-Ferguson	25	20	
1948	Ferguson	TO-20	Harry-Ferguson	25	20	
1951	Ferguson	TO-30	Harry-Ferguson	29	24	
1959	Fiat	411-R	Fiat		33	36
1959	Fiat	411-C	Fiat		29	37
1920	Flour City	18-35	Kinnard & Sons	35	19	
1920	Flour City	40-70	Kinnard & Sons	72	52	
1929	Four Drive	E, 15-25	Four Drive Tractor	28	18	

G

Year tested	Name	Model	Company	Rated belt hp.	Drawbar hp.	Rated PTO hp.
1949	Gibson	H	Gibson	24	22	
1949	Gibson	I	Gibson	41	36	

Tractor Test Results

Year tested	Name	Model	Company	Rated belt hp.	Drawbar hp.	Rated PTO hp.
1938	Graham-Bradley	503.103	Graham-Paige Motors Corp.	30	25	
1920	Gray	18-36	Gray	32	19	

H

Year tested	Name	Model	Company	Rated belt hp.	Drawbar hp.	Rated PTO hp.
1952	Harris	PH-53	Harris		50	
1954	Harris	FDW-C	Harris		43	
1954	Harris	FDW-C	Harris		42	
1920	Heider	C, 12-20	Rock Island	24	13	
1920	Heider	D, 9-16	Rock Island	19	11	
1925	Heider	15-27	Rock Island	30	21	
1920	Holt	T-11, 25-40	Holt	35	33	
1920	Holt	T-16, 40-60	Holt	57	51	
1922	Holt	T-35, 15-25	Holt	25	18	

I

Year tested	Name	Model	Company	Rated belt hp.	Drawbar hp.	Rated PTO hp.
1920	Indiana	5-10	Indiana Silo & Tractor	11	5	
1948	Intercontinental	C-26	Intercontinental	29	25	
1949	Intercontinental	D-26 / DE	Intercontinental	28	26	
1953	Intercontinental	DF	Intercontinental	33	31	

K

Year tested	Name	Model	Company	Rated belt hp.	Drawbar hp.	Rated PTO hp.
1962	Kramer	KL400	Kramer-Werke		28	32
1965	Kubota	RV	Kubota	8	6	

L

Year tested	Name	Model	Company	Rated belt hp.	Drawbar hp.	Rated PTO hp.
1920	La Crosse	G, 12-24	La Crosse	24	17	
1960	Land-Rover	88	Rover Co., Ltd.		29	30
1920	Lauson	15-30	John Lauson Mfg.	32	26	
1921	Lauson	12-25	John Lauson Mfg.	37	20	
1927	Lauson	16-32	John Lauson Mfg.	36	28	
1927	Lauson	20-40	John Lauson Mfg.	41	32	
1928	Lauson	20-35,S12	John Lauson Mfg.	40	29	
1949	Long	A	Long	31	28	

M

Year tested	Name	Model	Company	Rated belt hp.	Drawbar hp.	Rated PTO hp.
1949	Mercer	30BD	Farmaster	23	21	
1949	Mercer	30CK	Farmaster	28	25	
1920	Minneapolis	12-25	Minneapolis Threshing Co	26	16	
1920	Minneapolis	22-44	Minneapolis Threshing Co	46	33	
1920	Minneapolis	35-70	Minneapolis Threshing Co	74	52	
1921	Minneapolis	A ,17-30	Minneapolis Threshing Co	31	19	
1925	Minneapolis	B, 17-30	Minneapolis Threshing Co	34	23	

Tractor Test Results
Lesser-Knowns continued

Year tested	Name	Model	Company	Rated belt hp.	Drawbar hp.	Rated PTO hp.
1929	Minneapolis	27-42	Minneapolis Threshing Co.	48	34	
1929	Minneapolis	39-57	Minneapolis Threshing Co.	64	47	
1920	Monarch	18-30	Monarch	31	21	
1924	Monarch	D, 6-60	Monarch	70	53	
1925	Monarch	C, 25-35	Monarch	43	37	
1927	Monarch	F, 10 ton	Monarch		78	
1927	Monarch	6 ton	Monarch		50	

N

Year tested	Name	Model	Company	Rated belt hp.	Drawbar hp.	Rated PTO hp.
1958	Nuffield	3 Universal	Morris Motors	34	31	
1955	Nuffield	DM-4 Universal	Morris Motors	45	41	
1955	Nuffield	PM-4 Universal	Morris Motors	38	34	
1962	Nuffield	460	Morris Motors		50	54
1965	Nuffield	10/42	British Motor Co.		33	39
1965	Nuffield	10/60	British Motor Co.		47 55	

P

Year tested	Name	Model	Company	Rated belt hp.	Drawbar hp.	Rated PTO hp.
1920	Parrett	K, 15-30	Parrett Tractor Co.	31	20	
1935	Planet	Planet Jr. Garden Tractor	S.L. Allen & Co.	2	1	
1959	Porsche	L108 Junior	Porsche		9	11
1959	Porsche	L318 Super	Porsche		33	37
1920	Port Huron	12-25	Port Huron	28	20	

R

Year tested	Name	Model	Company	Rated belt hp.	Drawbar hp.	Rated PTO hp.
1927	Rock Island	18-35	Rock Island	36	30	
1929	Rock Island	G2, 15-25	Rock Island	29	22	
1929	Rock Island	G2, 18-30	Rock Island	35	25	
1922	Rogers	Rogers	Rogers Tractor & Trailer Co.	63	38	
1921	Russell	30-60	Russell & Co.	66	43	
1923	Russell	C,15-30	Russell & Co.	33	24	
1923	Russell	C, 20-40	Russell & Co.	43	32	

S

Year tested	Name	Model	Company	Rated belt hp.	Drawbar hp.	Rated PTO hp.
1920	Samson	M	Samson	19	11	
1927	Shaw	T-25	Shaw	1	.7	
1927	Shaw	T-45	Shaw	1	.7	
1921	Shawnee	Power Patrol		7		

Tractor Test Results

Year tested	Name	Model	Company	Rated belt hp.	Drawbar hp.	Rated PTO hp.
1925	Shawnee	30	Shaw-Enochs		13	
1936	Silver King	3 wheel R-66	Fate-Root-Heath	19	Steel - 13 Rubber - 16	
41953	Someca	DA 50	Someca	38	33	
1957	Someca	45 Som	Someca	40	37	
1920	Square Turn	18-35	Square Turn	36		
1920	Square Turn	18-35	Square Turn	32	23	

T

Year tested	Name	Model	Company	Rated belt hp.	Drawbar hp.	Rated PTO hp.
1952	Terratrac	GT-30	American Tractor	30	25	
1920	Toro	6-10	Toro	13	9	
1920	Townsend	15-30	Townsend	29	17	

U - W

Year tested	Name	Model	Company	Rated belt hp.	Drawbar hp.	Rated PTO hp.
1920	Uncle Sam	20-30	U.S. Tractor	32	22	
1957	Unimog	30	Daimler-Benz	28	25	
1961	Ursus	C-325	Zaklady Mechaniczne		22	24
1968	Ursus	C-335	Zaklady Mechaniczne		26	29
1968	Ursus	C-350	Zaklady Mechaniczne		37	43
1949	USTRAC	10-A	U.S. Tractor Corp.	21	15	
1958	Volvo	T425	A-B Bolinder-Munktell	23	22	
1958	Volvo	T55	A-B Bolinder-Munktell	63	59	
1957	Wagner	TR- 9	Wagner		87	
1959	Wagner	TR-14A	Wagner		155	
1964	Wagner	WA-4	FWD Wagner		97	
1927	Wallis	20-30	Wallis	35	27	
1920	Waterloo Boy	N, 12-25	Waterloo Gas Engine Co.	25	15	
1921	Wetmore	12-25	H.A. Wetmore	27	16	
1949	Willys Overland Motors	Jeep CJ-3A Universal	28	25		
1953	Willys	Farm Jeep	Willys Motors, Inc.	35	27	
1920	Wisconsin	E, 16-30	Wisconsin	31	22	

Z

Year tested	Name	Model	Company	Rated belt hp.	Drawbar hp.	Rated PTO hp.
1960	Zetor	50 Super	Zavody Jana Svermy		44	49
1962	Zetor	3011	ZKL-BRNO		27	33
1964	Zetor	2011	ZKL-BRNO		19	21
1964	Zetor	4011	ZKL-BRNO		41	48
1967	Zetor	5511	ZKL-BRNO		45	50

Cross-reference

A

Adapto-Tractor	Geneva Tractor Company
Advance	Henry, Millard & Henry Company
Ace	Horace Keane Aeroplanes, Inc.
Akron	Wellman-Seaver-Morgan Company
Albaugh-Dover	Square Turn Tractor Company
Allaround	Ford Motor Company
Allen A	Community Manufacturing Company
All-In-One	Pacific Power Implement Company
	Stroud Motor Manufacturing Company
All Purpose	Advance-Rumely Thresher Company
AllSteel	Bates Tractor Company
Allwork	Electric Wheel Company
American	Russell and Company
Americo	American Tractor & Foundry Company
Angleworm	Badley Tractor Company
Atlas	Lyons-Atlas

Auto Cat	A.J. Erstead
Autoplow	Hackney Manufacturing Company
Autopow	Autopower Company
Autotractor	Eason-Wysong Company
Auto Tractor	Hackney Manufacturing Company
Auto Tractor-Truck	Lombard Auto Tractor-Truck Corporation
Auto Tiller	World Harvester Corporation

B

Baby Creeper	Bullock Tractor Company
Baby Savidge	Backus Tractor Company
Ball Tread	Yuba Manufacturing Company
Bates	Foote Bros. Gear & Machine Company
Bear	J.I. Case Plow Works
	Mead-Morrison Manufacturing Company

Cross-reference

Bear Cat	Ellinwood Industries
Belt-Rail	Beltrail Tractor Company
Beaver	Goold, Shapley and Muir Company
Big Boss	Russell and Company
Big Bull	Bull Tractor Company
Big Chief	Waterloo Foundry Company
Big 4	Gas Traction Company
	Emerson-Brantingham Implement Co.
Big Mo	Minneapolis-Moline Company
Biltwell	Velie Motors Corporation
Blue	John Blue Company
Blue J	Dart Truck and Tractor Company
Bower City	Townsend Manufacturing Company
Boyer Four	Huron Tractor Company
Bulldog	Avery Company
	Heinrich Lanz

C

Cameco	Cane Machinery & Engineering Company
Canadian	Alberta Foundry & Machinery Company
Canadian Special	Gray Tractor Company
Capital	C.A. Dissinger and Brothers Company
The Cat	Four-Drive Tractor Company
Caterpillar	Holt Tractor Company
	Caterpillar Company
Centaur	Le Roi Company
	Central Tractor Company
Centiped	Phoenix Manufacturing Company
Chain Tread	Buckeye Manufacturing Company
Challenge	Challenge Tractor Company
Challenger	Massey-Harris Company
Champion	Chamberlain Industries Ltd.
Clarkat	Clark Tructractor Company
Cletrac	Cleveland Tractor Company
Cliff	John Minor Kroyer
Colby Plow Boy	Jones Manufacturing Company
Coleman	Winslow Manufacturing Company
	Coleman Tractor Corporation
Colt	Massey-Harris Company
Co-op	Farmers' Union Central Exchange Inc.
Corn Belt	Southern Corn Belts Tractor Company
Crawlerize	Hadfield-Penfield Steel Company
Creeping Grip	Bullock Tractor Company
Crop Maker	Hart-Parr Company
Cropmaster	David Brown

Cub	Wallis Tractor Company
	J.I. Case Plow Works
	International Harvester Company
Cub Jr.	J.I. Case Plow Works
Cultiplow	H.L. Hurst Manufacturing Company
Cultrac	Intercontinental Manufacturing Company
Cultractor	Baines Engineering Company
Cultitractor	United Tractors Corporation
Cultor	Continental Cultor Company
Custom	Custom Tractor Manufacturing Company
	Harry A. Lowther Company

D

Dakota	G.W. Elliot & Company
	Pope Manufacturing Company
Deering	International Harvester Company
Dixieland	Dixieland Motor Truck Company
DoAll	Advance-Rumely Thresher Company
Dreadnought Guide	Charles S. Whitworth
Du-All	Shaw Manufacturing Company
Duat	Clark Tructractor Company

E

Eagle	Eagle Manufacturing Company
Earthmaster	Aerco Corporation
Eclipse	Frick Company
Economy	Engineering Products Company
Espe	C.O.D. Tractor Company
EWC	Electric Wheel Company
E-Z Built	S. McChesney Company

F

Fair-Mor	Fairbanks, Morse & Company
Fairway	International Harvester Company
Farm Dozer	Isaacson Iron Works
Farm Horse	Farm Horse Traction Works
Farmco	Farm & Home Machinery Company
Farmer Boy	Columbus Tractor Company
	McIntyre Manufacturing Company
Farmall	International Harvester Company
Farmcrest	Gamble-Skogmo Incorporated
Farmers Tractor	Huber Manufacturing
	Sageng Threshing Machinery Company

Cross-reference

Farmford	Detroit Harvester Company
Farmmaster	Jensen Tractor Manufacturing Company
Farmobile	William Galloway Company
Field Marshall	Marshall & Sons Company
Fitch Four Wheel Drive	Four-Drive Tractor Company
4-Pull	John Minor Kroyer
	Wizard Tractor Corporation
4-Star	Minneapolis-Moline Company
5-Star	Minneapolis-Moline Company
Flexible	Franklin TractorCompany
Flex-Tred	Vaughn Motor Works, Inc.
Flour City	Kinnard-Haines Company
Fordson	Ford Motor Company
Four In One	Kardell Tractor & Truck Company
Four-Plow	Kinnard-Haines Company
Four Wheel Drive	Leonard Tractor Company
Fox	Fox River Tractor Company
Fuel-Master	M.A.N.

G

G, S & M	Goold, Shapley & Muir Company
Garner	William Galloway Company
Gas Pull	Advance-Rumely Company
Gearless	Union Iron Works
Geiser	Emerson-Brantingham Company
General	Cleveland Tractor Company
Giant	Russell and Company
	International Harvester Company
Giant Baby	Buckeye Manufacturing Company
G-O Tractor	General Ordnance Company
Golden West	Muscatine Motor Company
Graham-Bradley	Graham-Paige Motors Company

H

Heider	Rock Island Plow Company
H.M. & H.	Henry, Millard & Henry Company
Happy Farmer	LaCrosse Tractor Company
Hartsough	Gas Traction Company
Helping Henry	Auto Power Compnay
Heinze 4-Wheel	Traction Engine Company
Hoosier	Loe Rumely Tractor Company
Hudson	Evans Manufacturing Company

I

Ideal	Goold, Shapley and Muir Company
Imperial	Valentine Manufacturing Company
Imperial Super-Drive	Robert Bell Engine & Threshing
	Illinois Tractor Company
Ingeco	International Gas Engine Company
	Worthington Pump Company
International	International Harvester Company
Iron Horse	Samson Tractor Company
	General Motors Corporation
	Sweeney Tractor Company

J

Jerry	G.F.H. Corporation
Jet Star	Minneapolis-Moline Company
Jim Dandy	General Motors Company
Joy-McVicker	McVicker Engineering Company
Junior	Holt Tractor Company
	Pioneer Tractor Company
	Kinnard-Haines Company
	Nilson Tractor Company
	Russell & Company
	Porsche
	Buckeye Manufacturing Company

K

K.C.	Square Turn Tractor Company
K-C	Kansas City Hay Press Company
Kay-Gee	Keck-Gonnerman Company
Kerosene Annie	Advance-Rumely Thresher Company
Kingwood	Knickerbocker Motors Incorporated
Kinross American	American Tractor Corporation
Klear-View	Centaur Tractor Corporation
Klumb	Dubuque Truck & Tractor Company
Knapp Farm Locomotive	Gray Tractor Company
Knickerbocker	Knickerbocker Motors, Inc.

Cross-reference

L

Leader	Dayton-Dick Company
	Leader Tractor Company
Le Percheron	Societe Nationale de Construction
	Aeronautique du Centre
Liberty	Emerson-Brantingham Company
Lightfoot	Monarch Tractor Corporation
Light Four	Huber Manufacturing Company
Lindeman	Deere & Company
Line Drive	Line Drive Tractor Inc.
Lion	Lion Tractor Company
Little Bear	L.A. Auto Tractor Company
Little Boss	Russell and Company
Little Bull	Bull Tractor Company
Little Chief	Farm Engineering Company
Little Farmer	Will-Burt Company
Little Giant	Holmes Manufacturing Company
	Mayer Brothers Company
	Little Giant Company
Little Oak	Humber-Anderson Manufacturing Company
	Willmar Tractor & Manufacturing Company
Little Pet	Flinchbaugh Manufacturing Company
Little Red Devil	Hart-Parr Company
Little Traction	Adams Husker Company
Louisville Motor Plow	B.F. Avery & Sons Company

M

M.P.M.	Farmers Tractor Corporation
Macultivator	American Swiss Magneto Company
Macdonald	Cusman Motor Works
Major	Ford Motor Company
Master Four	Huber Manufacturing Company
Master Huffman	Huffman Traction Engine Company
McCormick	International Harvester Company
McCormick Deering	International Harvester Company
Merry Garden	Atlantic Machinery & Manufacturing Co.
Midget	Shaw-Enochs Tractor Company
Midwest	Wichita Tractor Company
Mid-West	Agrimotor Tractor Company
Mighty Man	Winter Manufacturing Company
Minnesota	Minnesota Tractor Company
Modern Farmer	Huber Manufacturing Company
Modern Four	Huber Manufacturing Company
Mogul	International Harvester Company
Mohawk	United Tractors Corporation
Montana	Montana Tractor Company

Morton Traction Truck	Lambert Gas Engine Company
	Ohio Manufacturing Company
	Pennsylvania Tractor Company
MotoRower	Hume-Love Company
Motox	Plano Tractor Company
	Wabash Tractor Company
Motrac	Minneapolis-Moline Corporation
Muley	C.L. Best Traction Company
Multipedal	F.C. Austin Company
	Austin Drainage Excavator Company
Multivator	Paul Hainke Manufacturing Company
Mustang	Massey-Harris Company

N

National	Denning Motor Company
	General Ordnance Company
	National Tractor Company
Neverslip	Commonwealth Tractor Company
	Monarch Tractor Company
New Elgin	Puritan Machinery Company
New Way	New Way Motor Company
Nu-Horse	William Stimson Barne

O

Oil King	Hart-Parr Company
Oil Pull	Advance-Rumely Threshing Company
Old Reliable	Hart-Parr Company

P

Pacemaker	Massey-Harris Company
Pacer	Massey-Harris Company
Paramount	National Pulley & Manufacturing
Pathmaker	Moon Tractor Company
Patriot	Hebb Motor Company
Peerless	Emerson-Brantingham Company
Planet Jr.	S.L. Allen Company
Platypus	Rotary Hoe Cultivators Ltd.
Plow Boy	Interstate Engine & Tractor Company
Plow Man	Interstate Engine & Tractor Company
	Plowman Tractor Company
Plymouth	Fate-Root-Heath Company
Pony	Pioneer Tractor Company
	Massey-Harris Company
	C.L. Best Gas Tractor Company
Powerbilt	General Tractor Corporation

Cross-reference

Power Horse	Harris Manufacturing Company
	National Implement Company
Power-Horse	EIMCO Corporation
Power Ox	Equipment Corporation of America
PowerRower	Hume-Love Company
Prairie Dog	Kansas City Hay Press Company
Princess Pat	Scientific Farming Machinery Company

Q

Quadpull	Antigo Tractor Corporation
Quincy All Purpose	Electric Wheel Company

R

Ranger	Southern Motor Manfacturing
Red-E	M.B.M. Manufacturing Company
Reeves	Emerson-Brantingham Company
Rex	Leader Tractor Manufacturing Company
Rigid Rail	Henneuse Tractor Company
	Hadfield-Penfield Steel Company
Road Hog	Hadfield-Penfield Steel Company
	W.A. Riddell Company
Road Layer	La Plant-Choate Manufacturing
Roadless	Richard Garrett Engineering Works Ltd.
Road King	Hart-Parr Company
Rototiller	Rototiller Inc.
Ro-Trac	Avery Company
Rumely	Advance-Rumely Threshing Company

S

Samson	General Motors Corporation
Sandusky	Dauch Manufacturing Company
Savidge	Backus Tractor Company
Seager	Olds Gas Power Company
Senior	Nilson Tractor Company
Shawnee	Shaw-Enochs Tractor Company
Shop Mule	Marsh-Capron Manufacturing
Sieve Grip	Samson Iron Works
Silver King	Fate-Root-Heath Company
	Mountain State Fabricating Company
Simplicity	Turner Manfacturing Company
Simpson Jumbo	Jumbo Steel Products
Skibo	Minneapolis Threshing Machine
	Union Iron Works
Special	Pioneer Tractor Manufacturing
Speedex	The Pond Company

Speedpull	Le Tourneau-Westinghouse Company
Square Turn	Albaugh-Dover Company
Steady Pull	Blumberg Motor Manfacturing Company
Steel-Clad	Denning Motor Company
Steel Hoof	Lambert Gas Engine Company
Steel King	Hart-Parr Company
Steel Mule	Bates Machine and Tractor Company
	Joliet Oil Tractor Company
Sterling	J.A. Hockett Company
Strait's	Killen-Walsh Company
Sunflower	Locomotive Finished Material Company
Sunshine	Sunshine Harvester Works
Super	International Harvester Company
	Oliver Corporation
	Vendeuvre Company
Super Drive	Illinois Tractor Company
Super Four	Huber Manufacturing Company
Super 4 Drive	Four Wheel Traction Company
Super Major	Ford Motor Company
Sure-Grip	Union Tool Company

T

T-C	Minneapolis Steel & Machinery Company
Tank-Tread X	Scientific Farming Machinery Company
Tank-Tread	Pan Motor Company
Terratrac	American Tractor Corporation
Thorobred	Allen-Burbank Motor Company
Tilsoil	Farm Motors Company
Tiger	Inexco Tractor Corporation
Tiger Pull	Gaar, Scott & Company
Titan	International Harvester Company
Toe Hold	Advance-Rumely Thresher Company
Tonford	Detroit Truck Company
Toro	Toro Motor Company
	Advance-Rumely Threshing Company
Tournapull	Le Tourneau-Westinghouse Company
Townmotor	The Townmotor Company
Tracford	Standard-Detroit Tractor Company
Tracklayer	C.L. Best Traction Company
Trackpull	Belle City Manufacturing
	Bean Sprayer Pump Company
Track Runner	Avery Company
Trackson	Geo. H. Smith Steel Casting Company
Tractair	Centaur Tractor Company
TractorRower	Hume-Love Company
Tractractor	International Harvester Company
TracTred	Miller Traction Tread Company
Tru-Draft	B.F. Avery & Sons Company
Trundaar	Buckeye Manufacturing Company

Cross-reference

Tructractor	Clark Tructractor Company
Twentieth Century	Minneapolis Steel & Machinery Company
Twin City	Minneapolis Steel & Machinery Company
	Minneapolis-Moline Company
2-Star	Minneapolis-Moline Company

U

Ultimate	Plantation Equipment Company
Uncle Sam	U.S. Tractor & Machinery Company
United	Allis-Chalmers Company
Unimog	Daimler-Benz
Uni-Tractor	Minneapolis-Moline Company
Universal	Advance-Rumely Threshing Company
	Elderfields Mechanics Company
	Lawter Tractor Company
	Minneapolis-Moline Company
	Moline Plow Company
	Northwest Thresher Company
Universal Farm	American-Abell Engine & Threshing Company
Ustrac	Federal Machine & Welder Company
Utilitor	Midwest Engine Company

V

Rotary Plow	W.S. Jardine Company
Vaughn Gearless	Eaton Gas Engine Company

W

Wadsworth	Detroit Engine Works
Wallis	J.I. Case Plow Works
Waterloo Boy	Waterloo Gasoline Engine Co.
	Deere & Company
Weber	American Gas Engine Company
Webfoot	Blewett Tractor Company
Wellington	Sterling Manufacturing Company
Westrak	General Tractor Company
Wheat	Hession Tiller & Tractor Company
	Wheat Tractor & Tiller Company
Whitney	Ohio Manufacturing Company
Winona Special	Pioneer Tractor Manufacturing Company
Wisconsin	McFarland & Westmont Tractor Company
Wizard 4-Pull	Kroyer Motors Company
Wolf	Star Tractor Company
Wolverine	Ypsilanti Hay Press Company
Wonder-Boy	Simplicity Manufacturing Company

X, Y, and Z

Yankee	American Tractor Corporation
York	Flinchbaugh Manufacturing

OTHERS

_____
_____
_____
_____
_____
_____
_____
_____
_____
_____
_____
_____
_____
_____
_____
_____
_____
_____
_____
_____
_____
_____
_____
_____
_____
_____
_____
_____
_____
_____
_____
_____
_____
_____
_____
_____
_____
_____
_____